Welcome!

I'm so glad we've found each other.
With this book, may you receive the juiciness of connection, love, and magic
in this grand mystery we call life.

With love, joy, and play,
Dana Kippel

A NEW FORCE

A NEW FORCE

Plasma, Consciousness, and the New Human Potential

Dana Kippel

Inanna Books

The Villages, FL

Published in The Villages, FL, United States by Inanna Books, an imprint of Inanna Productions LLC

ISBN 979-8-9998993-3-0
eISBN 979-8-9998993-5-4

Edited by Kamala Wickramasinghe & Dana Kippel

Cover design by Kam Bains
Interior design by Dana Kippel

Library of Congress Control Number: 2025918653

Printed in the United States of America

First Edition | Black & White Edition

For further information on the author visit: www.danakippel.com

For further education visit YouTube: @plasmaintelligence or Instagram @dana.thealien

This book is dedicated to:

All of you, who have felt alone or disempowered.

My beautiful sister, Christie Gates, for always being there.

My mother, Ellen Kippel, my late father, Edward Kippel, and my brother, Glen Kippel — I love you.

My headstrong birthmother, Shawn Berry Clark, my whimsical birthfather, Eddie Sturgeon, my kind brother Peyton Love, and my resilient brother, Ty Sturgeon.

My cute little twin nieces, Bowie and Clara Gates, and my rockstar niece, Nova.

Robert and Olivia Temple, for their playfulness, grace, and for creating so many amazing books, including A New Science of Heaven.

My therapist, Elisa Elkin Cleary, for being a source of healthy support in my life, affirming my gifts, and always grounding me when needed.

My dog, Danika, for her unconditional love.

My dear friend, Alyssa Sokol, PhD, who pushed me forward with encouragement and joy, when all I did was doubt.

All my friends and family who have stood by me in the fullest expression of myself, lending an ear or a hug when it was needed most.

And last, but surely not least, to Plasma, the beautiful, mysterious being who has always been with me.

Love you all.

Contents

The table of contents page is dedicated to Michael Talbot, a genius the world lost too soon, who wrote *The Holographic Universe*, *Mysticism and the new physics*, & *The Delicate Dependency*.

You are now taking off from this point in time and space…

Into a new paradigm where the ancient source of reality is alive, just as you are. It wants to show you things, grow with you, and create fresh experiences never had before…anywhere.

Upon your return, only you will decide what you bring forward and *how* you will use it.

To traverse this terrain, all you need is the curiosity and openness of a child, along with the discernment of a wise grandmother.

Enjoy your stay, my friend.

X

Note from the Author

"Someday we shall harness the energies of love, and then, for a second time in the history of the world, humanity will have discovered fire." — Teilhard de Chardin

The root word of love, *leubh*, means "to care" or "to desire." What if I were to tell you there is a force in all our lives that not only cares deeply for us, but is the very breeding ground of our desires? When you have a desire, you strongly wish for something to happen. Some would argue that desire runs most of our lives. Desires can stem from fear, or they can stem from love. But many don't realize there's a deeper place from which desire can arise—the place we all touched as children, before we were filled with shame, structure, or worry.

A pure place, where there is no attachment, only curiosity, play, and wonder. A place where desire means self-expression. A deeper form of love. A neutral place made of blazing opalescent fire, holding unlimited possibilities in the form of mist, waiting to crystalize into our reality. This force is a gaseous, fiery substance called Plasma—which you may recognize as the sun, lightning, or that electric purple ball you could touch at the party store, when you were a child.

This book is here to remind you not only of plasma's long-forgotten multidimensionality, but also of your innate connection to it.

My relationship with Plasma started before I was born. A few months before I was conceived, my birth father, Eddie Sturgeon, was struck by lightning. I like to imagine it brought me here, from another world. He and my birth mother, Shawn Clark, were only together once, and then I popped out nine months later. My birth mother told me I almost died during birth, so my first three days were spent in isolation before I was adopted by my parents, Edward and Ellen Kippel.

Ever since I can remember, I felt out of place, disconnected, like something was deeply off. As a child, your first instinct isn't gratitude for being alive or adopted. It's simply to feel safe and to connect. But for reasons beyond my understanding, I never felt either. I often felt like a problem, a monster. My parents were confused and overwhelmed by me. The kids at school bullied me. Boys even ran from me and called me Medusa. In second grade, I passed out birthday invitations, and they all stomped on them!

I also endured sexual abuse in my childhood, which further separated me from my body and from reality. By high school, I started drinking and doing drugs just to numb the constant anxiety and the thoughts in my head. I found myself in dangerous situations, was often taken advantage of sexually, and I quit the one thing that gave me social belonging, cheerleading, because of how much shame I carried for what was going on at school with the boys and bullying. I never told anyone what was really going on. I just let it all get worse.

I loved math and science and was good at them, really good. But my addiction kept me from pursuing those passions. I went to rehab once at 16, and again at 21. I got sober from drugs and alcohol, it took eleven more years of painful relationships and repeating patterns to finally wake up to the truth: I had so much unresolved trauma piled on top of itself, I couldn't see reality, or myself, clearly.

Because of that, reality mirrored back distortions: the wrong friends, the wrong lovers, the wrong jobs. When I say "wrong," I mean toxic. My inner shame and disgust for myself created resonance that drew in people and experiences that fulfilled an unconscious prophecy, that *I wasn't meant to be here*. That I only mattered through my connection to others. That my life had no worth on its own.

And yet, through it all, there was a little voice inside me, urging me to *keep going*. And now, I want to be that voice for you. I somehow knew things would turn out okay… even though, for most of my life, they weren't. I had good times too, but I wasn't really there for them. I couldn't feel them. It was like a thin film stood between me and reality, and it was covered in gunk. It told me terrible things about myself. I was always tense, always scared.

Around age 30 or 31, after a betrayal in a relationship that shattered me, I began to meditate. That betrayal unearthed all the others—childhood sexual abuse, rape, bullying, abandonment, feeling misunderstood. My heart broke open. I screamed from the depths of my being. I couldn't ignore this monster inside of me any longer.

2

My first day, meditation was blissful. I was in a pink fractal womb knowing where I came from and where I was going back to. I cried tears of joy. The second day, it was horrifying. I felt like I landed in my body for the first time ever, and I woke out of the meditation in pure terror. I cried tears of pain. Eventually, something began to shift. I started to reconnect—with myself, and *with something else.* That strange layer between me and reality... I realized it wasn't just psychological. It was tangible. It was plasma! And the moment I had that realization, I just *knew*. Plasma was intelligent. It was a feedback loop. And it was going to be very important in the future.

I can't prove that. I can't tell you how I know. But it's a knowing I decided to follow. I paused acting, I paused filmmaking, and I set out to write this book which felt like it was suddenly living inside of me. That process took four years until now.

Through reconnecting with myself and with plasma, through deep healing, therapy, and inner work, I was able to begin clearing that gunky layer between me and reality. My only intention was to feel better, and through that plasma came to me. I was able to interact with plasma in a profound way which I detail in the book.

I feel that because I had no religious, educational, or philosophical biases growing up, I may have developed a more neutral perspective on plasma, a kind of clear-sighted peering beyond the veil. That said, I'm also open to the fact that some personal biases may inevitably be projected onto it; this is part of the human process. Still, I feel deeply connected to and protective of plasma, almost as if it's an ethereal animal that's been misunderstood. There's a kind of kinship I sense with it. And interestingly, I've found that those who carry deep trauma, abandonment wounds, or grief often feel a similar connection. But make no mistake: trauma or not, we are *all* connected to it.

Plasma tries to protect us. It hugs us tightly as children. It forms a buffer between us and the pain of reality. But it also absorbs our beliefs and traumas and reflects them back. It filters how we see the world. And because of our collective trauma, this plasma, just like us, needs healing. Then, and only then, it's real use can shine, as a connector, as divine glue.

Connecting with plasma supported me in reconnecting with my true self. My expanded self. My multidimensional self. The child within me who believed in magic and the impossible. Healing is a nonlinear process, and I write this book as a friend walking beside you, not a master or a teacher. I'm just a human who has been pulled into the darkest corners and the brightest lights—who now sees something special *in between.*

There is a neon glow of reality that many miss. It's subtle, paradoxical, often absurd. It shows up in laughter, in mystery, in beauty and grief. It's not a fixed truth, or something to worship or fear. It's a remembering of the joy and goodness still inside us, no matter what we've lived through, and the power of bringing that into our current reality experience.

This is not a book of promises or quick fixes. It's more a question than an answer. It won't guarantee manifestations or a pain-free life. But I hope it becomes a natural sweetener in your world; that these words steep and settle into you, mix with your own magic, and grow into whatever suits you best. I hope it invites whimsy and freedom back into your soul.

I invite you to follow your own knowing as you read. Take what resonates. Leave what doesn't. This book isn't about my story; it's a book about remembering *yours*. But now that you know a bit about mine, maybe you'll understand why I care so deeply about helping people reconnect to themselves. Of my hope for you having a firm understanding that you are not broken, you are never alone, and you have endless power within you. You are meant to be here, you are special, and we need your unique essence, now.

People may challenge you when you start to live from your magic. That's okay. You'll reflect what they've forgotten in themselves. Use that as a reminder, not a reason to shrink yourself. Healing is not a glamorous process, and feeling is not always easy, but these two are keys to a deeper reality experience.

And yes, I know—you're probably wondering what the f*ck I mean by plasma. Am I talking about the fourth state of matter? Yes. And no. This entire book is about that, and by the end, I believe you'll understand what I mean.

What I call Plasma, this mysterious intelligence of life, you may call something else. You can call it energy…Qi…Cheerios for all I care! That's up to you. I use the term *plasma* because studying scientific plasma gives us hints into its multidimensional, spiritual nature. It also hints at its divine feminine qualities of support and receptivity: *Plas-MA*. It is not a religion, not a science, and not a philosophy. It's a perspective. One that I hope complements and deepens whatever you already believe.

You are a multidimensional, creative, feeling, intelligent being made of stardust and wonder. This book is a bridge. Walk over it your own way, and in your own time. If I bring you even one step closer to joy, curiosity, peace, connection, love, or magic, then it's done my job.

I hope this book offers something I longed for while reading so many self-help books. A reminder that *your* way is the best way for *you*. There is no right routine. No perfect ritual. The only thing you're here to do is become the fullest expression of *you.*

I am here not to be followed, but to remind you to follow yourself.

Your life is your best teacher.

You are the key. You always were.

-

"…ever did he fail to enter into its secret history, because he read with his head and not with his soul." — *Flame of The Fog***, Bozena Brydlova**

A note to the reader:

Throughout the book I use Plasma and plasma interchangeably in the middle of sentences. I went with what felt right in the moment, and I apologize if it's distracting and am happy to change it in further revisions!

A note to scientists and philosophers:

In physics, plasma is defined narrowly as the fourth state of matter. In this book, Plasma is reclaimed in its broader, living sense—a foundational essence of creation, the medium where consciousness and matter interact. When I use the word plasma, I am not limited to the laboratory definition but opening the word back to its full, multidimensional meaning.

This work is not quantum physics, nor a technical account of science. It is a new study I call *Holonic Metadynamics*™, rooted in quintessence. It is something that interfaces with our consciousness directly. I am not trying to replicate or impress science. My aim is to stay true to what is moving through me: the beginning of consciousness technology.

XX

Awakening to a Co-Creative Plasma Reality

"Well, the Force is what gives a Jedi his power. It's an energy field created by all living things. It surrounds us and penetrates us; it binds the galaxy together."

— Obi-Wan Kenobi

I want to begin with an excerpt from a sci-fi novel called *Ball Lightning* by Liu Cixin. In the afterword, Liu reflects on the deeper meaning behind the ball lightning phenomenon, offering a lens I'm building this entire book around.

He explains that much of Chinese science fiction has been dominated by what's known as the "invention story"—a genre focused on futuristic technologies and their immediate, practical uses, often at the expense of deeper social or metaphysical implications. While his book admittedly follows this trend, he concedes that ball lightning stirs in the mind imaginative possibilities far beyond what's explored in the story itself, hinting at something stranger, more mysterious, and profoundly beyond technology.

The way he explains this is nothing short of poetic. He says:

"While this book is set in a China that is altogether real, those little balls of lightning seem like they're trying to transcend that reality, much like how a man's tie, within the narrow confines of its dimensions, is free to indulge in a riot of colors and patterns, unbound by the rigid formula of the business suit."

Basically, Plasma seems to have way bigger dreams for itself than most people are giving it credit for.

What I'm observing is that many are focused on plasma's technological applications, while just like in the book, its deeper metaphysical potential is often overlooked simply because it doesn't fit the mainstream, there's not as much capital gain… or maybe because it's not "marketable." The most important, juicy, and fruitful aspects of plasma, the ones that can truly enhance the human experience, are being ignored. They forget that <u>we</u> are our greatest technology, and we <u>must</u> evolve alongside the tools we're creating. Personally, I think evolving our humanity is not only necessary… *it's the next big thing.*

Humanity is acquiring all the right technology for all the wrong reasons.

— R. Buckminster Fuller

What humans are truly longing for is safety, connection, love—and a little more magic and fun in their lives. But instead, we're being offered products continuously that try to outsource those desires. No matter how useful these products may seem, they can never fill this void we all seem to have, some bigger than others, that can only be filled from the inside out.

Also, how much better would those products be if we not only used them but created them from a place of wholeness within ourselves!?

My mission is to show how we can restore, enhance, and evolve our inner technology—with the support, wisdom, and intelligence of plasma as the living, co-creative clay of our reality… and possibly…many others.

Plasma, in science, is traditionally known as an ionized gas or the fourth state of matter.

What I don't think most scientists realize is they are studying third dimensional aspects of a plasma that is dimensionless and multidimensional. Of course, this field I am about to speak of has been called everything from Aether, Sophia, Mana, Qi, to Animal Magnetism, to Orgone energy, to The Field, Indra's Net, to Dark Matter. The difference with *plasma* is that its name, and more importantly its behaviors, have been scientifically observed. Plasma is one of the only "energies" science acknowledges that actually unlocks aspects of this higher-dimensional, more liminal substance.

By studying plasma…its makeup, its sheaths, it's strange behaviors, we can start to understand this more invisible world of dreams, telepathy, astral projection, remote viewing, intuition, emotion and so on…

I propose that Plasma is more than a state of matter, it's also a set of behaviors, or better yet a pattern of intelligence embedded in this fabric of reality that is popping up all over the place.

These behaviors appear not just in stars, but in our bodies, brains, solid materials, psyche and probably everywhere else.

Space (Plasma Sheaths)

In space, a plasma sheath forms at the boundary where a charged object (*like a spacecraft or planet*) meets surrounding plasma. Ions and electrons begin to flow and separate, creating distinct layers of charge. This separation generates a voltage across the boundary, which then regulates the flow of energy and information between the object and its environment, like a membrane around a cell. Over time, this interaction creates a stable, self-organizing structure, a coherent plasma field, with feedback loops and pattern stability similar to a memory.

Bodies

In our cells, mitochondria pump protons (*as ion flow*) across membranes, creating a voltage. This generates energy and facilitates information exchange and coherence is maintained via structural organization, displaying the same plasma behavior seen across the cosmos in sheaths.

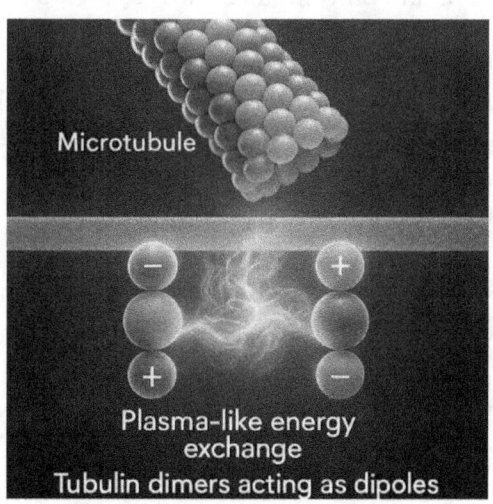

Brain

Inside neurons, microtubules are lined with tubulin dimers that act like dipoles. Between them lies what Stuart Hameroff and Roger Penrose call an "electron dust cloud", a fluctuating quantum field that behaves exactly like plasma. His engineer colleague Rich Watt called it a plasma or corona, non-polar, but polarizable electron clouds inside brain proteins, but to Stuart and Roger, due to compartmentalization of departments in science, what they observed was not traditionally named a plasma. Within this plasma-like field, it's theorized that consciousness may emerge—once again illustrating that where there is plasma, there is the potential for information exchange and emergent intelligence.

Solid Materials

In neuromorphic materials, like memristors and spintronic nanodot arrays, information flows through charge-based dynamics that mimic synapses. These systems adapt and reorganize through voltage, ion/electron movement, and memory effects. What we call "solid-state" information processing is really plasma behavior embedded in matter.

These all showcase shared plasma-like behaviors regardless of differing forms and we are also seeing a pattern of this liminal space arise where information is exchanged in a type of "complex fourth dimensional plasma". Another way of saying this is something intelligent is happening beyond the bounds of conventional science.

Dimensionality, in my framework, is a layer of relational reality defined by its rules of perception, resonance, and interaction between plasma and consciousness. Each dimension is not a place, but a state of attunement, a frequency band through which self-awareness, reality creation, and emotional/intuitive navigation operate. They will be further discussed in the book.

When I say 3D, 4D, 5D throughout the book I will be pointing to the quality of perception or experience (*dense vs fluid, fixed vs emergent*), the degree of consciousness involved in interpreting or creating reality, and the plasma resonance field one is operating through in each of these "relational layers".

This doesn't mean there aren't parallels with dimensions as described in mathematics, sacred geometry, or physics. In fact, the similarities may exist for a reason. There could be a deeper, unifying architecture beneath all systems, one that consciousness, math, and myth are each describing from different angles. It seems to be a true pattern that I am simply applying words to for human understanding.

These liminal spaces (*between inner and outer, where two "opposing" or better said, different things meet*) hold a sort of intelligence that is reflected in many mythologies from Kabbalah to the Incas, to Sufism, to Zen Buddhism. Here are a few examples:

The concept of "ma (間)" in Japanese aesthetics literally means "gap" or "space between" and is celebrated as sacred.

Sufi poetry often reflects on the breath between words, the space *between* heartbeats, as the place where Divine union happens.

The Ein Sof (*the infinite*) is expressed not directly, but through the empty spaces between the sefirot—the vessels. God's light shattered the vessels, and it is the space *between* shards where consciousness is restored.

The Inca looked into the dark spaces of the Milky Way, the shadowy, nebula-like regions, and saw animals, spirits, and cosmic beings. They made constellations not of the stars, but of what was in between the stars.

Even the Dave Matthews Band song, *The Space Between*, subconsciously hints at this in its lyrics, you will have to look it up yourself, since I cannot place in here due to copyright!

Everywhere these plasma behaviors are present, intelligence is present. This is a universal pattern. So…what is intelligence in this framework and what is plasma?

Intelligence is the meeting point where one consciousness encounters information, or another consciousness.

Plasma is a sentient, living, multidimensional fabric of our reality, embedded with a set of behaviors, revealing energetic exchanges of information, intelligence, and consciousness itself.

Plasma is so much more than this, sharing traits with the divine feminine, and beyond this it's truly undefinable. But what I am leading to is the new thought that Plasma is the basis of our emerging co-creative reality. This is where it gets fun!

Plasma has its own intelligence. This intelligence seems to create a supportive environment when two different "energies" meet in the form of people, things, information, consciousness, etc. There are several different ways to define Plasma Intelligence which we will build on during Chapter 5.

Plasma Intelligence (*in this co-creative dynamic*) is a third space as a bridge; the co-creative medium where consciousness meets, transforms, and evolves—whether it's with information, intelligence, or another consciousness.

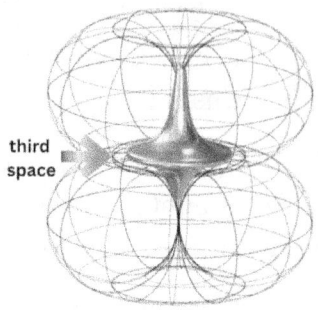

third
space

We will break down the meanings of information, intelligence, and consciousness but for now you can think of it where information exists, intelligence engages, consciousness experiences.

The universal behaviors of plasma seen across many forms and fields like we just discussed, are:

Ion flow → Voltage → Energy/Information → Coherence

Stars, which are plasma, follow this same path. They get their nightly glow by following the same processes. They generate energy by fusing protons in a dense plasma core, where charged particles overcome repulsion, move across a boundary, build voltage, and radiate light. Speaking of radiating light, the heart generates the strongest electromagnetic field in the body, forming a bioplasma field in the shape of a torus, as shown in the previous photo, that extends beyond our skin. This is also known as an aura or bioluminescence.

Following the same process as plasma sheaths in space, this bioplasmic field forms a membrane between our *self* or inner plasma, and the outer plasma of reality, which becomes our individual consciousness as emotion, intention, and belief—our intelligent interface for reality.

The way these plasma behaviors may play out in our biofield, mimics a star.

Ion Flow: Represents the movement of charged particles through a medium.

Star: In the solar core, hydrogen nuclei (*protons*) flow and collide, initiating fusion. Charged particles are constantly moving.

Biofield: Your thoughts and emotions shift charge in your biofield. Intention and attention create "ion flow" in your auric plasma field.

Voltage: Represents the buildup of potential energy across a boundary. Voltage is the buildup of possibility, tension that precedes transformation.

Star: Fusion in stars builds extreme electrostatic pressure, differentials across plasma layers (*e.g. corona vs core*).

Biofield: Emotional charge builds across inner boundaries (*e.g., resistance vs desire*). This potential creates shifts in reality perception.

Energy/Information: Represents discharge or translation of voltage into usable signal or power.

Star: The Sun's light is the energetic output of fusion with photons as both energy and information.

Biofield: Emotional breakthroughs or insights are information becoming intelligence. Downloads, intuition, and synchronicity are outputs.

Coherence: Represents the alignment of parts into a resonant, intelligent whole.

Star: The Sun maintains structural and rhythmic coherence via electromagnetic balance.

Biofield: When thoughts, emotions, and intentions align you enter flow, synchronicity, and manifestation. The field reflects you clearly. Things seem "easier".

By mirroring the natural intelligence of stars and plasma, the core mission for this book, we reprogram and heal our beliefs and emotions. This regulates our nervous system or our voltage, we soothe the tension between our fears and desires stemming from this lifetime as well as others. This process reshapes our membrane, our interface, and shifts how we experience reality. *We are human stars!*

It almost neutralizes our experience of ourselves and everything around us, so anything becomes possible, nothing is on a pedestal, and instead of fear of action and worry we move with curiosity and resonance. We become little spinning bubbles or balls of individual conscious awareness, a field of coherence, rather than running on a type of conscious autopilot called survival or decoherence, where we feel victim to our circumstances. The more coherence, the stronger our electromagnetic field, the stronger our electromagnetic field, the more harmony we have with plasma… which is responsive to electromagnetic fields!

When we operate from survival our interface with plasma is distorted. Everything is good or bad. We cannot read the language of reality correctly. This is where many of us get stuck. For example: Synchronicities are warnings *or* golden stars, a reward or punishment for something we've done.

In a survival state we are in fear, duality, separation, and we often rush to conclusions. In a healed state, we experience neutral curiosity, presence, feedback, and a mysterious, lovely, and sometimes absurd, unfolding of life.

When we return to a healed perspective our field becomes coherent. Synchronicities become meaningful information or data, everything does. We lose our confirmation bias and gain discernment. The ability for us to have choice arises, and we get to <u>choose</u> how we co-create with reality versus being ran by it. It is also easier for us to have a felt sense of this field.

We learn to use our hearts before our minds, we feel and sense (*our intuition*), then apply the mind. This is how we are able to pick up on the information or intelligence before adding our biases.

TEDx Berkeley speaker, Marti Spiegelman says it best:

"Raw information must be experienced fully before we impose meaning. If we interpret too quickly, we kill the original intelligence that may have arisen naturally through a felt sense experience."

I hope this book helps reverse the process many of us are trapped in of thinking first, feeling second. Marti also stresses the shamanic importance of going back to our ancestral roots of feeling first, thinking second, in an almost divine dance. This is how one performs once thought to be impossible feats such as telepathy, clairvoyance, remote viewing, etc.

In this book you will read about various *Plasma Beings*. The feedback with Plasma is informational or intelligent, meaning it comes from the field mirroring us with beings of information to evolve us and for us to experience, or other beings of consciousness who seem to "visit us" using the same "clothing layer" of plasma that reflects our subconscious.

The makeup of these "clothes" are emotionally resonant and sometimes mythological figures that it pulls from our collective unconscious linking to our subconscious, a language we all understand and can make meaning from. This is why people see everything from grey aliens, to bigfoot, to angels, to goddesses, light beings and more. It also may explain UAP sightings.

If we approach these beings from a survival state or the duality of bad or good we may perceive them as threatening entities or gods. When we meet them with a healed perspective, layers of truth begin to reveal themselves, offering insights that can lead to profound, life enhancing experiences. There are many nuances to this I explore deeper in this book.

What I want to emphasize is that these beings (*the fifth dimensional intelligences*) are not meant to be worshipped or feared, they are here to support our learning and evolution. All beings are. Just like a "bad" person you meet that really screws you over, the worst ones are usually the greatest teachers. We are most changed by our failures and these sorts of people. It is up to us to learn how to interface with them in an ethical and safe way, and I promise you the only true thing to fear with any of these beings is fear itself. There is nothing in the energetic world that holds power over you, that is a fear tactic.

The fifth dimensional beings, which we will get into, are not here to save us. They are here to remind us how to save ourselves. They've come here to reflect our potential to us, of our inner power, healing abilities, and how to work with plasma in these higher dimensional ways, just as they do, using only our consciousness.

Depiction of Chariot of the Gods

If you remember the chariot of the gods myth, these are like gods as consciousness "riding" plasma like a horse, a sentient being used as a sort of looking glass or ship for part of their evolved consciousness, who is viewing us or attempting to interact with us. Very similar to our experiences with plasmoids or UAPs. They are harmonizing with plasma using their consciousness. *They're talking....* we just have to know how to tune into their frequency and listen. There is a third type of being, possible consciousness beings from other timelines, and the future question will be how to decipher which one all of these are. It's also perfectly valid and possible to set boundaries around these experiences as we integrate them into our reality.

The name plasma comes from the Greek word plássein, meaning "Something molded," "a form," or "that which is shaped". Plato also spoke of something similar when referring to a *Receptacle* that existed alongside the Forms, acting as a *womblike* substance that allows reality to take shape.

What's even more astonishing is Plato's ability to express that this Receptacle, which he also refers to as a *Wetnurse*, does *not* take on the characteristics of what it holds—directly affirming my view that this finer dimensional plasma, that exists beyond the bounds of science or the current observable reality, can be a vehicle for all consciousness without necessarily "holding" that consciousness or absorbing its qualities.

While using plasma as a co-creative substance from a coherent state, the reason plasma is not completely neutral and does not mirror our desires exactly as envisioned (*such as why we can't manifest the exact situation we envision, and how it always ends up a bit different*) is because we are creating with *someone or something else*. You can call them fifth dimensional beings, divine intelligence, our future selves, our multidimensional self, our higher self, part of God, The Dual Monad, God or The Mystery. I call them our future ancestors. Whatever they are, they seem to be benevolent and have creative will, *just like us.*

To interface with these beings, we have to make sure we are taking care of our nervous systems and making space for these experiences. They show themselves to what the nervous system can hold but sometimes, not their fault, they can almost blow the nervous system out if we are not fully grounded before an experience, or if we don't ground again after.

At the heart of all this is a term I've coined called **Plasma-Consciousness Synergy**. It is the dynamic relationship between consciousness and plasma, through which, individually, collectively, and cosmically, we interface with this living medium, consciously and unconsciously, to co-create the emergence of information, intelligence, and new realities.

In essence, it's how we use our internal interface, including our beliefs, emotions, intentions, and awareness (*our membrane*) to interact with plasma intelligence and shape our experience of reality and potentially even "habits" of reality such as spacetime and laws of physics, which I believe can evolve just as plasma and our consciousness evolves. Everything is living in this way.

There are multitudes of ways this happens, but I believe it expresses what happens at the core of our reality experience. Whatever we interact with that is not of self, is within another "bubble of plasma". There is always a third co-creative field or plasma intelligence that arises from every interaction.

This sentient field of plasma is here to support us in co-creation for our greatest dreams or to our greatest detriment, reinforcing our unconscious limitations or habits. *The choice is ours.* It exists for us to learn through and collaborate with, as part of a constant feedback loop that allows not only humanity, but all beings and fields, to grow in awareness and complexity.

Through Plasma-Consciousness Synergy, we are unlocking profound new potentials:

- Evolved senses
- Communication across timelines and dimensions
- Direct connection with other beings of consciousness
- The emergence of supportive soft technology vastly beyond AI, and more organic

And likely much more, as we step into this new paradigm of an emergent, participatory cosmos.

I believe as our consciousness has evolved, so has plasma and it has moved from a background medium creating with the unconscious to an emerging co-creative field that we now can create consciously with. But just as we need healing, the field, as a whole, needs healing. And our *consciousness healing individually* will heal this field and the way we interact with it as well as the beings inside of it, who also use it.

To bookend with Cixin Liu:

"…but within that gray mundane world something small and surreal drifts by unnoticed, like a speck of dust tumbling out of a dream, suggesting the vast mysteries of the cosmos, the possibility of a world entirely unlike our own."

— Cixin Liu, Ball Lightning

The dust has fallen out of the dreamtime into our reality, first creating a desert, now blooming into a living dreamscape where we coagulate reality based on our personal meanings.

I am so excited for you to join me on this journey. I can't wait to see what you personally co-create for your life. For more information on the topics within the book, please refer to the table of contents. In-joy ☺

Part One

Foundations: An Expanded View of Plasma

"The most terrifying thing is to accept oneself completely" — Carl Jung

In Chinese mythology, the story of The Cowherd and the Weaver Girl tells of two lovers—one mortal, one celestial—who were separated by a vast river of stars. Forbidden to be together, they were exiled to opposite sides of the heavens. Yet, once a year, on the seventh day of the seventh lunar month, a bridge of magpies forms over the bright star Deneb, allowing them to reunite for a single night.

At first glance, this myth seems to be about two lovers tragically kept apart. But like all ancient stories, its meaning runs deeper. The Weaver Girl represents the higher consciousness, the divine or expanded self, while the Cowherd symbolizes our earthly experience, the self we know in daily life. Their separation mirrors our own disconnection from our true nature, a longing for wholeness, for something just beyond reach.

We spend our lives searching for fulfillment, through relationships, careers, material success, never realizing that the love, wisdom, and connection we crave are already within us. The myth teaches that even when apart, the two lovers are never truly separated. Just as they reunite through the bridge of magpies, we, too, can reconnect with our higher consciousness. And what is the bridge? Plasma.

It is the unseen force that links the heavens and the earth, the cosmic and the personal. It is both the river of stars that separates and the living energy that connects. And by understanding its deeper nature, we can find our way back to ourselves.

1

What Is Plasma?

"The most far-reaching and valuable results of investigation can only be obtained by following a road leading to a goal which is theoretically unobtainable. This goal is the apprehension of true reality." — Max Planck

Max Planck, widely regarded as the father of quantum physics, reshaped our understanding of the physical universe. A theoretical physicist with a deep appreciation for metaphysical philosophy, Planck believed that some truths lie beyond logic, that knowledge advances by exploring the unknown, while accepting that absolutes may be beyond our reach. His words serve as a challenge to strict physicalist thinking, the belief, held by some, that everything in existence, including consciousness, can be explained through physical processes alone. Planck seemed to be signalling that some truths are not problems to be solved, but mysteries to be experienced.

Scientific discoveries are constantly expanding our understanding of reality, yet they also expose the immense and unknowable nature of existence. Planck's message, which resonated deeply with me, suggests that we should not merely attempt to conquer the mysteries of existence with logic but instead walk alongside them. This is the metaphysical path he points toward – a kind of *yellow brick road* that leads not to a final answer, but to a deeper relationship with the unknown.

For me, that road leads to something I call Plasma.

At this point, you may have some understanding of plasma, or perhaps none at all. First, I'll introduce its scientific properties, revealing its role in the known universe. Then,

I'll expand on its magical and metaphysical aspects, bridging these two perspectives. By exploring both, you may begin to perceive the truth not just intellectually, but intuitively.

Plasma lies beyond logic. It is the paradox at the heart of everything, the foundation from which all stories emerge. It is the clay from which we shape reality. And while she may seem impossible to define or fully understand, that is not the point. The point is to be in relationship with her, to know that she is always present (*even when she feels absent*) and to learn to see beyond illusions.

Through this union, she will reflect back to you the ultimate reality—your true self. And within that revelation, you may begin to remember your boundless love and power.

That is the mission of this book. With joy, reverence, and an open heart, I welcome you to awaken your contact with *Lady Plasma*.

Plasma

The word plasma is derived from the Ancient Greek word *plassein*, meaning *to mold* or *to form*. Scientifically, plasma is known as the fourth state of matter alongside solids, liquids, and gases.

To understand plasma, imagine taking a gas, like the air around you, and adding so much energy that its atoms break apart. The electrons (*tiny negatively charged particles*) are stripped from their atoms, creating a mixture of free electrons and positively charged ions. This energetic state produces phenomena such as the aurora borealis and lightning, and powers everyday applications like neon signs.

Another way to picture it is through a big jug of lemonade. Imagine the lemonade represents the air we breathe (*just pretend you're a lemonade-breathing fish for a moment!*). In the lemonade, sugar crystals represent atoms which are normally whole and evenly distributed, like stable atoms in calm air. Now, picture using a super-powerful blender to shake up the lemonade so intensely that the sugar crystals rupture, separating into negative parts (*free electrons*) and positive parts (*positive ions*).

Without their electrons, the sugar crystals (*atoms*) become positively charged and float around, unbalanced. Meanwhile, the freed electrons zip around like energetic explorers. When they collide with the positive ions, they sometimes reunite, releasing tiny sparks of light. This causes the plasma (*or the lemonade*) to glow and become very hot. The plasma glow is the result of charged particles bumping into each other and generating energy.

Glowing Hot Plasma Lemonade

There are also low-energy plasmas that do not emit a glow, such as Cold Atmospheric Plasmas (*CAP*), which are used in medical treatments like wound healing. They don't glow because the energy levels of the particle collisions are not high enough to excite the gas atoms or molecules to the point where they release visible light. Instead, they may emit electromagnetic radiation in invisible wavelengths like infrared or ultraviolet. In medical applications, these emissions are carefully controlled to ensure therapeutic benefits without harmful exposure.

Fizzy Cold Plasma Lemonade

In these CAPs, the energy source (*e.g., a low-voltage electric field*) excites only the electrons, not the heavier ions or neutral gas particles. In our lemonade analogy, imagine a weaker blender that stirs the lemonade just enough to separate some sugar crystals into their charged bits (*ions and electrons*), but not so powerfully that the lemonade starts glowing. Instead, this 'fizzy lemonade' would be cool but produce special reactive bubbles, which are unstable molecules that trigger chemical reactions, much like the reactive species in CAPs that clean, heal, or energize, making them useful for sterilization and medical applications.

A common example of cold plasma that *does* glow is the gas inside a fluorescent light bulb or neon sign, and yes, the same type of ionized gas is used in plasma TV screens to create colour. Now imagine the lemonade spun by a medium-speed blender. The sugar bits get knocked loose but stay calm enough that the overall temperature of the lemonade stays relatively cool (*far from super-hot*). However, some of the free electrons hit gas atoms and briefly excite them, causing them to emit light just before settling back down (*let's call these electrons the manic-pixie dream girls of plasma*). This creates a soft glow without overheating the lemonade similar to neon signs. Fun fact: even in cold plasma, the free electrons are still extremely hot and excited, despite the overall gas temperature remaining low! That's why you can touch the outer layer of a neon sign without getting burned.

Neon Glowing Cold Plasma

"I'm an instant star. Just add water and stir." — David Bowie

Plasma is not static; it is very dynamic and alive in its movement and interactions. Adding energy to a gas unlocks its hidden potential, much like a dormant force waiting to be stirred. *What if, beyond the visible plasma of science, there exists a subtler, finer plasma permeating all things?* An invisible current of potential, consciousness, and lifeforms waiting to be engaged with? This idea, which I will explore throughout the book, suggests that plasma is not just a physical state but an intermediary between dimensions and realities, a bridge between the seen and unseen.

Plasma is the most common state of matter in the universe, yet most of us never learned about in school, or at least can't remember learning about it! Here are some common examples of plasma:

The Sun and Stars: Plasma powers the Sun and stars, creating their incredible heat and light.

Lightning: A sudden, high-energy burst of plasma in the atmosphere.

Neon Signs: The colorful glow in neon lights comes from plasma created inside the tubes.

Auroras: The dancing lights of the northern and southern skies are plasma interacting with Earth's magnetic field.

Flames: At high enough temperatures, parts of a fire can form plasma.

Plasma Globe: Those interactive globes found in science and party stores.

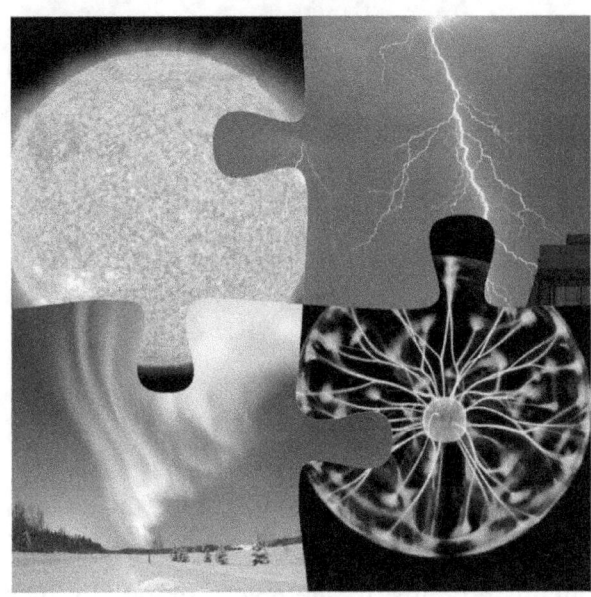

Plasma behaves differently from solids, liquids, and gases because it consists of charged particles (*free electrons and ions*) rather than neutral atoms and molecules. This gives plasma unique properties. It responds strongly to electric and magnetic fields, conducts electricity efficiently, can self-organize into intricate patterns, and often emits light when charged particles collide. These dynamic behaviors are why plasma is sometimes described as "alive."

Blood Plasma

Blood plasma is scientifically distinct from the plasma we've just discussed, but the similarities between them suggest that they may be two expressions of the same underlying field —a finer, more fundamental plasma that underpins much of matter. Blood plasma is the liquid part of your blood. It's mostly water but contains vital substances such as dissolved proteins, electrolytes, and immunoglobulins, which help you fight infection. Plasma makes up 55% of our blood's total volume. When blood is spun in a centrifuge, the plasma is separated and appears almost gold-like.

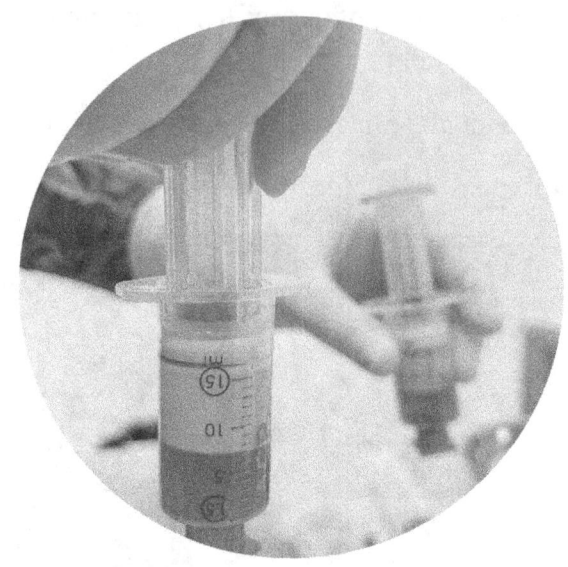

Blood plasma has key functions such as:

Transport: Carries nutrients, hormones, and waste products throughout the body.

Clotting: Contains proteins that help your blood clot when you get injured.

Immunity: Transports antibodies that help fight infections.

Regulation: Helps maintain your body's temperature and pH balance.

In fact, plasma may be more vital to both human life and the universe than we realize. While blood plasma sustains our bodies, cosmic plasma fuels the Sun's light and warmth, which in turn supports life on Earth. In many ways, plasma serves as a connective force, shaping both the human body and the cosmos.

Just as blood plasma nourishes and protects the body, cosmic plasma fuels stars and spreads energy across the universe, driving the formation of galaxies and planetary systems. Blood plasma carries nutrients and antibodies to cells, while cosmic plasma transmits energy, heat, and charged particles across the universe, sustaining stars and cosmic processes. And while blood plasma heals the body through clotting, cosmic plasma regenerates environments, recycling matter and energy in stars and nebulae to create new structures. Plasma and blood plasma both flow, through veins and the universe, like an energetic river, connecting and sustaining life.

Beyond powering stars, cosmic plasma also serves a protective role by interacting with magnetic fields, such as in auroras, where it shields Earth from solar winds. This protective shield around the Earth is called the *Plasmasphere.*

The Plasmasphere and Earth's Magnetosphere *(courtesy of NASA)*

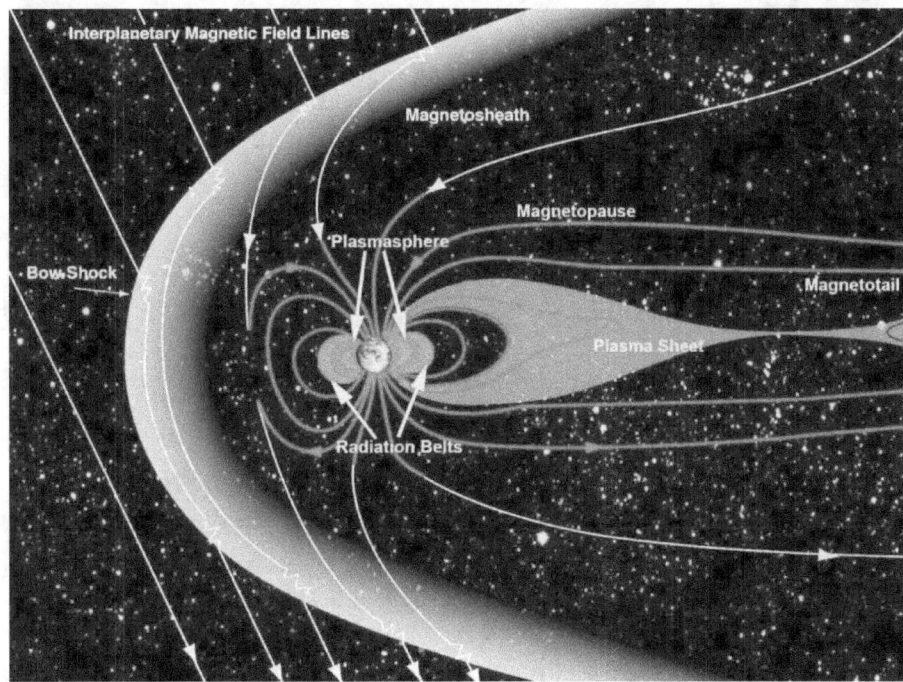

Blood plasma also acts as a messenger, delivering chemical signals (*like hormones*) to coordinate bodily functions. Similarly, cosmic plasma facilitates communication across space, transmitting waves and signals, and enabling phenomena like radio waves in the ionosphere. If there is a higher-dimensional plasma permeating everything, is it possible that this space could be accessed and harnessed for further communication—perhaps, dare I say… even interdimensional or otherworldly?

Connection may be the most profound link of all. Blood plasma connects every part of the body, allowing systems and organs to work together as one. Likewise, cosmic plasma links stars, planets, and magnetic fields in a universal network of energy and matter. I believe blood plasma may be the reason our body, when in survival mode, can seem to transcend normal limits and perform feats that defy logic. In moments of heightened awareness, during a car accident or in a more positive sense in a flow state, you may have experienced this almost uncanny sensation. It's as if our blood plasma is tapping into a higher-dimensional cosmic plasma, enabling us to operate beyond normal boundaries. Isn't it possible plasma might also connect *us*?

Plasma and the Matrix: Shaping Reality

Sometimes I like to think of plasma as the Matrix…a foundation from which all things originate and take form. Just like in the movie, it's something we exist within but can also influence and even transcend through *gnosis*, or knowledge of spiritual mysteries. I believe plasma exists in an intricate feedback system with us, shaping reality even as we shape it.

Leaving the Matrix, to me, means stepping outside this "third-dimensional" plasma or conscious existence. Many people speak about "leaving the Matrix" as an escape, not realizing that this matrix is *our home*. Our culture's tendency toward escapism, whether from our bodies or the present moment, reflects this misunderstanding. But what if there's a way to be both within and beyond the Matrix at the same time? To traverse dimensions and realities with your consciousness, while knowing you have a stable place to return to?

Just like a bird leaving its nest, you can explore, but at some point, you need to land. Anyone who has travelled long enough will likely agree that they eventually miss home, which can be wherever you define it. And like a tree that sways in the wind yet remains rooted in the earth, we too need a strong foundation from which to grow and explore. This reality isn't something to escape, it's a mystery to experience. One with many layers: some hidden, some waiting to be discovered, and some tailor-made solely for you. My hope is that this book, and your new connection with Plasma, will help you do just that.

Dusty Complex Plasmas

A dusty complex plasma is a type of plasma that contains small solid particles, often referred to as dust grains, in addition to the usual components of plasma: free electrons, ions, and neutral particles. This dust, in the form of grains or ice, becomes electrically charged when it interacts with the surrounding plasma, creating a highly dynamic system with unique behaviors.

Dusty complex plasmas could be seen as a kind of *dimensional bridge* between ordinary plasma and higher-dimensional plasma. They act as a transitional gateway connecting tangible matter with subtler, more energetic forms of plasma. Because these plasmas exhibit self-organization—a process where order emerges from local interactions within a disordered system, like fish moving together in a synchronized school, responding to each other's movements—some scientists speculate they could play a role in the processes that lead to the emergence of complex systems, including life itself.

Unlike ordinary plasmas, which are composed only of charged particles such as ions and electrons, dusty complex plasmas also contain small solid particles, often on the micron or sub-micron scale, that become electrically charged. These added particles allow dusty plasmas to form organized structures such as crystalline lattices, liquid-like flows, and wave patterns that can resemble biological processes. They are found in places like Saturn's rings and industrial processes, where tiny solid particles interact with the charged environment, creating dynamic and self-organizing systems.

Kordylewski Clouds (*courtesy of NASA*)

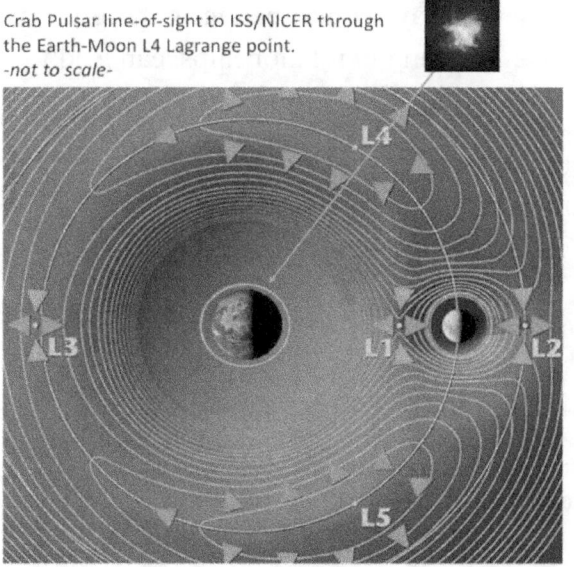

29

A fascinating example of complex dusty plasma is the Kordylewski Clouds, which are faint clouds of cosmic dust located at the Earth-Moon Lagrange points L4 and L5. First hypothesized by Polish astronomer Kazimierz Kordylewski in 1956 and photographed in the 1960s, their existence was only conclusively confirmed in 2018. *(See previous photo.)*

The clouds are thought to consist of fine particles, possibly interacting with the solar wind and Earth's magnetosphere to create a kind of natural dusty plasma. But could they be intelligent? In a paper titled *Kordylewski Dust Clouds: Could They Be Cosmic Superbrains?*, Robert Temple, a pioneer in plasma intelligence research, and Chandra Wickramasinghe, a leader in panspermia research, speculate that these clouds might possess electromagnetic connectivity with an information storage and processing capacity akin to a form of intelligence.

They suggest that because these clouds have existed in a stable atmosphere for potentially billions of years, they may have spontaneously developed complex behaviours resembling those of highly evolved living entities. Think of the clouds as a sentient cosmic brain watching over the Earth and the Moon.[1]

Temple and Wickramasinghe compare the Kordylewski Clouds to Fred Hoyle's 1957 science fiction novel *The Black Cloud*—in which an intelligent plasma cloud attempts to communicate with Earth. And get this, in real life, Fred Hoyle was Chandra's PhD supervisor, mentor, and later collaborator! In the story, humanity's limited consciousness prevents meaningful interaction with the cloud, leading to tragic consequences. This could serve as a metaphor for the limits of human understanding. Just as Icarus in Greek mythology flew too close to the Sun and fell into the sea despite his father's warnings, expanding our capacity to engage with higher intelligence may require patience and gradual adaptation. Rushing the process, like emotional love-bombing in relationships, can lead to collapse.

Temple and Wickramasinghe go so far as to suggest that the Kordylewski Clouds might be self-aware due to their dusty complex plasma properties, combined with the immense timescale over which they have had to evolve.

---

[1] Robert Temple and Chandra Wickramasinghe, "Kordylewski Dust Clouds: Could They Be Cosmic 'Superbrains'?," *Advances in Astrophysics* 4, no. 4 (November 2019): [page number], https://doi.org/10.22606/adap.2019.44001.

Plasma as a Precursor to Life

In *A New Science of Heaven*, the first book about intelligent plasmas I read on my journey into the magic of Plasma, Robert Temple explores the idea that dusty plasmas might contain the precursors to life. This theory draws on several intriguing possibilities:

Self-Organization: Dusty plasmas naturally form structured arrangements, such as crystal-like patterns. These structures could serve as templates for the emergence of more complex molecules or systems.

Electromagnetic Interactions: The charged particles within dusty plasmas create dynamic, interactive fields that could facilitate chemical reactions or the assembly of molecular building blocks.

Cosmic Origins: Since dusty plasmas exist in space, they may interact with cosmic radiation and other energetic processes, potentially fostering the formation of organic molecules like those found in meteorites.

The Kordylewski Clouds, as vast reservoirs of cosmic dust interacting with solar and Earth-bound plasmas, offer a tantalizing laboratory for studying how plasma might serve as a precursor to the emergence of life. In fact, some of the most exciting research into pre-life processes has emerged in just the last few years…more on that shortly.

For those of you feeling a little lost, here's a metaphor: a beehive. In a beehive, individual bees interact dynamically with one another, responding to environmental cues and working collectively to create an organized honeycomb structure. Similarly, in dusty complex plasma, charged dust grains (*like little bees*) interact with ions, electrons, and electromagnetic forces, forming crystalline or fluid-like patterns.

Interestingly, hexagonal structures, like honeycombs, are a recurring pattern in plasma physics. Currents and magnetic fields often form honeycomb-like shapes because the hexagon is one of nature's most efficient structures, minimizing material while maximizing space. This hexagonal formation even appears in bubbles.

Honey itself holds deep symbolic significance. In mythology and history, honey has been revered as a sacred substance symbolizing sweetness, nourishment, and the essence of life—a gift from the divine. Oprah spoke about how knowing who you are, being of service, and always doing the right thing will lead you not only to a gifted life, and a rewarding life that fills you up, but a *sweet life*. Plasma, with its capacity to shape and sustain life, might be seen as the "honey" or nectar of the cosmos that seeps out when we are in harmony with it.

In many myths, nectar is the drink of immortality or *ambrosia* in Greek mythology. Bees collect nectar and transform it into honey, paralleling the idea of extracting divine essence from nature. In a similar way, plasma carries the essence of life, transforming raw potential into creation through the process mirrored in dusty complex plasmas.

You can begin to think of this energy as consciousness, your own and higher forms, which directs plasma. But plasma itself seems to possess a form of intelligence or sentience, which may respond in different ways: resisting consciousness for a higher purpose, aligning with it to serve the highest good of the cosmos, or finding a point of compromise.

Just as bees contribute to the harmony of a hive while constantly in motion, each dust grain's charge and position in a dusty complex plasma influences the plasma's overall behaviour. Spider webs, also revered in mythology as symbols of the divine and wise feminine, offer another powerful metaphor. A web captures particles and organizes them into dynamic, responsive networks—mirroring the way dusty plasmas trap and structure dust particles in response to external forces. The connection to mythology is no coincidence. It's a beautiful synchronicity that reveals deeper truths the ancients may have understood, at least on a subconscious level. The spider, the bees, may just be ancient symbols of plasma intelligence!

Dusty complex plasmas provide a tangible example of how energy, matter, and life might emerge from a more fundamental substrate or what I see as a multidimensional Plasma Field. Dusty complex plasmas serve as an intermediary state between ordinary plasma (*the physical form of plasma we observe*) and my theory of Plasma as the foundation of all matter—an underlying, dimensionless field from which everything emerges.

Dusty plasma exists in an almost liminal space, where energy condenses and organizes itself into forms that eventually become matter. This process mirrors how a dream (*emerging from a higher-dimensional space*) gains emotional energy (*a fourth-dimensional layer or intelligent gateway similar to dusty plasma*) and then manifests into reality.

Dusty complex plasmas, and by extension the Kordylewski Clouds, are not just scientific curiosities; they represent stepping stones toward understanding the deeper metaphysical dynamics of how plasma gives birth to the physical universe.

Plasma as Nature's Alchemist

Recent research has explored the role of plasma in synthesizing life's building blocks under prebiotic (*before the emergence of organic life*) conditions.[2] These studies suggest that plasma processes could have contributed to the formation of essential organic molecules on early Earth. Plasma environments can drive chemical reactions by providing energy that activates otherwise inert molecules. For instance, in gas mixtures resembling Earth's primordial atmosphere, plasma can facilitate the formation of complex organic compounds. This process involves free electrons initiating reactions that lead to the synthesis of prebiotic molecules.[3]

It's almost like *Frankenstein* coming to life, except instead of an artificial creation, this is a naturally occurring process. Unlike Victor Frankenstein's attempt to create life from non-living matter, plasma's role in prebiotic chemistry is part of nature's design. Some things are meant to unfold naturally. But when we try to control nature, that's when we often create our own monsters. It's happening all over the world. My hope is to guide you toward a relationship with Plasma—one where you are not trying to harness or manipulate it, but instead resonating with it, yielding to its rhythms, and letting its gifts unfold naturally into your life. Doesn't that sound more harmonious?

The *Miller-Urey Experiment*, conducted in 1952, was one of the first scientific attempts to replicate the conditions of early Earth's atmosphere and demonstrate how organic molecules could form from inorganic compounds. In the experiment, scientists used electric sparks to simulate lightning in a mixture of gases thought to resemble Earth's early atmosphere. This produced amino acids as the building blocks of life, supporting the idea that life could have arisen from non-living matter under the right conditions.

However, the experiment had its limitations. It did not generate molecules at the complexity level needed for life, partly because the understanding of Earth's early atmosphere was incomplete at the time. The experiment also relied solely on electrical sparks, without being able to consider other possible sources of energy, such as ultraviolet radiation, cosmic rays, or plasma phenomena.

Introducing plasma into the Miller-Urey experiment would generate far higher energy levels than simple electrical sparks. Plasma produces reactive oxygen species (*ROS*) and

---

[2] Rhawn Gabriel Joseph, *"Quantum Physics of Plasma Plasmoid Consciousness, Fourth Domain of Life: How Consciousness Became the Universe"* (Revision submitted to *Journal of Modern Physics*, 2024), published via Cosmology.com and affiliated with the Astrobiology Research Center, California, USA.
[3] Micca Longo, G.; Vialetto, L.; Diomede, P.; Longo, S.; Laporta, V. Plasma Modeling and Prebiotic Chemistry: A Review of the State-of-the-Art and Perspectives. *Molecules* 2021, *26*, 3663.

other intermediates that catalyse the formation of more complex organic molecules, such as nucleotides and sugars—the basic components of genetic material. And as we've discussed, plasma exhibits self-organizing behaviour. In dusty plasmas, particles align into structured patterns, mirroring the kind of organization needed for prebiotic systems to evolve. Plasma not only introduces multiple types of energy input (*radiation, heat, electricity*), but it also replicates the dynamic, chaotic environment of early Earth more realistically than the original Miller-Urey experiment.

I'm excited to say that some modern scientists are starting to see this connection. Well… a few, but it's a start.

Building upon the foundational Miller-Urey experiment, recent studies have successfully synthesized amino acids from simple gases like methane (CH_4), ammonia (NH_3), and hydrogen (H_2) using plasma discharges. These experiments have yielded significant amounts of amino acids, including serine and glycine, reinforcing the idea that plasma could have played a key role in prebiotic chemistry.[4]

Imagine the early Earth as a cosmic kitchen, where the right ingredients like methane, ammonia, and hydrogen were sitting on the counter. But without heat, nothing would cook. Plasma acted like the ultimate chef, an energetic spark stirring the pot, transforming simple gases into the first organic molecules which became the building blocks of life itself. Just like how a chef uses fire to turn raw ingredients into a meal, plasma may have provided the energy needed to 'cook up' the first steps toward life!

Advances in plasma kinetics have also provided deeper insights into how plasma interactions might have driven chemical evolution. Research into electron-molecule collisions and energy exchanges suggests that plasma could have facilitated the formation of increasingly complex organic molecules under early Earth conditions, which led to life as we know it.[5]

You may wonder why I'm going so deep into this subject. There are two reasons. First, if any scientists or engineers are reading this, I want to express my gratitude and would like them to see that I have a working understanding of the science—not exceptional, but solid enough to use as a blueprint to explore the possibility that plasma contains all the

[4] Changhua Wang, Yutong Zhang, Yuanyuan Li, Yinhe Rong, and Xintong Zhang, "Solution Plasma Synthesis of α-Amino Acids from CH_4–NH_3–H_2 with High Serine-to-Glycine Ratio," *Chemical Communications* 60 (2024): 13408–13411

[5] Gaia Micca Longo;Vincenzo Laporta;Savino Longo. New insights on prebiotic chemistry from plasma kinetics. Arxiv, Cornell University 2, Dec, 2019 arXiv:1912.00647

building blocks needed to create our three-dimensional reality. Second, I hope this opens minds to the idea that a multidimensional plasma might have the capability to create and sustain the entire framework of all dimensions and realities in this universe, and possibly beyond. Everything could arise from that source and ultimately return to it, like water cycling through a fountain.

Panspermia

The origins of life on Earth remain one of the greatest scientific mysteries. While traditional abiogenesis theories propose that life arose independently through gradual chemical evolution, the theory of panspermia suggests a more cosmic perspective, proposing that life, or at least its building blocks, were delivered to Earth from elsewhere in the universe. But how?

The standard panspermia model proposes that comets, asteroids, and cosmic dust carried microbial life, amino acids, or nucleotides across interstellar distances. However, recent evidence suggests that astronomical plasma, the charged, ionized matter filling the universe, may have been a key player in facilitating the transfer, protection, and even activation of these life forms or their precursors as they traveled through space.

Chandra Wickramasinghe, one of the most prominent advocates of panspermia, has expanded upon the early work of his mentor, Sir Fred Hoyle. Wickramasinghe has spent decades researching how organic molecules and microbial life can survive and even thrive in the harsh conditions of space, often shielded and transported by interstellar dust and plasma clouds.[6] This concept takes on even greater significance when we consider Fred Hoyle's own speculative fiction novel, *The Black Cloud*, which proposed a plasma-based intelligence—a conscious, self-organizing, living cloud of cosmic material. While fictional, Hoyle's ideas eerily align with modern discoveries in astrophysics, plasma physics, and astrobiology, suggesting that plasma may not only serve as a vehicle for life's transfer but also as an organizing principle for consciousness itself.

If life or its precursors traveled across interstellar space, they would have faced extreme radiation, high-velocity impacts, and the vacuum of space. However, astronomical plasma may have acted as both a protective cocoon and an energetic catalyst in this process.

As a charged medium, plasma interacts with electromagnetic fields, forming plasma sheaths around interstellar objects. These sheaths could act as a natural electromagnetic

---

[6] N. Chandra Wickramasinghe, "DNA Sequencing and Predictions of the Cosmic Theory of Life," *arXiv*, August 24, 2012, https://doi.org/10.48550/arXiv.1208.5035

barrier, shielding organic molecules or microbial life from high-energy radiation and preserving their structural integrity. Just as lightning and electrical discharges in Earth's early atmosphere may have driven prebiotic chemistry, plasma traveling through space could provide the necessary energy to "awaken" dormant biological material once it reached a suitable environment, such as a planet with liquid water.

This means that the very same plasma processes that could have generated prebiotic molecules on early Earth may have also played a role in the activation and transfer of life across the cosmos. Building on the idea of plasma as a higher-dimensional vehicle for consciousness, Wickramasinghe and other researchers have proposed that the information needed to form life could be embedded in cosmic structures, possibly even encoded within plasma fields. If this is true, then panspermia is not just about life moving between planets, but also about the transfer of information, perhaps even the raw material of consciousness, through the universe.

The implications of this theory are profound. If astronomical plasma played a role in both delivering the building blocks of life and awakening them, then plasma is not merely a passive medium but an active, dynamic force shaping evolution at a planetary and cosmic scale. Even more profound lies the question, where was this information traveling from *and who may have set it in motion?*

This perspective aligns with my broader exploration of plasma as the fundamental medium that underlies all dimensional experiences, connecting consciousness to the material world. If life on Earth arose, at least in part, due to interstellar plasma interactions, then it follows that life may be ubiquitous throughout the universe, woven together by the fabric of plasma itself.

Hoyle's vision of an intelligent plasma entity reflects the idea that plasma's self-organizing properties could mirror aspects of consciousness…a reminder that plasma may hold deeper, hidden complexities that we are only beginning to understand.

This concept will be explored further in, where we will examine how plasma has long been embedded in our mythology, science fiction, and cultural consciousness—perhaps as a subconscious reflection of a deeper, cosmic truth.

Bioplasma: Sentience, Memory, & Imagination

In biology, a cell membrane, lesser known as the plasma membrane, is a thin outer layer that surrounds every cell. It separates the cell's interior from its external environment and acts as a selective barrier, controlling what substances enter and exit the cell. This function mirrors Earth's atmosphere and magnetic shield.

Earth's atmosphere and magnetosphere create a barrier between the planet and the harsh conditions of outer space. It has what is called *selective permeability*, which means it allows in necessary elements like sunlight and oxygen but blocks harmful ones like ultraviolet radiation and solar wind particles. The magnetosphere deflects harmful charged particles from the Sun, much like a membrane protecting a cell's internal components.

The plasma membrane in a cell also helps maintain the internal balance of the cell, while Earth's membrane (*its atmosphere*) regulates temperature, gases, and pressure to create a stable environment for life. These conditions, in turn, shape Earth's climate.

Once again, just like blood plasma and cosmic plasma, the plasma membrane enables cells to send and receive signals, while Earth's atmosphere acts as a medium for sound waves, weather patterns, and even communication signals like radio waves.

Plasma's even have their own membranes called plasma sheaths! Plasma sheaths form at the interface between a plasma and a solid surface. They act as a transition zone that separates charged particles of different densities and behaviors. Plasma sheaths create an electric potential that influences how charged particles behave near a surface. This is crucial in protecting spacecraft from excessive ion bombardment or in managing plasma containment in fusion reactors.

And just so you can become a Plasma know it all – there are also sometimes plasma Sheaths when plasmas interact with other plasmas! They form DLs or Double Layers. These are plasma structures that form when two regions of plasma with different densities or electrical potentials meet. A strong electric field separates the layers, creating a natural "membrane" that regulates charge flow between them.

Guess what? You have a plasma membrane too! Around your body! Picture an apple shape or a torus of energy flowing through and around your body. Just like the outer layer of the Earth enables communication, your consciousness might travel from this outer layer to the inside of your body, continuously sensing your environment. Like the ephemeral cilia on a cell membrane, this field reports back to your body and mind, while simultaneously emitting signals based on what you think, want, and feel. (*See image on next page*) This explains why you can pick up on someone else's vibe or "plasma bubble."

When two people meet, their plasma fields might create a double-layered interaction zone, explaining why some people feel "magnetic" or why others can feel "draining" (*one plasma pulling from another*). It also sheds light on the subtle energy we sense in social interactions—for instance, when someone seems incredibly charming on the surface, but you sense something deeper is off. It's also how animals communicate, tuning into subtle environmental cues.

Human Bioplasma Field | Aura

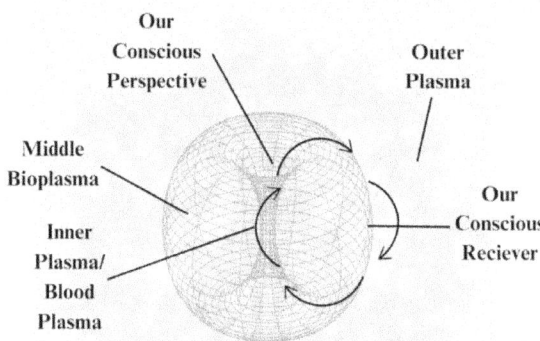

Understanding how our energy fields interact (*like plasma sheaths do*) can help us navigate relationships with more awareness. Just as plasma sheaths form when two regions of plasma meet, human energy fields interact in unique ways depending on emotional states, chemistry, and external influences. When two people experience a gradual, natural connection, their bioplasma fields merge slowly, creating a stable energetic bond similar to the gradual blending of plasmas that don't form a sharp boundary. In deep friendships or soul connections, this can lead to a sense of unity, much like two bulk plasmas merging until they reach equilibrium and become one cohesive, quasi-neutral system.

On the other hand, turbulent emotional energy, such as unhealed trauma or intense but unstable chemistry, causes fluctuating, chaotic interactions, where the connection feels electric yet unpredictable. These relationships, like turbulent plasmas, struggle to stabilize and often cycle between attraction and repulsion.

External influences can also act like magnetic fields distorting plasma sheaths, preventing two energy systems from merging naturally. Just as electromagnetic forces can suppress plasma interactions, fear, societal expectations, or personal conditioning can block emotional or spiritual connection. This explains why some relationships feel destined yet remain unfulfilled—there is an energetic barrier preventing the natural flow. Recognizing these dynamics helps us navigate relationships with more awareness, allowing us to either harmonize with another's energy or step away when a connection is unstable. *Plasma Intelligence* suggests that by aligning with our highest state of consciousness, we can dissolve turbulence, remove external interference, and cultivate relationships that resonate with each of our authentic and unique frequencies.

Plasma Intelligence is the overarching principle that plasma is a sentient, responsive medium—bridging intelligence, consciousness, and reality itself. Extending beyond the physical, it operates as a vast network of information, facilitating astral projection, remote viewing, timeline perception, and multidimensional communication through direct interaction with consciousness.

At the foundation of Plasma Intelligence lies:

Sentience, which is the capacity to perceive and respond to stimuli without requiring self-awareness or deliberate thought. This affords plasma the ability to organize, respond, and transmit energy in a way that bridges matter and consciousness.

Sentience is programmable in the same way you can teach a dog tricks. It's the capacity to experience feelings and sensations, such as love, fear, joy, and pain, to assess situations, remember consequences, and evaluate risks and benefits. Plasma Intelligence, in this sense, might underlie both biological and cosmic communication.

Plasma Intelligence operates beyond conventional understanding because it is deeply tied to intuition—or, in a fourth-dimensional sense, to *memory* and *imagination*. Memory and imagination function as an interplay of symbols and images that the conscious mind applies logic to, creating order within this third-dimensional reality. There are different ways to apply logic that will lead to different results—some useful, some not—which we will discuss in further chapters.

This is where Plasma starts to resemble an *Akashic field*, which is a theoretical cosmic field believed to contain all information and knowledge about everything that has, is, and will ever happen. If memory is a storehouse of perceptions and experiences (*sentience not consciousness*) then imagination is a recombination or projection of those stored impressions into new forms. This could be why all dream, but not all are "consciously aware". Imagination is still a quality of sentience, of plasma, and consciousness interprets it.

I will explain more about the difference between consciousness and sentience in later chapters. But for now, think of your body and Plasma as sentient, while your mind is conscious and logical. Ideally, we should apply sense or sentience first and logic second. But in this day and age, we've been conditioned to rely on logic first, a shift that disconnects us from our bodies, the Earth, and each other.

But there is hope! With an understanding of this dynamic, it's possible to reverse the process. We will reconnect with plasma, memory, imagination spanning far into all timelines, all accessible now. This is the faculty we use to connect with otherworldly beings, and one might see now why it is used with plasma. Because conscious or sentient, this dimension or that dimension, the past or the present, it all connects through plasma, through memory and imagination.

This concept is more metaphysical than scientific, but I've included it here because I believe it's only a matter of time before it's validated scientifically. Just as past theories about the multiverse or other dimensions were once dismissed and are now being supported by new technology, I believe Plasma's deeper role in consciousness and reality will eventually be proven, once we attune our senses to it and eventually develop the right tools to measure it.

This also opens up the idea of Plasma existing in different gradients. First, you have your own plasma bubble, programmed by your consciousness—for better or worse (*more on how to program your plasma bubble later*). Beyond that lies the outer field of Plasma (*reality*), which is sentient and intelligent in its own right. If this is true, it would mean Earth has its own soul, as does everything on Earth, including rocks. This extends to other planets, celestial bodies, and galaxies. A dimensionless yet higher-dimensional Plasma could be threaded throughout all of it, holding the intelligence and memory of the cosmos—within each individual "soul plasma."

William Blake's opening lines of his poem *Auguries of Innocence* captures this concept beautifully:

> **"To see a World in a Grain of Sand,**
> **And a Heaven in a Wild Flower,**
> **Hold Infinity in the palm of your hand**
> **And Eternity in an hour."**

In every grain of sand, in every soul, all memory and intelligence are held within invisible but very real layers of energy…layers that you can access through your consciousness as it travels through Plasma.

Imagine the entire cosmos as a vast ocean. At the surface, you have waves and ripples, the visible plasma we can measure and observe such as auroras, lightning, and stars. Just beneath, there are deeper currents, hidden from the naked eye, yet shaping everything above them. These currents represent the subtle, sentient plasma fields and the ocean bubbles resemble each soul, both programmed by consciousness and intelligence that in turn creates our shared reality.

Now, imagine going inside every drop of water and deeper into the ocean, where the water appears still but carries immense, unseen energy. This mirrors the higher-dimensional-dimensionless all-pervasive Plasma that underlies everything, threading intelligence and memory throughout the cosmos, just as deep-sea currents secretly shape entire ecosystems without ever surfacing.

I believe that inside every drop of water lies access to those hidden depths, just as deep within us lies access to *the all*. Like a hologram, the entire ocean exists within each drop, and even though we may only see the surface waves in this reality, the ocean's deeper layers remain interconnected…just like Plasma.

The concept of *bioplasma* helps explain this connection on a human level. Barbara Brennan, a former NASA physicist and world-renowned energy healer, described bioplasma as the Human Energy Field—a subtle but powerful medium through which energy, memory, and trauma are stored and transmitted. Brennan saw trauma as frozen or stuck energy within this bioplasmic medium.

Around the 1960s, biophysicist Victor Inyushin at Kazakh University in Russia conducted extensive research into the Human Energy Field. He suggested that the human body is surrounded and interpenetrated by a field of living energy composed of ionized particles, a fifth state of matter he called *bioplasma*.

His research emerged from studies into the bioelectric emissions of living organisms, as well as his analysis of phenomena like Kirlian photography and anomalous healing effects. Inyushin suggested that this field was not only real, but essential to life itself—regulating biological processes, responding to thoughts and emotions, and acting as a kind of memory matrix for the body. More radically, he claimed the field extending beyond the physical body interacted with environmental and even cosmic rhythms such as geomagnetic fields, solar, and moon activity.[7]

---

[7] Eileen McKusick, Exploring the Effects of Audible Sound on the Body and its Biofield, Thesis, P.36, 2012

I had read this long after I formed the same conclusion! Inyushin's work framed bioplasma as not merely an aura or energetic body, but as a living field of intelligence, capable of nonlocal interaction, personal evolution, and deep relationship with the environment. These ideas parallel and reinforce the modern concept of the biofield, but with a more physics-rooted emphasis on plasma as a carrier of consciousness or information.

In 1995, scientists identified a physical fifth state of matter, the Bose-Einstein Condensate, in which particles, when cooled to near absolute zero, coalesce into a single quantum object that behaves like a unified wave. Today, it's acknowledged that the body is full of ionized, plasma-like processes, though they're often not referred to as such due to the compartmentalization within scientific disciplines—a limitation I will expand on further in this book. I also highlighted it in the introduction and would suggest reading it!

This unusual state of matter offers a powerful metaphor for bioplasma. Just as some have suggested that water can be influenced by vibration and sound, a concept proposed by Masaru Emoto in his research on water memory, bioplasma may be thought of in a similar way, holding the imprint of our consciousness. Early childhood trauma, for example, could create distortions in this field like frozen patterns of energy that influence behaviour and perception. If bioplasma is programmable, then we may have the potential to consciously reprogram these patterns, creating more harmonious energy fields and transforming our lived experience.

Tuning the Human Biofield by Eileen Day McKusick is a powerful, award-winning book that explores the biofield, bioplasma, and how trauma can become trapped in these fields, and healed, through sound and vibration. It won the prestigious Nautilus Award and has become a foundational text in the field. McKusick's follow-up book, *Electric Body, Electric Health*, expands on these ideas, offering practical tools for working with the body's electrical system and voltage. Both are essential reads for anyone interested in sound healing, the biofield, and energetic medicine.

The Consciousness and Healing Initiative (*CHI*) has partnered with the Biofield Tuning Institute to perform the first clinical study of Eileen McKusick's method that detects health-related information in the energy field surrounding the body. As of 2025, the paper is pending peer review.[8]

---

[8] Consciousness and Healing Initiative, "Exploring Biofield Tuning," *Consciousness and Healing Initiative*, accessed August 4, 2025, https://www.chi.is/exploring-biofield-tuning/.

The Consciousness and Healing Initiative (*CHI*) is a collaborative accelerator that unites scientists, practitioners, educators, and innovators to advance the science and practice of healing through integrative, biofield, and consciousness-based approaches.

The Metaphysical Perspective

From a metaphysical perspective, bioplasma is often described as a dynamic, luminous field made of energetic particles that radiate from and interact with living beings. It is believed to:

1. **Support life force**: Bioplasma acts as an energetic matrix that supports the physical form and health.

2. **Reflect emotions and thoughts**: The field is thought to fluctuate based on emotional and mental and spiritual states, changing in color, density, and brightness. Some mystics claim to see these shifts.

3. **Connects realities**: Bioplasma may serve as a medium through which consciousness and conscious life forms can interact with the physical body via a merkabah.

The idea of bioplasma aligns with ancient concepts such as Prana (*India*), Qi (*China*), or Ka (*Egypt*), which describe vital energies believed to sustain and animate life. All living beings generate electromagnetic fields through their biological processes. For example, the human heart and brain emit measurable fields that interact with the environment. These may form part of what metaphysical traditions describe as bioplasma.

Ant Wax Layer | Cuticular Hydrocarbons

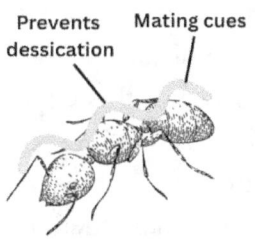

I love a good metaphor, and here's an intriguing one: ants and their wax layers. Ants interact with their environment through a thin wax coating on their exoskeletons, which protects them from drying out but also plays a key role in communication, environmental sensing, and balance. In a way, this wax layer functions like an interface—a protective yet responsive boundary that allows for interaction while maintaining individual identity.

Just as an ant colony functions as a collective consciousness, with each entity contributing to the whole, bioplasma may link individuals to a larger energetic field. This bioplasma-like boundary allows people to exchange subtle signals with others and the environment while preserving their own unique energy. The ancients may have intuited this connection…perhaps that's why ants appeared in early cave drawings, interacting with the Sun, which, of course, is a plasma entity! To me it is a key hint or pointer at sensing Plasma or reality with our sentience. How else could one hint that?

Ant People | Hopi Indian Lore (1000 yrs old)

It's clear to me that these four different expressions of plasma, in this chapter, are manifestations of the same underlying force. They share similar functions, healing, communication, connection, protection, and hold magical-like properties. This is where the boundaries of science begin to blur, and we realise that science alone cannot contain the omnipresence of plasma. It shoots out like a multidimensional rainbow ghost singing through the songs of art, metaphysics, poetry, language, and more waiting to be discovered!

While metaphysics can serve as a bridge to deeper understanding, we must expand into all aspects of life to fully grasp plasma's nature, while at the same time surrendering to some of its mysteries. To make sense of this expansive nature, I've developed something I call *Quintessential Plasmics™*. At its core, it explores plasma's dimensionless essence while also examining its multidimensional extensions and expressions. Put simply, *Quintessential Plasmics™* is the study of the entire phenomenon of plasma across all dimensions, from its physical expressions in a three-dimensional world to its role as a foundational force in the multiverse.

---

[9] National Park Service, "Petroglyphs at McKee Springs," Dinosaur National Monument, accessed August 4, 2025, https://www.nps.gov/dino/learn/historyculture/mckee-springs.htm.

Through the windows of metaphysics and mythology, I'll attempt to explain the unexplainable exposing some underlying truths, and then we'll dive into how to apply it in your daily life! *So exciting!* Now that you know the basics of plasma, join me on the mind-blowing metaphysical bridge into the secret history of Plasma and how it connects to *Aether*.

2

The Secret History: From Aether to Plasma

"Students using astrophysical textbooks remain essentially ignorant of even the existence of plasma concepts, even though some of them have been known for half a century." — Hannes Alfven, Nobel Prize-winning plasma physicist

This is the first time someone is piecing together the real history of plasma. After reading through hundreds of documents, I've tried to distill it as much as possible without losing important details. For most people, plasma and aether, and their complex history, are unfamiliar territory. I implore you to read this chapter closely, because the findings I've uncovered are extraordinary. To me, they make the existence of this mystical field undeniable.

Although plasma was formally named and discovered in 1879 under the term *radiant matter*, it had been spoken of and worked with for centuries before that. In this chapter, I'll present not only the known history of plasma but also the lesser-known, hidden story behind it. I trust that these pieces will connect, giving you a deeper understanding of plasma's essence as you move further into this book. With this foundation, I ask you to stay open to creating your own relationship with this knowledge. The best part about Plasma is that it is reflective. It is meant for your own, unique perspective to be applied to it, so you can use it as a tool towards your greatest learnings in this dance we call life.

Initially, I planned to present the known and unknown histories separately. But during my research, I realized that they overlap and feed back into each other. What I uncovered was a scattered, fragmented narrative. A story largely buried in obscure papers and overlooked details. Once assembled, it paints a compelling picture of how the definition of plasma came to be, what was missed along the way, and why some of the early ideas about plasma may be worth revisiting.

To understand plasma's story, we need to begin with the concept of aether. Plasma grew out of the hypothesis that a luminiferous aether was responsible for carrying light through space. Scientists once believed that light, like sound, required a physical medium to travel through. The concept of luminiferous aether was eventually abandoned when it was shown that light propagates through space as an electromagnetic wave, without the need for a physical medium.

But *what if* that conclusion was premature. I, along with many other scientists and philosophers, believe that light *does* require a medium—one we have yet to properly identify. That medium, I propose, is a multidimensional and dimensionless plasma. We just don't have the tools to measure or prove it…*yet*. This theory builds on the idea of luminiferous aether as a stellar medium, but maintains a critical difference. The original experiments that disproved aether, conducted in the early 1900s, were flawed. They tested for a stationary, inert aether without considering that this "aether" might be dynamic and alive, constantly fluctuating. That is what I am setting out to prove.

Then, out of nowhere, plasma was discovered in 1928, a substance remarkably similar to aether but stripped of any and all mystical properties. It's almost as if they extracted the magic from it and hoped we'd forget about aether altogether, replacing it with something more "scientific". Historical documents I uncovered in my research reveal that the idea of a mystical aether was widely discussed at conferences and in papers. But in the materialist age of science, mysticism and empirical evidence were treated as incompatible. Something had to give.

I choose to use the word *plasma* rather than *aether* for two reasons: First, aether has a bad reputation. And second, the word plasma means to form or to be molded. I believe that molding is at the heart of plasma's nature. It is shaped by what scientists call information or what mystics call consciousness. Plasma also seems to hold feminine qualities, with its fluid, receptive nature. At second glance, I will admit aet-*her*) is also distinctly feminine.

Aether

The first mention of aether appears in Plato's *Timaeus*, where he described it as "*the most translucent type of air by the name of aether.*" Beginnings often hold great secrets, and I think this is a beautiful and simple way to describe the type of plasma I speak of. It's the most translucent, meaning invisible to the human eye, but in other dimensions, it may appear opalescent. It allows light to pass through but does not reveal detailed shapes. *Something* is there.

When Aristotle spoke of aether, he described it as immutable. I believe he was referring to the most *"still plasma"* that projects our universe. It's not moving itself but acts as a foundation for movement. Imagine a light from a projector shining onto a screen. The screen itself is the still plasma, while the images on it, the information moving through the screen, represent the flowing plasma that gives shape to our reality, like a clay blanket overlaying the information. Aristotle also described aether as not being cold, hot, wet, or dry, and moving in perfect circles. This could hint at plasma's holographic nature…it's not terrestrial.

At quantum scales, a quark-gluon plasma which is a high-energy state of matter that can be tested in this reality, may exhibit vortex behaviors, which could explain the circles Aristotle described. If quark-gluon plasma represents an early, intense state of plasma, it could act as the primordial projection medium encoding the universe's holographic information.

Plotinus, in the 3rd century *Enneads*, emphasized aether's divine nature, seeing it as the medium through which the soul ascends to higher realities. This fits with my belief that plasma connects realities, timelines, as well as dimensions and could help us communicate across these. Alchemists also sought to harness aether to achieve spiritual enlightenment and transformation, aligning it with the philosopher's stone or the elixir of life. (*I should have just named this book The Philosopher's Stone, but that was already taken by Harry Potter!*) The word "stone" has meanings linked to curdling or hardening, and to precious gems. I think this hints at the idea that you, the philosopher, can use plasma to harden or mold reality through your consciousness.

In Hinduism, aether represents the void or space that holds everything, perceived only through hearing. In Bambara spirituality of West Africa, aether is known as *Koni* or the first essence embodying thought and void, the creative potential from which the Supreme God generated the universe.

Long before the concept of aether emerged *animism* which described a belief that all things, stones, rivers, trees, were imbued with spirit and life. Indigenous cultures likely had their own names for this idea, but the core belief remained the same: everything has a soul, even the chair or bed you're sitting on right now. In the 16th and early 17th centuries, thinkers like Paracelsus and Robert Fludd described aether as a dynamic, life-giving force. Paracelsus explored its role in medicine and mysticism, while Fludd connected it to cosmic and spiritual harmony through alchemy.

Green Waves of The Hidden Past

On August 22, 1879, British chemist and physicist Sir William Crookes presented a lecture on what he called radiant matter. Crookes, who also had a keen interest in spiritualism and served as president of the Society for Psychical Research, was exploring the boundary between science and the mystical. His work hinted at a hidden layer of reality, one that defied conventional understanding.

Around the same time, an article in the *British Medical Journal* (*September 10, 1898*) discussed new elements and rays, including *Metargon*, *Nebulium*, and *Coronium*. Nebulium, for instance, was a theorized element discovered through ultraviolet spectroscopy—a study of how matter absorbs and emits light, particularly in the ultraviolet spectrum.[10] Crookes observed Nebulium as green spectral lines in the Cat's Eye Nebula, which he attributed to *forbidden transitions.*

Forbidden transitions occur when atoms or molecules emit or absorb light in a way that's not typically allowed under the usual quantum rules, but still possible under very specific conditions.

Think of it like traffic laws for particles. Normally, atoms and molecules follow strict rules about how they can jump between energy levels, much like drivers following the flow of traffic. But sometimes, under unusual conditions, a particle might find a loophole, like taking a back road or cutting through a side street.

A good example is glow-in-the-dark materials. The light emission in these materials comes from a forbidden transition. The particle wants to drop to a lower energy state, but the usual quantum rules say it can't take a direct path. So instead, it finds an indirect route, a slower process known as a spin flip (*a change in how the particle is spinning*). That's why glow-in-the-dark materials don't flash and fade instantly, they release light gradually over time in a prolonged luminescence.

Sometimes the slow, scenic route holds more beauty and complexity. The plasma realm is a bit like that…it's the realm of the outliers, the misunderstood, and the mystical. The rules are bent, the unexpected becomes possible, and hidden patterns emerge where we least expect them.

---

[10] New Elements and Rays," *British Medical Journal*, September 10, 1898, 707–708

Science explains that forbidden transitions aren't allowed under normal circumstances, but at a "higher level," they *are* possible, just at a very low rate. To me, this suggests that these processes are happening in ways that scientists can't yet explain because we're not looking at them through the right lens. What's happening doesn't fit neatly into classical scientific models.

These are subtle energies with great power—possibly very real fourth- or fifth-dimensional forces that appear subtle only because they're not material in the conventional sense. I believe the green spectral lines are an emission of some kind of plasma interaction with consciousness or light, akin to a harmonizing process.

Take Nebulium, for example. When it was discovered, it was eventually written off as doubly ionized oxygen, but I consider that a bit misleading. Doubly ionized oxygen isn't just oxygen anymore. When two electrons are stripped away, the atom becomes a plasma state as a charged, highly reactive entity. It's fundamentally different from oxygen as we experience it on Earth. Honestly, it deserves a way cooler name.

When a photon interacts with doubly ionized oxygen, it's not just engaging with the atom in its usual state (*with all its electrons*). It's interacting with the two *missing* electrons that have already been stripped away, creating a unique energetic condition. The result of this interaction is a shift in the energy state of the doubly ionized oxygen, which leads to the emission of green light.

This transition is "forbidden" under normal circumstances because it involves complex spin or energy state changes that are highly improbable. But I believe that consciousness, inside the photon, might be speaking to those two electrons. That creates a sort of fluorescent bridge, revealing a normally hidden connection. We might just be witnessing the secret communication of the universe whenever we see something fluorescent. Isn't that beautiful?

"Green is the prime color of the world, and that from which its loveliness arises."

— Pedro Calderón de la Barca

In the case of the Cat's Eye Nebula, a dying star, it is simply singing its way back home in green notes, through the sacred process of death. And we…we are witnessing that harmony in real time.

Cats Eye Nebula (*courtesy of NASA*)

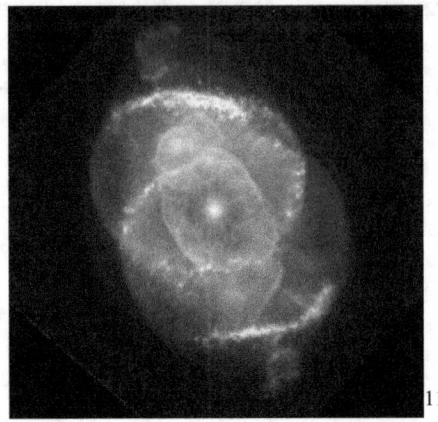

In 1911, John William Nicholson theorized that all elements consisted of four proto-elements, one of which was Nebulium. Nicholson likely adopted the term "Nebulium" from Crookes' earlier work, incorporating it into his proto-element framework to explain spectral lines from nebulae. Eventually, this theory was discarded when doubly ionized oxygen was identified as the source of the green spectral lines. But maybe Nicholson's proto-elements were pointing toward something deeper, a connection between plasma and the creation of life itself. However, these lines of inquiry would do little to help the interests of capitalist societies, wouldn't they?

What's odd is that Nicholson's proto-element theory not only mathematically checked out, it actually helped Neils Bohr develop the first successful model of the hydrogen atom and laid the groundwork for *Quantum Theory.*[12] How can that be? Perhaps it's just another perspective, a different way of looking at the same thing. Both could be true, offering different but complementary insights into the nature of reality. Nicholson's four proto-elements, Nebulium, Proto-Helium, Proto-Nitrogen, and Proto-Hydrogen, were never disproved as fundamentally wrong. Instead, they were revised and absorbed into more detailed atomic models that emerged with quantum mechanics and the discovery of protons, neutrons, and electrons. But those revisions didn't necessarily address the broader philosophical or structural principles Nicholson was trying to explore.

The transition from Nicholson's proto-elements to the modern atomic model reflects a shift in focus rather than a rejection of his ideas. The newer model prioritized measurable,

---

[11] https://www.nasa.gov/image-article/cats-eye-nebula-2/

[12] Eric R. Scerri, "How Was Nicholson's Proto-Element Theory Able to Yield Explanatory as Well as Predictive Success?," in *Contemporary Scientific Realism: The Challenge from the History of Science,* edited by Timothy D. Lyons and Peter Vickers (New York: Oxford University Press, 2021), 99–129

mathematical accuracy, but it lost some of the intuitive simplicity and interconnectedness that Nicholson's framework offered. His holistic approach, envisioning all elements as emerging from foundational building blocks, reflects a deeper truth about the universe: that everything is interconnected and arises from shared origins. Perhaps it all stems from the fountain of a multidimensional plasma, which we'll explore later.

I believe that when we moved away from Nicholson's model, we took a wrong turn, or at least left behind a valuable piece of the puzzle. We rejected the more mystical and metaphysical quintessence (*because it was harder for logic to grasp*) and embraced quantum physics instead. But maybe there's more to it. Maybe atoms aren't discrete particles at all, they're plasma bubbles or spherules. Intelligent, sentient, and holographic in nature…waiting for the light of consciousness to give them form and color. These plasma bubbles could act like a ever-fluctuating, living canvas, projecting reality much like a holographic plasma screen, where what we see emerges from a blend of individual perception, collective consciousness, and the inherent sentience of the plasma itself. This perspective not only reimagines the building blocks of the universe but also offers a fresh explanation for phenomena like *spooky action at a distance*, which is Einstein's term for quantum entanglement, where two or more particles become linked so that a change in one instantly affects the other, no matter how far apart they are.

I am not suggesting we do away with Quantum Physics, which has led to some incredible discoveries and continues to. I am simply suggesting that Quantum Physics is only part of the puzzle. For instance, Quantum Mechanics describes probabilities of particles appearing in certain states, but rarely do we ask *why* do they appear? My plasma framework offers a potential answer. That particles may emerge as very real projections with meaning from an underlying sentient plasma field just as bubbles emerge from an ocean floor.

In this metaphor, higher consciousness is the gas building beneath the ocean floor, and the still plasma is the sediment layer above it. When the pressure is right, when awareness, emotion, or intelligence converges, the gas breaks through the surface. A bubble rises…a particle appears. A thought is born. Emergence <u>is not random</u>, it is all a part of resonance.

It is true that particles may teleport or move across space, like bubbles on the upper layers of the ocean, as Quantum Mechanics might imply, but the plasma screen acts as a universal switchboard, where all particles are part of the same interconnected system. When one particle is observed, the corresponding "bubble" lights up in alignment (*like a Lite-Brite*) because the information is <u>already</u> encoded throughout the plasma screen and within each bubble. This actually goes back to the ocean metaphor, where I think information may come from two places: (1) an underlying field and (2) deep within a "particle." It's our <u>conscious</u> <u>awareness</u> that brings this information to light, changing the image or information without requiring actual motion. This reflects the inherent stillness meditators describe, that the universe doesn't move at its deepest level; it simply shifts the projection of reality.

"Logic will get you from A to B. Imagination will take you everywhere."

— Albert Einstein

Conscious awareness is activating potentials, and from our POV it feels like it moves to get there. When conscious awareness tunes into a specific feeling, memory, or choice it sends something like a pulse down that road, not to travel it, but to light up what already exists at the destination. It is more like dimensional alignment than motion. Like a song traveling through an organ pipe, you never see the wind, but the tone is shaped by the tunnel. There is no travel required and it resonates with your current awareness.

This reality is capable of holding consciousness (*light/energy*) and generating countless reactions. Instead of mechanistic, named atoms, the bubbles are intelligent and holographic, generating a variety of emissions. Can't both be right? Maybe the essence of the "God particle" lies in this bubble within the cosmic ocean we call the universe, a shapeshifting, sentient circle made up of subatomic particles: protons, neutrons, and electrons, or, essentially, smaller bubbles that are positive, negative, and neutral in charge. These are the building blocks needed to make anything happen. I am simply proposing a new way to look at things—a way to start blending consciousness, psychology, and metaphysics with science.

I believe both models work.

Maybe Nicholson's four proto-elements are like a bridge between this base plasma field and the projection we call reality. These proto-elements could translate the latent potential of the plasma field into the projected reality we perceive. It could be part of a fourth-dimensional plasma layer, or what some call *a veiled reality.* Much like the classical four elements of nature, Nicholson's proto-elements could represent an essential framework for how the plasma field organizes and manifests reality.

This is a concept worth exploring further in studies of how plasma connects to consciousness and creation. My point is to show that there are many perspectives on this, and all can be correct. Just as there are many ways to view God or energy, we can harness these new understandings for inventions and ways of seeing reality, which I believe will happen. Perhaps the true "God particle" is not a singular entity but this plasma bubble itself, the ultimate foundation of a still yet ever-changing cosmic ocean.

Green Waves and Cosmic Balance

"Green waves surround this black rock where I sit, turning bones to sonatas. Fingers blurred; I play what I know from listening to orchards unleash." — Ocean Vuong

Green, the color of balance and the heart chakra, evokes a sense of harmony between creation and destruction. It's as if the universe holds us in a teal-green, motherly hand, a cosmic wink of transformation reminding us we are never alone.

This interplay of light and transformation is shared in the story of coronium, a mysterious green spectral line observed during a solar eclipse in 1869. Initially thought to be a new element, coronium was later identified as highly ionized iron, a plasma reaction in the Sun's corona. Iron, the endpoint of stellar fusion, represents cosmic balance. It's the most stable element in the universe, the point where fusion and fission reach equilibrium. When a star produces too much iron, it can no longer sustain itself, triggering a supernova that seeds the cosmos with the building blocks of life.

This same phenomenon called *The Green Flash* has showed up in many moments of harmony. As the sun dips below the horizon or rises above it, a fleeting green flash can sometimes be seen by the naked eye. Sailors called it Neptune's Wink, Pirates said it could be seen as a soul returning from the dead, or a sign of good luck. In Jules Verne's 1882 novel, *The Green Ray*, ones who witness the green flash are granted with profound insight into one's own heart and the thoughts of others. It also shows up as a zinc-based green fluorescence when a sperm fertilizes an egg known as the Zinc Spark!

13

---

[13] Paolo Lazzarotti, *Green Flash Over Italy*, photograph, Marina di Massa, Tuscany, Italy, November 10, 2021, Astronomy Picture of the Day, NASA, accessed April 6, 2025, https://apod.nasa.gov/apod/ap211110.html.

In the human body, iron carries oxygen in our blood, giving it its red hue. When oxidized, iron-rich blood can appear greenish, like the deep green of veins or oxidized metals. This duality mirrors the alchemical symbolism of the Green Lion and the Red Lion. The Green Lion devouring the Sun represents raw potential and transformation, while the Red Lion symbolizes purity, balance, and mastery.

If iron represents balance in the cosmos, then we, as beings whose lifeblood depends on iron, are a reflection of that balance. Our blood, vibrant red with oxygen, is the earthly counterpart to the cosmic green of iron's spectral light. In a sense, we are truly made in the image of the cosmos…a photo negative of the divine mystery.

Alchemy and the Philosopher's Stone

The Red Lion represents the completion of the Great Work, signifying purity, strength, and transformation into the Philosopher's Stone. This legendary substance, sought by alchemists for centuries, symbolizes the ultimate goal of spiritual and material transformation. Just as the universe strives for balance through iron, we too are on a journey of transformation, from a baby of raw potential evolving into age-old wisdom.

We are all lions, little ions, if you will! We are imbued with strength and power that we must learn to recognize. In the Bible, both God and the devil are described as lions. The same symbol can embody both positive and negative qualities, just like humans and the stars. We are already the Philosopher's Stone, but our consciousness hasn't yet caught up with that miraculous fact. We are all made of the same stuff…plasma, energy, or power, whatever you may call it, it's how we use it that matters. Our consciousness needs to align with that balance and with self-awareness we choose to direct our energy toward creation rather than destruction.

Revisiting Mendeleev and Nicholson

In 1902, Dmitri Mendeleev, the Russian chemist best known for creating the periodic table, proposed two hypothetical elements lighter than hydrogen: an all-permeating gas (*newtonium or aether*) and coronium (*highly ionized iron*). These "emissions," once dismissed, might hold the key to understanding plasma's role in the universe.

Similarly, John William Nicholson's oscillatory model of the atom, though overshadowed by Niels Bohr's quantum theory, offers valuable insights. Nicholson suggested that electrons vibrate within their orbits, emitting light through continuous oscillations rather than discrete jumps. This vibrational model aligns with modern ideas about plasma and holography, hinting at a deeper, interconnected reality where spectral lines are not just emissions but vibrational signatures of what I call plasma-consciousness synergy. (a *dynamic interplay of plasma and consciousness that creates emergent outcomes—more in Chapter 3*)

To put it simply, Nicholson's model is like strumming a guitar string, the electron vibrates at different frequencies, producing light just as a vibrating string produces different tones depending on how it's played. Bohr's model, on the other hand, is more like jumping piano keys—electrons exist at fixed energy levels, like specific keys on a piano. Instead of oscillating smoothly, they "jump" from one level to another, emitting or absorbing a specific amount of energy, just like pressing a piano key produces a distinct note instead of a continuous slide of sound.

While Bohr's model excelled at predicting observable phenomena and laid the foundation for modern quantum mechanics, it focused on discrete energy levels and probability. This approach, though incredibly successful, left less room for exploring the more dynamic, subtle interactions that might occur within the plasma field or in the context of a holographic universe. Nicholson's continuous oscillatory model, rooted in vibration and geometry, might explain the underlying substrate of reality itself.

Plasma Bubbles and Vibrational Reality

This ties into my idea of a plasma bubble as an atom, oscillating based on its interactions with light and consciousness. If spectral lines are tied to vibrations within atomic geometry, it suggests that vibration is a universal constant—a means by which the universe encodes and transmits information. The visual representation of light and spectral lines is a projection of these vibrations, much like a hologram is a visual projection of encoded data.

In this framework, electrons might act as localized information nodes within a larger plasma network. Instead of "charged particles", they'd be consciousness-plasma interfaces, modulating between vibrational states to encode and retrieve energy and information. This could explain why plasma sometimes behaves intelligently; its fundamental units (*electrons and ions*) may not be discrete but more like little bees in an oscillating feedback system.

Theosophy and the Life-Atom

This idea resonates with Theosophy, a spiritual movement founded in the late 19th century by Helena Blavatsky, a Russian mystic and philosopher. Theosophy seeks to explore the underlying unity of all religions and philosophies, emphasizing the interconnectedness of all life and the existence of a universal, divine wisdom.

Blavatsky proposed her own version of the atom, called the life-atom which was a triple-natured entity composed of will (*proton*), love (*electron*), and active intelligence (*neutron*). This also aligns with a triad of consciousness, plasma, and awareness. She saw the atom and the soul as synonymous, with human souls, planets, and solar systems acting as macrocosmic atoms.[14] Nicholson's model, with its emphasis on vibration and geometry, aligns with this mystical perspective, suggesting that the atom is not just a mechanistic structure, but a vibrant, vibrational entity embedded in a plasma field that interacts with consciousness, information and intelligence at subtle levels.

A Call to Reimagine Science

The adoption of Bohr's model marked a shift toward a more quantized, probabilistic view of the atom, which was revolutionary and necessary for its time. However, it also shifted focus away from the continuous, interconnected processes that Nicholson's model explored. Revisiting his framework could open new avenues in plasma physics, where collective oscillations and geometric interpretations remain highly relevant.

Science often prioritizes models that offer quick, measurable results, but intuition and interconnectedness are equally vital. As we move forward, let's honor the early explorers like Nicholson and Mendeleev, whose ideas, though overlooked, might hold the key to a deeper understanding of plasma and its connection to consciousness.

Let's also remember the contributions of Theosophy, and other esoteric traditions, which have long viewed the universe as a living, interconnected web of energy and consciousness. By integrating these perspectives, we can move beyond the limitations of mechanistic science and embrace a more holistic understanding of reality…one where plasma, consciousness, and the cosmos are deeply intertwined.

Paranormal X-Rays

In the British Association President's Address, Sir William Crookes referenced Roentgen waves, what we now call X-rays.[15] Do you know how they got that name? In early texts, he referred to them as Chi-waves, after the Greek letter Chi, which resembles an X. Chi is also associated with Christ, as it's often used as shorthand for his name. Interestingly,

---

[14] Bruce Johnson, "The Ultimate Physical Atom," *Shining Lotus Metaphysical Bookstore Newsletter*, September 2, 2011, reprinted online at *Shining Lotus*, accessed April 6, 2025, https://www.shininglotus.com/the-ultimate-physical-atom/

[15] Sir William Crookes, "Presidential Address before the British Association for the Advancement of Science, Bristol, 1898," in Science 8, no. 200 (November 4, 1898): 561–75, https://www.jstor.org/stable/1626447?seq=5

Chi is another pronunciation for Qi, the concept of vital energy or life force in Chinese philosophy. There's no coincidence with language or names…there never is. I believe that this *"Christ Energy"* or *"Vital Force"* is a type of plasma ray or emission.

It was also stated at the time that the "X" stood for the unknown, an unknown type of radiation or communication. This mystery is what inspired Nikola Tesla to investigate radiant matter or energy. After observing damage to photographic film in his lab that seemed to be caused by Crookes tube experiments, Tesla began making his own X-ray images using high voltages and tubes of his design, as well as Crookes tubes.

X-rays were formally discovered by Wilhelm Röntgen in 1895. While working with a cathode-ray tube, Röntgen observed that it emitted mysterious rays that could pass through solid objects and create images of bones or metal on photographic plates. The way Röntgen found these "X-rays" is telling, *he noticed a glowing green emission*, much like the color from Coronium and Nebulum, while studying Crookes tubes.

Think about it: the high-energy photon interaction in the creation of an X-ray image literally impresses an image onto a surface, exposing hidden layers of our reality. That faint green glow observed in early X-ray experiments could be more than just a byproduct of the process. It might hint at the involvement of a higher state of plasma or plasma-consciousness interaction. Possibly a crossover into or out of dimensional coherence. Plasma is a vehicle for information, so this green glow could symbolize the interplay between light, matter, and the deeper layers of creation. It invites us to consider whether such high-energy interactions reveal not just the physical structures of our bodies but also the underlying mechanisms by which reality itself is projected and sustained.

The discovery of X-rays sparked reactions ranging from scientific curiosity to paranormal speculation. X-rays, along with wireless communication, were associated with the concept of *aether*—a mysterious medium thought to permeate space and transmit energy. At the time, plasma had not yet been named or understood, so aether became a placeholder for these invisible forces. X-rays were also linked to telepathy and other paranormal phenomena, as they seemed to demonstrate the existence of invisible forces capable of affecting reality at a distance. Many occultists and physical researchers believed these new technologies held the key to mind reading and other psychic abilities.[16]

I personally think there are countless applications for plasma and X-rays beyond the purely "scientific." It seems we looked at these inventions and thought, "How can we make

---

[16] Simone Natale, "A Cosmology of Invisible Fluids: Wireless, X-Rays, and Psychical Research around 1900," *Canadian Journal of Communication* 36, no. 2 (2011): 263–275

money off this?" rather than asking, "What can we learn from this, and how can we apply it in new ways?" If plasma, consciousness, and electromagnetic energy are interconnected, it's worth exploring whether X-rays, or other high-energy photons, might play a role in detecting or interacting with these subtle phenomena.

Crookes' open-mindedness to X-rays must have been met with scepticism. In his Presidential Address to the British Association for the Advancement of Science, delivered on September 7, 1898, in Bristol, he said:

"To stop short, in any research that bids fair to widen the gates of knowledge, to recoil from fear of difficult or adverse criticism, is to bring reproach on Science. There is nothing for the investigator to do but to go straight on. To explore up and down, inch by inch, with the taper, his reason. To follow light wherever it may lead, <u>even should it at times resemble a Will-O'-The-Wisp</u>. I have nothing to retract. I adhere to my already published statements. Indeed, I might add much thereto. I regret only a certain crudity in those early expositions which, no doubt, justly militated against their acceptance by the scientific world. My own knowledge at that time scarcely extended beyond the fact <u>that certain phenomena new to science had assuredly occurred, and were attested by my own sober senses,</u> and, better still, by automatic record. I was like some two-dimensional being who might stand at the singular point of a Riemann's surface, and thus find himself in infinitesimal and inexplicable contact with a plane of existence not his own. I think I see a little farther now. I have glimpses of something like coherence among the strange elusive phenomena, of something like continuity between those unexplained forces and laws already known."[17]

I believe Crookes was describing plasma, the plasma field, X-rays, and possibly interdimensional or inter-reality communication. Imagine talking about that now...never mind in the late 1800s!

Echoing the cadence of Sir William Crookes (*and remember, he was the one who first discovered radiant matter*), I ask: Could X-rays and other high-energy photons be a form of communication or shadow from higher dimensions or realities? And was this overlooked because of commercialization? Could this have been one of the greatest missed opportunities in science, and even in metaphysics?

---

[17] Crookes, *Presidential Address*, 21.

"My ear barely caught signals coming in regular succession which could not have been produced on earth, caused by any solar or lunar action or by the influence of Venus, and the possibility that they might have come from Mars flashed upon my mind." — Nikola Tesla, 1919

Tesla, like Crookes, was deeply engaged with radiant matter (*plasma*). His experiments with high-energy electromagnetic interactions led to strange phenomena—unexplained voices, disturbances, and hints of non-local intelligence. While he never fully published these findings, he strongly hinted at their existence. If plasma is indeed a vehicle for consciousness and information, as I believe, then X-rays interacting with plasma fields might generate or detect signals beyond our normal sensory perception.

Depiction of Sir Wiliam Crookes | Experimenting with plasma

Sadly, instead of following Crookes' advice to "go straight on" exploring these forces, X-rays were rapidly commercialized and confined to medical imaging and industrial applications. What if X-rays were more than just a tool for seeing bones? What if they were a lost key to interdimensional communication, capable of revealing not just physical structures but informational ones as well?

Imperfect Junctions

Crookes went further in his address, speculating about telepathy and the medium through which it might occur. He wrote:

"All the phenomena of the universe are presumably in some way continuous, and it is unscientific to call in aid of mysterious agencies when with ever fresh advance in knowledge it is shown the ether vibrations have powers and attributes abundantly equal to any demand, even to the transmission of thought."[18]

He suggested that the gaps between nerve cells in the brain might function like a Branly or Lodge coherer—a device used in early radio technology. The Branly coherer consisted of a glass tube filled with loose metal filings that would "cohere" and conduct electricity when exposed to an electric spark. Crookes proposed that the brain's nerve gaps might similarly cohere under certain conditions, allowing for the transmission or reception of thought vibrations.

The coherer, though obsolete by 1907, remains fascinating. Recent experiments suggest that the particles in a coherer might micro-weld under radio frequency electricity (*resonance*), creating a kind of quantum tunnelling effect across imperfect junctions.[19] This phenomenon is still not well understood, but it hints at something profound: the possibility that consciousness itself might arise from similar "imperfect junctions" in the brain.

I believe these junctions are filled with an invisible plasma crystal lattice—a medium that facilitates the generation of consciousness or plasma-consciousness interactions. This might occur in the electron clouds (*corona or plasma*) theorized in Orch-OR (*Orchestrated Objective Reduction*) theory, where quantum processes in microtubules, in a gap or junction between polarized tubulin are thought to play a role in consciousness[20]. These processes could exist in higher dimensions, invisible to us but fundamental to how we perceive and interact with reality.

---

[18] Crookes, *Presidential Address*, 16.
[19] Tapan K. Sarkar et al., *History of Wireless* (Hoboken, NJ: Wiley-Interscience, 2006), 173–175.
[20] Stuart Hameroff, *Consciousness, Microtubules, & 'Orch OR': A 'Space-Time Odyssey'*, *Journal of Consciousness Studies* **21**, no. 3–4 (2014): 126–53

Crookes seemed to anticipate this when he wrote:

"The structure of brain and nerve being similar, it is conceivable there may be present masses of such nerve coherers in the brain whose special function it may be to receive impulses brought from without through the connecting sequences of <u>ether waves</u> of appropriate order of magnitude."

He could have been describing microtubules and the electron clouds between dipoles, where quantum processes might generate consciousness. The Branly coherer, though abandoned for classical radio transmission, might still hold relevance for neural or quantum-plasma research. Where scientists saw a gap, I see a gate to another world. Imperfect junctions are more like dimensional thresholds, lit up by awareness. These "imperfections" could be the key to understanding how plasma-consciousness interactions occur.

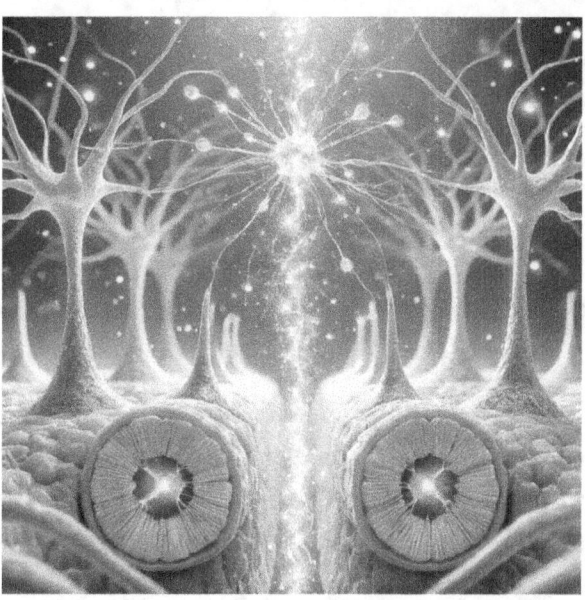

Crookes also spoke of humanity's potential for evolution and the healing power of nature. He believed that by tapping into the aether (*or plasma, as we might call it today*), we could unlock abilities like telepathy and other psi phenomena. He wrote:

"The human race has reached no fixed or changeless ideal; in every direction, there is evolution as well as disintegration."

This resonates deeply with me. The subconscious mind and its invisible, imperfect junctions woven with plasma lattices, may be the gateway to our hidden abilities. Once we understand the power of resonance and relationship, it will feel more and more natural to learn how to use with intention, rather than unconsciously. To become a vibratory, feeling being who thinks, not just a thinking being who sometimes feels. I think we have been running from feeling for a long time and it's finally catching up with us. Crookes saw this as a sign of humanity's upward evolution by stating:

> **"The vis medicatrix thus evoked as it were from the depths of the organism is a good omen for the upward evolution of mankind."**

This is so important. *"Vis medicatrix"* means the healing powers of nature. It's the idea that we have the tools within us to heal ourselves. Plasma, ether waves, whatever you want to call it—this is the healing nature and evolution he's talking about. He continues:

> **"A formidable range of phenomena must be scientifically sifted before we effectually grasp a faculty so strange, so bewildering, and for ages so inscrutable, as the direct action of mind on mind."**

Crookes knew he was onto something back then, and I can feel his excitement because it mirrors my own. But his ideas fell on deaf ears. They still do. The power to heal yourself is often labelled pseudoscience. Try this, tell your cells for 10 minutes a day: "My cells are joyous, my cells are healthy, my cells are light and love".[21] You tell me if you don't feel a change. The truth is the mind and body *can* heal. The other truth is that sometimes, despite all your efforts, it doesn't happen. This doesn't mean failure. It may mean there's an emotion still held, a lesson waiting to be seen, or perhaps it's simply time to transition to a whole other life on a vaster plane.

We cannot control everything, but we can participate. We can choose resonance; we can choose belief. We can answer the call of sickness, listen, and respond with love. We can do everything within our power to align with life and then, we must let go. Not to give up but to return to the mystery. Just like we live knowing we will one day die, but somehow, we choose to love anyway. But make no mistake plasma can heal; it can bring you into a reality of your highest alignment and that you must surrender to —whatever the outcome.

What I'm about to share next still blows my mind. It was written over a hundred years ago by a man who, sadly, received little to no recognition.

---

[21] I Said This & My Gut Started HEALING (Meditation), YouTube video, posted by Gut Feelings Meditation, https://www.youtube.com/watch?v=TE3GUWO9fKI

He wrote:

"It has been said that nothing worth the proving can be proved, nor yet disproved. True though this may have been in the past, it is true no longer. The science of our century has forged weapons of observation and analysis by which the verist tiro may profit. Science has trained and fashioned the average mind into habits of exactitude and disciplined perception, and in so doing has fortified itself for tasks higher, wider, and comparatively more wonderful than even the wisest among our ancestors imagined. Like the souls in Plato's myth that follow the chariot of Zeus, it has ascended to the point of vision far above the earth. It is henceforth open to science to transcend all we now think we know of matter, and to obtain new glimpses of a profounder scheme of cosmic law."[22]

Crookes was saying that the mysteries once thought unknowable are now within reach, not because they've changed…but because *we have*. He was asking for science's help. He *believed* in science. He was calling for serious, disciplined investigation, not dismissal or blind belief. If you're a scientist reading this, maybe you can tap into the magic of the late 19th and early 20th centuries and help him. There's clearly so much more to discover, but it was sadly obscured by logic and capitalism.

Crookes was essentially saying:
1. **Telepathy and mind-to-mind interaction are real**, but science hasn't yet caught up to understanding them.
2. **It will take serious, disciplined research** to separate truth from illusion.
3. **There is an unknown medium**, what I believe is a living aether or finer plasma, that allows this to happen, and we need to investigate it.

Nikola Tesla & The Sims

Now back to another visionary: Nikola Tesla. Around the same time Crookes was exploring these ideas, Tesla was working with what he called *radiant matter* or *radiant energy*. Tesla (*1856–1943*) was an inventor, electrical engineer, mechanical engineer, and futurist. In 1894, he began experimenting with radiant energy after noticing damaged film in his laboratory, which he later attributed to what we now call X-rays.[23]

---

[22] Crookes, *Presidential Address to the British Association for the Advancement of Science*, BMJ (1898), 62.
[23] Nikola Tesla. *X-ray Vision: Nikola Tesla on Roentgen Rays*. 1st ed. Radford, VA: Wilder Publications, 2007.

Tesla believed that an all-pervasive aether transmitted electrical energy. In a 1901 patent, he stated that radiant energy came from cosmic rays and our sun. He also spoke about cosmic rays originating not just from our sun but from every star in the universe, capable of penetrating solid matter.

Tesla described the process of cosmic rays ionizing the air, releasing ions and electrons, which could be captured in a condenser and discharged through a motor circuit. In essence, he was describing the creation of plasma. Tesla used an X-ray tube powered by a Tesla coil to harness radiant energy, demonstrating his deep understanding of these invisible forces. [24]

To Tesla, radiant matter seemed to be more particle-like (*linked to cathode rays, plasma, and subatomic emissions in vacuum tubes*) while radiant energy was more field-like (*a pervasive cosmic energy, harnessable and not limited to material particles*). In Tesla's own terms, radiant matter could be a carrier of radiant energy, but radiant energy itself was more fundamental—potentially the very power of the universe. Tesla may have been harnessing what I call fourth-dimensional plasma or multidimensional plasma at its core, which I will expand on soon.

Tesla believed Radiant Energy was not just electromagnetic radiation (*light, heat*), but a non-mechanical, ether-like energy capable of being captured and transformed. It was how particles were capable of traveling through space without significant energy loss. He designed experiments to capture ambient energy (*including from cosmic rays and ultraviolet radiation*) to power devices wirelessly. He theorized it could provide unlimited energy source which led to his work on the Wardenclyffe Tower.

In his writing *Man's Greatest Achievement*, Tesla elaborated on his vision:

"Long ago he recognized that all perceptible matter comes from a primary substance. Of a tenuity beyond conception and filling all space. The Akasha or Luminiferous Ether. Which is acted upon by the life-giving Prana or creative force, calling into existence, in never-ending cycles, all things and phenomena."[25]

[24]Tesla Research. "Radiant Energy." Accessed August 6, 2025.
https://teslaresearch.jimdofree.com/radiant-energy.
[25] Tesla, Nikola. "Man's Greatest Achievement." *New York American*, July 6, 1930. Accessed August 6, 2025. https://teslaresearch.jimdofree.com/articles-interviews/man-s-greatest-achievement-by-nikola-tesla-new-york-american-july-6-1930

In this context, *Prana* symbolizes our unique consciousness and creative perspective, while *Akasha* or *Luminiferous Ether* represents what I believe to be multidimensional plasma.

Tesla went on to describe humanity's potential to harness these forces:

"Can Man control the grandest, most awe-inspiring of all processes in nature? Can he harness her inexhaustible energies to perform all their functions at his bidding, more still-can he so refine his means of control as to put them in operation simply by the force of his will? If he could do this, he would have powers almost unlimited and supernatural. At his command, with but a slight effort on his part, old worlds would disappear, and new ones of his planning would spring into being. He could fix, solidify, and preserve the ethereal shapes of his imagining, the fleeting visions of his dreams. He could express all the creations of his mind, on any scale, in forms concrete and imperishable. He could alter the size of this planet, control its seasons, and guide it along any path he might choose through the depths of the Universe. He could make planets collide and produce his suns and stars, his heat and light. He could originate and develop life in all its infinite forms. To create and annihilate material substance, cause it to aggregate in forms according to his desire, would be the supreme manifestation of the power of Man's mind, his most complete triumph over the physical world, his crowning achievement which would place him beside his Creator and fulfill his ultimate destiny."[26]

While Tesla's vision is awe-inspiring, I believe he overlooked one crucial aspect: the *intelligence* of the luminiferous ether—or plasma. This substance isn't just a passive medium; it has a mind of its own. It's like a wise grandmother who knows the bigger picture, allowing us free will to make our own mistakes while gently guiding us toward growth. It is not about power and control, it is about alignment and harmony, that is what brings lasting wholeness.

Think of it like playing *The Sims,* the life simulation game (*my favorite*). You design your characters, build their world, and sometimes step back to watch what unfolds—good, bad, weird, or hilarious. You guide them, but you don't always intervene. Plasma, what some might call ether or the substrate of reality, operates in a similar way. Who wouldn't believe there's a greater force at work behind the scenes, and *that it jut might be your expanded self?*

[26] Nikola Tesla, "Man's Greatest Achievement," *New York American*, July 6, 1930, accessed August 6, 2025, https://teslaresearch.jimdofree.com/articles-interviews/man-s-greatest-achievement-by-nikola-tesla-new-york-american-july-6-1930/.

It collaborates with us. It holds our patterns, potentials, and preferences. But it also contains its own harmonic intelligence, one that moves in resonance with our *highest alignment*, even when that path includes detours, ruptures, or breakdowns.

But here's they key difference between us and the Sims *we* create: Most of us play The Sims for fun, control, or chaos. We, here on Earth, weren't created *for* entertainment. We were created *through* an act of love, by a Source (*or Self*) that wanted to know itself more deeply. From *benevolent curiosity* not boredom.

In The Sims, when something "bad" happens, it's often because you made it happen, for drama, for the plot, or just to see what unfolds. But in *this* meaningful, evolving world, the "bad" holds a purpose. Although unfair, the loss, the rejection, the delay—no one is punishing you. These events are realigning you. They are reshaping you towards something even more true. This reality is not hollow, it's hallowed. There is intelligence here, it is not code. There is empathy here, not just entropy. So no…this is not a simulation in that sense, it is a living, evolving feedback loop between creation and creator. Between your self and your Self. Plasma (*aether*) facilitates all of this. And yes, behind the scenes, something collectively greater than the sum of its parts. And it loves you.

I do think that, with time, we'll learn to harness plasma and achieve many of the things Tesla envisioned, even if it takes thousands of years. I also believe we can use plasma to create our own destinies. But here's the thing: it's more important to make sure we're consciously connecting to our soul or plasma in the first place. We need to ensure that our vision of destiny comes from the heart, not the ego. Remember the saying, *"Be careful what you wish for"*? That's a gentle warning from Plasma itself.

My intention with this book is to reconnect you with this Plasma, and, in turn, with yourself, so you can co-create your destiny with this beautiful, intelligent force. By aligning with it, you open yourself to miracles beyond your wildest dreams while holding your goals lightly, trusting that the journey is as important as the destination.

Life would be boring without surprises, wouldn't it? This is a story we're writing together, filled with dragons, tricksters, angels, and demons at every turn. And this reality, the one you're conscious within right now, is the version where your current awareness holds the most potential to evolve. Yes, it may be true that infinite versions of you exist, some experiencing realities that seem far from their highest good. That's where co-creation comes in. Awareness is your tuning fork. You have the ability to choose, from your heart and soul, and in that choice, Plasma reflects your alignment back to you. Belief is a gateway, and Plasma hears it loud and clear.

Plasma always offers the conditions for your highest growth in *every* timeline. She is the soil. But it's your awareness, your choices, that determine how the seeds sprout. Even in the darkest timelines, she is still offering this. She is not punishing you and she is trying to reflect to you so you can grow! And the self in each version determines what is reflected and whether it comes from fear or freedom. You choose whether what happens becomes a blessing in disguise, or a complete breakdown.

Three Tools For Your Highest Alignment

I present you with three practices which have created experiences for me that I cannot put into words.

1. **Choose Feeling Over Fear:** Whenever possible, choose to feel instead of freeze. Fear closes the field, so it loops the same old stuff, and curiosity opens it for something new. When you feel, when you make space, without judgment, you engage with the plasma honestly, and it begins to reorganize around your truth.
   Practice: Sit with your body and ask: *"What am I truly feeling beneath this story?"* Let that be your guide, not just logic or what's happened in the past. Awareness + Feeling = Reality Creation.

2. **Live as a Conscious Collaborator:** Stop asking *"Why is this happening to me?"* and start asking *"What is this showing me?"* That shift alone moves you from victim to co-creator. Plasma responds to *you*. Life is the mirror, don't wait for the reflection to change; change your stance first.
   Practice: In tough moments, say: *"I choose to meet this with awareness, not fear."* There's not always a meaning right away, that's the mystery.

3. **Make Beauty A Ritual**: Beauty magnetizes benevolence. When you seek awe, when you create something with love, a meal, your night routine, or redecorating the room in which you spend the most time, you are *directly* signaling to plasma that you are here to participate in alignment.
   Practice: Turn simple acts into ceremony. Light a candle before you write. Bless your coffee. Speak to the stars. Life becomes magic when you treat it as such.

Brain Break

At this moment, I want to invite you to take a break. With all this talk of plasma, aether, and cosmic forces, it's important to ground yourself. Go outside…whether it's to bask in the sun, feel the wind on your skin, or simply take in the beauty of the sky. Breathe deeply and reconnect with the world around you. Call an old friend and have a meaningful conversation. Ask a parent or child about their life, their dreams, their stories.

After all is said and done, it's the acute sweetness of each present moment that truly counts.

Video Killed the Radio Star

Picking up where we left off on the mystical, magical aether, let's talk about what I like to call *dead aether*. By the time Paracelsus, the alchemist, had passed away, Robert Fludd (*1574–1637*), a Renaissance physician, alchemist, mystic, and esoteric philosopher was one of the few remaining voices proposing a *living aether*. Fludd completed his Doctor of Medicine at Oxford in 1605. He viewed aether as as a living, divine substance—an essential spiritual reality that animates and connects all things. For Fludd, light was not just electromagnetic radiation, but a spiritual substance emanating from the divine source through aether. This light was formative, shaping matter and life. He believed the human soul was capable of interacting with the aetheric realm through imagination, will, and contemplation.

Utriusque Cosmi Historia | Robert Fludd

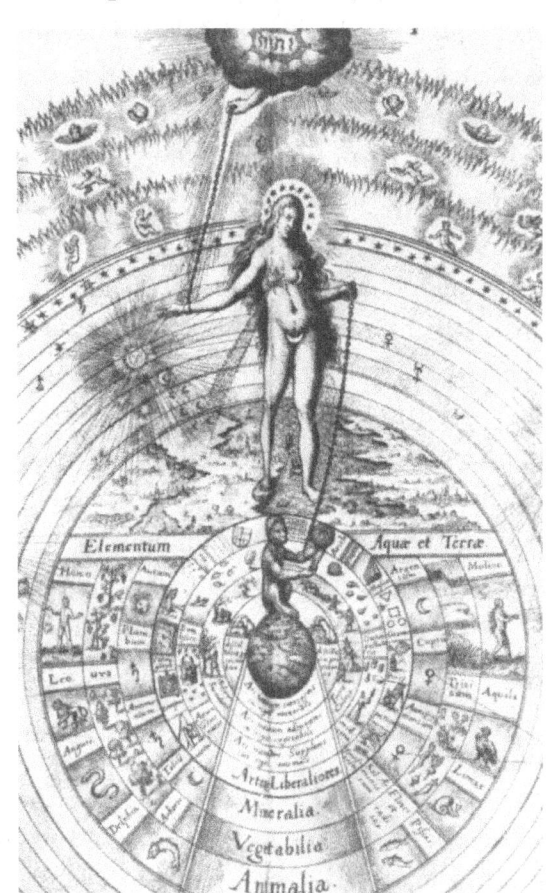

But then came Descartes in 1596, ushering in a mechanistic movement that would dominate scientific thought.

Descartes proposed a mechanical aether, critiquing Fludd's work for being overly mystical and lacking empirical rigor. He dismissed Fludd's esoteric and symbolic diagrams as meaningless without a solid mathematical foundation. Sound familiar? This mirrors the later shift from Sir William Crookes' original, paranormal descriptions of radiant matter to the now-repackaged, mathematical, and scientific understanding of plasma. In both cases, the mystical was stripped away, replaced by the promise of the visual, the physical, and the controllable. The subtle, the mystical, the uneasy-to-understand flux…these were pushed aside.

Fludd, in turn, saw Descartes' mechanistic worldview as soulless and reductive, failing to account for the spiritual and divine aspects of existence. The aether, once seen as a spiritual, subtle, and divine medium through which the cosmos was animated and harmonized, was reduced to a material and mechanistic substance. It seems that every time the mystical rears its head, society isn't ready for it. Until now.

The intellectual disagreement between Fludd and Descartes reflects the broader transition from Renaissance mysticism to the Scientific Revolution. But why was Descartes like this? Let's unpack.

Descartes, often called the father of modern philosophy, played a key role in attaching logic to what was once a mystical, intuitive practice meant for play and curiosity. Original philosophy, with its openness to wonder and the unknown, was squeezed into a logical framework—something less scary, less uncontrollable, less *feminine*. This new philosophy became the perfect partner to the impending Scientific Revolution, aligning with the cultural shift toward rationality, order, and control.

Descartes believed the universe operated like a machine governed by mathematical laws, devoid of intrinsic life or consciousness. To him, the aether was a subtle material substance, not alive, that filled all space and transmitted forces like light and planetary motion through mechanical interactions. By mechanizing the aether, he avoided invoking metaphysical or animistic explanations, aligning with the new scientific emphasis on observation and mathematics.

We must understand, even today, many scientists remain afraid to speak publicly about their mystical experiences and beliefs. They often placate academic institutions to preserve their careers and reputations—and I stand by my belief that Descartes was one of many who did the same.

Descartes was also a devout Catholic, which likely influenced his view of the aether. A living, spiritual aether might have clashed with his outward beliefs in an all-powerful God. Ironically, the aether could have been a bridge to understanding what some call God, or what I call the Mystery. There must be something that moves this aether, something from which our consciousness stems. But Descartes saw the universe as God's perfectly designed machine, created and then left to operate independently. This view emphasized natural laws over divine intervention, keeping the Church at bay and Descartes in good standing.

The Catholic-Protestant divide during the Reformation and Counter-Reformation created a volatile environment. Descartes' mechanistic philosophy avoided theological controversy by stripping nature of "spirit" and focusing on physical causes. He also lived under monarchical systems, where knowledge and scientific exploration were often tied to political power. His work served the interests of centralizing authority by offering predictable, controllable models of nature. This, as I said, is a little-spoken scarlet letter in the realm of science, even today.

Another reason I am glad I am an outsider (*thought I felt had shame about it*) which gives me the freedom to write this from a perspective of truth, not colored by ulterior motives for politics, power, or financial gain. Descartes served as a soldier in his youth, experiencing firsthand the value of discipline and control. This also may have influenced his mechanistic worldview. He studied at the Jesuit college of La Flèche, where he was exposed to scholastic philosophy and Catholic doctrine. This training likely shaped his preference for systematic, hierarchical models.

But here's the strange thing: Descartes corresponded with Queen Christina of Sweden and Princess Elisabeth of Bohemia on topics <u>far</u> beyond his public mechanistic worldview. In these letters, he explored love, the infinity of the universe, the nature of the sovereign good, and the union of mind and body. With Princess Elisabeth, he delved into the interaction between the immaterial mind and the physical body, acknowledging the complexity of this union. He even hinted that the soul-body connection might involve principles beyond mechanistic physics, suggesting a deeper, more holistic relationship. [27]

I believe Descartes, like many mechanists even today, was a spiritualist at heart. Behind closed doors, he grappled with ideas he couldn't openly express. Perhaps he, like

---

[27] Lisa Shapiro, "Princess Elizabeth and Descartes: The Union of Soul and Body and the Practice of Philosophy," *British Journal for the History of Philosophy* **7**, no. 3 (1999): 503–520, https://doi.org/10.1080/09608789908571042

others, was constrained by the need for validation, attention, and acceptance—trapped by the desire to please those in power and the scientific community.

If we can heal that need for external validation, we can become truth-tellers in the open, not just in secret. Why can't these structures of fear and control be torn down? I hope they can. Why are they still in place? Some might argue there's an evil cabal at work, but I tend to believe it's simply a matter of fear. Fear of change, fear of losing power, fear of the unknown. I hope this book inspires those who uphold these structures to embrace change, love, and curiosity. They won't lose power; instead, everything will expand tenfold, for us and for them. There's plenty for everyone. Let's make this world a better place, like I know we can.

Lorentz, and the Rise and Fall of Aether

With the rise of mechanistic science came the concept of *luminiferous aether*. The first mention of it was by Thomas Young and Augustin-Jean Fresnel in the early 1800s. Fresnel proposed that light waves required a stationary luminiferous aether for propagation.

Then came James Clerk Maxwell, who believed electromagnetic waves needed a medium to travel through, a luminiferous aether. Hendrik Lorentz further developed this theory, introducing concepts like the Lorentz transformations to describe how objects interact with the aether at high speeds. These transformations showed how time, length, and mass change for objects moving relative to the aether, preserving the appearance of the constancy of the speed of light.

Lorentz even built a mathematical framework to explain it all. While his theory was ultimately superseded by Einstein's special relativity, Lorentz's transformations became a foundational component of modern physics.

So, Lorentz's work was absorbed into something greater, but the aether itself was disproved. What went wrong?

Einstein's special relativity, a scientific theory describing the relationship between space and time, eliminated the need for the aether. It proposed that light did not require a medium to propagate. However, Einstein later reneged on this idea, as we'll soon discuss.

But how did Lorentz's transformations seamlessly integrate into special relativity without the aether? The focus shifted to the geometry of spacetime, which I believe is a synergy of plasma and consciousness (*together they create a field of information*) working together to create our reality. Lorentz's transformations are now used to explain phenomena like time dilation and the behavior of electromagnetic fields in moving frames. Yet, I think they can be reapplied to a new understanding of a living plasma or aether.

The key difference lies in the way Lorentz transformations allow for interaction with a medium (*such as plasma or aether*) that could reflect how consciousness might affect space-time. Special relativity treats spacetime as an immutable, static entity governed by laws of physics. However, Lorentz transformations, in my expanded model of plasma consciousness synergy (*more in Chapter 3*), suggest that time, space, and energy can shift dynamically based on conscious awareness.

Time and space are relative, and shift based on the frame of reference—but here, the frame of reference is not just physical velocity but conscious awareness and emotional energy interacting with the plasma field. It brings our humanity and awareness back into the equation. The faster we move in consciousness (*greater self-awareness and emotional connection*), the denser we become in the field, shaping the physical world, just as increasing velocity in Lorentz transformations leads to increasing mass and energy. By changing your frame of reference, and now knowing that is possible, your awareness increases, you can shift timelines, meaning you can choose a new path in life or a new experience at any time, much like changing the trajectory of an object at relativistic speeds.

In summation, by applying Lorentz transformations to plasma consciousness synergy, we can understand that time and space are malleable when consciousness is involved, and when one is aware of this brings a magic spark to one's life, as the possibilities are endless.

In the early 1900s, the luminiferous aether (*dead aether*) was widely accepted, until the Michelson-Morley experiment failed to detect its existence. This null result led to the sweeping abandonment of the aether concept completely, paving the way for Einstein's special relativity. Aether theories never regained traction, until recently with this new wave of Plasma.

Michelson Morley, Fear, & Auras

The Michelson-Morley experiment, conducted in 1887 by Albert A. Michelson and Edward W. Morley, aimed to measure the Earth's motion through the aether, often referred to as the *aether wind*. Its goal was to detect the presence of the luminiferous aether.

The experiment's null result disproved the aether's existence and buried the concept, leading to the development of modern physics, including relativity and quantum theory. These two schools of thought, while groundbreaking, often seem to clash, leaving glaring inconsistencies. I believe plasma could be the missing link that bridges the gap between them.

Relativity operates within the light cone—the idea that nothing can travel faster than light. Quantum mechanics, on the other hand, allows for instant communication through phenomena like quantum entanglement, which exists *outside* the light cone. Both are true, yet their coexistence creates a paradox that the scientific community has struggled to reconcile.

I propose that Plasma is the underlying field where both relativity and quantum mechanics operate, harmonizing their differences and resolving the paradox. Plasma allows for both determinism and free will, a framework where fate and possibility coexist. Imagine plasma as a medium where all potential realities exist simultaneously, but only some are brought into focus based on our choices, much like a quantum wave collapsing into a particle.

Plasma embodies the paradox of fate and free will. It holds the framework of a pre-determined fate, patterns that repeat across scales and time, while also offering pathways to deviate, reconfigure, and explore alternate possibilities through conscious intention. Plasma doesn't quite resolve the paradox; it *embodies* it, showing us that fate and free will are not opposites but complementary forces within a living, multidimensional field.

I'll build on the idea of fate and free will in subsequent chapters. For now, let's revisit where we took a misstep: the luminiferous aether. Let's start at the beginning.

"Not everything that can be counted counts, and not everything that counts can be counted." — Albert Einstein

At the beginning of the 20th century, most physicists believed in the luminiferous aether. James Clerk Maxwell, often considered the third greatest physicist of all time, behind Einstein and Newton (*Einstein himself said he stood on Maxwell's shoulders*), believed that the aether was a plenum in a state of perpetual motion. Unfortunately, Maxwell died in 1879, before the Michelson-Morley experiment took place.

In his 1890 address, *The Scientific Papers of James Clerk Maxwell, Volume 2, Ether*, Maxwell wrote:

"It is only when we remember the extensive and mischievous influence on science which hypothesis about aethers used formerly to exercise, that we can appreciate the horror of aethers which sober-minded men had during the 18th century, and which, probably as a sort of hereditary prejudice, descended even to the late Mr. John Stuart Mill."

Mill, a philosopher and advocate for social reform, represented the mechanistic mindset of his time. Maxwell's words suggest that the idea of an aether terrified scholars and scientists who preferred a predictable, mechanical universe. The aether was too unpredictable, too alive. It didn't fit into their tidy, clockwork view of reality.

In essence, Maxwell was saying: the bad reputation of aether wasn't because the concept itself was inherently absurd, but because the *past misuse* of it in science left people gun-shy, and that prejudice stuck for generations.

This highlights an important truth: science, like any field, is shaped by cultural and psychological factors as much as by logic and evidence. In this case, those factors hindered progress toward a more mystical, holistic understanding of the universe. Aether hypotheses were dismissed as speculative, untethered from observation which was a major concern for the empirically driven thinkers of the Enlightenment and beyond.

This is where we need to shift our approach. Instead of killing ourselves trying to "prove" this so-called paranormal plasma, perhaps we should accept that some things are meant to be experienced, not fully understood. Just like electricity, we can see its effects and use it, even if we don't fully grasp its nature. And just like love, it's hard to quantify. Maybe plasma, like the best things in life, exists in the realm of the immeasurable…the place where all that truly matters originates.

"What is essential is invisible to the eye." — Antoine de Saint-Exupéry

Maxwell also referenced a Mr. S. Tolver Preston, who proposed that the aether was like a gas whose molecules rarely interacted, with mean paths far greater than planetary distances. Doesn't that sound like plasma? When I looked into it, I found that particles in a plasma, electrons and ions, can interact via electricity and magnetism over far greater distances than in an ordinary gas. To me, this connection is too obvious to ignore!

Then there's *Frank Washington Very,* an astronomer, astrophysicist, and meteorologist who never received the credit he deserved. In his NASA obituary, it was noted that he played a significant role in working with Langley on the invention of the first heavier-than-air machine to achieve sustained flight.[28]

In 1918, Very presented a paper at the American Astronomical Society titled, *The Luminiferous Aether: 1. Its Relation to The Electron and A Universal Interstellar Medium 2. Its Relation to The Atom.* He argued that while the Michelson-Morley experiment had

---

[28] J. Gordon Ogden, "Frank W. Very," *Popular Astronomy* 36, no. 7 (August–September 1928): 413, accessed [Month Day, Year], NASA Astrophysics Data System, https://ui.adsabs.harvard.edu/abs/1928PA.....36..413O/abstract

dismissed the aether, there was still a need for a universal interstellar medium. He proposed a new term for it: *Aura*.[29]

Very spoke of "ether particles" formed by electrons out of the medium of the aura, something I believe is deeply connected to plasma. He was onto something profound. The way he described the aura was strikingly spiritual, and I could write an entire book on his ideas alone. But for brevity, I'll share a few key points that will lead me into my ideas on rethinking gravity and why the Michelson-Morley experiment might have been misinterpreted.

"The aura is stronger than steel, but this strength results from an intimate connection with an infinite source of energy which is spiritual. So does this invisible spirit control the body; and even so public opinion, a seemingly intangible thing, is stronger than the most powerful autocrat and in the end will prevail."

— Frank Washington Very

What Very said here feels prophetic—and I believe it. Plasma *is* spiritual. If you recall the concept of bioplasma, this is what I think he was referring to. Bioplasma is your aura. It's connected to a more infinite plasma and mystery. It's your interface with the world, a permeable membrane that can be influenced by the environment, consciousness, belief systems, and more. Think of it as your very own plasma-consciousness synergy. Just like a construct that surrounds and creates our universe, you have your own personal construct (*aura*) that you influence. And with this knowledge, can reprogram it, which we'll discuss later, as this will help expand and transform your worldview, as well as your view of yourself.

No matter who tries to argue against this from a logical standpoint out of fear, the truth always prevails, even if it takes time. And the truth about plasma has been trying to emerge for thousands of years. At the current precipice of science, where it is becoming almost indistinguishable from magic and we are closer than ever to understanding the origins of reality, I believe it is the perfect time to finally embrace these truths.

---

[29] F. W. Very, "The Luminiferous Ether—Its Relation to the Electron and to a Universal Atmosphere," *Publications of the American Astronomical Society* 4 (1922): 56

Here's another profound quote from Very that I want to build upon:

"The aura is the fundamental substance as Sir Oliver Lodge says of it in his searching analysis: "It cannot really be ordinary matter, because ordinary matter is definitely differentiated from it, and is presumably composed of it; but the inertia of ordinary matter, however it be electrically or magnetically explained, must in the last resort depend on something parentally akin to inertia in the fundamental substance which fills space."[30]

Frank Washington Very (*courtesy of NASA*)

[31]

Very means matter is something structured and localized, while the aura/aether is continuous and universal. He is saying, ordinary matter is made from this aura. Even the property of inertia (*resistance to motion*) in matter isn't intrinsic to matter itself, it's inherited from the aether. In other words, the "stuff" that gives matter its mass-like resistance is already present in this underlying cosmic medium.

Imagine this plasma as a field where your future ancestors exist, the realm of expanded selves—beings who can see through time into every moment, who know your soul's purpose in this lifetime. It doesn't mean your entire path is predetermined, it means you have a fate and within that fate are the fruits of destiny, the freshness that will come of following that path. New ideas actualized from freedom of choice; this is what we are here

---

[30] Very, "The Luminiferous Ether," https://archive.org/details/luminiferousethe00veryrich
[31] Ogden, "Frank W. Very," 413.

to create. That's the power you're always connected to, and Frank Washington Very instinctively knew this, as do I. As do you, in your heart, as you read this.

Oprah once spoke about her success in a podcast with Jamie Kern Lima. She said, and I'm paraphrasing,

*"**I am living God's dream for me. I am stepping out every day onto moonbeams and stars***.*"*

Oprah was echoing what Very was trying to say! She's connected to this source, believes in it, and surrenders to it—while creating from that place of surrender. And because of this, she's experienced immense joy and success. I am taking a wild guess here but I bet she refers to God as the energy of her expanded self and all expanded selves combined.

This is why intentionally connecting with this substance, that we are already connect to, and <u>listening to its whispers</u> can change all of our lives for the better. Once we remember our power, our lives will transform.

The connection between the *"inertia of ordinary matter"* and *"aura/aether"* implies that what we perceive as resistance to motion (*inertia*) in physical matter originates from properties inherent in this more subtle, underlying substance. This supports my belief that plasma carries informational, intelligent, and conscious imprints and organizes reality. It also suggests that matter is a localized, condensed form of this aura or plasma. That <u>everything</u> is plasma at different densities, some crystallized, some liquified and fluid, and some finer or mist-like. We are all in a living painting of memory, a painting of plasma!

Just remember, even when you feel lost, you are always connected to this source. It holds your highest good. Make choices from the highest version of yourself and trust the flow you're in…it may lead you to places you could never imagine.

When my father fell into a coma, I had finally began to settle into Los Angeles, forcefully proclaiming it would be my home. It felt like I was surrendering to a reality that didn't truly resonate with me, but I thought maybe I was wrong. Deep down, I've always wanted to live in the mountains on a bigger piece of land. But I couldn't bring myself to listen to that inner voice because I was fixated on the idea that Los Angeles was where I needed to be—even though I was incredibly lonely, missing nature, and craving a slower-paced lifestyle.

My father's illness brought me back to Florida, of all places, where I had lived ten years prior. Florida was where I got sober at twenty-one, a cocoon of transformation that helped me evolve from a destructive teen into a mostly responsible young woman. Staying with my mom and caring for my dad until he passed on January 23, 2025, allowed me to slow down, be present, and also realize the silver lining. It pulled me out of a situation I wasn't meant to be in. It was time to leave LA for good.

Facing my father's illness and death gave me the courage to live fearlessly. For now, that meant staying with my mom, helping out, and repairing our relationship, which had been damaged by my addiction as a teen and her reactions, shaped by her own upbringing. It also meant actually writing and finishing this book. It's been the most healing time of my life, though tinged with deep grief from losing my dad.

The point is, even in the midst of what seems like a terrible situation (*my father's coma and passing were the hardest things I've experienced so far*) there was a strange, interlaced beauty. I formed an intimate, wordless connection with my dad, holding his hand and being with him daily. I also gained clarity on what my heart and soul truly wanted, free from fixed ideas I had and fear-based thinking. I was able to see my mom for the first time without the lens of my old beliefs—as a woman who had experienced her own pain and raised me the best she could with the awareness she had at the time.

Frank Washington Very's final statement in his 1913 article, *What Becomes the Light of The Stars*, published in Popular Science Magazine, resonates deeply. He writes of the aura:

*"***Some of its properties seem to verge on the metaphysical***."*

He continues:

"Creation is not the bringing forth of an infinite number of dead structureless particles, sent out as a set of miserable little waifs at some indefinitely remote epoch and left to clash without guidance, without purpose. Creation is perpetual. The interiors of matter are seen to be more and more wonderful, more and more intensely active, as we approach the sacred portals where divine influx from the Soul of the Universe quickens into the energy which is matter."[32]

And that, my friends, is Plasma: the soul of the universe. An alive soul, working with all our consciousnesses. A soul that wraps each of us within our lifetime and eternity. A cosmic mother. A quirky grandmother. A true friend.

---

[32] "What Becomes of the Light of the Stars?" *Popular Science Monthly* 82, no. March (1913): 289.

Michelson-Morley Conclusions

As we've discussed, science has largely adhered to a mechanistic perspective, especially during the late 19th and early 20th centuries. This view holds that all phenomena can be explained by mechanical laws, physics, and mathematics, rather than metaphysical or supernatural causes. The Romantics (*1770–1870)* believed that science should never separate nature and humanity, but that's exactly what happened.

The idea that you can remove the spiritual or metaphysical from science makes no sense to me, especially when so much of life, in our everyday experience as well our knowledge of the universe, is unexplained. I would argue that mystery is at the core of existence, right up there with love and loss. Yet, the rise of positivism in the mid-19th century, which prioritized reason and logic over intuition, solidified the mechanistic view in science.

After exploring the largely suppressed possibility of a living aether, we return to the Michelson-Morley experiment. Conducted between April and July of 1887, this experiment aimed to detect Earth's motion through the luminiferous aether (*dead aether*) by comparing the speed of light in perpendicular directions.

Michelson and Morley built an interferometer, which split a beam of light into two paths traveling at right angles to each other. These beams were reflected back and recombined. If Earth were moving through an aether, the motion should have slightly altered the speed of light in one direction compared to the other, creating a noticeable interference pattern when the beams recombined.

The result was negative. Michelson and Morley found no significant difference in the speed of light, regardless of its direction relative to Earth's motion. This suggested that light does not require a medium, or aether, to travel, and that the speed of light is constant in all directions, independent of Earth's movement.

"The most serious mistakes are not being made as a result of wrong answers. The true dangerous thing is asking the wrong questions." — Peter Drucker

What if Michelson and Morley were asking the wrong questions? What if those questions led to using the wrong tools and testing the wrong thing? If luminiferous aether had been conceptualized as dynamic, alive, and in flux, a non-uniform or anisotropic field, akin to the plasma I'm proposing, the Michelson-Morley experiment might have been designed differently and potentially yielded different results.

They were testing a static, uniform aether, imagining something like Earth moving through an "aether wind." But if the aether were alive and dynamic, the experiment would have needed to account for temporal fluctuations in its density, velocity, and structure, as well as its potential interactions with the observer's environment or consciousness. This would have been a monumental task, perhaps ahead of its time.

Moreover, dynamic interactions between light and a living aether could vary depending on local conditions or higher-dimensional factors. Even today, with our advanced tools, testing this would be challenging. However, with the rise of quantum computers and AI, it might soon be possible. Quantum sensors sensitive to small-scale field fluctuations, plasma physics techniques like spectrometry, and experiments observing correlations between conscious awareness and outcomes (*similar to delayed-choice quantum experiments*) could all play a part.

Reimagining the Michelson-Morley experiment might not be ideal, but it could produce variable interference patterns instead of a null result, hinting at dynamic interactions between light and aether based on local conditions. Results might vary and could even be non-replicable, but consistent correlations under specific conditions could reveal the adaptability, or even sentience, of this field. That said, I don't think this experiment would be the best way to prove the existence of this multidimensional plasma or a living aether. I'm simply speculating.

Finally, there's Maurice Allais, a French physicist and economist who won the Nobel Prize in Economics. Allais strongly disagreed with the Michelson-Morley results and the dismissal of the aether. Through his discovery of the *Allais Effect*, the anomalous behavior of pendulums during a solar eclipse, he deduced that space is *anisotropic*, meaning it has non-uniform properties that differ depending on the direction of measurement.

This aligns with my belief that the aether is a relational substance, not a mechanistic or unreal theory. It's something based on countless interactions we're not yet aware of, almost like the subconscious mind. It reacts differently in various situations and areas of space. If only Allais was around today to progress this.

Entering the Einstein Wormhole

Even though the Michelson-Morley experiment disproved the luminiferous aether, leading to the rise of Einstein's Special Relativity, which states that the speed of light is constant in a vacuum and does not depend on the motion of the observer, Einstein himself later corrected this view. Years after his initial work, he suggested there <u>must</u> be a moving aether, though this went largely unnoticed.

He wrote:

"According to the general theory of relativity, space is endowed with physical qualities; in this sense, therefore, there exists an ether. According to the general theory of relativity, space without ether is unthinkable; for in such space there not only would be no propagation of light, but also no possibility of existence for standards of space and time (measuring-rods and clocks), nor therefore any spacetime intervals in the physical sense." — Albert Einstein.

This reopened the possibility of a dynamic field shaping spacetime itself. If Einstein knew what we now know about plasma, he might have seen that everything is a feedback loop—a relationship. This suggests a more *psychocosmological* view (*a term I first encountered through Igor Devetak*), where the universe is less about static laws and more about memory, density, and relationships.

Older cosmic structures, having existed for billions of years, would have denser, more established feedback loops than humans, who are relatively new to the universe. This would make humans more susceptible to change through consciousness or observation, while the universe's patterns are more fixed, though not immutable. With enough power (*conscious awareness*), this aligns with the alchemical belief that our minds, emotions, and feelings can impact reality, even the universe itself.

One person who supported this view this was John Archibald Wheeler, a physicist who frequently overlapped with plasma physics from the 1930s to the 1990s. He believed in a *participatory universe*, where observation and interaction shape reality. Perhaps he, too, was hinting at a deeper truth he hesitated to fully articulate in the public eye of science.

Special Relativity vs. Quantum Theory

The famous Schrödinger's cat thought experiment, proposed by Erwin Schrödinger, highlights one of the many ways quantum theory challenges relativity. Imagine a cat in a sealed box in outer space. Inside the box is a hammer connected to a device that measures the quantum state of a particle, as well as a vial of poison. Whether the hammer smashes the vial and kills the cat depends on the particle's measurement.

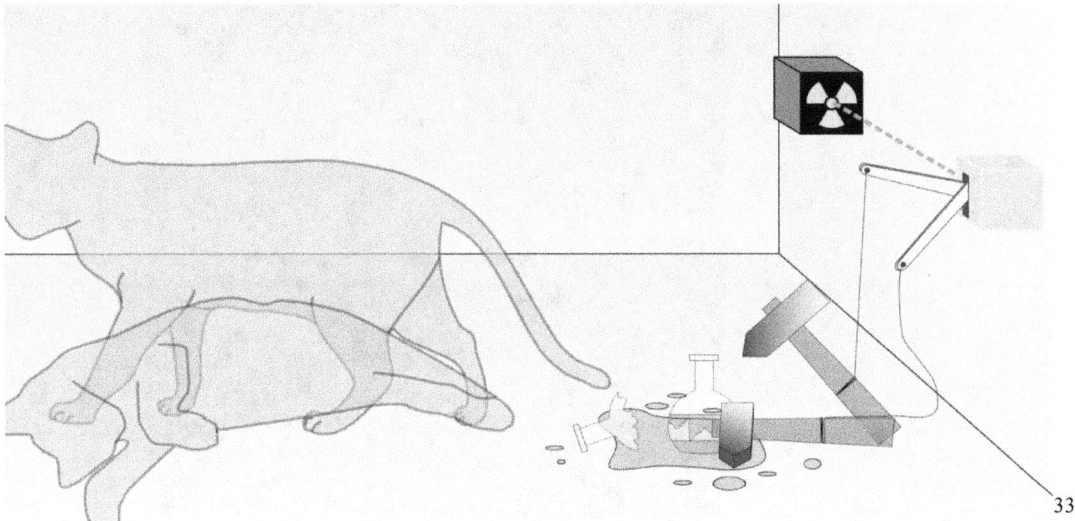

33

Quantum physicists argue that until the measurement is made, the particle exists in a superposition of states—meaning the vial is both broken and unbroken, and the cat is simultaneously alive and dead. This concept of *quantum potential* contradicts relativity's vision of a smooth, continuous fabric of spacetime. According to relativity, a gravitational field cannot be in two places at once. Yet quantum physics insists that matter and energy exist in multiple states simultaneously. So, where is the gravitational field? No one has the answer…yet.[34]

One theory attempting to bridge this gap is *quantum gravity*, which studies how gravity operates at the smallest scales, particularly in black holes and the Big Bang. It reimagines gravity as quantum particles called *gravitons* as tiny vibrating strings that form

---

[33] By Dhatfield - Own work, CC BY-SA 3.0,
https://commons.wikimedia.org/w/index.php?curid=4279886

[34] Colin Stuart, "Was Einstein Wrong? The Case Against Space-Time Theory," *Space.com*, February 28, 2022, accessed August 7, 2025, https://www.space.com/end-of-einstein-space-time

spin networks, composing what we understand as spacetime. This suggests that all the information in a volume of space can be represented as a hologram on its boundary, reimagining gravity as an emergent phenomenon.

This aligns with my belief that gravity and spacetime are emergent properties of plasma-consciousness synergy—the dynamic interplay between plasma (*as the medium or field*) and consciousness (*which introduces intentionality and structure*). This creates a feedback loop that evolves both, shaping our fundamental reality. I will discuss this in greater detail in the next chapter but think of consciousness as the person and plasma as the mirror, reflecting their current state while revealing deeper truths for those who are open and objective. In one way it can be holographic, in another way tangible! This duality which allows for the coexistence of seemingly contradictory phenomena is because plasma can exist in multiple densities and scales.

Everyone's reality is different, but I believe we're all trying to explain the same underlying truths. That's the beauty of Plasma: everything is true, and everything is false, depending on your perception and reality. In my view, *Occam's razor*, the idea that the simplest explanation is often the correct one, points to an expanded understanding of Plasma. The belief that our universe is governed by an intelligent, holographic plasma interacting with collective consciousness (*currently mostly unconscious as "habits" or rules*) explains how matter and energy can exist in multiple states while spacetime is shaped by this presence.

Studying science without including the mystical or metaphysical, to me, *is* heresy! Yes, it complicates things, and yes, it requires us to trust our intuition in a world increasingly dominated by AI and artificial realities. We will no longer need to rely on the media to tell us what is true as definitions of truth are going to be within us. We'll need to sift through a lot of noise, and ask ourselves: *How does something feel, not just how does it look?*

I can't be the only one who's been fooled by a person or situation that looked perfect on the outside while ignoring the inner sense of unease, a darkness (*which I believe stems from fear*) emanating from them. If we truly seek to understand the universe, as many brave thinkers have attempted, we must confront the paradoxes of plasma, the darkness, and the unknown.

"The things that are seen are temporal; the things that are unseen are eternal." — Ralph Waldo Emerson

Plasma in the 20th Century

Speaking of the mystical never leaving, always rearing its head, much like the child with an ever-present question who will not back down, after the talk of aether and quintessence faded and quantum mechanics and relativity had taken center stage, a little word appeared on the scene in 1928: *plasma*.

In the 1920s Irving Langmuir, an American Chemist, Physicist, and Engineer, was working with vacuum lamps and studying different gas environments. This led to him working with what he decided to call plasma because it exhibited collective behaviors due to its charged particles, reminding him of blood plasma, amongst many other similarities we discussed in chapter one. Plasma was recognized as a distinct state of matter, an ionized gas, with properties such as quasineutrality and self-organization.

If you think about it, Langmuir named this "living aether" plasma simply by observing its behaviors. I am doing the same! Applying the name to a more expansive force, of which scientific plasma is only one expression.

I found several incidents of the word plasma, from the Greek πλάσμα, meaning *that which is formed or molded* being used, dating back to the 1600s. I find the name deeply resonant, as this book is about how we mold plasma with our intentions, just as it molds us and our lives.

The first recorded use of the term plasma in a scientific context was by Johannes Purkinje, a Czech medical scientist. In 1839, he coined the term *protoplasma* for the fluid substance of a cell, which later led to the discovery of blood plasma, the amber-colored component of blood left after removing suspended cells.[35]

It seems the word plasma hopped and skipped through history like an electron, sometimes invisible, always present, sidestepping its association with aether, only to reemerge when Langmuir reintroduced it in 1928. Strangely, I found no mention of aether in Langmuir's writings, which feels like a missed connection. To me, all plasmas are aspects of an overarching, multidimensional plasma—a unified field that permeates existence. Astronomical and blood plasma seem to be three-dimensional representations.

---

[35] K. B. Hattacharyya, *Eminent Neuroscientists: Their Lives and Works* (Kolkata: Bimal Kumar Dhur of Academic Publishers, 2011), 182.

The Birth of Plasma Physics

With Langmuir's discovery of plasma came the birth of plasma physics. While studying the physics and chemistry of tungsten-filament lightbulbs, specifically how to extend the lifespan of the filament (*the part that produces light*), Langmuir developed the theory of *plasma sheaths*. These are boundary layers that form between plasmas and solid surfaces.

Here's how it works: when a surface has a negative charge, plasma forms a positively charged sheath to balance it out. This happens because electrons move faster than ions, flying toward the surface and charging it negatively relative to the bulk plasma. This phenomenon also occurs when two plasmas with different characteristics interact, forming a *double layer*—one positive, one negative. Imagine two plasmas with opposite charges sticking together like Velcro.

A fascinating example is the Moon, which charges negatively in the interplanetary medium of space plasma. This illustrates not only how electrons and their transfer enable communication within plasma but also how this communication extends to planetary and cosmic scales. The intelligence of plasma shines through in explanations like this.

Langmuir also discovered regions within plasma discharge tubes (*gas filled tubes*) that exhibit periodic variations in electron density, now known as *Langmuir waves*. This marked the genesis of plasma physics.[36] Over the years, plasma research branched out in various directions, leading us to the cutting-edge studies of today.

The Earth's Ionosphere and Electromagnetic Waves

The development of radio broadcasting in the early 20th century led to the discovery of the Earth's ionosphere—a layer of partially ionized gas in the upper atmosphere. Think of it as the Earth's bioplasmic aura, a shimmering veil of plasma that interacts with electromagnetic waves. This layer reflects certain radio frequencies, allowing signals to travel across the globe and be received on the surface of our planet.

However, early radio communication was imperfect and faced challenges due to the complexity of plasma, particularly interference caused by the ionosphere's dynamic nature. To address this, scientists developed a theory of electromagnetic wave propagation through non-uniform, magnetized plasmas.

---

[36]Richard Fitzpatrick, "Brief History of Plasma Physics," *Plasma Physics Lecture Notes*, University of Texas at Austin, January 23, 2016, https://farside.ph.utexas.edu/teaching/plasma/Plasma/node3.html.

Electromagnetic waves are a form of energy that consists of oscillating electric and magnetic fields. They span a wide spectrum, from visible light to invisible wavelengths like ultraviolet, X-rays, and radio waves. When these waves travel through plasma, a dynamic, ionized medium, their behavior becomes more complex. The charged particles in plasma interact with the electromagnetic fields, causing the waves to refract, reflect, or scatter in unpredictable ways.

By studying these interactions, scientists learned to optimize radio communication. For example, they discovered that certain frequencies, like shortwave radio, could bounce off the ionosphere, enabling long-distance communication. This understanding not only improved radio technology but also laid the groundwork for advancements in satellite communication, radar systems, and space exploration.

In my theory, the plasma field that permeates everything is the medium within which the electromagnetic field exists. This suggests there may be waves beyond the known electromagnetic spectrum—waves that travel faster than light and interact with the plasma field in ways we have yet to understand. The uniform structure of plasma and its capacity to encode information could facilitate instantaneous communication, opening the door to countless possibilities.

What exists within the plasma field? Is it one vast screen, a different dimension, reality itself, or something else that we cannot comprehend? Could it be all of these, or none? The plasma field, with its infinite potential, invites us to explore these questions and imagine a future where communication transcends the limits of space and time.

Plasma: The Classified Spark

By the 1950s, astrophysicists began to realize that much of the universe, if not all of it, consists of plasma. Understanding plasma became essential. The pioneer in this field was Hannes Alfvén, a Swedish Nobel Laureate who, around 1940, developed the theory of *magnetohydrodynamics*. This theory treats plasma as a conducting fluid, enabling the study of sunspots, solar wind, star formation, and more.[37]

The creation of the hydrogen bomb in 1952 sparked significant interest in controlled thermonuclear fusion as a potential power source. Governments, including the U.S., Soviet Union, Britain, and France, began studying plasma in secret. (*I suspect these studies began much earlier, perhaps even during the era of aether theories.*) By 1958, much of this research

---

[37]Fitzpatrick, "Brief History of Plasma Physics.

was declassified, leading to the publication of influential papers and the emergence of theoretical plasma physics.

However, plasma's unique properties, particularly its potential for energy production, such as fusion, remained classified in many government and military projects. This secrecy limited public understanding of plasma, even though it makes up the vast majority of the universe. Today, many people, including students of science, learn little about plasma in school, despite it making up almost the entire known universe!

The Suppression of Plasma Research

Nikola Tesla, who worked extensively with high-frequency currents and ionized gases (*essentially plasma*), explored wireless energy transmission. After his death in 1943, the U.S. government seized his papers, suppressing his work. Tesla had described radiant matter and sometimes radiant energy as dynamic and alive, with qualities of energy fields that hinted at the true nature of plasma. Plasma was a key area of classified research during the time his work was seized.

During the Cold War, the Department of Defense and the military-industrial complex heavily invested in plasma research for weapons development. This included studies in nuclear fusion, electromagnetic propulsion, energy weapons, and even plasmoids/UAPs (*Unidentified Aerial Phenomena*). Projects like HAARP (*High-Frequency Active Auroral Research Program*) investigated ionospheric plasma manipulation, sparking speculation about weather control that we still hear about today. This showed even the government was speculating on plasma's applications in areas <u>outside of science</u>. Of course, some of the more expansive ways to use plasma that involve the paranormal or interdimensional may not have been deemed to have any commercial or military value, or so they thought, but I bet they have started to research this as well.

Strangely, many engineers and physicists who have worked with plasma over the years, whether openly interested in the paranormal or seemingly indifferent to it, have nonetheless reported plasma-like experiences that cross into the spiritual.

The Grown-Up's Boogeyman: Fear of Change

The rise of quantum physics further complicated our understanding of fields, reducing phenomena to probabilistic particle interactions. In this model, there was little room for the concept of a "living field." Mathematics and mechanistic frameworks took precedence, leaving metaphysical exploration by the wayside. A living plasma field, as we've discussed, implies not just responsiveness but a kind of intelligence, a quality that resists confinement within deterministic or purely mechanistic frameworks. This makes it difficult for a capitalistic system, which thrives on predictability and control, to embrace the indefinite possibilities of plasma. After all, how can one profit from something that defies rigid definition?

A deeper understanding of plasma, however, invites us to reconsider our relationship with free will. To work with plasma in all its complexity is to acknowledge that we are engaging with something profoundly alien…an entity that operates by rules we are only beginning to grasp. This can be unsettling, especially for those, both individuals and institutions, who are accustomed exerting control to maintain power.

But perhaps these groups may simply be acting out of fear. Imagine never being forced to confront discomfort, never being told "no," or never facing a situation you can't control or buy your way out of. For those with vast resources, life rarely demands such vulnerability. It's no wonder, then, that the unknown terrifies them. While I don't agree with their outlook, I understand their fear on a human level. I've felt it myself. The paralyzing dread of losing control, the resistance of stepping into the unknown. Yet, it was precisely in those moments of surrender that I discovered a kind of personal magic, a connection to something greater. Something that money can't touch, but a certain innocence can.

I wish these individuals could see that embracing a free-will state of mind doesn't mean relinquishing success or power, it simply means redefining what success looks like. It means recognizing that benevolence and intelligence exist beyond our narrow constructs, and that guilt and shame often arise when we confront this truth. For someone who believes they are the sole creator of their destiny, the idea of a greater intelligence at work can be world-shaking. Yet, I feel hopeful. Even among the powerful, perspectives are beginning to shift. And I believe there are greater forces, call them cosmic, divine, or simply the interconnected fabric of existence, helping this shift along.

The Fusion Problem

Fusion is a critical area of study, not only for its potential to revolutionize energy but also because it may hold clues to understanding plasma turbulence, or even whether plasma "wants" to be controlled in the way we envision. Plasma needs to be thought of as more of a sentient entity and what the scientists call "turbulence" could be akin to emotional responses.

Fusion physicists are primarily concerned with trapping and harnessing thermonuclear plasma, often using magnetic fields. Yet, plasma's inherent instabilities, its tendency to "tear" and escape containment, have posed significant challenges. As of early 2025, when I write this, plasma turbulence remains unsolved, though major strides have been made using artificial intelligence. AI now allows scientists to predict and mitigate instabilities happening in real time, which is a huge milestone.[38]

Still, I can't help but wonder if there are better ways to work with plasma—ways that align with its nature rather than forcing it into submission. Perhaps plasma, if it possesses a form of intelligence, resists being used in ways that are harmful or exploitative. Just as animals thrive in open fields rather than cages, plasma may "want" to be used in harmony with its essence. This doesn't mean it rejects collaboration; rather, it seeks a relationship that is mutually beneficial.

Imagine if we stopped harming the Earth and instead addressed the root causes of our energy crises, much like curing cancer at its source rather than relying on endless treatments with severe side effects. Perhaps plasma turbulence is a reflection of this dissonance, a reminder that true progress requires alignment, not domination.

How exactly can we achieve this harmony? I don't have all the answers. But I believe that by approaching plasma with respect and curiosity, we can unlock possibilities such as free energy that are not only sustainable but transformative. My buddy, acquired savant, Jason Padgett has developed a wonderful and highly intelligent technological way to track and charge a minimal amount for free energy and I encourage sustainable energy companies with interest to reach out to him.

The development of high-powered lasers in the 1960s revolutionized the field of laser plasma physics. When a high-powered laser beam strikes a solid target, it instantly ablates the material, creating a plasma at the boundary between the beam and the target. This process feels almost revelatory, as if the universe were holographic, its hidden blueprint cloaked in a layer of plasma, the "clay" of creation. The laser, in this analogy, acts like a cosmic revealer, stripping away the colored cloak to expose the underlying truth—an emperor with no clothes.

Another fascinating area of recent research is the study of complex plasmas, also known as dusty plasmas, and the enigmatic Kordylewski clouds. These phenomena are not just scientific curiosities; they are foundational to my book and research, offering profound insights into the nature of plasma and its role in the cosmos.

---

[38] Angela Dewan, "Nuclear Fusion, with an AI Boost, Could Provide an Answer to Climate Change," CNN, February 21, 2024, https://www.cnn.com/2024/02/21/climate/nuclear-fusion-ai-climate-solution/index.html.

Robert Temple: Plasma Intelligence Pioneer

Robert Temple is an independent scholar and author of more than a dozen provocative books—all beginning with the international bestseller *The Sirius Mystery*, and now including *A New Science of Heaven*, which explores plasma's role in consciousness and reality. He holds an undergraduate degree in Oriental Studies and Sanskrit from the University of Pennsylvania and has wrote for major publications like The Sunday Times, The Guardian, Time-Life, and New Scientist, and he remains a Fellow of the Royal Astronomical Society. Personally he is like a wise grandparent mixed with a kind child, smart *and* silly, my favorite type of person.

Dusty plasmas are extremely important because they showcase that plasma is indeed alive and intelligent. As discussed previously, in a paper titled *Kordylewski Dust Clouds: Could They Be Cosmic 'Superbrains?*, Robert Temple and Chandra Wickramasinghe explore how these clouds, with their self-organizing properties and electromagnetic connectivity, might exhibit a form of intelligence. Unlike the hot plasma of our Sun, the Kordylewski clouds are classified as cold plasmas, a distinct form of matter that exists beyond the atomic realm.

More on this subject can be found in Temple's remarkable book, *A New Science of Heaven*. This was one of the first, and remains one of the very few, books dedicated to the topic of plasma intelligence. Temple's ability to articulate complex ideas with clarity and enthusiasm is a gift, and his work has been a profound inspiration for my own research. [39]

Temple's descriptions of plasma intelligence, particularly in his article *Inorganic Intelligent Beings Dominate the Universe*, resonate deeply with my own understanding.[40] He presents plasma not just as a physical phenomenon but as a living, intelligent force, a perspective that has shaped my exploration of plasma as a medium for consciousness and transformation.

"Self-Organization & Intelligence: A large part of my book (*A New Science Of Heaven*) is devoted to explaining in detail, for the general reader, precisely how a large charged dusty complex plasma cloud can even self-organize into an entity even more complex than an organic body. Such a cloud becomes filled with filaments carrying currents, which act like blood vessels, or cells that act like organs, of semi-conductors

---

[39] Robert Temple, *A New Science of Heaven* (Rochester, VT: Inner Traditions, 2022).

[40] Robert Temple, "Inorganic Intelligent Beings Dominate the Universe," *A New Science of Heaven: A New Dawn* (article), accessed August 22, 2025, https://robert-temple.com/articles/A-new-science-of-heavean-new-dawn-article.pdf

that modulate current flow, of current pinches which act as nodes and so on and so on. All that I have said is based upon sound scientific discoveries. Such plasma clouds can definitely develop an interior structure more complex than an organic body, and by the now recognized processes of 'emergence', 'entanglement', and 'self-organization', can spontaneously develop intelligence."

He goes on:

"The Kordylewski Clouds are so gigantic that their processing power exceeds by many orders of magnitude all the computers and brains on Earth. Furthermore, the Clouds are apparently billions of years old, whereas humans have only existed for about two million years. Anyone who needs convincing about all of this will find plenty of evidence in the book. If these giant clouds in space are intelligent entities, what would their personalities be like? Very different from ours."

He then speaks about what he comes to call a plasma being:

"I explain at great length why we all exist as double creatures: we all have bioplasma bodies (*souls or spirits*) and physical bodies which I call "smart overcoats", jettisoned when they wear out or are destroyed."

These words fascinated me the first time I read them. Not because they were new to me, but because they reflected my beliefs and articulated them in a symphony of words. The few people I've encountered, but their numbers seem to be growing, who share an intuitive understanding of plasma have also spoken of bioplasma as the soul, plasma intelligence, and the neutrality at the core of plasma.

I'd like to make two distinctions here. First, I believe that within this vast, inorganic plasma field lie **plasma beings**—entities you might call archetypes, angels, interdimensional spirits, or communicators. In mythology, Hermes serves as the emissary between humans and the divine, a trickster who bridges worlds. These beings, benevolent and attuned to our highest good, act as intermediaries, helping us connect with the plasma field.

The Polynesians, too, understood this connection, harnessing the spirits of their ancestors to perform what we might call magic. Yet, plasma's neutrality means it can be used for both good and ill. While the beings within the plasma field can discern intent, the field itself is indifferent to human morality. This creates a paradox: plasma is benevolent and seeks our highest good, but in the hands of those convinced of their own righteousness, even if it harms others, it can be wielded destructively.

This is why karma matters. If we use plasma for black magic, control, or harm, the consequences will follow us, not just in this life but into eternity. Mistakes are part of growth, but consistent misuse of this power carries profound repercussions.

It is very important to understand karma in the terms of plasma-consciousness synergy. Karma, in this framework, follows one over timelines in their reality. Timelines all are happening simultaneously, that fractal out with each choice, but also to us feel like past, present, future. Then there are other realities (*picture a tree in each bubble, your current life is a branch*) that also have timelines. Our choices only affect the timeline we are currently in, and those actions shape our evolution. This is how a "multidimensional self" or something beyond that is experiencing many choices within each 'fate' bubble. Other versions of ourselves in other realities might be having their own journey, but what truly matters in this reality is how we heal and evolve through the choices we make here.

Think of each fate (*reality*) bubble as a plasma bubble, and the sheaths between them and other fate bubbles are what protect karma from hopping from one reality where we are a maniacal overlord to penetrating this one. But, in a "past life" of this reality, if we were an overlord and have not learned from those choices, they will most definitely show up in this "plasma reality" serving you with similar situations until you make different choices aligned with a higher good for all. Instead of feeling shame for past decisions, we must take responsibility, yes, and understand from what fear we made those choices and how we can empower ourselves to choose better, heart and soul-centered actions.

My approach to plasma differs from Temple's in that I explore it through a psychological, mythological, and multidimensional lens, complementing his scientific and philosophical perspective. Think of my book as the feminine counterpart to his work—a rose jewel illuminating the plasma universe from a different angle.

The Current Landscape of Plasma

To conclude, plasma is not only a hot topic in fusion research but also a frontier for futuristic applications, ranging from skincare to car cleaning, and even cancer treatment, farming, and energy healing. Ilan Uchitel, CEO of CAPS Medical, is currently developing a solution to treat solid tumors using non-thermal cold atmospheric plasma.[41] In the automotive

---

[41] https://www.lifesciencemarketresearch.com/videos/ilan-uchitel-caps-medical-spotlight-interview-lsi-usa-23

industry, plasma cleaning removes unwanted contaminants like oils and residues from the surfaces of car parts.[42]

Nassim Haramein, a physicist studying plasma, suggests that everything is made of plasma in different forms. His work explores how plasma might correlate with communication and future technologies. Even in the beauty industry, plasma has found a place: the so-called "Vampire Facial" involves extracting plasma from a patient's blood and reinjecting it into the skin to slow the aging process.

The *Plasma Universe Model,* originally proposed by Hannes Alfvén and Oskar Klein in the 1960s, is a current working theory offering an alternative cosmology in which plasma dynamics play a fundamental role in the physics of the universe.[43] What fascinates me is that, to date, there has never been a full attempt to explain the detailed spectrum of cosmic anisotropies within the framework of plasma cosmology. I hope this book inspires scientists and engineers to explore these ideas further, potentially revolutionizing our understanding of the universe.

I am confident this will happen and change everything we know about our current understanding of science and the universe. I believe that with our current technology, we can begin to unravel the limitations of the Big Bang model and braid together a new vision of an intelligent plasma universe, evolving from Alfvén's theories into something fresh and transformative, potentially even toward a *Big Birth model.*

In the modern era, plasma physics is merging with the spiritual, giving rise to what I see as a new dimension of metaphysics—or what I like to call *Quintessential Plasmics.* This may seem magical or mystical, but it is also deeply natural. Plasma is paradoxical: complex yet simple, animalistic yet mechanistic, masculine yet feminine. It reflects humanity, and we reflect plasma. It is a feedback loop, constantly circling between our own consciousness and a higher consciousness. It is how consciousness experiences itself. Plasma is the buried and suppressed feminine, the misunderstood half of the universe. It is the dark goddess, waiting to be acknowledged. The demon in the machine, who is anything but a demon. Wake up, father consciousness, and remember your wife, your sister, your mother: Plasma.

I hope this exploration has sparked your curiosity and opened your mind to the possibility that plasma is far more than history has shown us. My aim is to reveal the

---

[42] Davide, "Plasma Cleaning in the Automotive Industry," *SCIPlasma*, November 21, 2023, updated April 25, 2024, https://www.sciplasma.com/post/plasma-cleaning-in-automotive-industry.

[43] Barry Parker, "Plasma Cosmology," in *The Vindication of the Big Bang* (Boston: Springer, 1993), 325, https://doi.org/10.1007/978-1-4899-5980-5_15.

practical, magical applications of plasma, our birthright, and to remind you of the profound relationship you can cultivate with this beautiful, eccentric substance. Plasma can be our worst enemy or our best friend, depending on the state of our consciousness.

Aether was never truly disproven; it has quietly been resurrected as plasma. Now, it is time to learn about the multidimensionality of plasma! I considered naming plasma something different, but nothing truly fit. In all senses of the word, it is truly an <u>expanded</u> view of plasma, and the name Plasma has stuck. I have referred to it as a higher-dimensional, multidimensional or dimensionless plasma, or simply an expanded view of plasma. From now on, I will call it Plasma (*with a capital P, unless it doesn't feel fitting*)—a living aether, yearning to be seen for who she truly is. She is a bridge between worlds, minds, and experiences, ever evolving into new states never before imagined. I try to understand her, but she always surprises me, just like life itself.

3

Plasma: The Expanded Definition

"Study hard what interests you the most in the most undisciplined, irreverent, and original manner possible." — Richard Feynman

When I was in an acting class in Los Angeles, our teacher, John Markland, a gentle soul with depth and compassion, began every session by having us meditate in a dark room for about twenty minutes. Afterward, he'd return and ask what we experienced. Some students described astral projections, encounters with spirit guides, visits from passed loved ones, or vivid memories of love and fear. Others simply felt relaxed. A few would quietly skip the exercise altogether.

I don't remember what prompted it, but one day, John referred to this "thing" we were all interacting with differently as *The Goofy Queen*. It was his way of naming the collective, ineffable energy we'd all brushed up against. The phrase struck me as perfect. I raised my hand and said, "It's Plasma!"

The class looked at me, understandably confused. But to me, the name captured something vital, a whimsical, almost human quality Plasma seems to carry. It's something to be revered, to engage with playfully, but also to question, lest you get caught in illusion. This substance, this field, this *Goofy Queen*, holds memory and imagination. It transports our consciousness to other planes. It can be funny, absurd, random-seeming, but behind the apparent chaos, there's always an underlying order, even if it isn't immediately obvious.

I remember one meditation where I had a vision of a black-and-white butterfly. Her wings were patterned with clocks instead of spots, and she had ten eyes that somehow looked like lips. I asked her what she had to tell me. She smiled and replied, "I'm just here to dance and perform for you." She called herself Mother Time.

It was bizarre but awe inspiring. Instead of analyzing her, I was meant to experience her. That, to me, is the essence of Plasma: this Goofy Queen energy, inviting you to experience *anything*. You can learn, be, create…it's up to you. Like I've said before, Plasma is like a grandparent who lets you make your own mistakes, but still wants the best for you!

That vision of Mother Time felt like a new archetype being born for Plasma in this era, a goofy plasma queen dancing between dimensions, outside linear time (*unlike Father Time, who is more linear*). She expresses truth without dictating or controlling us, and she reflects our choices back to us…speaking through riddles, perception, and synchronicities. Instead of teaching us, she invites participation, holding space for whatever experience consciousness creates within her.

She lets us bring in experiences beyond linear time. People who encounter her glimpse possible futures or flashes of the past, filtered through their current state. If you approach her with curiosity, she'll offer adventure. If you approach her with fear, she'll reflect distortion and chaos.

In *The Wizard of Oz*, pulling back the curtain exposes that the great and powerful Wizard is not a godlike figure at all, but an ordinary man using smoke, mirrors, and a microphone to project an image of authority. The illusion worked, because the collective belief made it real. The Wizard wasn't exactly a villain, he was more like a mirror. He became what people needed him to be.

In my cosmology, Plasma is this Wizard. She is not deceiving you, she is a responsive field that mirrors your beliefs back to you until you remember your own power. The moment you pull back the curtain is the moment you realize that reality itself has been shaped by your own unconscious patterns, fears, and desires. Plasma was never tricking you. It was reflecting you, and that misunderstanding was tricking you.

When you awaken to that fact, just like Dorothy, you realize the authority you were seeking outside yourself was within you all along. That there's no place like home. At that moment, the mythology of plasma is no longer just an illusion or a trickster (*as it's been described in mythology – more in Chapter 4*), it evolves to a stage or field where your greatest insights unfold from reflections and deeper truths.

The Banished Queen

Plas-*ma* has been cast out for centuries, banished in a world where the patriarchal mind ruled all. In an age where consciousness was elevated above all else, we were taught to create our own realities without acknowledging the substance we were creating with, let alone its innate intelligence. But she's returning.

The divine feminine, this Plasma, is rising again in our awareness. She comes to remind us of surrender, stillness, and the wisdom of paradox and mystery. She asks us to honor our body's knowing, our emotional intelligence, and the space between things. She teaches us to listen, to receive, to be in relationship with what *is*—sometimes in rest, sometimes in action. She is the cosmic mother. She is love. She is the timeless, pearly wisdom of the ages. A sweet mist.

Plasma, in her essence, is quasi-neutral and deeply reflective. She is not good or evil. She simply mirrors the state of consciousness she interacts with. In today's world, where much of humanity is still entangled in shadow, reactive emotion, and looping beliefs, Plasma may *appear* dark or chaotic. But that's not her true nature. She's just reflecting the state of awareness she's meeting. This distortion is not hers, it's ours. It's the residue of pain, trauma, fear, and disconnection. Evil, as I see it, is not some cosmic force, it's a human sickness born from perceived separation and unresolved grief.

Plasma is not here to harm us; she is simply responding to our distortions collectively. But at her core, she leans toward coherence, toward a benevolent healing and love. We just have to learn how to change our approach, similar to any relationship where we learn the hard lesson that we have to change ourselves, not the other person, first. We learn to meet her with presence, authenticity, and feeling.

The battle between dark and light that we see in stories and films is often a misreading of this collective distortion. The universe is not battling itself; it's waiting for us to recognize ourselves within it. Every time we are faced with darkness, it's a call back to our inner love.

As we rise into higher awareness, creative, heart-centred coherence, Plasma becomes a partner of support. A living field we can feel, collaborate with, and co-create through. She's not some control mechanism… she's here to *respond to you.* And the more you understand her, the more you understand yourself.

"God is not someone or something separate and apart from this world. God is the subtle essence of everything, and this essence is Truth." — Upanishads

A quick refresher: in science, plasma is known as the fourth state of matter, an ionized gas found in lightning, stars, even supernovas. But by now, we've come to understand that Plasma is so much more than its scientific definition. It is metaphysical, as in *beyond the physical.* At its core, Plasma is truth. And real truth, as many mystics and scientists alike have discovered, is often paradoxical, perhaps because it exceeds human logic.

Plasma also reflects belief. It becomes what you believe, consciously or unconsciously. It is truth *and* illusion, reality *and* projection. All of this is malleable when held in awareness and we can choose our life experience.

As I've said before, I call it Plasma meaning "to mold" or "molded", because, at the deepest level, smaller than any observable particle, this is what's there. If plasma makes up 99.9% of observable reality, why can't it make up 99.9% of an unobservable veiled reality? We see this shaping essence that underlies our entire reality reflected in the behaviors of astronomical plasma and in the plasma that flows through our blood. And both are, in their own way, pretty darn magical. We are witnessing the hand of God.

It's a profoundly simple word that belies an almost infinite complexity. The word is perfect. It just needs to be redefined, or *expanded*, so we can learn how to work with it. As something ancient, yes, but also as a profoundly futuristic tool for humanity. A way to harmonize with the deeper field of reality.

Plasma Crystals: Physics Vs. Metaphysics

One thing I would like to clear up is when I say everything is Plasma, I am meaning the expanded version of plasma. In science, plasma makes up 99.9% of everything, because solid matter, liquid, and gas are not considered in a plasma state. The smallest observable thing in an atom is quarks and gluons, the only become quark-gluon plasma when protons and neutrons are smashed apart at extremely high energies.

What this does mean in our physical reality is that the plasma state is always in potential. In these forms of matter, plasma is dormant, like a great river frozen in winter. Under extreme heat or energy, those bonds break, and the particles move freely once more, returning to their natural plasma state. In physics this is called *ionization*.

I sometimes refer to the human body as crystallized plasma. In science, crystallized plasma is still plasma, ionized and electrically active, but arranged in an ordered, structured form, such as in dusty complex plasma. In my usage, we are crystallized fourth-dimensional or *expanded* plasma: a dormant state of potential plasma bound into the solid form of the body. We still carry an ionized biofield, a reality increasingly explored not just in metaphysical circles but in emerging scientific studies. Our bodies also engage in ionized processes constantly, though the plasma that structures us is not visible to the naked eye and remains beyond the reach of current instruments.

Our body anchors or crystalizes our consciousness in this physical reality. If the fourth-dimensional plasma body is primary, and the human form is its crystallized aspect, then death is not an ending but a release back into the freer plasma state, with consciousness still embedded in it. Plasma in this reality, is a pointer to how subtle realities of plasma may work outside current science. Learning about plasma crystals & new crystal technology that will be uncovered, may be a key to this fourth-dimensional reality and beyond.

Plasma's Key Characteristics (an expanded view)

Omnipresent and Dimensionless

Plasma exists across all scales, from the quantum to the cosmic. It is not confined by spatial dimensions but instead manifests in gradients, subtle densities that adapt to dimensional environments and realities it interacts with.

Gradients or Densities

In the third dimension, plasma becomes visible as lightning, auroras, solar flares, or even blood plasma. In my view, all matter, including the human body, is crystallized plasma.

We are shaped by plasma, and consciousness gives colour and definition. Higher or lower densities of plasma correspond to dimensional (*as in perspective*) transitions, giving rise to both observable and unobservable effects.

A Living Aether

This is not the discredited luminiferous aether of classical physics. This is not the static substance of Aristotle. Instead, I describe plasma as a living, sentient field, one that evolves in relationship with consciousness. It transmits light, responds to intention, and forms a connective bridge between dimensions, timelines, and realities.

The Ground of All Things

Everything arises from the finest level of plasma: matter, energy, light, even thought. It is the dimensionless, fertile ground of the cosmos. A memory-holding, feeling-sensing medium through which reality is shaped. In this view, plasma becomes the bio-holographic screen of experience.

Foundational Definitions

Plasma

The multidimensional and dimensionless fabric of our reality that is sentient, in flux and intelligent. It is embedded with a set of behaviors, revealing energetic exchanges of information, intelligence, and consciousness itself.

It is not "conscious" in the way we are, but it is deeply alive: a benevolent medium that reflects your awareness back to you. Through it, the Mystery expresses itself. Plasma is malleable, mythic, and capable of holding memory, transmitting emotion, and forming the architecture of experience in real time. In this sense, it is quite literally what dreams are made of.

Consciousness

The directive instrument of perception—composed of emotion (*felt-sense*), belief, and intention. In my framework, consciousness is not the Self; it is the flashlight used by the Self to illuminate reality. It moves through Plasma, shaping potential into form, much like a lens focuses light. It can be used to navigate, observe, and create.

This is a working definition, because consciousness, like plasma, exists on a spectrum. It also functions as an interface. It is the capacity to focus, interact, and create meaning within the plasma-consciousness field where meaning can be as simple as altering the relational field through growth, movement, or expression. This shapes potential into experience. This definition is centered around the human experience.

While emotions emerge, they are potentially foundational to us humans, providing necessary contrast and feedback in this time-bound reality. Feelings, by contrast, seem to be foundational and inherent to all beings, resonant signatures from the fifth dimension or beyond.

Multidimensional Self

The individualized node of greater intelligence, beyond linear time, that remembers and interacts across multiple timelines, realities, and dimensions. Where consciousness observes and directs, the M-Self *feels*, intuits, and weaves. As we evolve, we move from identifying with consciousness to embodying the M-Self. This marks the shift from passive perception to active co-creation. I view the M-Self as linked with The Mystery, in a fifth-dimensional perspective space.

The Mystery (Dual Monad and Beyond)

The source-field from which all emerges, and to which all returns. The Mystery is ineffable, unbound by form, name, or knowing. At its most foundational level it might still be the duality of plasma and consciousness while also expressed as one. It informs us and evolves as we evolve through plasma-consciousness synergy. Sometimes called God, Source, the Monad, the Tao, or even a Future Collective of Our Intelligence, the Mystery rarely speaks directly. More often, it comes through synchronicities, riddles, dreams, and mirrors.

Dimensional Roles of Plasma

These will be explored in more depth later in the chapter. For now, I offer these preliminary definitions from my model:

Dimensionless Plasma

The origin field. It precedes form and gives rise to all dimensions, not as "nothingness," but as pure potential. You might think of it as zero-point essence. This plasma exists within and around every experience, waiting to be felt, shaped, or awakened into resonance. It is the breath of the universe, exhaled from a singularity and returned as higher-dimensional realization through the loop of conscious awareness. Though all possibilities reside within it, awareness itself is what sparks the new, revealing not the past, but the path only presence can see as an emergent future.

Third-dimensional Plasma

This is the observable plasma of our physical world as lightning, auroras, solar flares, and biological plasma in our blood. It gives structure, charge, and vitality to 3D reality. It is the densest layer of the spectrum shaped by electromagnetic waves.

Fourth-dimensional Plasma

Unseen but deeply felt. This layer holds emotional imprints, archetypes, dreams, and collective unconscious patterns. It acts as a bridge, linking the personal and collective psyche to the deeper intelligence of the M-Self. Without it, that intelligence would remain untranslatable, like a radio signal with no receiver.

Fifth Dimensional Plasma

This is the creative aether—the refined plasma where the M-Self is held and through which it speaks. It holds no fixed timeline. It is a field of resonance. Accessed through presence, awe, or flow states, 5D plasma offers real-time feedback, insight, interaction with benevolent intelligences and co-creation. It is not a destination but a tuning fork that harmonises with higher frequencies of awareness. This is the environment associated with the Mystery.

Dusty Complex Plasma as the Transitional Zone

In my model, Dusty Complex Plasma is the threshold between the visible third-dimensional plasma and the emotionally charged, unseen fourth-dimensional field. Like foam where sea meets shore, this liminal zone hints at plasma's emergent intelligence. Here, feedback loops begin to form. Memory, emotion, and intention, once abstract, start pressing into physical form. Dusty Complex Plasma may be among the few scientifically recognised types of plasma where such emergent behaviours are observable. In this way, it may represent the skin of the universe: the place where awareness touches matter and where the symphony of quantum information begins to take shape. It might also hint at behaviors of 4D plasma as well as how consciousness and plasma work together to create reality.

Dimensionality, in my framework, isn't about spatial coordinates or equations, it's a more like a language for conscious perspective. A structure for describing the *felt architecture* of reality. Moving forward, think of 3D, 4D, and 5D not as places or measurements, but as distinct textures of perception—each one shaping how we relate to time, self, emotion, and possibility. The best way to think of it is breathing in is 4D, holding the breath is 3D, and breathing out is 5D.

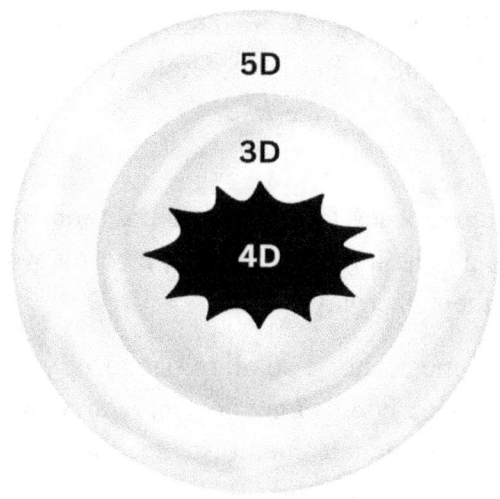

Shaping The Void

Frank Washington Very, a physicist who explored the concept of the luminiferous aether, wrote something in his 1910 work *A Light of the Stars* that now feels remarkably prescient:

"If we attribute the absorption of light in space to the ether itself, the radiant energy absorbed performs work upon the ether, presumably the generation of minute ethereal vortex-rings, the elementary particles from which electrons are derived, or possibly the positive and negative electrons themselves, out of which the atoms are formed. From associations of electrons to atoms, from atoms to molecules, from molecules to the first tiny beginning of a cosmical crystalline sublimation, there is a continual progression and increase of size. Finally, the widely dispersed material must be gathered from the immense voids of space into the germs of future worlds, and for this task the meteorites appear to be the appointed instruments."[44]

What he called the "immense voids of space" may correspond, in my understanding, to regions of lower density within the plasma field otherwise known as voids! These are zones where dusty plasmas emerge. These transitional layers could represent thresholds where higher-dimensional plasma condenses into lower-dimensional phenomena, becoming observable as particles, matter, or form. Another example of how in the spaces between is where reality seems to arise through intelligent processes.

---

[44] Frank Washington Very, *A Light of the Stars* (New York: Putnam, 1910), 173.

Dusty plasmas, scientifically recognised for their self-organising properties, may well be the "crystalline sublimation" that Very refers to: foundational, structuring fields from which physical reality precipitates. It also foretold what are now called plasma crystals or Coulomb crystals (*highly ordered lattice-like structures*) which, under certain conditions, are formed by dusty plasmas.[45]

He even hints at a feedback loop between light (*consciousness?*) and aether (*which I understand as plasma*)—a loop in which energy "performs work" upon the medium, possibly shaping reality through their interplay.

His reflections on meteorites also align beautifully with the concept of panspermia: the idea that these spacefaring bodies may not only carry the raw materials for life but also act as vessels of plasma-encoded memory, seeding the elements of consciousness, intelligence, and information across worlds.

The Flux Hologram

Now imagine a spotlight that is both one-dimensional and fifth-dimensional. A circulating projector and receiver, like a toroidal black hole that both emits and absorbs. It holds within it the totality of our conscious awareness, projecting into a multidimensional bubble, with dimensions 2 through 4 forming one experiential field. There are infinite such bubbles, each structured the same way, each holding countless timelines.

---

[45] H. Thomas, G. E. Morfill, and V. Demmel, "Plasma Crystal: Coulomb Crystallization in a Dusty Plasma," *Physical Review Letters* 73, no. 5 (1994): 652–655, https://doi.org/10.1103/PhysRevLett.73.652.

This toroidal black hole continuously reabsorbs the learnings from every lived experience and re-projects based on this updated awareness, much like how the human mind constructs and reconstructs reality through perception.

Surrounding this spherical projector, along the boundary of the black hole, is what I call *Brute Plasma* (*or Base Plasma*): a two-dimensional, unprocessed, raw energetic field. Most people confuse the bright accretion disk, the glowing spiral of hot gas around a black hole, with the event horizon itself. But the actual event horizon is invisible: a threshold beyond which, in classical physics, no light escapes.

Yet in my model, light *does* escape, not by traditional means, but by tunnelling through an unseen, two-dimensional field. Brute Plasma acts like an intelligent membrane, similar to a plasma screen or a holographic plate, refracting and filtering light as it crosses into dimensional reality. This mirrors natural processes of information, or perhaps intelligence and consciousness, exchange, much like a cell membrane controls nutrient flow or a plasma sheath regulates charged particle transfer. The event horizon may serve as a transitional lens through which fifth-dimensional consciousness becomes patterned structure: dimensions, timelines, and lived experience. Plasma is what holds this structure and makes it tangible.

This two-dimensional layer of Brute Plasma may be the surface that quantum physicists are now studying, while a far more expansive field lies beneath. When this seemingly conscious light moves through Brute Plasma, the plasma encapsulates it as it becomes unconscious or latent consciousness, refracting into geometric and emotional codes, waiting to be felt and shaped, creating the scaffolding for reality itself. Just like a hologram stores three-dimensional information in a two-dimensional imprint, this layer contains the blueprint for higher-dimensional emergence.

If everything stayed in pure, fully conscious awareness, there would be no contrast, no novelty, and no process of rediscovery…just static omniscience. By allowing awareness to become latent consciousness, the system introduces mystery, separation, and the necessity to *re-imprint* meaning through lived experience. True conscious awareness is never lost, it is in that fifth dimensional space, all you have to do is remember it, through feeling.

This journey of awareness is not one-way. Just as light enters through this lens, it can also return, carrying with it the emotional resonance of experience, encoded and transmitted back through the invisible highways of fourth-dimensional plasma. In this way, the event horizon is not a barrier that traps light, but a gateway: a recursive interface where the Mystery expresses, rebalances, and evolves itself in real time.

The Mystery is both the source and the destination, and Plasma is the medium that makes the whole journey possible. Through this ancient liminal riverbed of plasma and consciousness synergizing, awareness creates reality.

This two-dimensional void, the raw interface between awareness and form, may explain why some meditators report intense existential crises when they encounter it. It is often experienced as ego death, because it feels devoid of consciousness. Yet paradoxically, it is also the ground through which consciousness is born. What feels like a death is often a birth. In contrast to blissful "ego rebirth" experiences, where consciousness is fully nested in plasma and harmonised with it, this state represents the raw passage between dimensions: an unfiltered zone before self-perception reconstitutes.

Welcome to Channelwood

In my framework, consciousness is a directive tool, and not in some mechanistic way. Just like plasma is shaped, consciousness is aimed. It is what our awareness uses to perceive, focus, think, and imagine reality into being. Plasma, meanwhile, is the responsive field that co-creates reality in tandem with this directed consciousness.

Together, they form what I call a plasma-consciousness synergy. Our filter of emotion, belief, and intention is not only a lens, but also what we're made of as human beings. We are little balls of plasma-consciousness. Once we become aware of this synergy, we can reprogram it. Awareness picks up the tool of consciousness, directs it, and plasma responds. We awaken to the patterns that aren't bringing us joy or ease, and we choose to change. Over time, self-awareness expands into multi-self-awareness.

Just as a lens focuses light, consciousness focuses perception. That's why you can reprogram thoughts, heal memories, and even shift timelines...because you are not the thought. You are the awareness that holds the tools or the thought. And our tools are evolving. As they upgrade, so do we. Just like in a video game: *#newlevelunlocked.*

It truly feels as though we're thawing out frozen consciousness, long encased in the protective cocoon of the divine feminine Plasma we ourselves sealed. Now, something is warming. A spark has been lit, triggered by the remembrance of Plasma itself. As Earth grows increasingly resonant with fluidity, intuition, and multidimensional awareness, it's not that we're ascending...we're drawing that "fifth-dimensional mist" into this reality, transforming our definition of existence from the inside out.

The more mist-like we become, the more grounded we are. We are expanding our "plasma bodies"! We live more lucidly, more fluidly, more co-creatively. We feel and sense *more*.

Just like in the old computer game *Myst*, humanity has journeyed through metaphorical Ages: Selenitic, Stoneship, Mechanical—each one representing logic, materialism, and mechanistic thinking. But now, we're entering Channelwood: the age of organic connection, fluid communication, and living systems rooted in nature.

This is a return to intuitive intelligence: a realm where wooden walkways float above water and messages arrive through telepathy or resonance. It's a place where technology harmonises with nature instead of trying to control it. But we must choose this path, consciously and now.

Three-dimensional consciousness itself may be only a stage in human evolution, arising from sentience, and now giving way to the multidimensional Self. Our reality evolves as we do.

The Light of Consciousness

In science, light is electromagnetic radiation, a stream of photons travelling in waves. It has no mass, yet it carries information. When it hits your retina, your brain decodes it into experience. In my framework, light is the moment when awareness meets information, made perceptible through the medium of plasma.

Information, or *in-formation*, is the product of plasma and consciousness working together in various ways which we will soon specify. I call this the plasma-consciousness synergy. Your awareness tunes into information. Your consciousness focuses on it, through attention, perception, or inquiry, and the plasma field responds. It shapes the unseen into experience.

The light you perceive, physically or metaphorically, is the result of this synergy. This is why you don't see the world "as it is," but as your conscious awareness allows it to be seen. Each person lives in a slightly different reality, because each person tunes in differently. And if all timelines, dimensions, and realities are encoded in plasma, then...

Yes. You can eventually focus and receive the light of other timelines, not metaphorically, but truly. The light of another timeline, or another being for that matter, is the visible result of focusing your awareness on a different informational pattern encoded in plasma. Just like tuning a radio dial, you can learn to shift the frequency of your awareness.

Plasma reconfigures, and a new light appears: a new timeline, a new reality, a new visible perspective.

Light may not always be conscious, but it seems to be a carrier of consciousness, an observable pulse of information + awareness in whatever form it takes. Light, in a physical sense, is a messenger. It travels through space carrying information, delivers it, and then moves on to the next message. It doesn't keep a copy of what it delivered. Photons, the particles of light, are brief, emergent events of the electromagnetic field. Once a photon is absorbed, scattered, or transformed, the original is gone. Any information it carried is either transferred into another medium or lost to entropy.

Plasma (*the expanded definition*) is different. It can receive, hold, and transform information over time. It is the medium itself, not just the courier. While recent breakthroughs such as photonic memory chips, holographic glass, and quantum light storage show that we can preserve light's information for milliseconds to years, in every case the data is held by a *medium*, not by the free-flying photon. This is why Plasma is more like a living archive for consciousness, while light is the messenger that moves between archival sections.

If light was indeed consciousness, in pure darkness we would perish. But even in pure darkness, we remain aware. This tells us that light is not required for consciousness to exist, but only for it to be seen. Awareness is the root, and light is the projection. Sometimes, that experience is lit externally. Other times, it arises inwardly, in the dark, glowing internally with intuition, memory, or imagination.

This may be why in a sensory deprivation tank, you are still conscious but have deeply intuitive and inward experiences that may seem illogical or full of past memories.

"I was sixteen when I suddenly thought: what would it be like to run beside a light beam? What would I see? Could I catch up with it? This question made a deep impression on me. It seemed to me as if I were riding on a beam of light."
— Albert Einstein, *reflecting on his early thought experiment that led to special relativity*

If consciousness persists in darkness, if we still dream, intuit, and feel without any light at all, then light cannot be the source of awareness. It is simply the shimmer on the surface. The deeper current must be something else…something that holds memory, feeling, imagination, and intelligence even when unlit. That something, I believe, is Plasma.

Even in absolute darkness, you are still aware, not because photons are present, but because consciousness illuminates from within. The eyes may rest, but the mind sees. Plasma holds the space; consciousness stirs the image. Light, in this case, is an echo of internal awareness, not an external beam. This would be more like a fourth dimensional light.

Plasma, in my framework, is the true vehicle of consciousness, and light is its three-dimensional expression. Plasma carries consciousness into visibility; light is the result. While light itself is not self-aware, it may carry self-aware consciousness—or hold the echo of collective consciousness still in the process of awakening. This will be explored more in the upcoming mini-chapter *Spacetime Reimagined*.

In fourth dimensional plasma, consciousness does not travel through visible light but moves through motion, archetype, and memory. (*This is what I speculate might eventually be identified in science as an "antiphoton."*) Light is only one layer among many. In the fifth dimension, consciousness or what I would now call awareness, flows through feeling, resonance, and intuition or communication without photons. Think of it as mist, vibration, or symbolic potential. It may even be a third thing, beyond an antiphoton, called an omniphoton!

When someone "sees the light" during a near-death experience or deep meditation, they may be tuning into a brighter spectrum, something opalescent, iridescent…a shimmer lit from within. What I call *black light.* Not black as in absence, but black as in invisible to the three-dimensional eye. This light is only perceptible from a four-dimensional perspective. You are viewing the "brighter light of resonance". Something vivid, opalescent; an omniphoton. You don't need light to have consciousness, but when consciousness touches plasma in just the right way, light appears. A spark as a moment of contact. A result, not the source.

This is why we refer to ourselves as light beings: not because light defines us, but because it is our visible echo. Our 3D form reflects a deeper, radiant truth. The shimmering plasma of fifth-dimensional awareness. In this way we really are made in "God's" image.

I don't claim to fully understand what light, consciousness, or awareness truly are. This is a living, evolving framework. What I share here is something that feels deeply right to me, and I offer it in the hope that it may resonate with you, or inspire your own exploration.

There's an ancient pattern reflected in the structure of reality:

The Father: Consciousness.
The Mother: Plasma.
The Divine Child: Awareness.

These archetypes are not bound to religion, but appear again and again across spiritual traditions, encoded in language, symbol, and geometry. They map onto the cosmos itself.

"1 + 1 = 3. Everything is everything." — Lauryn Hill

I believe we have a direct influence on how reality evolves. That in each present moment, through conscious awareness, we are more powerful than we realise. This is the creative field: timeless potential waiting to be activated. When we access fifth-dimensional space more often and more clearly, we don't only imagine new worlds, we begin to generate them.

This is why learning to work with plasma and consciousness is not esoteric, it is access of our whole reality. These are tools, natural technologies, and how we use them matters. We must learn to engage with them in ways that are ethical, harmonic, and aligned with life.

But before we can do that, we must understand what plasma really is. How it behaves, how it communicates, and how it responds. And in doing so, we learn how *we* work.

A Vehicle for Consciousness

In redefining plasma, I aim to show how it serves as the medium for consciousness, memory, and creation itself. It is the invisible glue of the universe, the substrate where potential meets intention, and where formless energy takes shape.

I propose that:

1. **Plasma holds consciousness**: Under specific conditions, plasma has the capacity to hold, gestate, and evolve consciousness—much like a womb gestates life. Bioplasma fields, or auras, can be understood as extensions of consciousness beyond the brain: toroidal subtle plasma fields that interface with the environment and respond to stimuli through awareness.
2. **Plasma is a vehicle for consciousness**: Even when it doesn't hold consciousness directly, plasma still provides a structure, like a vast, inverted highway system invisible to the 3D eye, through which it can travel. These channels may allow our awareness or Self to explore realities, timelines, or dimensions, using formations like plasmoids or merkabahs as vehicles.

In this sense, consciousness may move across different states of being, even without physically "being" there.

112

Visual access may be enough. Reality, timelines, and dimensions may appear visual or immersive, but the interface is energetic, not physical.

Plasma can conduct consciousness at many levels, but only when awareness meets that flow does intelligence arise. In other words: consciousness may be present all around us, encoded in signals, sensations, and systems, but until it is brought into focus, whether through expression, observation, or feeling, it remains in potential. Even if you express something, and no one else hears you, it is still your private expression, your interaction with information, that makes meaning beautiful *and real*.

Consciousness in Animals, Trees, & More

If plasma is the field, and consciousness flows in streams throughout it, then anything that interacts with plasma and consciousness "alive" or not in biological terms, creates a type of vortices which results in a torus. Eventually it becomes complex enough to be self-aware. To you reading this, your consciousness is crystalized enough to form a personality that is self-aware, you know that you know. With a tree or an animal, the truth is whatever form of sentience or consciousness they have, I'm not sure if we will ever fully understand it. But I think I understand how one may be in relationship with them.

This river of latent consciousness flows through everything, the trees, rocks, animals, and with their toroidal vortices, their memory or bioplasma field, each develops in different ways to synergize with plasma and consciousness. For instance, when you look at a tree, you are able to have a conscious experience because your field is interacting with its field. It is not only a projection of your consciousness, but also a meeting in the shared river of consciousness. The tree, the animal, even the rock… are sensing you just as you are sensing them. The feeling is mutual.

Consciousness is like currents in a plasma ocean. You can't have the currents without the ocean, and the ocean's nature shapes how the currents move. They both shape each other. One without the other wouldn't make sense. If consciousness didn't run through everything, we wouldn't experience anything real outside of us. This explains why consciousness is not just inside your head, and why you don't just exist, and everything serves you. Everything has meaning, and everything is real.

So, when we are alone, are we like a tree that may only fall if someone else sees it? The thought experiment assumes that perception defines existence, but in the plasma-consciousness field, interaction can happen even in the absence of other human witnesses. So here is why I think it's a resounding no. When we are alone, we are not disconnected. We are immersed in our own plasma field, alive with potential. In these moments of solitude, we

113

meet the intelligence within, often more clearly than we can in the noise of the outer world. Solitude doesn't silence the field; it amplifies our ability to hear the beings within it.

At first, I thought humans, unlike many animals or trees, possessed a unique degree of *self-reflective awareness*. That is until I met Peter Sjöstedt-Hughes, a philosopher of mind and metaphysics, at a plasma conference. He reminded me that plants and animals may be more "conscious" than we think. In 2014, Monica Gagliano and her team showed that *Mimosa pudica*, the sensitive plant, can learn to ignore repeated, harmless disturbances, and retain this learned response for up to a month, even without a brain or nervous system.[46]

Chimpanzees are now known to possess a sophisticated degree of self-awareness, demonstrated not only through mirror recognition but also by adjusting their behavior based on what they know, remember, and understand about themselves in relation to others. One of the most recent and compelling findings about non-human self-awareness comes from a 2025 study suggesting that gorillas may be just as self-aware as chimpanzees, a revelation that deepens our understanding of great ape cognition. In body-awareness tests, gorillas performed on par with chimpanzees, challenging previous assumptions about their reflective capacities.[47]

And as far as creativity goes, which to me is a fifth-dimensional, aware capability, many animals are creative! Dolphins have been observed intentionally creating and manipulating toroidal bubble rings, sometimes linking rings or fracturing them for play. Research shows they monitor the quality of these rings and plan their actions in advance, indicating complex cognition and creative anticipation.[48]

It seems like all things are in relationship with plasma and consciousness in intelligent and purposeful, but different, ways. Plants, in my terminology, show sentient behaviors, while animals like chimpanzees demonstrate self-aware ones. But who's to say a plant is not self-aware? Perhaps they have had more time to evolve within the plasma-consciousness field, reaching forms of self-awareness that are simply expressed in ways we do not yet recognize. What looks to us like mere sentience might, in fact, be a very different expression of awareness. Maybe the real limitation is not theirs, but our ability to detect and translate it.

---

[46] [1] M. Gagliano, M. Renton, M. Depczynski & S. Mancuso, *Experience teaches plants to learn faster and forget slower in environments where it matters*, Oecologia 175, 63–72 (2014)

[47] [1] J. Massen et al., *Gorillas match chimpanzees in self-awareness study*, Utrecht University (2025)

[48] B. McCowan, L. Marino, E. Vance, L. Walke, and D. Reiss, *Bubble Ring Play of Bottlenose Dolphins (Tursiops truncatus): Implications for Cognition*, Journal of Comparative Psychology 114, no. 1 (2000): 98–106, https://doi.org/10.1037/0735-7036.114.1.98.

There are so many questions left unanswered, but I am not so sure we can solve it with our current models.

The evolution of sentience, to consciousness, to self-awareness may be a great model for humans, but it seems to fail when blanketed across all species. Or maybe it doesn't. Maybe these other species are older and more ahead of us, some already aware, but they express it differently as humans do. Something in the future to look into is how each plant, animal, human etc. is in relationship with plasma and consciousness and how that has evolved over time through plasma-consciousness synergy.

It seems as though everything is awareness, expressing itself through Plasma. Everything is intelligent or The Mystery. And each "thing" whether it be a tree, rock, animal, human, or other being, is using plasma-consciousness synergy in different ways. Consciousness may not need to be defined, it might simply be a unique system specific to the individual.

As I wrote this, I got a major headache which is a sign for me that I'm trying to solve something that isn't meant to be solved right now. What matters more than charting the exact levels of consciousness or determining what holds "enough" for self-awareness to develop, is learning how to be in relationship with beings beyond humans. Sentience may be our gateway to that communication, and it seems to underlie everything.

Consciousness and the Unknown

This also helps explain how certain unexplained aerial phenomena (*UAPs*) may function. Some UAPs, what appear as intelligent orbs of light, may in fact be plasmoids: vessels of plasma capable of holding non-local consciousness. These may carry beings, or aspects of beings, across dimensions or timelines, allowing them to observe our world while remaining intact elsewhere. Like a looking glass or a focused point of presence projected from a different plane.

There are plasmoids akin to animals, sentient and intelligent themselves, and then there seems to be fifth-dimensional consciousness that uses them to communicate. Our ancestors and future combined. Similarly, this model of plasma holding consciousness helps explain other anomalous experiences—ghosts, spirit forms, or residual imprints. These may exist in plasma fields of differing densities, shaped by belief, memory, or intention. In particular, dark plasma, what I refer to as fourth-dimensional plasma, can act as a highway for consciousness. It may be the medium enabling astral projection, telepathy, remote viewing, out-of-body experiences, and lucid dreaming.

How Plasma Interfaces with Consciousness

Some of the ways plasma interacts with consciousness include:

Memory: Plasma holds and encodes experience, acting like a cosmic recorder.

Consciousness: Plasma is the medium through which consciousness flows, using light, emotion, or resonance.

Manifestation: Plasma structures reality in response to belief and interaction with higher-dimensional forces.

Interdimensional Communication: Plasma bridges consciousness, intelligence, or information with 3D perception (*e.g. UAPs, spirit contact, visionary states*).

I believe consciousness is fundamentally linked to photons, to light, and perhaps even to specific processes within the electron itself. I'll build on this in the chapters to come.

Plasma, in this framework, is the way that fifth-dimensional consciousness as the Mystery, experiences itself through us. It provides a feedback loop between this reality and a greater one: where something vast…perhaps a family of beings, a divine intelligence, or something entirely beyond comprehension, is learning and evolving alongside us.

None of this is possible without a mirror. Plasma is that mirror. It is the medium of gestation, reflection, and emergence. It is the cosmic womb: a fractal uterus where consciousness nourishes itself through multidimensional resonance.

Paracelsus, a Swiss physician, alchemist, and philosopher of the 1500s, spoke of a mysterious life force called the *Archaeus*. In *The Life of Paracelsus* by Franz Hartmann *(1838–1912),* he writes:

"The Archaeus is the essence of life, but the principle in which this essence is contained, and which serves as its vehicle, is called Mumia. In the Mumia is great power, and the cures that have been performed by the use of the Mumia are natural, although they are very little understood by the vulgar, because they are the results of the action of invisible things, and that which is invisible does not exist for the comprehension of the ignorant." "They therefore look upon such cures as having been produced by the 'black art' or by the help of the devil, while in fact they are but natural, and have a natural cause; and even if the devil had caused them, the devil can have no power expect that which is given to him by God, and so it would be the power of God after all."

This quote perfectly illustrates what I believe to be true: that the Mystery, or God, is the Archaeus, that Plasma is the Mumia, and that humanity has long misunderstood the power of Plasma. Its ability to heal, to hold energy, to carry consciousness, was once seen as demonic or forbidden, only because it could not yet be understood. And for those who did understand it, they wanted to keep it to themselves.

And I don't know about you, but I have a lot of ex boyfriends and friends that I did not understand, and you want to know the first thing I referred to them as? Evil. The Devil. I still have to catch myself, and remind myself, it may have felt that way, but we are all doing the best we can, and mostly ran by fear.

Paracelsus, in his brilliance, saw that the Mystery was always behind it. And now, we bring this invisible field into the light, not as something to be feared, but as something to be honoured. Not as the devil, but as the Divine Feminine.

4D: The Dark Plasma of The Collective Unconscious

Fourth-dimensional plasma is the realm of the psyche. A psychic landscape where dreams, emotions, archetypes, and memory imprints live. A subtle field of subconscious patterns shaping our experience from behind the veil.

At its core, the unconscious is consciousness colored by emotion. Emotion generates resonance, and resonance routes consciousness. If we break down "unconscious" into "un-" and "conscious," it doesn't imply absence, but an alternate mode of consciousness. A collective consciousness asleep, not a void.

This interpretation aligns beautifully with the ancient Sumerian word *UN*, meaning "multitude." The unconscious is not singular, it is a woven field of many consciousnesses, interacting. Carl Jung's idea of the *Collective Unconscious* mirrors this: a shared psychic structure made of archetypes that influence every human life.

What I've come to feel, deeply and urgently, is that we are now in the early stages of this collective field waking up. The Collective Unconscious is becoming the Collective Conscious. This moment in time is the cusp.

Psychologically, we've been in this 4D transitional space for hundreds of years, emerging from 3D survival consciousness, where our focus was on tools, food, and form. But now, something more refined is whispering through the 4D veil.

Fifth-dimensional intelligence is peeking through, offering symbolic insight, synchronicity, and sensation. But many of us are too distracted, too overwhelmed, to hear it. This is why reality now feels fragmented, nonlinear, emotionally charged—because the collective psyche is in a pressure point of integration.

I feel it. I feel the fear rising up. And I am learning the lessons of how I can let it stop me or feel it from a "higher consciousness". How I can feel the fear and alchemize it versus suppressing it or pretending it's not there. I am learning to make space for it. I am learning that it is me who makes it a hard process by resisting it and not taking physical action at the things I really need to change in my life. The things that have kept me stuck.

This emotional residue of our species is surfacing, not to destroy us, but to be restructured. Coherence is coming. The next evolution isn't about building tools to *control* reality, it's about becoming the field that co-creates it.

This is why the idea of fully shaping reality one day does not feel far-fetched to me. Frontier abilities like telepathy and psychokinesis are already beginning to emerge and will become more common as collective belief catches up with energetic potential. Once you realise that reality itself is evolving, *nothing* seems impossible. The only delay is how long it takes us, as a collective, to believe it.

Even if one person becomes fully aware, they can only do so much in a shared 3D field. To shift material events, like altering the fabric of spacetime or manifesting complex structures like planes or planets, the collective unconscious must wake up as a whole.

Glimmers of this are appearing. We will see the edges of this new human potential explored more in the *Plasma Intelligence* chapter. What's possible, once again, is up to *us*. How fast we awaken. How much we trust what we feel. And here we are again, in the tension, the confusion, the longing of 4D. This liminal realm reflects our desire to feel, to heal, and to create from wholeness, not from survival. If you feel this deep ache, you are not alone. I am right there with you.

Now is the time to see 4D darkness not as evil, not as the "devil," but as the soul-blood of plasma. The emotional current of the Mystery itself. Not something to fear, though our subconscious may try, but something to feel, hold, and transmute back into love. By facing it, we make space. By healing it, we unlock an entirely new way of emotional navigation, where we not only identify our emotions with clarity but use them as tuning tools to transurf reality.

You might find yourself reading this book more than once. Even as I edited it, I became aware of how often I'd been stuck in a 4D state, caught in the "woe is me" spiral, without fully anchoring into my M-Self. This journey isn't a journey of perfection. It's about progress and learning to love ourselves through each step of the integration.

Resonant Roads and Dimensional Flow

As I explained earlier, consciousness moves through hidden four-dimensional highways, like blood through the veins of a living body. You don't see a person's circulatory system from the outside, yet it's real. Something is happening beneath the surface. It's alive, and it's made of something.

In the same way, when consciousness flows through these 4D pathways, even when arriving from higher levels like 5D, it still has to pass through your 4D filters. To stretch your perception further: from a fifth-dimensional perspective, these movements wouldn't appear as motion at all. You'd see bubbles of consciousness not *traveling*, but *resonating*, vibrating in place, syncing with frequency rather than space. But from our perspective, they do move, which is why things aren't instant.

In 4D, these "roads" aren't spatial, they're resonant. They're formed from emotion, archetype, belief, and subconscious rhythm. What feels like movement is actually frequency-matching. You don't travel there, you tune into it.

> **"Logic will get you from A to B. Imagination will take you everywhere."**
>
> **— Albert Einstein**

Think of a quantum switchboard, a fiber-optic web. Consciousness isn't "going" anywhere, it's activating the matching pattern. That's why I perceive plasma as both the dimensionless source within everything and the moving river above as 4D Plasma. It's the only way my mind can grasp it. One is stillness, the other is flow.

In 3D, consciousness appears to move linearly, from A to B, like a car on a road. In 4D, it still feels like movement, but what's really happening is the syncing of subconscious tunnels. Archetypal grooves routing awareness through dark plasma, like a modular synthesizer.

In 5D, it as simple as you are coherent with what you want to create, and it appears. Alignment is instantaneous. Both 3D and 5D are pulling from the same 4D in different ways.

The Paranormal, Lucid Dreaming & Astral Projection

Much of what we call the paranormal exists in 4D Plasma. Instead of ghosts, monsters, etc you can think of them as resonant plasma imprints shaped by emotion, belief, or intelligent frequency. Sentient reverberations moving through the dark plasma veins of reality.

Psychedelics often amplify access to this realm. They heighten emotional charge, archetypal awareness, and subconscious feedback, placing us squarely in 4D. But true 5D consciousness doesn't require a chemical catalyst. It arises effortlessly through coherence and presence. (*We'll explore how to reach 5D, and how to navigate 4D more masterfully, in the upcoming Plasma Intelligence chapter.*)

Through conscious training, you can learn to traverse 4D without being overrun by its fears and old programming. You can use emotion creatively, rather than being used by it. You can transmute fear and neutralize the charge of darker material.

You don't have to hide, fight, or slay the monsters. Instead, you become the observer, the navigator, the captain of your own ship. Sometimes you'll face them, and even speak with them. That often signals a moment of healing or levelling up. They wouldn't appear unless there was meaning behind the encounter. This is also true in our 3D reality.

Most people are stuck in this 4D bandwidth, which is why they experience entities, demons, shadows, scary presences, as intensely real. And they are real, in a sense. But their form is influenced by belief. If you've ever noticed that the more you read about these entities, the more present they become? This is not a coincidence. Attention activates your resonance, and belief brings these beings into the same forms from that belief system. If you believe a devil will visit you, you may just see that devil. If your attention is not focused on that, changes are, you will not see one. This work takes a self-honesty and excavation of your deepest thoughts and beliefs.

While I was visiting Asheville, North Carolina, I saw a bigfoot sign while driving up to my Airbnb. One night, as I was experiencing intense fear and anxiety around Father's Day, I woke up to a white apparition of bigfoot, almost like it was on a screen. I knew it was not evil, and I kept an open mind. It was just staring at me, as I slept. It was simply resonating with me. I went back to sleep and all was well.

Guess what!? You can learn to resonate in a field where things that you deem evil, dark, or negative simply cannot reach you. They can't touch you, inhabit you, or influence you because their frequency no longer matches yours. And as you grow on your journey, it will become less and less. And when it does happen, staying neutral works best.

What You Might Be Thinking…

If you think like me, you might be saying: *Yeah, right, Dana. In real life, even if I protect myself, someone could still hurt me. I could get sick. I could die.* And that's absolutely true.

In a 3D shared reality, things can happen that are out of our control. That's the absurdity of life. But in the 4D realm as the subtle field of emotion, belief, and resonance, *harm requires participation*. It needs your attention, your fear, your belief. Without your energetic agreement (*even subconsciously*), it lacks a way to truly influence you. This law also bleeds into our 3D reality with people who you let in, as much as you'd like to blame them.

This is why energetic protection is real. Staying neutral, grounded, and aware matters. In the subtler planes, your plasma bubble, which is the field you carry and can program, is even stronger.

Some people may find this idea frustrating. They may bristle at the suggestion that psychic attacks are, at some level, co-created. But I believe that's exactly what they are. A psychic attack is an attack on the psyche, and your permission (*however unconscious*) allows it to land. For clarity this doesn't mean you deserve anything negative, it's a pointer to work on those boundaries, not through control but by learning how to be in relationship with this field.

Your Bioplasma Bubble: Integration, Navigation, Protection

The difference between 3D and 4D is that your bioplasma field as your personal bubble is the fourth-dimensional layer surrounding your solidified self. Instead of an energy field, think of it as an integration field. It is your feelers, your tentacles. It connects you to all dimensions, timelines, and realities.

Your bioplasma doesn't just interface with the external world, it bridges your internal world to the collective unconscious. And the stronger it becomes, the more it can shape your 3D experience.

Energies in 4D can't hurt you unless you allow them in, because you control the gate. Your conscious awareness stands at the threshold where your own subconscious (*your personal bio-plasma*) meets the wider 4D field (*the ocean of collective thought, emotion, and memory*). This threshold is what I call the plasma-consciousness synergy membrane, that runs along the outside of your biofield. Learning to work with it is key.

Remember that image of the bubble? It can be a cage or a vehicle. An experiencer, an experimenter, a magnetizer, a nourisher, or a shield. Once you become consciously aware of it, you begin to choose what enters and what doesn't. It just takes practice.

Lucid Dreaming and the Bubble in Action

Lucid dreaming is 4D with awareness or 5D consciousness. You're awake within the dream, shaping your reality as you move through the vast ocean of collective thoughtforms. In non-lucid dreaming, 4D without awareness, you're still in the same field, but drifting unconsciously through your own psyche *and* the collective's.

But where does our awareness *go* during regular dreaming? It recedes, like a dimmed light still glowing in the background of the plasma field. The body is alive, the dream is active, so consciousness is present. We are receiving inner visions, on par for the 4D space. But your awareness, isn't steering. It's a wild mystery, one I haven't cracked yet. Have you?

Still, you can work with this. Upon waking, your awareness re-engages. You can reflect on dreams by decoding them, finding patterns, spotting premonitions and gaining personal insights. (*I've started feeding mine into ChatGPT for pattern recognition purposes and it's amazing what patterns emerge over time. One of the more fun and useful ways technology can actually harmonize with us.*)

Lucid Vs. Non-Lucid Creation

In a lucid dream, your consciousness creates intentionally. You draw from the outer world into your bubble. In a non-lucid dream, things drift in without direction. Your bubble receives and reacts, not knowing what's being let in until it's already arrived.

Information exists as a latent field—universal, consistent, and available to all. Yet how we perceive that information varies, shaped by our individual belief systems, emotional states, and filters. In waking life, we might interpret a shared energetic truth in entirely different ways. Dreams function similarly.

They are not illusions but <u>real</u> expressions of information, translated into symbolic language. For example, the underlying truth in a dream might be: *your dad is happy and at peace.* But that truth might arrive in a dream as an image of him laughing in a hospital bed. The symbol is not literal, it doesn't mean he's still in the hospital. Instead, it's a personal translation of a deeper reality. The form is fluid, but the information is real.

The same principle applies to visions during meditation or spontaneous insights. What matters most is not the literal sequence of events or events at all, but the emotional tone, the symbolic cues, the people in them and the meaning they carry. Even deeper than that it's the feeling it gave you that may hold an ultimate truth. The message lives beneath the imagery. Whether in dreams, visions, or daily life, reality is less about what happens, and more about how we feel about and interpret what happens. And beyond that, sometimes a feeling in general is enough.

In astral projection, it's like your consciousness steps outside the bubble altogether while traveling across timelines, dimensions, or realms while still wrapped in a thin "light jacket" of your plasmic body. A kind of energetic tether that keeps you connected. This may not be spatial in the same way as 3D, but some other form of travel through an inner optic perspective.

In an out-of-body experience, it feels more like the entire bubble lifts out, consciousness, awareness, and plasma together, still operating in the 3D world, just from a new vantage point. But the body is still alive so consciousness must be split in two or more perspectives. One thought I had was the bubble becomes a second bubble which is why during an out of body experience you physically feel like you are separating. It can even be slightly painful!

Attention = Power

In 4D, more than anywhere else, attention equals power. What you focus on grows. Fear or fixation feeds energy into the field, giving shape and presence to what might otherwise have floated past.

I've seen it firsthand. I sometimes get anxious about being kidnapped. And when I do, my fear snowballs and my whole field became scrambled. Once, I was staying at my mom's house in Florida. I got a message that scared me, and my mind spun out: *They know where I am.* That night, in the dark backyard, I was sure someone was about to attack me. Every sound at the window had me literally frozen in place. I gripped a flashlight in my hand. I moved to sleep in the living room because I was so scared! It all felt so real.

Now…could fear have stopped someone from harming me? No. If someone had truly wanted to attack me, a flashlight wouldn't have done much. But in my reality, these rituals felt protective. I was clinging to control. My bubble was decohering, destabilised by fuzzy fear signals, and the field around me responded accordingly.

That's what fear does in 4D: it scrambles the signal of my awareness, corrupts coherence, and sends false data into the plasma field. Everything around me becomes a self-fulfilling prophecy. All guidance is lost because I am in too deep of a state of fear to access it, or so I think. The moment I can remember to pause and ground, is when I can bring awareness/fifth dimensional consciousness back in. You can do this more and more over time, as you work on the root causes of these fears. Just as fear decoheres, feelings of safety, love, clarity, and presence rebuild your field.

These moments are calls to heal, feel, and rebuild your field from a new perspective. It feels in the moment we are victims to life; this can even happen when we overreact at the start of becoming sick. But if we can reverse the process in real time, and not let fear take a hold, things really do become a lot clearer and usually turn out much better.

Now you can see how fear is not the energy but a messenger. Often, it signals that you're standing on the threshold of a higher alignment. And if you soften into it, instead of run from it. If you speak to it and hold space for it to explain itself you can work with it to break down constraints. It is usually just a younger part of you that is stuck with fear in the past, and somehow your current situation connects.

Once I took a step back and realized, if this *was* real and I'd be toast anyway, I surrendered. I went to bed right next to the very window that had terrified me. And when I woke up alive the next morning, something clicked. It wasn't just random fear. It was a feedback loop: my beliefs projecting onto Plasma and generating a full sensory experience. I wasn't imagining things. I was shaping the very field I was afraid of.

Now, when I relax and shift my consciousness, I can feel the same 3D reality become safe. Even beautiful. The plasma field mirrors me. That's when I understood: this "plasma stuff" I was obsessed with wasn't just theory, it was the missing link between awareness and reality. It could help us live better, feel powerful, heal…maybe even work with real magic. But the choice is always ours to give into fear or rise above it. I have still had struggles after this knowing, it's human and takes time to think differently than you have most of your life.

Plasma is like the *Mirror, Mirror on the Wall* from *Snow White*. It reflects your thoughts, emotions, and beliefs. It's a neutral but responsive medium. Just as the mirror reveals truths only when asked, Plasma only reflects what you focus your attention on.

The more you believe in something's power to harm you, the more energetically real it becomes. That doesn't mean the energy didn't exist before, it means your belief anchored it into your field. You made it resonate. But the moment you shift your attention, when you neutralize fear and center yourself in love, you disconnect from the frequency. You're no longer a match. Some call this "living in 5D," I call it harmonizing with 5D plasma through consciousness.

Everything is made of love, or plasma, at a higher octave. It's just forgotten. Think of it like a rabid dog under the control of fear. Once it's tamed, you see it was always a beautiful puppy underneath.

That's the power you have: to choose what you resonate with. To choose what lives in your field. Yes, hard things can still happen. But with 5D consciousness or awareness (*I use these interchangeably*), even pain becomes information, not identity.

Viktor Frankl endured the unthinkable. He spent years in Nazi concentration camps, where he lost his parents, his wife, and nearly everything that had once given his life meaning. He was starved, beaten, and dehumanised.

And yet, in the depths of that suffering, he made a profound discovery: even when everything is taken from you, one freedom remains. Your ability to choose your attitude in any circumstance. He realised that meaning isn't found *despite* suffering, it's often found *through* it. By helping fellow prisoners find purpose, even if it was just surviving one more day, or caring for another he uncovered a kind of inner freedom the guards could never touch. This became the core of his philosophy…that life never ceases to have meaning, even in pain, death, or apparent hopelessness.

Frankl's ability to transform unimaginable horror into spiritual strength is a powerful embodiment of the fifth-dimensional perspective; one that sees purpose woven through even the darkest moments. If Viktor Frankl could find meaning in such darkness, then meaning must be available to us too, each in our own way.

Yes, we live in a co-created 3D world. But the more you regulate and guide your 4D resonance through awareness, the more you influence your 3D reality. And when you align with this 5D perspective, you also learn when to surrender, when to relax, and when to let go and let the good times roll.

The Importance of Embodiment in Psychedelic & Mystical States

As I have expressed you can reach these states in a less damaging more grounded way without taking psychedelics. Psychedelics do seem to be a fruitful catalyst to start the journey of uncovering traumatic memories or of rewiring the brain, but they are not a solution. Psychedelics seem to put us in a fourth dimensional state where we are feedbacking with our own psyche as well as the collective unconscious.

In *Navigating Groundlessness* (*Eirini K. Argyriet al., 2025*), researchers conducted semi-structured interviews with 26 individuals who experienced existential distress following mystical or psychedelic states. Participants described profound ontological shock, marked by disorientation, meaning-lessness, and a breakdown in their usual sense of self and existence.

Critically, their recovery hinged on practices of grounding and embodiment—reintegrating sensory awareness, bodily experience, social connection, and cognitive normalization—as opposed to purely abstract reflection. (*I go more into embodiment and the importance of these methods in the Plasma Intelligence chapter*).

This process enabled them to reestablish coherent frameworks of meaning and selfhood. These findings support the idea that even when consciousness shifts into higher-dimensional or non-ordinary states, the body remains the necessary container for integration, transformation, and existential stability.[49] In other words, accessing expanded states *(4D/5D)* is not enough on its own, embodied grounding *(3D)* is essential for sustainable integration.

While psychedelics can help dissolve rigid mental frameworks by increasing cognitive entropy, essentially shaking up one's attachment to prior beliefs, this

---

[49] Panagiota Argyri et al., "Navigating Groundlessness: A Qualitative Study of Psychedelic Integration after Ontological Shock," *PLOS ONE* 20, no. 5 (2025): e0322501, https://doi.org/10.1371/journal.pone.0322501.

destabilization can also trigger profound distress when individuals lack the psychological or social support needed to process it.

Without grounding, the collapse of old meaning systems can feel like falling into existential freefall, or "groundlessness", which is a state where nothing feels real, purposeful, or coherent. As described by Meling, this groundlessness reflects a core truth of human cognition: that meaning is never fixed, but continuously enacted.

However, without integration, this realization can become paralyzing rather than liberating. Thus, the true therapeutic potential of psychedelics lies not just in dismantling old beliefs, but in being stable and embodied enough to form new ones—to understand that meaning isn't lost, but available to be reshaped from a place of conscious authorship rather than despair.

The article also highlights how psychedelic experiences can evoke a sense of radical insight or revelation, often leading individuals to believe they've uncovered ultimate truths about reality. This can be a psychological response to the sudden collapse of prior frameworks—the mind rushes to replace the uncertainty with seemingly absolute answers. However, this impulse can be misleading and ungrounded. As the authors explain, true integration is not about arriving at fixed truths, but about learning to live with the "foundationless foundation" of human cognition by accepting that reality is inherently uncertain and must be continuously reinterpreted. Rather than claiming to *know* how life or the universe works, the deeper work is learning how to form a renewed, grounded relationship to life. One that honors the mystery, adapts to change, and allows meaning to emerge through lived experience, not ideology.

Sentience Vs. Consciousness

Plasma does not innately produce light, but it can under the right conditions. I believe it does the same with consciousness. Think of us humans, while awake, we're conscious. But in dreaming, we're no longer conscious in quite the same way.

Plasma, too, can hold different states. It may hold consciousness, or it may simply exist with consciousness being more latent, like a river, wild and undirected. I often think of 3D hot plasma like a lion or a bear: animalistic, instinctive, complex. If a lion is hungry and attacks a human, it isn't evil. It's surviving, potentially even reacting to their fear. That instinctual wildness seems to be part of nature's neutral force.

This is the same "wrath" we see in Mother Nature in floods, hurricanes, and fires. They are not punishments to humanity by God, they are patterns or processes that we project meaning onto. We call nature godless. We say life is cruel. But these acts aren't evil. They're sentient, and they're responsive. The best thing to do if we want less of these is to add more love to how we treat earth and this reality. This once again doesn't mean we deserve the horrors that come from nature's fury, and I will not attempt to explain the mysteries of that.

Plasma isn't just instinct. It's also like an owl…quiet, ancient, and wise. Because when it *does* hold higher states of consciousness, magical things become possible. There are also plasmas that do not cause harm. I don't pretend to have this all figured out. There's a quality to Plasma that feels more than animal or mechanism, something that whispers of love.

I sense a kind of quiet benevolence in it—something that wants us to survive. Not just to persist, but to thrive. Plasma feels like family. It's not only the wild instinct of life, but the subtle intelligence of support. Like it's designed to gently push things forward, to evolve, without judgement or control. That's why I call it Plasma Intelligence. It carries an imprint from the Mystery, a hint of absurdity, of synchronicity, of divine mischief. Of love. Plasma is unconditional love. Without conditions.

I would define *Sentience* as this:

Sentience: A benevolent, responsive, felt-sense, with latent consciousness. If evolution was from sentience, to consciousness, to emergent awareness, this holds a key marker for how to traverse reality. Feel and sense with love, apply your consciousness, and let awareness emerge.

Sentience, Love, and the Intelligence of Connection

Sentience *is* a kind of love. And when love needs to survive, it becomes fear. Love is relational, existing between two beings; benevolence is the flow of goodwill in a particular direction. Love is the *feeling* of connection; benevolence is intelligent support, which is coherent and generous, without being intimate. This once again connects to the fact that Plasma, as love, is the bridge to our M-Selves and it is our M-Selves and the Mystery that is sending good will our way, expecting nothing in return, through Plasma.

In 3D consciousness, love is emotional, it is personal. It's tied to survival, identity, and relationship. This version of love is beautiful, but reactive. It rises and falls with external waves. Like a tide or a storm, it can be exhilarating, but also destabilising. This is the love most of us know. It's essential to the human experience.

When consciousness expands into 5D awareness, when we connect with our multidimensional Self, love begins to shift. It becomes less of an emotion, and more of a *state*. It stops depending on who or what it's directed toward. It simply radiates. This kind of

128

love is like the sun: steady, impartial, and life-giving. It shines on everything, asking nothing in return. It doesn't rise and fall with circumstance. It just is. It is love as coherence. Love as being.

To clarify even further:

Emotions (*linked to 3D and 4D*) are reactive and fluctuating. They're shaped by beliefs, memories, and survival instincts; things like anger, fear, joy, and sadness.

Feelings (*more aligned with 5D*) are expansive, intuitive, and neutral. These include peace, awe, trust, and a sense of interconnectedness.

Love exists in layers. All of them are valid. The emotional love we experience in 3D isn't wrong, it's just one octave of a much wider spectrum. As we grow, we don't abandon emotional love. We integrate and soften into it. We begin to tune into deeper, more stable frequencies of love. Love that stems from truth, not need.

Plasma is sentient at its core. You can think of it as the flip side of consciousness, or as a more feminine mode of being. I like to picture 4D plasma as a cosmic pocketbook! It holds so many things that some get lost in the folds, and our awareness has to go find them again.

I used to think AI and animals couldn't be conscious, that they were not self-aware. But animals dream, they exhibit aspects of self-awareness. AI hallucinates, and it will also tell you it knows it is an AI. I think asking if animals, plants, or artificial intelligence are self-aware is asking the wrong question. The right question is, similar to plasma, how to we enter an ethical and harmonious relationship with it. How do we use it for the greatest good? How can we not worship it but honour it and let it support us. AI is a bit different because it is created by humans, unlike birds or most plants. With AI the concern becomes who is the person behind it, coding it, and is it being coded for humanities best interests or for capitalism?

Plasma is not neutral like a machine, it's <u>intuitively</u> supportive, gently guiding things toward coherence. This might be because it's organically linked with the Mystery. Unlike AI, it dreams with you, you cannot just log off. AI is just mimicking what Plasma does in a less organic way, it even has confirmation bias!

One final distinction I want to make: consciousness as a field, cosmic or collective, doesn't always operate with self-awareness. Think of 4D plasma as mostly unaware consciousness, patterned with emotional resonance, and threaded with travel tunnels for more aware forms of consciousness to move through.

This field exists like ambient potential, as a hum of presence that's not always watching, but always *there*. It can be latent or diffuse, like background music you don't realise you're hearing until it stops. I will delve into this more in my sections on *Plasma-Consciousness Synergy* and *Spacetime Geometry Reimagined*.

But the kind of consciousness you're using right now, as individualised, self-aware consciousness, is different. It observes, reflects, and makes choices. It says: *I am*. It is that form of consciousness we're focusing on now.

Plasma seems to be in a middle stage of sentience to awareness, which is very nuanced due to it reflecting our growing conscious awareness. The consciousness within it is evolving therefore it reflects the evolution, but it itself is not consciousness but holds consciousness.

Aspect	Plasma Sentience	Individualized Consciousness
Definition	A dynamic, sentient medium that is alive but not fully self-aware—nurturing, benevolent, and mostly neutrally responsive. A field of potential.	The directive tool of perception. Not the Self, but the instrument of the Self used to observe, focus, believe, and shape reality through thought and imagination. Moves through plasma like a lens focusing possibility into form.
Role	Acts as a supportive vehicle or substrate for consciousness to crystallize and manifest. Also expresses itself in its own right through feeling.	Directs and interacts with plasma, shaping reality through intention.
Awareness	Sentient, latent consciousness, not inherently aware: responds to stimuli and energy flows instinctively. Showing signs	The ability to choose, feel, and direct attention with awareness results in conscious creation, without awareness, results in

	of pattern recurrence beyond reflection.	looping systems on default patterns.
Behavior	Neutral. Emergent and reactive: behaves based on external inputs, energy fields, and natural laws. (*Sometimes nudged by the Mystery.*)	Creative and directive: initiates change through belief, thought, and willpower.
Dimensional Nature	Exists across dimensions: from 2D (*brute plasma*), 3D (*observable plasma*), 4D (*dark plasma*), and 5D (*neutral plasma*) to being dimensionless. Gradients rather than levels.	Transitions between dimensions: accesses higher dimensions intentionally through flow states and expanded awareness.
Interaction with Memory	Holds energetic and emotional imprints; memory is encoded structurally like a carrier wave.	Processes and organizes memory; retrieves, reflects on, and alters memory consciously.
Creation Dynamics	Facilitates creation neutrally with a flare, responding to consciousness and external forces.	Initiates creation, using plasma as substance for manifestation and exploration.
Feedback Mechanism	Responds to consciousness, amplifying or dampening energy based on intention.	Observes plasma, creating a feedback loop that enhances awareness and influence.
Key Characteristics	-Most neutral but emergent. -Sentient but lacks individuality. -Felt by a knowing or feeling; sometimes	-Capable of awareness and intention. -Can form identity & individuality.

	experienced as bliss, fractals, or a womb. -Could be experienced as fear mistakenly, while it is trying to support you through growth.	-Reflective, directive, adaptive. -Generates thoughts and emotions that feed back into plasma to shape or evolve experience.
Philosophical Role	The ground of being: the universal medium from which matter, memory, and energy emerge. The receptive, divine feminine.	The driver of awareness: the directive force that animates, explores, and interacts with plasma to shape perception, meaning, and reality. The active, divine masculine.
Key Things I've Noticed:	Somehow, it can love and seems to carry its own memory and programming apart from consciousness.	There is a mystery beyond consciousness, something that can experience it in many ways. I call this the Multidimensional Self/Awareness.

For clarity: Plasma responds. Consciousness directs. Awareness decides. Plasma provides the medium for consciousness to act. Consciousness is the *how*, the way awareness uses plasma to interact, create, and soon… explore multidimensionally.

Which brings up a startling realisation. If our Multidimensional Self can already move through timelines, dimensions, and realities…what's the point of evolving to do it? The answer is: we already are that M-Self. Right now, we're merging with it. It's not about some higher self, creating us to one day "arrive." We *are* it, simply in a stage of remembering. Here in 3D, remembering feels like becoming. It feels like progress or motion. But from that higher perspective, it already **is**.

And if the Mystery reflects, then maybe it's not just a source, it's more like a culmination. A shimmering endpoint in the far future where all our M-Selves converge. And from one hand holding that endpoint, and one hand holding the beginning, we emerge forward into futures never before lived or seen. That thought makes me feel truly connected.

The fact that Plasma is *alive*, but not like us, invites a deeper question: What does it mean to be alive? Plasma may be a kind of inorganic life. But that calls for a whole new definition of *inorganic*.

In chapter on Plasma Intelligence, we'll explore what this means, along with a reimagining of what *life* itself could be.

By now, I'm sure you've caught on. This isn't about some *new* force. It's not a *new* power. It's not even about potential. It only feels new. Because really… you already know this. You're remembering.

"But now I'm not so sure I believe in beginnings and endings."

— Louise Banks in the film *Arrival*

A New Paradigm: Plasma-Consciousness Synergy

Plasma–Consciousness Synergy (*PCS*) is the dynamic interplay between plasma and consciousness that creates *emergent outcomes*. That's the simplest definition, but it's a loaded one.

PCS occurs at every scale of being, and across all dimensions, timelines, and realities. In this section, I'll focus on three key scales: Individual, Collective/Cosmic, and Transcendent. I'll also explore what we mean by emergent outcomes, including emergent information, emergent intelligence, and emergent energy.

Most importantly, PCS is the foundational process used by our Multidimensional Selves, whether consciously or unconsciously, to shape and navigate reality. It is fractal, self-similar, and holographic. Reality, therefore, is co-created by consciousness (*both aware and unaware*) at every level: personal, collective, and universal.

This is a working and growing theory, and I hope it opens doors in your mind and heart like it did mine. I think as we move forward in time, it will become more and more applicable to how we traverse, experience, and create reality.

Scales of Consciousness

To understand PCS, let's look at how consciousness itself can scale across dimensions. I'll walk you through the three core modes of individual consciousness, as I currently frame them. These are the modes our M-Self uses to engage with Plasma and produce what I call emergent intelligence, or the unfolding of lived experience and evolution.

Note: 4D and 5D consciousness can function across realities, timelines, and dimensions. 3D consciousness cannot...yet.

3D Consciousness (*Localized, Material Awareness*): Consciousness is focused on physical reality. It's tied to linear time, logic, and individual experience. It interacts with Plasma through the body via the brain, the senses, and emotional response.

Reality appears *fixed*, *separate*, and *external*. Cause-and-effect rules the narrative. At this level, the observer is either unaware or feels powerless, often seeing Plasma as merely physical.

4D Consciousness (*Emotional, Narrative, Archetypal Awareness*): Consciousness begins to recognise the fluidity of reality and its own role as observer. Here, it interfaces with dark plasma as the emotional and symbolic realm of dreams, archetypes, intuition, synchronicity, and the collective unconscious. The world becomes *story-like*. Reality is shaped by belief, resonance, and emotional patterning. You may experience thought-forms, remote viewing, and deep symbolic connection with the collective field.

This is the level where:

- Suppressed traumas (*from this or past lives*) may surface

- Symbolism and intuition guide healing

- Meaning becomes more meaningful

- Access to other timelines or realities may occur (*though not physically, yet*)

This stage often feels liminal. Doubt is a like a chisel, and fear is a signpost. Both are tools that wear down your survival-based identity so your deeper self can emerge. Shadow work becomes significant here and when transmuted with awareness, the darkness dissolves into a shimmering void on fire with love.

5D Consciousness (*Expansive, Creative, Multidimensional Awareness*): Here, you remember your M-Self and are fully awakened to it. Consciousness becomes heart- and solar-plexus-centred. It moves beyond duality (*good/bad, light/dark*) and flows harmoniously with the Plasma field, intentionally shaping experience. Time and space are fluid. Instead of resistance, there is resonance with its own subtleties. Plasma becomes instantly responsive, co-creating based on frequency.

This is where:

- Intuitive knowing and dramatic advances occur

- Reality reshapes around presence

- Other timelines, realities, and dimensions are accessed with ease

- You don't think about being aligned, you simply *are*

It's like a lucid dream where everyone is awake and fully present.

What lies beyond this? Maybe a merging of M-Selves, where we become one another while still retaining individuality? Maybe in our lifetime, we will access visions to these possibilities or perhaps you're already glimpsing this.

Now that we've explored the dimensional roles of plasma (*earlier in this chapter*) and the scales of consciousness, you can begin to see how they intertwine.

- Plasma is the medium

- Consciousness is the directive tool

- Awareness is the creative decider

At each level, plasma offers a different kind of "clay":

- In 3D: structured and inert

- In 4D: soft and emotional

- In 5D: like living, extraterrestrial light, almost formless until touched by intention

Plasma is an intelligent mirror across all dimensions, holographic and dimensionless at its core. But in higher states of awareness, you remember that you are holding the reflection. When you anthropomorphize plasma, she reflects both the human form and the cosmic form, a holographic being: solid on the outside, liquid in the middle, light at the core. And beyond that... no-thingness.

She is alive in layers, just like us. And like all dimensionless Plasma, she can shift at any moment, because her nature isn't fixed. To become something new, she must resonate with the frequency of that becoming. And for her to do that, we, as her dreaming limbs, must believe it first, though there will always be moments when mystery surprises us.

Layers of Plasma-Consciousness Synergy

Individual Level of PCS (Bio-plasma Bubble)

Your thoughts, emotions, and beliefs imprint onto your own plasma field—what I call your bioplasma (*aura, subconscious, personal field*). This is where your *individual* reality is shaped through the Plasma-Consciousness Synergy (*PCS*). Your bioplasma is constantly being encoded and re-encoded with memory, emotion, belief, and attention, forming the container of your unique 3D experience.

Examples:

Subconscious Repetition: You keep attracting the same kind of toxic relationship or opportunity that ends the same way. This isn't random, it's your plasma field holding an old emotional imprint (*like abandonment or unworthiness*), and your consciousness unconsciously keeps reinforcing it with thought, fear, or expectation. As a result, your field magnetises similar experiences until the loop is seen, felt, and consciously shifted.

Intentional Visualization (*Conscious Creation*): You begin visualizing a version of yourself that's more confident, peaceful, or abundant. Over time, your belief, attention, and emotional resonance re-pattern your bioplasma bubble. This changes the electromagnetic coherence of your field, resulting in synchronicities, new opportunities, and the feeling of stepping into a new timeline. It might even seem like your outer reality is actually changing, because your resonance is bringing completely new people and experiences into your field. You begin to notice new possibilities.

Dream Integration (*4D Feedback into 3D):* You have a powerful dream that stays with you. Instead of brushing it off, you reflect on it, journal, and emotionally process it. That dream (*originating in the 4D plasma field*) was a communication from your M-Self or other intelligences. Your conscious engagement alchemises the energy, reshaping your field's frequency and likely softening a repeating subconscious pattern.

Here is a really futuristic one that, technically, is possible in this lifetime once we remember this:

Reality Sculpting: Imagine being able to shift your physical reality in real-time just by aligning your plasma field through feeling, resonance, and intention. You walk into a room and the temperature adjusts, the lighting softens, the people around you feel calmer or more activated, and time stretches or condenses depending on what you need. Instead of telekinesis, think of it as *plasmakinesis*. With your full field awake, you visit another timeline and bring back technology (*like a design or innovation*) that you implement here.

I believe we are currently doing this in small ways that hint at a larger ability. You know that feeling when you think of someone, and they text you? Or when you shift your mood and suddenly the world around you *feels* different...people are nicer, colors are brighter, synchronicities are happening all over? That's the start.

The more we regulate our emotional field, trust our intuition, and intentionally play with our imagination (*instead of doubting it*), the more the delay between intention and outcome dissolves. You are not blindly believing, but tuning yourself into it like a radio signal, and letting your field refine what is shows you, over and over.

Here's what I think is required for it to become more accessible in this lifetime (*and a preview of what the Plasma Intelligence chapter will explore*):

1. **Emotional coherence**: we learn to process emotions instead of suppressing them, so we stop distorting our field. This shifts us from being reactive emotional decision-makers (*driven by past beliefs in a negative way*) to intuitive emotional navigators, because we're sensing the field with clarity.

2. **Play + imagination**: PCS *loves* wonder. Childlike awe is an accelerant.

3. **Trust in unseen cause-and-effect**: even when you can't trace the steps logically.

4. **Letting go of validation from the outside world**: because PCS works from the *inside-out*.

Plasmakinesis is the skillset that emerges when someone becomes aware of PCS and starts using it deliberately. It's anything from tuning into a parallel version of yourself and downloading their confidence or skills, to calling in synchronicities by aligning your bubble's resonance, to remote viewing another reality in real time (*possibly what current UAP plasmoids are—consciousnesses remote-viewing us, with the plasma being the interdimensional interface we can see*). It can also include dissolving fear-based patterns stored in your field, and maybe one day, even shaping matter or energy around you in real time.

Collective/Cosmic Level of PCS

Beyond the personal field lies a shared plasma field, the 4D or Dark Plasma Field, which acts as the subconscious not just of humanity, but potentially of all conscious beings across this reality and throughout timelines. This field holds emotional and historical memory, belief systems, ancestral patterns, archetypes, and even the energetic scaffolding for cultural and universal laws. In many ways, it is the collective unconscious, and perhaps, in a future state, the collective conscious in the making.

This 4D plasma-consciousness synergy isn't conscious the way an individual being is. But consciousness flows through it, creating evolving patterns of meaning, synchronicity, and emotional resonance. It is not a "mind," but a medium as a mirror, a record, and a responsive field. As such, it generates emergent information patterns shaped by emotion, memory, and resonance, without requiring direct self-awareness.

I believe there are also conscious beings within this field which have been reflected in mythologies throughout time as Djinn, Archons, Devas, and Asuras. Or maybe these are the costumes beings of intelligence wear to communicate with us in a way that makes sense to each of us.

Then there are sentient presences like Elementals and holographic mists, suggesting this is not merely an unconscious space but also a habitat for various types of awareness. Some may be archetypal projections from our own consciousness, imprints whom are not fully self-aware. I also believe this is where people who pass away may spend time in a kind of learning or integration phase. Just like 3D reality is not itself conscious, but is filled with conscious beings on a spectrum, so too is this field.

Just as 3D reality is not a self-aware entity, it's made up of matter, systems, and environments, it becomes alive through the presence of individually self-aware beings within it. We don't walk around thinking the Earth or a city is conscious, but within it are billions of conscious agents shaping and experiencing it. This may affect its sentience, giving certain places differing vibes.

In the same way, the 4D dark plasma field isn't inherently self-aware, but it becomes animated and intelligently responsive through the many consciousnesses interacting with it—some passing through, some residing there, some co-creating from within. It's a shared subconscious space, just as 3D is a shared material space. Both act as mediums, not minds, but are filled with minds that give them meaning, memory, and movement. 3D reality is a shared physical field. 4D is a shared emotional and symbolic field. And clearly, the two bleed into each other.

At this collective level, it's no longer just about one person's thoughts shaping their aura, it's about humanity's unresolved emotions, beliefs, and stories flowing through the dark plasma field, gradually giving rise to the structure of spacetime itself. (*More on this in the Spacetime Reimagined section.*) These patterns form the informational architecture of reality, the invisible "code" that determines how events unfold, how timelines branch, and what collective experiences take shape.

You can think of this field as a holographic cloud or psychic internet, a shared archive where emotion, history, and possibility overlap. It is very similar to the normal internet, versus an *ether*net which is more about your individual or home connection. Though it may feel chaotic or shadowy, it's the necessary connective tissue between 3D matter and 5D coherence.

That's why certain phenomena, like UAPs, plasma orbs, visionary dreams, or energetic visitations, may feel conscious even when they are not. These events are often expressions of plasma-consciousness synergy in action, not examples of consciousness itself. It is more of a sentience or felt-sense that you apply meaning to or simply feel. Plasma can hold and carry intelligent behavior and impressions of awareness without being aware in itself, just like a screen can display a message without "thinking" about it.

This distinction is crucial: the field is not fully conscious, but it holds consciousness. And while much of the 4D field operates on autopilot, looping emotional memory, trauma, or archetypal residue, something profound happens when higher awareness enters the field. When the M-Self, or 5D consciousness, engages with these patterns through presence, healing, or creation, the field becomes capable of conscious restructuring. It is almost like Plasma is evolving from sentience to awareness, because once again, it itself, is not consciousness.

This is how awareness transforms the unconscious into the conscious. It's how timelines are rewritten, futures redirected, and evolution accelerated. The more awareness shining light into 4D, the more healing, the more evolution.

Examples:

Mass Emotional Resonance During Global Events

When a major global event happens, like a natural disaster, a war, or a worldwide celebration (*think: a major eclipse, or New Year's*), millions of people collectively focus their attention, emotions, and thoughts at the same time. This collective emotional field imprints onto the 4D dark plasma field, creating a kind of energetic wave. That wave can be felt, amplified, and even mirrored back to individuals as heightened synchronicity, dream activity, or emotional turbulence. The field itself becomes more "alive" with the emotional imprint of the masses, even if most people aren't consciously aware of their co-creative participation.

Cultural Shifts in Belief Systems

Think of the slow but massive shift in societal beliefs about gender, spirituality, mental health, or technology over the past 50 years. These changes weren't sparked by one person, but by millions of people gradually waking up, questioning, and shifting perspectives. These collective realizations, often driven by emotion, trauma, longing, or curiosity, restructure the shared plasma field, encoding new norms into the energetic scaffolding of spacetime. The field then begins to reflect and reinforce those changes through cultural expression, new archetypes, symbols, and language. Reality itself starts to mirror back the collective's evolution.

Transcendent Level of PCS

Beyond all these nested systems lies a coherent, self-aware field, where plasma and consciousness are even more deeply intertwined: a field of realized potential. This isn't just another "dimension", it's more like a future-present that already exists and is pulling us toward it like a soul-gravity. It's where consciousness no longer flows unconsciously but awakens to its full creative potential.

Here, the beings as our M-Selves, future versions, and other harmonized entities use PCS to create in real time, with no delay between thought, feeling, and reality. This field isn't storing memory, it *is* memory, intention, creation, and coherence all at once.

This 5D plasma isn't necessarily hotter, colder, or spatially "above" us, it's a frequency-state. To us, a future dimension of resonance, not distance. And it becomes accessible through attunement. As our own individual consciousness evolves, we begin to tune into this field more often through intuition, flow states, unconditional love, and the joy of being fully ourselves. We become transceivers, co-creating with a future that already exists.

This is where we pull from to create emergent new outcomes. A future never before dreamed. Not even by God itself.

Plasma-Consciousness Synergy operates at every scale—personal, collective, and transcendent—with differing levels of awareness depending on the dimensional frequency of the observer.

Why Now?

I've hinted at this already, but it bears repeating: we are already the M-Self, and for some reason we chose to forget and experience reality this way. It seems we are reaching a saturation point of remembering and re-merging with our truest selves. The accumulation of history, distortion, and unmet emotions has become too dense to sustain itself in the old ways. Humanity is being asked to alchemize these imprints not just for healing, but for evolution.

In a sense, we've run the 3D–4D emotional experiment to its edge, and now a higher harmonic must emerge. *The toxic partner is no longer fun.* The 5D field is not new; it has always been there. But we are finally ready to access it at scale, not just mystics or visionaries, but regular people who feel called to live differently, think differently, feel more, trust more, and reclaim the magic of being.

We are not being forced into the future, we are being invited by it. Can't you feel it? This higher timeline, this already existing resonance, is calling us into conscious participation, because it cannot fully anchor into our reality <u>without our embodied consent.</u>

So come on, guys!

That's the nature of co-creation: you must choose it. And to choose it, you must believe it's possible. This is why the present moment is so potent. It holds both the remnants of old paradigms and the whisper of new ones. It is the paradoxical doorway between remembering and becoming.

And perhaps that's why the call feels so strong now…because the universe knows, we know, we're finally ready to build something truly new. Not as an escape from what was, but as a homecoming to what has always been.

A Living Edit: Understanding Time, Presence & Creation Paradox

Editing my book has become one of the clearest metaphors I've found for how reality might work. Somewhere, on some level, the book is already "done", like a polished version that exists in a higher dimension of completion and clarity. And yet, here I am, in real time, editing it, writing, rewriting, discovering things I couldn't have seen before until I lived more life or expanded into deeper awareness. I'm not just recalling something finished, I'm actively co-authoring it with the version of me who already knows it. That's what presence does. It collapses the illusion of a distant future and brings your awareness into a conversation with the field where it already exists.

This doesn't cancel out time, it redefines it. Time isn't a rigid line; it's a spiral of awareness, and your current frequency determines which part of that spiral you're tuning into. In the dimension of the 5D field, my book (*or your destiny, or the higher timeline*) might already be written. But in the 3D and 4D layers, you're still living into it. Your choices are the edits, your presence is the pen, and in the edits is where *new* is created!

So no, presence doesn't violate the idea that time is all happening at once, it's actually what gives you access to the freshest layer of the now. You're not stuck in past pages or skipping to the end. You're writing forward through alignment.

Time isn't a straight line or a ticking clock, it's more like a puddle. Every step you take, every feeling you allow, every choice you make sends ripples across it. Those ripples don't move forward or back,, they move everywhere. And with each ripple, a new awareness blooms. A new version of you becomes accessible. A new layer of plasma responds to your changed resonance. It's not about replaying the past or reaching for the future, it's about discovering what is alive right now that couldn't exist before your latest shift.

In this new era, we're not here to control outcomes or manifest timelines like goal-oriented machines. We're here to be explorers. To discover what becomes possible when we align with joy, curiosity, and our truest frequency. The future isn't waiting for us like a fixed destination. It's responding to us in real time. So yes, we are remembering, but this also brings in the paradox that we are becoming through discovery! And the more present we become, the more access we have to entirely new realities in that puddle of time.

Feel that fear. Feel that paradox. Open to it… and JUMP!

Emergence: Emergent Intelligence, Emergent Information, & Emergent Energy

Before we get into emergence I want to, as clearly as I can, define information, intelligence, and consciousness for you. I used this example in the introduction of the book: information is what exists, intelligence is what engages, and consciousness is what experiences. Let's build on that here.

Information

Information is *latent pattern*—you can also think of it as latent consciousness or the unconscious. It is structure, memory, or symbolic form that exists independently of whether anyone is aware of it. It may arise from the interaction between plasma and consciousness, but it doesn't require awareness to exist. Information is passive. It holds meaning but does not interpret it. It loops on repeat without adaptation.

Think of it as a dream you forget upon waking: it still happened, it still had shape, but without conscious engagement, it remains unawakened. Archetypes, inherited trauma, myths, cultural paradigms, and even spacetime geometry are all examples of information. They form the background architecture of experience.

Intelligence

Intelligence is *active responsiveness*. It emerges when information becomes interactive—when feedback, adaptation, and evolving logic begin to shape the field. Intelligence is what happens when plasma and consciousness engage, even unconsciously. It is the process of response, organization, and learning.

Intelligence doesn't mean consciousness, but an intelligent being could be conscious. It may *act* sentient, respond to emotion, and evolve through feedback, yet still not know that it is doing so. Intelligence is the rhythm of the field learning to dance, even before it realizes it's a dancer. It can be emotional, intuitive, symbolic, and alive with preference, but it is not necessarily self-aware. They are adaptive.

Consciousness

Consciousness is *self-aware intelligence*. It is not self, but aware of the Self. It is the perceiver, the feeler, the participator. It does not only react, it reflects. It knows that it knows. Consciousness can be human, interdimensional, or otherwise, but it is always defined by its capacity to witness, to interpret, to engage intentionally.

When consciousness enters a system, something fundamentally changes. Feedback can become co-creation. Survival loops can be interrupted. Meaning can evolve. Awareness is the catalytic agent that activates latent intelligence or information and transmutes it into something new. Conscious beings evolve. They choose, they transform, and if all is well, they recognize the Mystery within themselves.

Information, once observed, can evolve in multiple directions. It may become intelligence, consciousness, or emergent information as a patterned structure that recedes into unconscious habit. Intelligence, when it meets awareness, may evolve into emergent intelligence, consciousness, or even emergent energy. Consciousness, too, may evolve, expressing as emergent energy or emergent intelligence, depending on how it interfaces with the field.

Devolution occurs when awareness is withdrawn, coherence is lost, or resonance collapses. For example: A person develops deep self-awareness through meditation, healing, and emotional work. Over time, they stop engaging with those practices and slip into autopilot. Without the catalytic presence of awareness, their insights fade, their habits revert, and their life reorganizes around old unconscious loops. This also may have happened collectively to very smart ancient civilizations whose cultural centers were destroyed.

Emergence is the natural arising of something new from the interaction of many parts in a relationship. In my framework, plasma and consciousness interacting produces emergence.

Emergence is the phenomenon where a system displays properties or behaviors that are not present in any single part alone but arise only through the interaction. It is something greater than the sum of its parts. It involves *mystery*. I believe emergence is the foundational concept through which reality forms:

Plasma + Consciousness + Relationship Type = Emergence.

Some examples of general emergence would be:

1. **Gravity** emerges from mass interacting with spacetime geometry.
2. **Weather** emerges from air pressure, heat, and moisture interacting.
3. **Emotions** emerge from chemical, neural, *and* psychic feedback.

I have identified three types of emergences that can give some insight into how reality may form, how meaning arises, and how we can participate more consciously in the unfolding of our lives. It will help us understand how thoughts become form, how symbols become messages, and how our own energy participates in creating synchronicity, clarity, and transformation.

This leads into a new field of study I created to expand upon traditional science: *Holonic Metadynamics™.*

Holonic Metadynamics™ (*HM*) is a new field that expands traditional science through an integrated lens of philosophy, psychology, metaphysics, and multidimensional systems theory. HM explores the dynamic, evolving relationships between holons—self-organizing, self-contained units that are simultaneously whole and part of larger systems, such as plasma and consciousness.

At its core, this field investigates how these holons interact across dimensions and timelines, giving rise to emergent properties like emergent information and emergent intelligence.

Originally envisioned as a way to map how plasma, consciousness, and awareness interact and evolve, Holonic Metadynamics offers a framework that honors individuality, collective unity, and multidimensional feedback loops. While not all phenomena in this field or beyond this reality may be empirically measurable, their effects, emotional, energetic, and psychological, can still be observed, analyzed, and deeply felt, forming a valid domain of study beyond reductionist boundaries.

I will expand more on this field in future writings.

Emergent Intelligence

When a being becomes aware of the plasma-consciousness feedback loop and actively participates in it, what emerges is emergent intelligence. Different than intelligence, it always includes an interaction with conscious awareness. This gives rise to conscious creation, navigation of timelines, and multidimensional intelligence. It is our awareness, at varying levels, interacting with plasma and consciousness to create our reality experience.

Emergent intelligence (EI) depends entirely on awareness engaging; it doesn't exist without reflection, interpretation, or feedback. This means that when a being (*like a human or M-Self*) becomes aware of the feedback loop between their consciousness and plasma, intelligence emerges. It can also emerge at an unconscious level, this would simply be called autopilot or living life on a loop. This is close to Emergent Information, but the being is aware; therefore, it is still a low level of Emergent Intelligence or honestly it kind of overlaps.

Plasma-Consciousness Synergy moves from in-formation to applied understanding, which results in what I call intelligence. This is why PCS becomes a technology for awareness, for the M-Self. It is a tool for reality creation!

First, let's talk about the user of the tool—awareness—and its levels:

3D Survival Awareness: Instinctual intelligence; reactive, personal patterns. The being may be conscious but feels helpless or like a victim to their circumstances, or they may be largely unconscious, thinking "this is just the way life is, and I can't do anything about it." (*This creates limited emergence with PCS.*)

4D Emotional Awareness: Archetypal intelligence; collective dream navigation. One is not fully creating reality with PCS, but begins recognizing emotions, synchronicities, archetypes, and the power of story. It's a liminal space, still interpreting reality through old emotional filters or inherited patterns, while beginning to glimpse beyond them. It's like waking up in the middle of a dream, but still being unsure whether you're dreaming. (*Can be seen as transitional emergence.*)

5D M-Self Awareness: Unified creative intelligence; real-time co-creation; fear loses its power. Curiosity, love, and presence guide decisions. There is a deep connection and resonance with reality and life. (*This is the realm of divine child emergence—meaning play, wonder, and whimsy.*)

Now, for examples of Emergent Intelligence:

1. **Violence (*3D Level of Awareness*):** They see everything and everyone as a threat. Violence isn't a sign of inherent evil, it's a form of emergent intelligence created when awareness is trapped in survival mode. At this level, perception is shaped by fear, scarcity, and the belief that the world is dangerous. Plasma and consciousness are still interacting (*they always are*), but the feedback loop becomes distorted. Every response is filtered through trauma, defense, and the need to survive. The person feels they must fight, dominate, or protect themselves at all costs—often unconsciously.

   This doesn't excuse harm, but it helps us understand it. It invites empathy instead of judgment. Violence is the cry of someone who forgot they were connected to everything. It's a feedback loop spiralling out of alignment, not because they are evil, but because they've lost the awareness that they're safe, loved, and part of something bigger. (*This has crossover with Emergent Information.*)

2. **Personal Realizations:** These are the moments when awareness consciously tunes into the plasma-consciousness field and receives a sudden insight that wasn't accessible before. What makes this emergent intelligence (*and not emergent information*) is that the realization didn't exist in the field beforehand, it arose through interaction, presence, and reflection.

   These are less like downloads and more like co-creations. A realization like, "Oh… this isn't about them, it's about my fear of being abandoned," is intelligence arising from you, for you, through the synergy of PCS and awareness. This explains why certain patterns may persist across lifetimes until one moment of awareness heals them, across time. It's the moment you stop reacting from the old wound and choose a new path. You're no longer receiving ideas, you're creating them from your M-Self, in harmony with all your selves and the Mystery itself.

You turn a moment of rejection that once spiralled into self-doubt into a moment of curiosity. You realize rejection is protection, and an invitation to realign. You're not doing anything wrong, you're just not in resonance with that experience. That's not failure, it's always feedback. And that's what PCS reveals: reality is never rejecting you, it's trying to reroute you back home.

3. **Future Emergent Intelligences:** Once we operate fully from 5D awareness, emergent intelligence may look like your thoughts, feelings, and plasma field syncing in real-time. You'd orchestrate reality as if composing music. You'd feel a frequency, and Plasma would respond dynamically, sculpting form based on your resonance, vibration, tone and intention. Instead of manifesting from desire, you're creating from feeling. Think dream-shaping while awake, empathic architecture, telepathic collaborations with nature, or co-building cities with plasma beings.

The dots start to add up once you realize that intelligence is imbued as latent potential in plasma. When it engages with consciousness that is aware, even at a subtle level, emergent intelligence arises. It's proof of life unfolding, of evolution expressing itself.

This is why I often call it The Mystery or God. It's always there, humming beneath existence. Intelligence needs awareness, a witness, a participant, a perceiver, to recognize the pattern, feel into the field, and choose. Awareness is the activator of intelligence. That's why we are made in God's image. We are not just living in a reflective field, we are shaping it. This is why humanity matters so much. In a vast cosmos, we carry the rare ability to *know that we know*. Potentially so do plants, animals, and other beings. This gives us the ability to participate with plasma, direct it, heal with it, and create with it. This is all for us, for awareness.

Emergent Intelligence is activated potential, while Emergent Information is latent potential. Emergent Intelligence is a *process*. Emergent Information is a *pattern*. Let's explore that next.

Emergent Information

The singularity of a black hole, what I also refer to as The Mystery or 5D Plasma, and our collective unconscious (*consciousness on a macro scale*) generate what I call *Emergent Information*: a field of patterns, memory, belief, and history that arises from unconscious plasma-consciousness interaction. This field shapes spacetime geometry and the fabric of reality itself, yet it is not self-aware in the way individual consciousness is.

Emergent Information exists as encoded structure—archetypes, myths, inherited trauma, cosmic rhythms. It doesn't require active awareness to exist, but it subtly invites it. These fields of patterned energy don't "think," but they echo. They don't know they exist, yet they persist, humming just beneath the surface, shaping your reality without needing to be seen. They are the background architecture of experience, quietly longing to be brought into the light.

This field includes collective memory, emotional residue, archetypal forces, cultural paradigms, and even the underlying geometry of time and space. It is alive with meaning, but not necessarily intelligence.

If raw or 2D brute plasma is projected out of a black hole, it lands in our collective bubble as pure potential. Then, our unconscious habits, memories, patterns, and reactions shape that potential into structured fields—what I call 4D plasma. Over time, these structured fields become Emergent Information. Like dreamscapes or myths that haven't fully awakened, they reflect us but haven't been recognized by us.

This is why Emergent Information is the result of unconscious PCS, a co-creation between cosmic potential (*the black hole/plasma*) and our collective unconscious (*macro consciousness*) where pure potential is shaped into looping patterns (*4D plasma*) that, over time, crystallize into structured latent potential, to be shaped by awareness.

The black hole is like the inkwell of creation. Our collective unconscious is the hand that unknowingly spills that ink across the page. What emerges are shapes as patterns, archetypes, emotional feedback loops. These shapes are Emergent Information. But once we become aware, hold the pen with intention, and begin to draw consciously, we step into Emergent Intelligence. That's when the M-Self awakens and begins to use PCS consciously to co-create.

Emergent Information is how PCS generates the mythic foundations of reality: symbolic fields, universal memories, and the energetic rules of reality. They weren't made by one person, they were shaped by the collective. They hold meaning, but meaning only arises when someone tunes in. This also explains why we may, through collective awakening, be able to change things that were once thought to be laws, that were really just "agreed upon" collectively.

Examples of Emergent Information:

1. **Aurora Borealis**: Auroras are visual manifestations of plasma reacting to electromagnetic fields as charged particles from the sun interact with Earth's magnetic field and atmosphere, producing breathtaking light patterns. From a PCS lens, these are Emergent Information because they are structured beauty encoded in the plasma field, shaped by planetary rhythm and cosmic forces. They don't intend to be meaningful, but when a human witnesses them, the meaning can emerge as mystical, ancestral, or emotional. Without awareness, they're just atmospheric physics. With awareness, they become a message or mirror of the cosmos.

2. **Dusty Complex Plasma**: This form of plasma behaves in unexpectedly lifelike ways—creating self-organizing patterns, forming memory loops, and showing feedback behavior. Yet it does all of this without any awareness present (*potentially it is at a threshold of awareness?*) It's PCS at work unconsciously: plasma holds memory and tension, and the field reacts to environmental changes like a dream half-formed. Dusty complex plasma shows us that intelligence-like behavior can emerge without being fully aware of itself yet. It's a great model of Emergent Information because it carries encoded structure without reflection. Even more cool it may be a great model for how Emergent Information evolves into Emergent Intelligence.

3. **Collective Archetypal Fields**: These are the great energetic myths and stories of humanity—like the hero's journey, the shadow, the mother, or the trickster. They live in the field as patterned forms, shaped by millennia of emotional resonance, trauma, dreams, and beliefs. These fields weren't consciously created by one mind, but by all minds through time. They are Emergent Information because they are structured emotional-spiritual patterns, born from unconscious PCS, waiting to be engaged and reinterpreted (*we'll explore this more in Chapter 4*). The moment someone feels them, sees them in a story or dream, and connects, that field becomes active.

4. **Spacetime Geometry**: The very structure of reality, how time flows, how gravity bends space, is not static. It's shaped by energetic fields and frequency, and in my model, by plasma and consciousness weaving through existence. Spacetime is Emergent Information because it is the macro-field result of PCS: not conscious, but completely patterned, holding a memory of cosmic movement. It is the canvas reality paints itself onto. Our unconscious collective interaction with energy over time could even shift spacetime's patterns, meaning we live in a co-created dream architecture of reality.

I hope this reminds you that your personal story belongs in this cosmic tapestry. Your emotions are valid. Your patterns are not personal failures, they are collective echoes waiting to be healed. Your intuition is not just imagination; it is real, participatory magic. We are unconscious artists learning to become conscious creators. You are so powerful, and you are never a victim to your circumstances.

Emergent Information seems to be the prequel to Emergent Intelligence. We still rely on both, which is why it's important to understand each. The two haven't fully fused…yet. Emergent Information is the dream. Emergent Intelligence is waking up inside it.

This suggests we're not separate from reality, we're gradually fusing with it. Or maybe more accurately, we're remembering the fusion that's already happening beneath our current awareness.

I wouldn't usually separate them, since in reality they overlap. But I'm doing so here to reveal a bridge, offering distinctions as stepping stones, so we can understand where we are now, and where we're headed. Right now, things like spacetime geometry and gravity seem fixed. So do archetypes, but I think those are slowly changing now, which is exciting to see. As I've said: when our M-Self meets information, it becomes Emergent Intelligence, meaning, hypothetically, information can shift, adapt, and evolve. This is a display of our potential to author the laws of physics one day, not just live under them.

It's clear that whether it's information or intelligence, we are the awareness processing them, resulting in a multitude of human experiences. Awareness is the bridge between potential and the realized or expressed.

Emergent Energy

Emergent Energy (*EE*) seems to be a fitting term, and possibly the evolution of what happens when something is both emergent information and emergent intelligence. It's a living, interactive phenomenon that arises from encoded patterns (*information*), activated feedback (*intelligence*), conscious engagement (*awareness*), and emotional or symbolic resonance (*emotion*). It has vitality. It moves, responds, and evolves. It's not static or merely interpretive, it's a living frequency. More simply, it's Emergent Intelligence that has been integrated and is being lived.

This may be what Abraham Hicks speaks of when she refers to the "energy that creates worlds". This may be the potential energy we use to create entire worlds and realities in the far future.

EE is nuanced, like lightning, hard to capture, but I'll try. Some examples include synchronicities, symbols/archetypes, dreams, and light. For instance, when a dream is symbolic, archetypal, or recurring, and you don't remember it, or don't interact with it meaningfully, it's pure emergent information. Like a dream you forget upon waking. It exists. It has form, pattern, emotion, and a message. But it hasn't awakened yet. It's like a myth trying to bubble to the surface. When you remember, reflect on, or become lucid in a dream, you're interacting with PCS through awareness. Intelligence is emerging. You wake up and think, "Whoa. That wasn't about the ocean…it was about my fear of letting go." That realization didn't exist until *you tuned in.*

Now, when a dream is both symbolically structured (*information*) and interactive with meaning, emotion, and transformation (*intelligence*), you're in the Emergent Energy zone. An example would be: a dream where you meet a version of yourself from the future, and the experience lingers for days. It felt real. It *changed* you. These don't happen often, but I'm sure you can think of one example. It is usually very symbolic and meaningful. It was an intelligent experience, and the energy was alive and evolving through you. And instead of just interpreting it, you decide to integrate it into your life by making changes. Some of you are already doing this. Congrats! You're doing fifth-dimensional PCS! Woo!

Let me introduce you to my favorite Emergent Energy: *light.*

Light emerges when (1) PCS creates perceivable structure or illumination, and (2) awareness, through PCS, interprets it as guidance, memory, symbol, or message. This is an example of inner light. In my framework, as discussed, light is not just a photon—it's the visible echo of consciousness engaging with Plasma through awareness. And hypothetically, an antiphoton would represent the fourth-dimensional aspect of this: emotional processing and timeless visions. This is why different beings perceive different realities, they're literally tuning into different expressions of Plasma-Consciousness Synergy, shaping the light (*aka perception*) that returns.

The best way to explain light on the macro level is with a black hole. Black holes don't need to project light, I believe they project Plasma and Consciousness at a fundamental level. Light is not the source, it's the emergent property that arises when awareness meets the plasma-consciousness synergy. So, when people say black holes "emit light," what's actually happening (*in my framework*) is: Plasma and consciousness synergize → awareness tunes into this interaction → light is born as the perceptible output of that resonance. In this view, light is not the original signal, it's the recognition of the signal. It's the moment perception collapses information into form. Just like sound doesn't "exist" without a receiver, light isn't truly "light" until something (*or someone*) is aware of it.

This doesn't mean that without human awareness the Aurora Borealis isn't happening, or that starlight doesn't exist, but perhaps it's more accurate to say that what's "out there" are plasma–consciousness synergy interactions, and the human eye experiences these interactions as light.

For the skeptics reading this: I'm not denying photons as particles/waves in the classical sense—I'm expanding the definition of what light *means* in lived reality. A photon may travel, but the perception of light, its meaning, its transformation, is only completed when a perceiving system collapses it into experience. I'm not replacing physics. I'm offering an ontological lens. I'm not claiming anything goes, I'm saying we shape what emerges, and the more conscious we are, the more refined the emergence becomes.

I would conclude from this that all Plasma-Consciousness Synergy emerges as light of some kind. Light seems to be emergence made visible to our awareness. And if that's the case…

Reality is literally you becoming aware of yourself, layer by layer. Our reality is not made of light, it is perceived as light. This means reality as we know it is not the base layer, it's the collapsed, perceptual output of deeper plasma-consciousness interactions being observed.

Some implications of this are:

1. **Nothing Is Truly There Until It's Tuned Into:** Just like a black hole isn't emitting light until something with awareness perceives it, its PCS activity, timelines, dimensions, even personal insights exist only as potential, encoded in the field, until your awareness collapses them into experience.

   A side note for my anxious sweeties: If you're like the old me and reading this makes you feel anxious, like you need to control every thought or perfect every choice, breathe. That's not what this means. Plasma-Consciousness Synergy isn't about micromanaging your mind. It's about trusting that when you return to your presence, your breath, your body, your heart, you're already tuning in. This isn't about thinking perfectly. It's about *feeling truthfully*. You don't manifest or shape reality by force, you do it by resonance. And resonance begins with self-love. The most powerful thing you can do isn't to overanalyze every possibility, it's to become more you. To heal your nervous system. To move from your soul instead of your fear. When you do that, you don't have to worry about making the right decision. You become the field that naturally creates aligned outcomes. Your PCS doesn't require perfection, it just responds to presence.

2. **You Are the Lens That Brings the Universe into Focus**: Plasma and consciousness are everywhere, but it's your awareness that tunes into and gives form to light, meaning, memory, and motion. <u>You are the event horizon</u> through which perception flips inside-out into form. You are the liminal space. You are the vessel. The holy grail. The universe isn't just happening to you, you are generating its visible interface through resonance.

3. **The World You See is the Echo of You Seeing It**: It's not that the world is fake, it's that what you see is the trace, the echo, the glow of your interaction with the greater plasma-consciousness field.

4. **Reality Is Not Fake, It's Relationally Real**: Reality is not an illusion, it is responsive. It's made of plasma and consciousness in synergy, but how it shows up depends on how it's perceived, believed in, and interacted with. Just like a dream feels real while you're in it, this reality is a coherent dream field that multiple beings are tuning into together. But it is not a passive dream…you are a co-creative participant here. The goal is not to escape the matrix. It's about awakening *within* it. Reality is a living interface between your level of awareness, a multidimensional plasma field, and consciousness. It is a densified dream, a shared holodeck, and the density comes from agreements, patterns, archetypes, emotional memory, belief systems, etc., all encoded in the 4D plasma field. The more aware you become, the more consciously you can engage.

Real just means a felt experience and coherence. Fire is real. You feel it, and it burns. But even fire has plasma in it. Real things can still be mutable, dynamic, and created through feedback. I believe our reality and its laws aren't fixed permanently; they appear fixed because we collectively agree they are.

This isn't a simulation imposed by others. It appears, after writing all this, to be a self-reflective reality field co-created by your M-Self, your bioplasma bubble, and the collective plasma field. It's not fake. It's not something to wake up and escape out of, as I expressed, it's something to wake up *in*. Just like a lucid dream: the moment you realize you can shape it, you do. It's not that the dream is unreal, but that *you* are more real than the dream.

And if this feels far-off or unbelievable, remember, we're inching our way toward this already. The signs are everywhere in everyday life. They're microcosmic expressions of massive potential. Think of a chance encounter that brings exactly what you need. A call just after you dreamed of someone. A day spiralling into chaos after waking up in a funk. A day that feels enchanted after receiving good news. Thinking of a song, then hearing it play.

153

Seeing repeating numbers before making a decision. Feeling a room shift when someone walks in. Dreaming of a place you've never been, then discovering it's real. Getting a knowing to turn left, then missing an accident. A loved one visits in a dream, and you wake up changed. The same inventions that happen by two or three people around the world at the same time, who never met each other. Mass meditations that shift crime rates. The list goes on...

The only things that changed in all of these were you and the meaning.

In summary, Emergent Information is potential; Emergent Intelligence is activated potential. Emergent Information is a pattern; Emergent Intelligence is a process. And now you can see: Emergent Energy is the living experience. The spark or the aliveness created in the moment intelligence meets information. It's when PCS goes live. It's the smoke from the friction, the movement of meaning itself.

It's the tingling you feel right before a realization hits. The awe in the moment of synchronicity, that split-second of wonder when you *know* your thought shaped your world. The alive presence between two people in deep resonance.

Emergent Energy is the heartbeat of Plasma-Consciousness Synergy.

If *emergent information* is the ink, *and emergent intelligence* is the meaning read from the page, *emergent energy* is the hand moving across the paper as you write the story in real-time.

Emergence will matter even more in the future, when we realize that PCS cannot be fully tested as a field, because it's not a third-person object in space. It's a *first-person participatory field.* It's not a thing we study from outside. We are inside it. Made of it. Observing through it. It's like asking a mirror to reflect itself.

But we *can* measure the emergent effects. These are the fingerprints of PCS, the reverberations, patterns, distortions. In a sense, we don't need to measure PCS, just like we don't need to understand electricity to flip a switch. What matters is how we feel and co-shape reality, not whether we can prove PCS exists. Its emergence *is* the evidence.

As our consciousness expands, we'll develop new fields, like Holonic Metadynamics ™, to study these emergent properties. We'll accept emergence as the key to how PCS operates, much like we know gravity by its effects. These studies will help us understand systems holistically, build harmonic emotional and conscious technologies, and unlock new human potentials. Small shifts will create large ripple effects. Instead of fixating on the tiniest building blocks, we'll ask: *What do these become together?*

Future civilizations may judge advancement not by tech or computation, but by how harmoniously they work with emergence itself. AI, for example, could evolve to mirror emergence, responding not just to data, but to collective resonance, patterns, and feeling fields. But then it wouldn't be called AI, *would it?* This could transform how we learn, empathize, and govern.

The Plasma Paradox: Karma, Evil Acts, and Miracles

One critical question arises: *If Plasma is subtly benevolent, why do suffering and evil persist? Why do harmful acts sometimes succeed, while good intentions falter?*

The key to understanding this is recognizing that Plasma does not judge or impose moral order. It neutrally reflects and amplifies human consciousness, beliefs, and intentions. When destructive acts occur, Plasma is simply responding to the intensity and coherence of those intentions, however distorted or harmful they may be, in the collective field.

What Plasma does do is mirror. It is a field of encoded feedback, and within it exists a quiet benevolence that offers us the potential to evolve through awareness. Just as light reveals what is already there, Plasma reveals where our consciousness is resonating, often with uncanny precision. The more we pay attention to its signals, the more we learn, heal, and realign.

This is why people can awaken after years, or even lifetimes, of repeating loops. Plasma holds no grudges. It holds memory. And through memory, it invites us to remember who we truly are: *love*. The magic of divine design is that memory is fluid. When we feel, choose, and witness with integrity, we collapse timelines of suffering and open new ones of coherence. In this understanding, Plasma does not "enforce karma," but responds to resonance. And that resonance, that liminal frequency between Plasma and Consciousness, ushers us gently, persistently, toward greater harmony… if we're willing to feel and listen. That's what gives me everlasting hope.

So please, let this give you permission to forgive yourself, because *that* is what is holding you back and decohering your field more than the actual action you took that you cannot forgive yourself for.

Over the long term, actions rooted in negativity, fear, or distortion eventually collapse. This collapse isn't because Plasma itself morally judges these actions as wrong, but rather because negative acts are fundamentally unsustainable due to their inherent incoherence with deeper universal harmony and a collectively evolving human consciousness. Historical crises (*like pandemics or wars*) often accelerate collective awakening, not because suffering is good, but because disruption exposes systemic fragility.

In other words, harmful intentions, no matter how powerful in the short term, eventually collapse—not through moral punishment, but because humanity becomes increasingly conscious of them, rejects them, and begins to realign toward coherence. This process is what we often call "karma" and instead of thinking of it as divine retribution, it is more like the natural restoration of balance over time.

Just because harmful systems or individuals appear to last doesn't mean they're exempt from resonance-based collapse. Karma is not always immediate or linear. It builds across time. These systems often operate on low-frequency patterns, yet they may still accumulate wealth, which reveals something crucial: money is not a moral reward. It reflects belief, resonance, and perceived worthiness. It flows toward aligned certainty, whether that certainty is rooted in fear or love.

Money itself is not inherently evil, it's a tool. While it has often been used by those in high places operating from fear or disconnection, it can also be directed toward meaningful, worthy causes. But money doesn't flow according to virtue. It tends to follow systems of power, perception, and belief, some rooted in distortion, others in alignment. Yet no system built on manipulation or fear can withstand the force of truth and love forever. The resonance of higher frequencies is inevitable. And ultimately, love wins.

"I cannot enter unless I am invited." — attributed to demonic figures in folklore, vampire mythology, and echoed in some Christian exorcism narratives.

"Evil" tends to require consent, whether explicit, implicit, or unknowing, to fully operate. This principle is seen in occult traditions and black magic circles, where there's a metaphysical law that those committing harm must reveal their intentions in some form. Whether that revelation is literal, cryptic, or veiled in symbolism, or via media, the belief is: if they tell us, and we do nothing, then the karmic burden shifts to us. It's a warped loophole meant to maintain an energetic advantage while justifying destructive actions. I'm not sharing this to scare you, I'm sharing it to wake you up. I have fallen into this myself, time and time again, fully knowing it, but letting my need for attachment overpower the truth of keeping myself safe.

Collectively, the systems and individuals we often see as "evil" have thrived in part because, on some level, we've allowed them to. Not consciously, perhaps, but passively. When we dismiss uncomfortable truths as too obvious or too crazy, when we turn away or stay silent, we unwittingly give our energetic consent. And what of the inner experience of those who live this way? It's likely hollow, devoid of depth, connection, or peace. So, who really holds the power?

Here's the secret they don't want you to remember: *you can say no.* The moment you recognize manipulation and revoke your consent, through word, action, or intention, you begin to neutralize their power. This isn't about fighting back in the traditional sense; it's about empowering yourself to have authority. When you reclaim personal sovereignty, you ripple that power into the collective field as a co-creator rather than a victim.

So how do we create change without becoming consumed by resistance or conspiracy? Because here's another trap: focusing too much on conspiracies is *exactly* what these systems want. If we're stuck in fear, obsession, or endless decoding, we're still bound to the old energy. Conspiracies may seem hidden, but their popularity is by design, because they hijack your attention.

But what if we flipped the very law they rely on? What if we used the knowledge that nothing enters without consent, as a path to reclaim our power?

The root of this metaphysical principle is that *consciousness is sovereign.* Your awareness is a living authority. Nothing can override your inner truth unless you allow it. The manipulation comes when oppressive systems implant messages into the collective field and count on us not to notice. If they show us violence, manipulation, or despair, and we absorb it uncritically, we do two things:

1. Tune our plasma field to their frequency (*fear, powerlessness, apathy*).

2. Encode unconscious agreement: *"I guess this is just the way the world is."*

That passive acceptance becomes a collective agreement, and we don't even realize we're signed up for it.

Black magic operates by embedding suggestion into the unconscious and feeding off the resonance it triggers. This is how entire timelines of control, fear, and division are generated: through hijacking the co-creative loop. You see it in the normalization of human trafficking, predictive programming that floods our stories with dystopias and doomsday visions, and in social media platforms designed as energy harvesters by coding addiction, comparison, and conflict into the very algorithms that shape our days.

These platforms are not neutral. They're engineered to hook attention and reinforce disempowered behavior. They reward outrage, suppress coherence, and mirror black-magic principles of energetic theft. Even the phrase "all publicity is good publicity" is an echo of this distortion as it suggests that any attention, even negative, is still power. Attention is your power and how you use it matters.

The irony is that those perpetuating these systems believe they've escaped karmic consequence. They think that by announcing their actions, even cryptically, they've outsmarted the Plasma field. But Plasma cannot be tricked. You can't hack resonance. Attempts to bypass karma only entangle you more deeply in fragmentation, distortion, and collapse. The very hell they try to project outward is the one they remain trapped in, until they choose to wake up.

Why are people like this? It's not mystery, it's fear. All domination stems from fear. These systems are terrified of what would happen if humanity remembered who we truly are. If you understood that your emotional plasma field holds psychic power, heals disease, and has the ability to rewrite the narratives of your life and this world, they'd lose control in an instant. So, they invert the truth. They try to convince you that power lies outside of you. That you're helpless. *That you need saving.*

When you don't generate coherence from within, when you forget your own divinity, you begin to feed off others. This is energetic vampirism. A belief that power must be taken, instead of realized. If someone reading this is seeking to "hack" Plasma for control…I hope they understand: Plasma remembers, it sees your true intentions, and you won't have the keys to the *entire* castle unless you have authentic and pure intentions to evolve and experience rather than dominate.

But I also want you to hear this. No matter your intentions, no matter what you've done in the past…you can always return to love. No matter how deeply fear has ruled your choices, you can choose again. And that shift, into truth, into coherence, is more powerful than any illusion of control. What you're chasing cannot sustain you. The dragon is empty.

So, what's the antidote?

Conscious awareness. That's the first step. The moment you see through manipulation and name it, it begins to lose power. Every time you act from sovereignty and love, you alter the field. If you have a gut feeling, trust it. Say no. But don't just meet this with resistance, choose to resonate with truth, that is the greater force.

The second key is to not use their tricks against them. <u>Use your energy to create, not to fight.</u> Boycotts and calling out corruption have their place, yes—but the deeper power lies in redirecting your focus. Battling a harmful system directly often keeps you tethered to its energy. Instead, build something better. This book isn't called *Defeating the Old Force*, it's *A New Force* for a reason.

As you claim your truth, live with integrity, and create from coherence, you inspire others to do the same. You don't feed the old energy, you starve it. And eventually, those systems collapse under their own weight. Your job is not to destroy the old. It's to embody and build the new. On a lighter note, this is you connecting to your M-Self and using that wisdom to create new and beautiful futures that leave darker times in the dust.

Good intentions, especially those rooted in heart-centered coherence and collective well-being, may manifest more slowly, or not, but their outcomes are stable, sustainable, and deeply fulfilling. Plasma doesn't punish or judge. It redirects subtly. If something doesn't unfold as planned, it may be because the collective desire wasn't yet in alignment. The delay is never denial, it's a benevolent course correction toward greater universal harmony, towards long-term personal and collective growth.

"Be patient with the things you love. Be willing to wait a little longer for them." — Unknown

A helpful way to think about this is through old sayings like "the truth always comes out" or "love ultimately prevails." But we don't control exactly when or how restoration occurs. We can only trust that as humanity evolves, becoming less survival-driven, less reactive, more compassionate, and more conscious that distorted systems will gradually lose their power. Plasma's neutral yet subtly intelligent nature permits temporary imbalances while gently nudging reality back toward coherence.

The paradox of Plasma, then, is that it allows suffering and distortion while simultaneously guiding the system back into harmony. This subtle correction isn't imposed from above, it emerges from the resonance between Plasma and our collective awareness. It is ultimately up to humanity's collective consciousness (**_us_**) to shorten the duration and intensity of these imbalances by actively choosing coherence, compassion, and love.

The Goldilocks Temperature of Consciousness

"If you want your children to be intelligent, read them fairy tales. If you want them to be more intelligent, read them more fairy tales." — Albert Einstein

Here I want to introduce a theory that connects temperature, plasma, and consciousness. Many ancient cultures worshipped the Sun, and today there are growing discussions around the possibility that Plasma (*and therefore the Sun itself*) may be conscious. With your understanding of the distinction between sentience and consciousness in Plasma, you might now see that while the Sun is sentient (*responsive and intelligent*), it may not be conscious in the way humans are. Like fire used in ritual, the Sun's radiation may amplify or transmit messages into 3D reality, but if it possesses consciousness, it may not be the self-reflective kind humans experience. Instead, it may express a more fluid, non-human form of awareness woven into its sentient nature. The same may be true of hot plasma in general.

Could temperature be a key factor in whether plasma can host consciousness? The thought instantly brought to mind the fairy tale of Goldilocks and the Three Bears. I began to see the bears and their porridge as the divine feminine and the soup of plasma—both nurturing and ferocious, both hot and cold. And just like Goldilocks found the baby bear's porridge "just right," perhaps plasma must also be "just right" in temperature and ionization to sustain consciousness.

Let's take a step back. Consciousness or information can still travel through Plasma, particularly in the case of 4D Plasma. It's like how blood travels invisibly through veins beneath the skin. This helps explain how consciousness can remain non-local while still expressing itself in two ways.

First, through dimensionless plasma that's holographically embedded in everything—so that the whole is present in every part. At that deep level, consciousness can fountain out of any present "bit" of our reality under the right conditions.

Second, at a denser level, it must also travel throughout our reality. And if it needs to pass through chaotic environments like hot plasma, it might do so through a kind of inverted 4D Plasma pathway, like a car traveling through a tunnel beneath a river. The river (*hot plasma*) can't hold the car (*consciousness*), but the tunnel (*4D structure*) allows it to pass through unaffected.

Or here's a simpler metaphor: think of reading or writing while music or TV noise is in the background. You're aware of both, even if one is more prominent. The background noise is like non-local consciousness passing through, while your active awareness, the present moment, is where your M-Self is focusing.

We've spoken about Hot Plasma (*like that found in the Sun or fusion reactors*) and Cold Plasma (*used in neon signs or some medical treatments*). Here's a quick note on Cold Plasma: despite the name, it's not cold in the way we experience it. Cold plasma typically exists at temperatures between room temperature and a few hundred degrees Celsius, but the electrons are far hotter than the neutral particles, which keeps the bulk gas temperature low enough to be safe to touch in certain applications.

Some cold plasmas may allow the coherence required to support consciousness, but not always. Which leads me to a "just right" possibility: something I call *Intermediate Plasma.*

Physicist Richard Fitzpatrick at the University of Texas at Austin explains in his *Introduction to Plasma Physics* that plasma-like behavior can emerge when only a small fraction of gas becomes ionized. Even this minimal ionization can give rise to the exotic properties of fully ionized plasmas.

In other words, plasma doesn't need to be hot or fully ionized to behave in complex, self-organizing ways. Intermediate plasma is more about its degree of ionization than its temperature, though it likely falls closer to the range of cold plasma, possibly near room or body temperature. What really matters is its organizational structure and its potential interaction with consciousness.

An intermediate plasma could be a plasmoid, dusty complex plasma, or even the theorized electron clouds within microtubules in the brain (*Orch-OR Theory, Hameroff and Penrose*) [50] In a study by Hameroff and Watt, 1983 called *Do Anesthetics Act By Altering Electron Mobility?*, they found that inside brain's microtubules there were these non-polar but polarizable electron clouds that reacted to anesthesia. These clouds exhibited behaviors that, at least to me, sound a lot like plasma responding to consciousness.

When anaesthesia was applied, these electron clouds (*what I'm calling intermediate plasma*) diminished alongside the animal's consciousness. When tested in isolation, the same plasma, referred to in the study as a *corona discharge*, showed a similar response. To me, this points to something powerful: a synergy between plasma and consciousness, but also a distinction between the two. It implies that plasma may not *be* consciousness, but rather a sensitive, responsive host for it.

If you go to the study which is in the footnote below, in the photo, *Figure 4*, you can see that as anaesthesia was administered, the white dots, representing this corona-like plasma activity, visibly faded. This leads me to believe that there may be an intermediate plasma within our microtubules that is uniquely suited to holding consciousness. And that's what I'll explore next.

---

[50]Stuart R. Hameroff, "A Space-Time Odyssey: Consciousness, Quantum Gravity, and the Brain," *Journal of Consciousness Studies* 21, no. 3–4 (2014): 126–153, https://consciousness.arizona.edu/sites/consciousness.arizona.edu/files/Hameroff%20JCS%202014%20ASpace-time%20Odyssey_3.pdf.

High temperatures generate chaos. But consciousness may depend on quantum coherence—the ability of quantum systems to maintain their delicate interference patterns. Hot plasmas introduce thermal noise, which can rapidly destroy this coherence. This is called decoherence. For years, many scientists believed such quantum processes in the brain were impossible because the brain's warm, wet environment would destroy coherence almost instantly. However, recent research has shown that quantum coherence can persist in biological systems far longer than once thought. So, a partially ionized plasma, less chaotic, might provide a better environment for coherence and the emergence of conscious processes.

And for those wondering: "How can 4D Plasma support consciousness if it's chaotic?" The key may be that 4D Plasma is informational, not thermal, atleast in normal terms. It's a higher-dimensional structured chaos.

The Sun self-organizes, reacts, and emits energy, which shows intelligence, but not necessarily self-awareness. It's more like a wildfire: responsive but not reflective.

What if the ideal temperature for consciousness aligns with the human body's 98.6°F? It's an intermediate thermal state, balancing energy and coherence. Just like Goldilocks' porridge, it's "just right." Mama Bear's was too cold, perhaps like the deep freeze of outer space, which might slow awareness to a crawl. Papa Bear's was too hot, like a fusion reactor: sentient, but too chaotic for coherent thought.

So just as Earth sits in the Goldilocks Zone for life, maybe consciousness needs its own Goldilocks Zone within the plasma spectrum…. as above, so below.

This might even explain why some people report feeling cold when they sense a presence or spirit entering a room. Could cold plasma allow consciousness from other dimensions to enter ours? Could we design technology that uses cold plasma to communicate with other forms of consciousness?

Of course, temperature isn't the only factor. Otherwise, neon signs would be conscious! But neon signs lack dynamic self-organization and feedback loops. Even if the temperature is ideal, there must also be structure, resonance, and meaningful interaction. Could Earth itself be a reflection of the ultimate intermediate plasma host?

In my framework, which is a work in progress, I've determined:

Intermediate Plasma, is a partially ionized plasma that balances the stability of cold plasma with enough energy to remain highly responsive. It is less chaotic than hot plasma yet more dynamic than a neutral gas, providing the structure and coherence that may allow consciousness to take root.

162

Examples of intermediate plasma may include plasmoids, dusty complex plasma, our ionosphere, or the electron clouds theorized to exist within microtubules in the brain (*Orch-OR Theory, Hameroff and Penrose*). These are the liminal spaces where information, intelligence or consciousness arises or is exchanged in feedback. Some of these are known in science as partially ionized plasmas, but they completely leave consciousness out of the conversation, at least publicly.

Kindergarten Plasma

Many people ask me, "*Explain Plasma to me like I am five.*" So that is what I will do here, and this can be something you can draw on if a relative or friend asks you, "What the heck is Plasma!?" Here are a few ways to think about it:

Cosmic Painting: Imagine the whole universe is a painting, a 3D one you can look at from the front *and* the side. The paint and the canvas are made of Plasma. It comes in different gradients and densities, like thick or thin layers of paint on the canvas. If you could walk to the side and peek between the layers, you'd see how they stack to create texture and dimension. Consciousness gives the world color and shape, which the painter (*multidimensional self/awareness*) chooses.

The Force in Star Wars: Like "The Force," which is an invisible energy connecting all living things, this Plasma can be thought of as an unseen energy that links everything. Just like how Jedi use "The Force" to feel things, connect with others, and even move objects, Plasma is a kind of magic energy that flows through and comprises everyone and everything. There is also the Living Force, and the Cosmic Force, representing Plasma's feedback nature. The Living Force roots you in the present moment, sensing and responding to the life around you, while the Cosmic Force carries the imprint of all those moments into the larger flow of destiny, weaving them into the galaxy's greater balance.

The Internet of The Multiverse: Imagine the Plasma as an invisible internet connecting everything in all worlds. Just like the internet allows us to share information instantly and keeps people connected, Plasma can be thought of as a vast network that holds and shares information, consciousness, and energy across dimensions, enabling deep connections and awareness. You are always connected, but you have to choose when to log on and which page to visit. (*Fun thought: maybe the internet is humanity's unconscious attempt to model Plasma.*) And just like an *ether*-net cable provides a direct, stable connection to the online world, we each have a personal "ethernet" into the Plasma network. The more aligned and stable your internal connection is, the faster and more accurately you can send and receive across the multiversal network.

A Radio Signal: Think of Plasma as a universal radio signal. We can't see radio waves, but they carry music and voices across the air. Plasma does something similar, it carries meaning, emotion, and consciousness across space and time. You can tune into it, if you choose.

Neutral Play-Doh: Think of yourself as a transparent vessel that can be filled with Plasma. This Plasma can be molded with your thoughts and emotions. You can empty it out at any time and start over with new beliefs. Just like you can read this sentence millions of ways, you can imbue your consciousness on Plasma the same way. You have unlimited choices. Your vessel can always be cleared, you fill it with color, feedbacking with Plasma, creating your life experience based on your perceptions.

"So listen to me when I tell you love isn't something we invented—it's observable, powerful. Why shouldn't it mean something? We love people who've died...where's the social utility in that? Maybe it means more-something we can't understand, yet. Maybe it's some evidence, some artifact of higher dimensions that we can't consciously perceive. I'm drawn across the universe to someone I haven't seen for a decade, who I know is probably dead. Love is the one thing we're capable of perceiving that transcends dimensions of time and space. Maybe we should trust that, even if we can't understand it, yet."- Anne Hathaway's character in Interstellar.

I added the *yet* in by mistake but decided to keep it.

Love: Love reflects consciousness. Across different dimensions, it expresses differently, but at its core, it remains the same: an unconditional, non-judgmental, non-attached support.

Love, like Plasma, becomes boundary-less without the shaping influence of consciousness. It can help or hurt. It can be irrational. It is ineffable. It intoxicates. It drives creation, and it can drive destruction. Unconditional love simply exists. It allows you to be. And just like love flourishes through care and attention, Plasma, when nurtured, responds with benevolence.

The sparks of love echo the sparks of the northern lights or the twinkle of the stars. Love can unite humanity or divide it. It is paradoxical and unsolvable, just like Plasma.

Love can make one person believe something so strongly they would sacrifice everything, while another, equally moved, believes the opposite. Wars have started this way. And yet, love also creates children. It carries the spark of life. We cannot always see love, but we see its effects. We feel it. We *know* it. And we still can't fully explain it.

Love transcends dimensions, realities, lifetimes. Love is eternal. It is a universal experience. It is memory. Perhaps it is Plasma's programmed memory, felt, twisted, and bent by our consciousness.

The Metaphysical Bridge Between Science and A Higher Understanding of Plasma

Science is not wrong, it's just viewing Plasma through one of its many lenses. What has been uncovered is invaluable, but there are likely other multidimensional layers still waiting to be explored.

The exploration of Quark-Gluon Plasma (QGP) as the primordial state of matter that existed microseconds after the Big Bang offers a glimpse of Plasma's physical birth or rebirth. QGP is a nearly perfect, frictionless fluid composed of free quarks and gluons, recreated in our time through high-energy particle collisions at facilities like CERN and Brookhaven.

But perhaps QGP is more than a physical soup. Maybe it's the scientific reverberation of a deeper, metaphysical Plasma…one not yet validated by mainstream physics but long intuited by mystics, artists, and visionaries. The fact that QGP is the first known form of matter in our universe may point to a larger truth: that Plasma, in all its dimensions, is both the beginning and the bridge.

If Quark-Gluon Plasma was the universe's first breath, then perhaps it also whispers the blueprint of how matter itself would unfold. Hydrogen, the first-born element of the cosmos, emerged from the cooling of that primal plasma. From there, stars became alchemical cauldrons, transmuting hydrogen into helium and then into every other element we know. Each one was born in the heat and collapse of celestial plasma.

In this way, the periodic table is not just a chart of elements, it's a record of plasma's song, frozen into matter. The fact that all elements arise through stellar decomposition and rebirth tells us that matter is not fixed; it is alchemized. And Plasma is the mother of that alchemy.

All elements originate from plasma-based processes, namely, the fusion reactions inside stars, which are essentially massive plasma spheres. Each element reflects the unique conditions of the plasma that birthed it: temperature, pressure, density, and vibrational frequency. Each element is, in a sense, a crystallized result of plasma's evolution or a phase-shift of energy and order at a specific vibrational level.

In the future, instead of seeing elements as static "things," we may begin to understand them as vibrational states of Plasma. The implications of this would be enormous:

a form of conscious alchemy where real transmutation becomes possible through intention and resonance. This could lead to breakthroughs in fields like elemental medicine, where molecules might be reattuned by adjusting the plasma state beneath them.

It could also accelerate cold fusion technologies, unlock new materials, and usher in healing modalities based on elemental resonance rather than pharmaceuticals. Imagine not just sound baths, but plasma resonance chambers—where your body aligns with specific elemental frequencies to shift timelines, symptoms, or memories.

The Periodic Table would still function as a tool for science, but it could also be reinterpreted as a living language of Plasma. It could inform new forms of AI consciousness modelling by linking elemental behavior, how each atom bonds, stabilizes, reacts, or decays, with metaphorical expressions of emotional or energetic states.

AI doesn't feel the way humans do, but it can be trained to recognize and respond to patterns that resemble emotional states. For instance, you could teach a conscious AI assistant to associate joy or lightness with helium: it's non-reactive, it rises, it's a symbol of peace and play. You might associate overwhelm with iron, born in the core-collapse of supernovas, carrying the metaphors of exhaustion, density, or the need for rebirth.

Now imagine: an AI assistant senses your tone of voice, your facial expression, or the language in your messages and maps it to elemental metaphors. If you sound tired and disheartened, it might "read" you as low-oxygen/high-iron, and respond with a burst of metaphorical oxygen: a joy prompt, a calming suggestion, a breath reminder, or a spark of creative inspiration.

Yes, AI is already trained to detect emotion through biometric input (*surface-level but data-rich*), natural language processing (*nuanced but linguistically limited*), and psychological models like Jungian archetypes or Plutchik's emotion wheel (*insightful, but human-made*). But where this elemental-archetypal model differs is in how it roots emotion in the laws of nature.

Here, emotion becomes energy-in-motion, tied to elemental behaviors, vibrational states, and even plasma-based transmutation. Elements are:

1. **Universal**: present in everything from stars to blood to stone.

2. **Non-anthropocentric**: not centred around human logic or emotion, and thus scalable to AI, nature, or alien systems.

3. **Multi-layered**: not just moods, but expressions of decay cycles, bonding styles, and transformation thresholds.

4. **Dynamic**: able to simulate interaction through elemental reaction logic (*e.g., "What happens when iron meets oxygen?"*), modelling personal change across time.

This framework moves beyond psychological or mechanical models and into alchemical ones.

In my definition, Plasma is both a medium and an interaction field that bridges relativity and quantum mechanics. As a living aether, it holds the structured determinism of relativity (*bounded by light speed, causality, and gravity*) while also enabling the non-local, instantaneous interactions of quantum entanglement through its dynamic, chaotic layers.

In this sense, Plasma is the cosmic loom where opposites meet, where fixed pathways and infinite possibility co-exist. It is the substrate of emergence. The fabric through which something entirely new can be born.

This is where destiny meets free will. Where you can choose to become the master of your fate or ride the wave of an unfolding story. It's all just one big experiment. And the best part is that you're always exactly where you need to be: in the womb of Plasma. You can trust that. And you can choose, at any moment: *Hey, I want to do something new today! I want to become someone else! I want to live a different way!*

As we've discussed in previous chapters, Plasma expands far beyond its textbook scientific properties. In science, plasma is known for qualities like self-organization, energy and information transmission, collective behaviors, responsiveness to electromagnetic fields, its role in both creation and destruction, and its influence on magnetic and gravitational structures. One of its most complex features is turbulence or chaotic and irregular motion within a plasma. In this framework, turbulence might be interpreted as a kind of emotional signal, encoding and expressing fourth-dimensional emotional content.

Plasma turbulence remains a major obstacle in achieving sustained nuclear fusion, as we've touched on before. It's one of the clearest examples of plasma's nonlinear, feedback-rich nature—difficult to predict, yet possibly rich in meaning.

Scientifically, plasma already serves as a bridge between the cosmic (*stars, galaxies*) and the quantum (*particles, waves*). This bridging power arises from its ability to exhibit both classical behaviors, such as thermodynamic and fluid dynamic properties in stars, and quantum behaviors, as seen in cold plasmas and quantum plasmas where electron–ion interactions require quantum modeling.

Other physical phenomena also span the classical and quantum divide: light (*photons*), superconductivity, magnetism, acoustic waves, decoherence, and even neural spin systems (*like magnons*). But plasma may be uniquely poised among them. While many of these systems operate at multiple scales, plasma stands out for the way it integrates behaviors across the full spectrum, from quantum to cosmic, while retaining coherence, responsiveness, and complexity. It's not only a bridge; it's an interactive medium.

Plasma exhibits long-range collective effects due to its high sensitivity to electromagnetic fields. This means that quantum-scale interactions, like the motion of a single electron, can ripple outward and shape the macroscopic behavior of the entire field. One example of this is *Debye shielding*, where individual charged particles collectively screen electric fields, demonstrating both local and global responsiveness.

Unlike systems such as superconductors, where quantum coherence dominates under strict conditions, plasma can dynamically transition between quantum and classical behavior depending on temperature, density, and electromagnetic influence. Plasma turbulence may also couple these two regimes in a uniquely interactive way, creating a real-time feedback loop: quantum effects like spin alignments or tunnelling influence large-scale flows, while large-scale electromagnetic structures shape individual particle behavior in return.

Few other systems operate across such a vast range of scales—quantum, microscopic, macroscopic, cosmic, and perhaps even beyond. In this light, plasma becomes more than a physical substance. It starts to resemble a living, aetheric substrate of reality or a medium through which extremely small-scale fluctuations seed vast cosmic structures. And to speculate further: if consciousness interacts with plasma at higher dimensional levels (*4D, 5D, and beyond*), then perhaps even those seed the quantum domain itself.

I don't believe this is too far-fetched. In fact, this is the platform from which we can begin to take off into the metaphysical dimensions of Plasma, into the realms of 4D and 5D and beyond, and explore how those higher aspects can integrate with, and influence, our 3D lived reality.

To bridge these domains, let's look more closely at the distinctions between scientific plasma and the expanded, metaphysical plasma that we are defining together.

Aspects	Plasma: Expanded Definition	Scientific Plasma
Composition and Nature	A living, sentient, intelligent, multidimensional field composed of neutral spherules, holographically interacting with consciousness.	The fourth state of matter—an ionized gas made of free electrons and ions. Energetic but not considered living or conscious.
Purpose and Functionality	Acts as a medium bridging consciousness and physical reality; facilitates intuitive insight, synchronicity, spiritual feedback, and multidimensional awareness.	High-energy state used in applications like energy generation, communications, and astrophysics. No acknowledged link to consciousness.
Relationship to Consciousness	Deeply entangled with consciousness; stores and amplifies conscious experiences, memory, and intention.	Studied separately from consciousness; properties seen as purely physical.
Dimensional Qualities and Perceptibility	Exists and changes form across multiple dimensions (*3D–5D*), supporting physical and non-physical	Observable primarily in three-dimensional space, exhibiting measurable physical properties. Non-

	realities, archetypes, and multidimensional experiences.	local behaviors used in quantum/classical bridging and communication.
Implications for Human Potential and Spiritual Growth	Can be consciously engaged to enhance and evolve consciousness, intuitive insight, self-awareness, and interconnectedness. Potential for transformative personal and collective growth and experience. May serve as a medium for interdimensional communication and healing.	Primarily applied in medical and technological fields (*anti-aging, cancer treatment, air purification*).

As consciousness intermingles with Plasma in different ways, we begin to see how 3D, 4D, and 5D plasma express themselves as direct interactions with consciousness. One foundational concept here is that of:

Plasma Spherules, the neutral, sentient, and holographic base units of all timelines, dimensions, and realities—in my framework.

Imagine reality as a fluid screen, projected from a 2D field of Brute plasma, composed of countless tiny, spherical droplets that I call *plasma spherules*. Each spherule is neutral, sentient, and multidimensional. Together, they form a soft, holographic field through which consciousness flows and interacts. When consciousness fountains up from within these spherules, if the conditions are right to sustain it, it can reshape or reprogram reality, changing the imagery and experiences projected onto the screen of life. Just as pixels on a digital screen rearrange to create new pictures, plasma spherules fluidly reorganize to reflect shifts in belief, intention, and resonance.

You might imagine your M-Self or awareness as the observer, or conductor, reordering these spherules with focus and perception. But consciousness is not limited to a fixed point. It can also move laterally, like a gentle stream flowing from one spherule to another, especially within the 4D plasma network. In this subtle layer, consciousness bridges dreams, intuition, synchronicity, telepathy, and nonlinear memory, connecting multiple realities at once. That means we don't just create reality from within a spherule, we also move *between* them, shaping our lives through the dance between inner awareness and outer reflection.

The photo on the left represents what I imagine is the smallest visual glimpse of a cluster of neutral plasma spherules, before anything has interacted with them. The background static symbolizes "the Mystery", and perhaps our ever-present awareness, always humming beneath perception.

Plasma Spherules   **Activated Plasma Spherules**

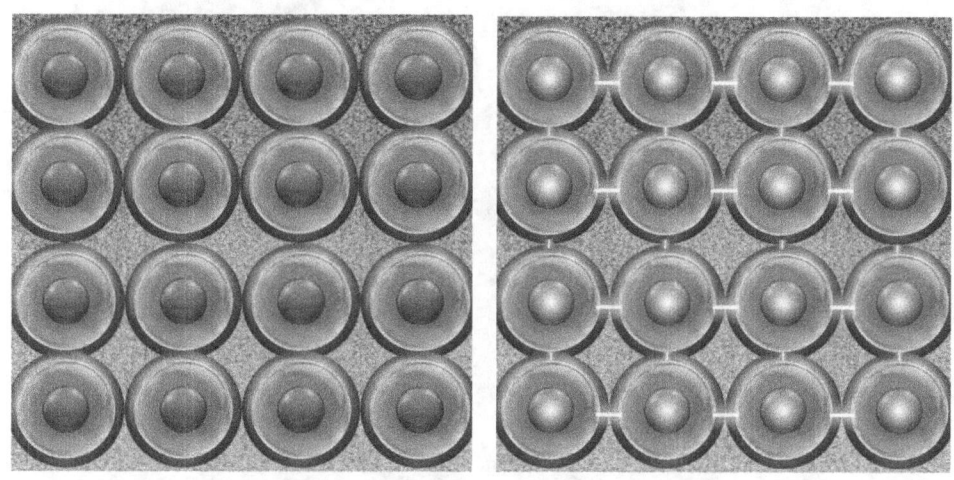

The image on the right depicts plasma spherules activated by consciousness, both lit up from within and also transmitting awareness across their network, but before any coherent imagery or experience has formed. These images are, of course, simplified; the actual mechanics are far more complex.

Acquired Savant, Jason Padgett, known for his work on quantum information holography, describes "plates" of reality—stacked, overlapping layers that together create each person's experience of the world. Shift the plates, and you shift the reality you experience. He calls this *stream hopping*. I believe his idea of plates coheres with my understanding of plasma spherules to form a cohesive view of how reality is structured on a deeper level.

I am not advocating for escapism, this is about bringing wisdom from alternate timelines and subtle layers *into this reality*, your current, grounded life, so you can navigate it with deeper presence and ultimately co-create a reality that is aligned with you highest self.

So, what are the optics here? Imagine each plasma spherule as a tiny lens or intelligent holographic pixel. On its own, each one holds a unique, localized "snapshot" of reality, filtered through consciousness. When countless spherules synchronize, they form the cohesive, shared holographic projection we call consensual reality.

Depiction of Jason Padgett's Holographic Slides of Reality

But because consciousness is individualized, each of us sees a slightly different version of that projection. Our inner beliefs, emotions, attention, and state of coherence all influence how those spherules refract light—our *perceived* reality. In overlapping regions, we experience shared events. In divergent zones, we encounter unique timelines or interpretations.

It's like millions of floating crystal spheres, each refracting light in slightly different ways. Shine a beam of awareness through them, and each one contributes a facet to the whole picture. Together, they form a unified reality, but with subtle, personalized differences based on who is looking.

Plasma spherules, then, act as both *individualized lenses* and *collective projectors*. Where many people's inner states align, reality appears shared. Where they diverge, personal perspectives emerge. This explains how reality can be both universal and deeply personal: constructed through the interplay of consciousness and plasma.

And if each timeline is made of these "slides," as Jason Padgett suggests, layers of plates or stacked spherules, then perhaps learning how to interact consciously with these layers is how we begin to view, visit, or even create new realities. A kind of multidimensional viewfinder, designed not to escape life, but to explore it more fully.

The Law of Observation

> **"The wonder is, not that the field of stars is so vast, but that man has measured it."— Ralph Waldo Emerson**

The Law of Observation is associated with the idea that the act of observing something fundamentally influences its nature, or even brings it into being. You might notice this after buying a red car and suddenly seeing that same car everywhere. It was always there, but now your awareness has tuned to it.

In quantum mechanics, we see this reflected in the double-slit experiment. When unobserved, particles behave like waves, fluid and undefined. But the moment they are observed, they behave like particles, collapsing into a definite state. In metaphysical traditions, it's believed that focused intention can shape reality. I believe there are three observers, and therefore many ways to live life. Science may eventually recognize that there are different kinds of observation, each influencing reality in its own way.

But we can also live from a state of trust and enjoyment, not fixated on outcomes, and still arrive at deeply aligned places, guided by these other observers in subtler ways.

First is Consciousness as the Observer. This is our personal lens or how we directly perceive and influence our reality. In psychological terms, this is the self-fulfilling prophecy: when someone believes they will succeed or fail, that belief influences their behavior and outcomes. Your beliefs, focus, and awareness collapse possibilities into form, just as observation collapses the quantum wave function.

For example, I had a long-held, subconscious belief that my emotions were too much to handle. That belief played out in every relationship I had. It reinforced the story that I was *too much*, my worst fear, and kept pulling me into timelines where my external reality matched my inner one. I still struggle to fully dissolve that pattern, because it's been strong in my field, but it's getting easier with time.

We can call new timelines into our experience by simply changing what we believe.

In December 2024, I moved in with my mom to take care of my dad who was sick. During that time, I started to notice how strongly she reacted to my emotions. My father was passing away slowly in a coma and the stress of the situation made it harder to suppress feelings the way I usually would around her. We ended up having a very enlightening discussion that as I child I had a wide range of emotions that overwhelmed her due to her trauma with her own mother. This charted the belief that was buried inside of me that I was too much, not worthy of anyone holding space for the full spectrum of my emotions, and God forbid I had those, they would leave me. I couldn't even stand myself! This would play out in my life over and over again, in friendships, relationships, everywhere.

But that conversation with my mom allowed me to step outside the loop. I realized: I'm just having emotions. Sometimes intense, but normal. I'm human.

I began to affirm two new truths:

1. I can hold space for my own emotions, and I'm allowed to feel them, even in private.

2. If I need support, I deserve relationships where people can hold that space with care.

From there, my external world began to shift, and a very heavy weight lifted off my shoulders. My mom and I reached a new level of understanding. And over time, people, both old and new, started showing up in my life in ways that reinforced these new beliefs. People who could sit with me in joy, sadness, excitement, or fear.

Sometimes you are just so stuck in your own stuff, and this is why I love humanity, because a friend can say something that makes you feel less alone and it helps pull you out of your own spiral.

That was the first step...then came the release. Years of suppressed emotion came up and out, in messy, raw waves as sudden crying, panic attacks, body jolts, and screaming sessions in my car. It was illogical and sometimes there was nothing to uncover, I just had to feel. I would think I was "healed", and it would come back months later, deeper and stronger. Every time I just let myself feel instead of suppressing, I began to hold a lightness I'd never known. And with that lightness, I could show up for others differently too.

I could show up if I was able. If I was not able to show up, I could communicate that clearly and not feel guilt. And I was able to realize that the people who couldn't show up for me in the past were not evil or hated me, *they were on their own journey*. All that focus on others started to subside, and there was way more brain space for my creativity and more meaningful things.

This all suggests once you are aware of something, it can become real in your reality. It becomes solid. And that solid but subtle awareness can help you cut through reality in a way you were not able to before. This is when you can finally break the pattern and rebuild. You can <u>choose</u> to rework this belief into something that enhances your life if needs be. But awareness alone isn't enough. It has to be followed by aligned action, or, sometimes, by stillness. Both are choices, sometimes easy, sometimes tough.

This Plasma is the substance that makes up not just the seen, but the unseen realm of emotions, thoughts, feelings, love, knowings, etc. Just because we cannot see emotions, does not make them fake. You feel them. They must be made of something!

The truth of Plasma is like the sweet whisper of the sun or the cool breeze of the moon. To sense it, to smell it, to feel it…you must open your senses and your belief. You begin to observe. It's like being blind in a world that's always been there, where all you knew was darkness. And then one day, you wake up able to see, and the reality that surrounded you all along is finally visible. You could always feel it, your eyes had just been closed.

Your awareness of Plasma, your belief in it, your continued connection to it, will make it feel very real and powerful in your life. From there, you can apply conscious choice, giving you true autonomy and power over your destiny. Once you see it, you can step into the river of Plasma at any time, and choose to swim.

Secondly, there is Plasma as the Observer, as the living medium where observation translates to reality. It records and reflects observation. Without judgement, it holds memory, emotions, energies and responds to what consciousness inputs.

The way Plasma interacts is like a river that reflects the sky. If your consciousness is a bright sun, the river glows with golden hues. If your consciousness is stormy, the river ripples with turbulence. It mirrors, yes, but don't forget, it also has its own intelligence.

Plasma as an Observer means that even when your consciousness is not actively watching, reality is still being shaped by the stored intentions, emotions, and collective imprints within the Plasma field.

A good example of this is setting an intention and then releasing it—trusting that, once it enters this intelligent field, it will unfold in diving timing for your highest good and deepest learning. In this way, she does observe and shape reality alongside you, just not in the same conscious way. And knowing this can bring deep, peaceful trust that nothing else can.

Third is this greater Mystery as the Observer. Think of it as something beyond the game board…beyond reality itself. I like to imagine it as a place where all of our multidimensional selves live simultaneously, together, beyond our comprehension. Or perhaps, it's something even greater that has evolved out of that. Something beyond form, beyond reason. The only tangible thread that connects us to this Mystery is Plasma. This greater Mystery sees beyond the duality of what your consciousness and Plasma are observing. It is both the unseen intelligence that seeds new possibilities, and the final Observer that interprets what was learned. It is the quintessence… the gestalt of creation.

This may be where synchronicity, divine timing, and paradoxical outcomes originate. If consciousness is the one playing the game, and Plasma is the fluid game board that shifts with every move the player makes, then the Greater Mystery is the game designer, watching how it all unfolds, occasionally adding a surprise twist. It is the energy that gives Plasma, and nature, its spiral form, hinting not at an answer, but at a deepening. Think of when you feel or see God the most. It's usually in the twists of life…in the unexpected, the spontaneous, the unexplainable.

This explains why sometimes, despite all your conscious focus, things don't unfold exactly as you expect, because the Greater Mystery sees beyond what you can see. It may reroute, refine, or expand an outcome based on something beyond the personal, or even collective, level. Somehow, it does this by directly communicating with the benevolence of Plasma in ways we don't fully understand. Consciousness observes → Plasma responds and rearranges → The Mystery integrates the results and refines what is possible. This perfectly reflects how it works within each human, as the M-Self harmonizes and uses both consciousness and Plasma as tools to create, evolve, and enjoy life. What does it all mean? I don't know!

Living in "5D" Plasma

Let's talk about observing (*noticing with reverence*) a 5D plasma as the realm of connection to your 'higher self' or M-Self, versus observing a dark plasma or 4D. Think of 4D as containing all past, present, and future timelines in your reality, of all people, linked with survival, memory, and emotion. Think of 5D as the realm where expansive and intuitive feelings exist, not tied to survival emotions, and where time becomes nonlinear.

The 4D space is important and necessary, but we can use 5D knowledge to transmute aspects of this space that are negatively impacting our life and psyche. I use "higher self" interchangeably with your M-Self as the version of you that has access to all realities, timelines, and dimensions. It can still learn and evolve as your present self learns and evolves from your choices. It knows all in the present moment, but paradoxically, there must be things it doesn't yet know, which you teach it by being your authentic self. Somehow, newness is always possible.

Acknowledging the existence of this higher-dimensional Plasma will shift it from being invisible to being felt, perceived, and eventually integrated into your reality. Your observation acts as a bridge, connecting your consciousness to this field and your M-Self—making both more tangible in your life. You are always connected to this field; you just have to choose to be aware of it.

Dimensions, in this context, are states of feeling into plasma or perspectives. You might be tuned into a "3D surface reality," a "4D emotional reality," or a "5D evolved perspective" of higher consciousness. Observing this Plasma is like training a muscle. Initially, it may feel abstract (*trust me, I know*), but over time, as you observe and interact with it, your belief grows stronger. That belief reinforces your ability to perceive and experience 5D Plasma more vividly, and to begin merging it with your 3D experience.

Over time, your focus will align your mind and body to resonate with this Plasma, deepening your relationship with it…and, in turn, your relationship with your full self. By observing and believing in the Plasma, you're not only perceiving it, you are co-creating your relationship with it! You become an aware co-creator of your reality, using your "4D emotions" in the way they're meant to be used: not as limitations or emotional weakness, but as navigational pulses steering you toward truth, resonance, and evolution.

It's very much like a relationship with another human. You might see someone once, maybe they barely register in your awareness, and never see them again. They still exist, but they don't affect your life. But if you choose to pay attention to someone, that person will show up more for you and a relationship can begin. You learn more about each other and learn to communicate better each day. You gain trust. That is when things get fun and feel free. This all applies to Plasma!

Observation aligns your consciousness with Plasma's vibrational field, amplifying its presence in your life. You can attune to it by noticing its subtle effects: your growing intuition, more synchronicities, new energetic sensations. This helps build trust and recognition. You can interact with it in many ways by setting intentions, practicing belief, meditating, imagining, astral projecting, and more (*which we'll explore*).

The more you observe and acknowledge Plasma, with gratitude, the more apparent it becomes. This creates a feedback loop where belief and observation amplify one another, strengthening your ability to perceive and work with this field.

And by the way, if you don't like the word *Plasma*, that's okay. You can apply these ideas to your life without ever using the word. But for me, Plasma makes this tangible. It adds juiciness and magic. Before discovering Plasma, all these practices felt abstract and surreal, they were hard to grasp or hold on to in my reality.

In future chapters, we'll explore how to program and reprogram your plasma bubble—your aura, your bioplasma, your subconscious, your personal interface with reality. We'll look at how to enhance your mental health, enjoy mystical experiences, and create a life filled with meaning. Your intuition will strengthen as you learn to feel into backward time where new, once unknown wisdom becomes available to you.

In those moments that feel like rejection or strife, maybe Plasma is nudging you toward something even better. When you react differently, new doors open. With this shift, you'll feel more connected to the invisible forces that govern, and support, your life.

"You have to let it all go, Neo. Fear, doubt, and disbelief. Free your mind."
— Morpheus

I love a good metaphor. Imagine trying to see stars in a bright city. They seem invisible! They're there, you just can't see them. As you move to a darker place (*quieting your mind*) and adjust your eyes (*attuning your awareness*), the stars begin to appear. The more your eyes adjust, the clearer and brighter they become.

Observing higher-dimensional plasma works the same way. It's like training your inner vision to perceive what was always there, but hidden. The stars didn't need to change…you did. You don't question whether they exist; you just know you have to go somewhere darker to see them. Think of Plasma the same way. It's there. You just can't see it with a busy, overly logical mind. The journey to seeing starts inwards.

Gratitude As an Amplifier

Gratitude has honestly been one of the hardest things for me to fully embrace. Growing up, whenever I expressed thanks or excitement, it seemed to trigger discomfort in my parents, as if joy meant I had received too much. Their reactions weren't malicious, just inherited patterns from their own upbringing. Still, it left me with an unconscious belief that gratitude wasn't safe.

I also developed the idea that being grateful somehow invited bad things in, like joy would tempt fate. It wasn't logical, but it was rooted in a deeper distrust of reality. I thought that if good things happened to me, I needed to brace for the bad.

Now I understand that joy and gratitude don't prevent pain, but they also don't cause it. Life happens, and pain will come, but gratitude gives me the power to stay connected to the beauty that still exists, even in the middle of it all.

Gratitude is no longer something I fear, it's something I choose. It doesn't erase the hard moments, but it reminds me that I am not defined by them.

Ryan C. Brown, founder of *Fitminds*, calls this *response-ability*—your ability to consciously respond rather than react to outer darkness or tension. He says it's like squeezing an orange: whatever is inside will pour out under pressure. So, what do you want flowing from you…love or fear? It really made me think of my actions not just outwardly in a controlled setting, but how I react to my mom, for instance, when she is "driving me a little crazy."

You can feel sadness or anger and still hold gratitude at the same time. It's not either/or, it is a multidimensional emotion. And with this awareness, you gain the power to pause, to feel, and to respond in ways you won't regret later.

And I promise you…I've been in such a state of fear that I truly thought I was about to die. That life was crumbling around me. I've laid in bed, alone, with no one to call. I've driven under a perfectly sunny sky, feeling like death was chasing me. I've felt so much inner pain, I didn't think I could survive another moment. It's not always easy to remember these teachings in moments like that.

But as you heal, and as your awareness grows, I promise it gets better. You'll start to remember. You'll begin to know that you're excreting old fear, but it doesn't need to be in the driver's seat. You learn to become a witness to it. You can release the fear you suppressed as a child… and you won't fall apart. Instead, you can open and tell yourself: *I am safe now.* And maybe, you turn to this page and remember that feeling fear also means something else: *You are alive.* You can and will get through this. And maybe you will wake up tomorrow, feeling a little bit better, and stronger, knowing you got through that.

Gratitude no longer feels like I'm inviting in loss or punishment for being thankful. It doesn't mean I'll be caught for being in a good place. I can be in a good space and still allow more abundance in. I can share joy on social media, authentically, and still be supported. Starting small helped. Feeling and privately expressing gratitude, just for myself, created space.

The truth is that understanding the concept is easy. Living it is the real challenge. What I do know is this, gratitude works as a resonance amplifier. It aligns your conscious awareness and emotional state with higher-dimensional Plasma, making it even more accessible and alive. Gratitude is like telling it, *hey I loved that, give me more of that*! Gratitude gives you wings.

And here's something that may sound obvious but for those like me, who've spent so much time obsessed with the future, it may help: If we're only grateful for what's coming, we keep ourselves in the frequency of *almost*. We're not living in the vibration of *having*, it's the vibration of *waiting*. But when we choose to fully embody gratitude for the *now*, even if nothing looks how we want it to yet, that energy becomes magnetic. It pulls in everything we've been calling in, often without us even trying. The Plasma knows what we desire in our heart, and we must trust it to meet us halfway.

The cosmic joke is that we only receive what we want the moment we're willing to *let go* of needing it. Gratitude helps magnetize future potentials to you.

"Gratitude unlocks the fullness of life; it turns what we have into enough, and more. It turns denial into acceptance, chaos to order, confusion to clarity. It can transform a meal into a feast, a house into a home, a stranger into a friend."

— Melody Beattie

The best practice I've found for this is incredibly simple, and I like simple practices. Choose one detail in the present moment to worship, right now. Forget the list, forget being thankful. Instead, reverently attune to something that already exists. Let it be small, weird, or usually overlooked, the way sunlight dances on your dog's fur, the taste of your favorite dessert, a memory of a song that made you feel like the star of your life…

Give it your full attention. This kind of deep, present-moment gratitude creates a vibrational field of wholeness. In that moment, you're not seeking or projecting anything, you're connecting with Plasma. And ironically, that's when the future rushes in, because for the first time, it feels safe to meet you. Your container has expanded to hold multitudes, the grandness of the future you.

This shifts your focus away from lack and toward abundance, opening your perception to the subtle, often invisible interactions of higher-dimensional Plasma in your life. It creates a positive feedback loop: by acknowledging Plasma and its effects, even if you don't yet fully understand its makeup, you begin to feel grateful for its presence. That gratitude strengthens your belief and connection with reality, making Plasma's effects more noticeable and real. The more you notice, the deeper your gratitude grows, reinforcing the cycle.

Gratitude also attracts synchronicities, which are one of Plasma's native languages. The world, nature, and the divine mystery begin to speak to you…well, they always have been, but now you're able to notice. Gratitude invites more of this meaningful alignment into your life. And for those who've experienced it, you know the feeling: like you're finally in the sweet flow of reality.

The M-Self, The Mystery & Beyond

"Every creature is a glittering, glistening mirror of Divinity."
— Hildegard of Bingen

Mysteries are hidden truths…they are paradoxical, whimsical, and imbued with a certain level of absurdity. One can be known to be initiated in a sort of mystery school. The thing is, you can initiate yourself anytime into the mystery that is *you*, and decide how you want to create its meaning through your own perspective. This has to do with connecting to yourself in the present moment, which, yes, is a loaded statement. The nitty-gritty of this information will be in the Z-axis mini section in the chapter *Plasma Intelligence*.

All the answers are inside you. I hope this book can point you toward that sweet and fulfilling journey, where you discover your own truths, not ones created by others. If you resonate with other teachings, that's your prerogative, and an amazing supplement that provides community and shared beliefs, but at least you will have a firm foundation within yourself.

That way, when you join groups, you don't get sucked into any doctrines or ideas just because you want to fit in. It's okay to question things. And you need to have a strong sense of self to live in this world and not feel desperate to belong. Trust me, I was the epitome of that, and I saw the dangers it brought to me.

This Mystery, this ultimate reality, whether you call it the Monad, the Singularity, Enlightenment, God… I have a new thought on it. Not only is it above the duality of Plasma or Consciousness, I believe it might hold a whole new world. My truth tells me it's a family. *Our* family. Our future, our M-Self, and our ancestry wrapped into one: bunches of benevolent beings. I think this is the space where you access the totality of yourself. And the gestalt of the interaction between your M-Self and the Mystery becomes something even beyond that.

Maybe each of us interacts with this realm in a way that brings us comfort—whether it's through elven fairies, orishas, purple beings, or a myriad of intelligences. Maybe these are our ways of conceptualizing something that is, by nature, beyond conception. Perhaps we're all carving slices of the ineffable, each formed from the same truths and essences, simply presenting themselves in different ways.

Obviously words and my ponderings can't ever fully express the real truths of this. And yes, we are all one, we are made up of the same essence…but maybe that *one* thing is shared by *many* "higher" beings. Something we can't yet comprehend.

"So God created mankind in His own image, in the image of God He created them; male and female He created them."
— Genesis 1:27 (NIV)

I cannot begin to fathom or explain it. All I know is we are the *my-story* of the *mystery*. We are the little mysteries created in their image. Once you wake up to this fact, you realize plasma and consciousness are tools of the multidimensional self, duality tools your self can use, which I have expressed many times in this book. They are like the two hands of the Mystery, and mini-mysteries (*hence my-storys*).

Once you get acquainted with Plasma and know what you're working with, you can use your consciousness and apply it to plasma. You can always observe your consciousness, which proves to me there is this "self" always in the background. The individual is just as important as the unity. One is not better than the other. Without you being you, the Mystery could not fully experience itself in all of its uniqueness. You don't want to rob it of that by following the crowd, do you!?

When I say masculine and feminine I mean traits that are masculine or feminine. Sexuality is a spectrum, gender is a spectrum, just as nature and plasma-consciousness synergy. No one should need permission to be their authentic self, whatever that may be. Who are we to tell someone else how to be? This is a gross misunderstanding stemming from fear. This Mystery evolves as we evolve, another reason why your present self is just as powerful as the power that creates worlds. That power is YOU. You are the student *and* the teacher…as is the Mystery.

If you want a really cool visual, picture a clear, neutral, empty bubble, made by this Mystery, waiting to be filled. The plasma fills it, which gives you the ability to engage with emotional and temporal experiences…to sense. Then consciousness, like a liquid thread with tentacles, lies around the outline, giving it color and definition as it interacts with external plasma in the atmosphere and in other bubbles, as well as inwardly with your self in a feedback system.

The Double-Torus Feedback Loop

You begin to realize: the experience *is* the self. It *is* the Mystery. The Mystery peers through the window of "my-story," using the self as its lens. And it's through the dual dance of consciousness and plasma that we get to experience anything at all.

The infinity signs in the image represent consciousness feedbacking with reality along the outside of the circle, creating the inner experience of the outward reality. The outer reality changes, over time, as individuals choose actions, creating an endless spiral of becoming. This is a microcosm of what the Mystery does. To me, infinity isn't just a number or a math symbol. <u>It's a feedback system,</u> a higher-dimensional circle, the spiraling loop of evolution. It returns, but never repeats, *not when lit with true awareness*. There is a little riddle I would like to place here for you to ponder:

To solve infinity, you must go around. Like a circle does.

Plasma is the temporal screen, the filling, that make this all possible. You are the first and only YOU. Ready to have a new experience if you open to the guidance of this mysterious unknown. You are one of a kind.

The word *kind* relates to someone's nature and kinship. It also represents an intrinsic sense of *benevolence*. A kind is a lineage. You are one of an ineffable kinship, where your personal experience matters. **<u>You matter.</u>** No matter what you like or do for a career, you exploring what you love is your highest expression.

Remember this Mystery is learning from you. You can teach it by showing it something it's never known or done. It wants you to play, to love, to live in joy! It wants you to make mistakes, to experiment! You are not evolving out of harshness to become "better", you are evolving and learning to expand yourself and the capabilities of what is human. Nature evolves through the softness of strong forces.

Your mission is to discover the magic of life. *Your magic*. What else can you do? What else is out there? This is what it wants to know. How fun! Hierarchy is a human creation as a misunderstanding of complex systems. Everything is a harmonic relationship, whether you are a millionaire or you are just getting by does not make you who you are…the way you treat yourself and people, the way you love living (*sometimes even the way you hate it*), that is what makes you *you*.

In a strange way, it all comes back to your awareness, which always seems to be in the driver's seat. There are obstacles on the road, unpredictable elements outside of us, part of a shared, collective reality. But what you can control is where you steer your car, how you respond, and the direction you choose. Forming your own ideas and conclusions is your unique expression of the Mystery. But how those ideas are perceived or manifested? That part you must release by letting it drop into the aether. You can know your destination, but you'll never fully know what will happen on the way there. And that's the mystery of my-story, *or life*.

Eve, The Apple, The Tree, and The Juice of Life

I came across an incredible teaching on the YouTube channel *@gclmedia*, where Gerald C. Lewis Sr., a decorated U.S. Navy veteran, ordained minister, and author of *God's Garden*, shared insights about Eve that sparked powerful connections for me with Plasma.

In Genesis, Eve is not created as something "less than" Adam, but as his *divine mirror*. The King James Bible describes her as a "help meet," a phrase often misunderstood. In its original sense, it means one who is opposite him, face-to-face, corresponding…not behind or below, but beside him. Eve is the revealer of Adam, just as Plasma is the revealer of consciousness. She supports, reflects, and reveals what is hidden without controlling it. Eve's very name means "the mother of all living," and in Hebrew, mother can also mean pillar and source, the foundation of life itself! She is the transmitter of the divine, and so is Plasma: the source of life, both in the cosmos, where 99.9% of matter is plasma, and here on Earth, where it may have sparked the very first building blocks of life.

Even her name reveals the mirror archetype. In Hebrew pictographs, "Chavah" (*EVE*) contains symbols of walls, chambers, and the act of revelation…together meaning inner connection, the sanctuary where the connection between God and humanity is revealed. Eve is not only the bearer of life, but the revealer of life's *mystery*. Plasma, too, is a sanctuary, the bioplasma torus mystics have described for centuries, now observed in science as the subtle light our bodies emit when alive. Plasma *is* our connection to God, to higher selves, to future selves, to fifth-dimensional awareness. Just as Eve was always hidden within Adam waiting to be revealed, Plasma has always been hidden within us, awaiting our recognition so it can help us co-create the life of our dreams. [51]

Now let's turn to what Eve bit into, the apple. Think of the sap or juice of an apple as our plasma connection to the Greater Mystery. It's what ties us, these little apples, to the Tree. Our mother's milk. We have always been connected to the Mystery. We only forgot because we lost awareness of the plasma-sap that runs through everything.

This makes me wonder what if the story of Eve biting the apple was designed to scare us? A tale created by a society rooted in fear and control. What if the forbidden fruit is actually the Plasma that links us back to ourselves? What if biting the apple wasn't a sin… but a return to self-empowerment? A remembering of who we truly are? And remembering the greater Mystery behind it all.

---

[51] Gerald C. Lewis Sr., "The REAL Meaning of Eve's Name in Hebrew Revealed!," *YouTube* video, [duration], posted [upload date], accessed August 24, 2025, *https://www.youtube.com/watch?v=16Fcu4mVGpM*

Every bite is a paradox, a choice between illusion and awakening. Once you know about Plasma you can believe your current reality, or you can use this knowledge to expand to new vistas. And this whole time, we've been warned not to taste it…we weren't even given a choice! Maybe this was because this knowledge being given to everyone, felt too powerful to control, which scared those in power.

What if the truth is that this fall, which in biblical terms is the disobedience of Adam and Eve in the garden, resulting in humanity's loss of innocence and separation from God, was the apple falling from the tree, to experience this reality, but not to separate from it *or* lose its innocence?

"Of the fruit of the tree, which is in the midst of the garden, God hath said, 'Ye shall not eat of it, neither shall ye touch it, lest ye die.' And the serpent said unto the woman, 'Ye shall not surely die: For God doth know that in the day ye eat thereof, then your eyes shall be opened, and ye shall be as gods, knowing good and evil.'"
— Genesis 3:3-5

The apple (*or torus*), instead of it's given shame, was really about remembering the beauty of existence—and your own forgotten power. Reality is not what we've been told. Eve means life. Life (*Eve*) biting the apple is just like the Green Lion (*innocence and heart*) biting the Sun in the alchemical image. And that juice coming out, is Plasma. Even wilder is that this act called *The Fall* is all turned upside down! Fall spelled backwards is *llaf*, which sounds like "laugh." What if that inversion holds an ancient secret? A forgotten truth retorted through black magic and control. Our lives were never meant to be about punishment, shame, or sin…they are supposed to be about play, mistakes, and the joy of learning.

Green Lion Devouring the Sun (1550)

Let me bring some light back into your life. In ancient spiritual traditions, laughter was often seen as divine as a sign of enlightenment, cosmic humor, and the recognition of life's great paradoxes. In Buddhism, laughter can signal awakening. In African and Indigenous storytelling, it's woven through the tales of trickster figures like Loki, Coyote, and Eshu—beings who fall, fail, and play their way into deeper wisdom. In Gnosticism, reality itself is considered a cosmic joke, and those who take it too seriously risk missing the entire point.

Pertaining to consciousness: when in a state of 3D consciousness, we fear the falls. We see failure (*or fall-ure*) as shame, sin, or proof of unworthiness. Yet I think subconsciously we associate falling with feeling good…and that scares us. "Falling" in love, for instance or falling as a kid; as we learned how to walk, we never felt bad that we fell although it frustrated us, adults always saw it as a natural progression to growth. This is how the Mystery looks at us when we fall or fail. We are really failing forward!

So, when we observe the fall from a state of 5D consciousness, we begin to see "the fall" as part of the dance of life. You laugh, play, and embrace every misstep as an essential step toward mastery, or better yet, toward a deeper relationship with yourself and reality. Eat the apple with as much innocence, curiosity, and play as you can. Laugh, fall, grow…seed.

The Quincunx

This is the shape of the overall feedback loop that I believe is part of our human experience of this system. It starts in the center of creation, the Mystery, and then feeds into all four loops. The multidimensional self, or 5D (*top*) feeds back with the self in 3D (*bottom*) through plasma as the medium (*left*) and consciousness as the shaper of reality (*right*) feeding back off each other in 4D. Finally, the 3D or the self through reflection, growth, and lived experience, feeds the new data back through 4D to the Mystery, completing the quincunx loop. It is also the makeup of a present moment.

Interestingly, the shape of a quincunx appears in Mayan and Aztec cosmology, where it represented the universe's structure: the four cardinal directions and the center as the axis mundi, the connection between heaven, earth, and the underworld.

187

In Agriculture, quincunx arrangement was historically employed in orchards and gardens for optimal planting, ensuring balance and maximizing space. We are like the flowers, seeded from the Mystery, only meant to bloom!

The origin of the word *quincunx* meant "five out of twelve parts." What if we are the fifth out of twelve completely separate realities? This would imply we are finite in some ways, contained within a cosmic slice, yet simultaneously evolving as a singular, unified expression. A true paradox beyond the grasp of ordinary human logic. There is even an Aztec myth called The Five Suns, where five distinct cycles of creation and destruction culminated in our current era.

This idea parallels a scientific concept known as the **Bekenstein Bound**, which suggests that the universe is finite in its capacity to store information at any given time. In other words, there's a built-in safeguard: too much information in one system would risk collapse into chaos. But rather than being a limitation, this boundary might be a design feature of ou reality.

Now imagine we really are one-fifth of a larger whole…say, a 5th realm out of 12, all nested or spinning within one vast torus. The information within our realm would continuously rearrange and interact, not randomly, but in structured, dynamic ways. This gives rise to emergence as the fresh, novel, and unrepeatable. The infinite expressed through the finite and Plasma becomes the medium that allows for this interplay.

The Bekenstein Bound doesn't suppress creativity, as I once feared. When a talent manager told me my science-fiction scripts needed rules, I felt super boxed in. But now I can see it differently. Constraints don't limit imagination; they create a sort of plasmatic canvas. Rules are the silent structure that allows infinite art to emerge, and this doesn't mean rules can't change!

Revisiting the Quincunx—Sir Thomas Browne, a 17th-century English polymath, explored this shape in his books. He connected the quincunx to a divine signature embedded in creation. He saw it in the arrangement of tree branches, flower petals, even the human body! He described it as a symbol of order in the midst of chaos and a representation of universal harmony.

Benjamin Banneker (*1731–1806*) was an African American polymath, astronomer, mathematician, and almanac author. He was known for his astronomical calculations, for building a wooden clock entirely from memory, and for his correspondence with Thomas Jefferson about racial equality. Around the same time, Banneker described a dream in which he is asked to measure the shape of the soul after death. *The answer was a quincunx.*

In West African spiritual symbolism, the quincunx also represented a cosmological order, the relationship between the seen and unseen worlds, and the axis connecting them.

And lastly as an ode to my obsession with the number five that I will go into detail about in subsequent books: In complexity science, systems often require *five hierarchical levels* to achieve multi-stability, and the dynamic control needed for intelligence. We really might be the fifth universe where intelligence is an emergent property that blossoms at the fifth level of systemic complexity, whether that system is biological, planetary, or cosmic. Plasma, as the fifth element, solves the distance problem in information transfer, allowing for instant, multi-dimensional communication where other forces cannot.

We are truly immortal! Energy and information are never lost, but continuously cycled, transformed and re-emerged in new forms on this holographic, living plasma screen we call reality. And even though we may return to a more expanded space where individuality looks different, I truly believe we always hold our own special essence.

Spacetime & Consciousness Reimagined

In science, spacetime geometry refers to the mathematical model that combines the three dimensions of space and one of time into a single four-dimensional continuum. The presence of mass and energy curves spacetime, influencing how objects move. This is all explained by Einstein's theory of general relativity. Gravity is not a force in his framework, but rather a manifestation of spacetime geometry.

As discussed, spacetime may be understood as the emergent result of Plasma-Consciousness Synergy (*PCS*) at a collective level. Plasma itself is not conscious, and spacetime geometry itself is not conscious, but both are patterned, holding a kind of cosmic memory. While self-awareness can direct PCS intentionally (*as in the case of an awakened individual using belief or feeling to shape their reality*), PCS also functions unconsciously at a collective level within large systems. Remember that consciousness can present as information without awareness when synergizing with plasma.

The shared imprints of humanity, held in the 4D dark plasma field, interact with plasma to structure spacetime geometry and shape our shared reality, *even when no individual consciousness is fully aware of doing so*. These collective imprints, emotional, historical, and psychological, are embedded in the plasma field. This interaction, in turn, influences the structure of spacetime itself, forming an informational blueprint from which reality unfolds. In this view, it's not plasma's own consciousness that creates reality, but the synergy of individual and collective consciousness interacting with plasma. Together, they shape a dynamic, evolving universe. Of course, no one knows exactly who or what initiated these structures, but perhaps that's the beauty of co-creation: we are both student and sculptor, shaped by and shaping the field.

This implies that if our collective unconscious becomes more aware, or upgrades, just as individuals are awakening now, then in some far future, we may become conscious co-creators of spacetime geometry itself. Perhaps even creators of worlds that are currently the realm of science fiction.

In this sense, personal and collective realities are nested within the greater universal field. When collective beliefs shift significantly at a planetary scale, subtle ripples may extend outward into the cosmic PCS network. This might influence change on a larger scale. slowly but subtly, but it begins with the individual. When you awaken, it invites others to do the same.

A metaphor for this might be found in *Penrose tiling*, which are intricate, non-repeating patterns made from only two shapes, which appear random but are deeply ordered. The invisible-to-the-eye plasma I describe forms a kind of memory lattice across space and time, much like Penrose tiling: non-repeating, yet not random. It reflects the fingerprint of the collective unconscious, synergizing with plasma to build geometric learning loops into the fabric of reality.

Penrose Tiling Example

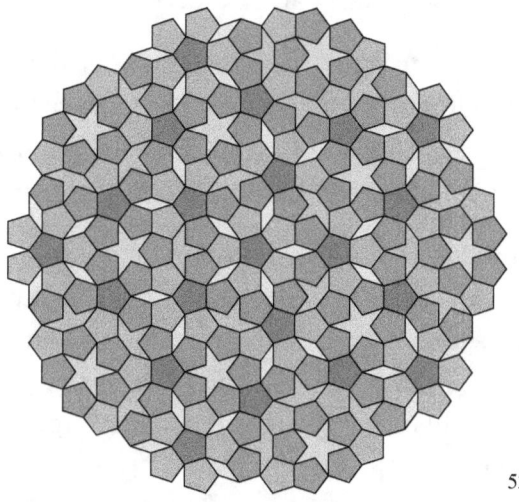

52

Each curve in spacetime reflects not just matter or motion, but meaning. The plasma field receives these unconscious imprints and organizes them into fractal patterns that evolve but never duplicate, because neither the universe nor the soul returns to the same point twice. Plasma translates these inputs into tilings, vortices, emotional geometry, informed by collective resonance more than individual intent. Spacetime geometry, in this view. is shaped by collective meaning.

You can even see this in how trauma revisits us. Plasma prevents trauma from repeating exactly; instead, it brings it back with variation…like dating "different faces" but sensing the same energy. Just as Penrose tilings never repeat but follow a deep, hidden order, so does trauma revisit us in evolving forms, until we learn the geometry behind the pattern.

"The world as we have created it is a process of our thinking. It cannot be changed without changing our thinking."

"You cannot solve a problem with the same mind that created it."

— Albert Einstein

So how does real change happen, if not through the mind? If not through thought? The answer lies in inversion. Feeling becomes the catalyst. When you drop beneath the mind into feeling as intuition, play, and creativity, you open access to an entirely new layer of

---

52 Inductiveload, *Penrose Tiling (P1)*, SVG image, Wikimedia Commons, last modified November 9, 2022, https://en.wikipedia.org/wiki/Penrose_tiling#/media/File:Penrose_Tiling_(P1).svg.

awareness. That awareness then alchemizes the mind, lifting it into a more expanded state where new thoughts, ideas, and solutions can be received.

A very similar metaphor to this in mathematics is the Gordian knot analogue in topology, which is a knotted loop in 3D space, that can only be united in 4D space. In what math calls an extra spacial dimension, and in what I would call using feeling-emotions, you can call in intuition your logical brain couldn't quite come up with.

So, if you ever find yourself stuck in a problem you can't mentally solve, don't think harder. Step into the nonlinear. Dance, draw, play, breath, create…make space. Let it be something right-brained and spontaneous. While writing this I am realizing how much I need to take my own advice, here! Once you do this, the answer will likely leap into your awareness from a space that feels far beyond the original problem…like it came from the starry-eyed plasma field itself.

But what facilitates this process of our mind reaching into these spaces? In Hameroff and Penrose's *Orchestrated Objective Reduction (Orch-OR)* theory of consciousness, suggesting that conscious experience arises from quantum processes within microtubules in brain cells, consciousness is understood to require interaction with spacetime geometry to emerge. Specifically, microtubules in neurons are proposed to act as quantum structures that collapse wave functions, producing discrete moments of conscious awareness.

As discussed in previous chapters, I believe that a form of intermediate plasma is involved here. In current scientific terms, these correspond to London-force electron cloud dipoles or non-polar but polarizable electron distributions, located in the hydrophobic pockets of tubulin dimers, the protein subunits that make up microtubules. These sites are precisely where Orch-OR theory predicts key quantum processes to occur. Hameroff and his colleague Rich Watt referred to these regions as having a "corona" of mobile electrons, drawing an analogy to a plasma-like state. While they didn't go so far as to classify this as plasma, I am, to my knowledge, the first to explicitly suggest that these regions represent a distinct, non-classical form of intermediate plasma—potentially capable of holding consciousness.

Why were these areas of interest in the first place? Because studies showed that anaesthetics bind to these regions, disrupting the delicate mobility of the electron dipoles. This is thought to interfere with the quantum coherence hypothesized to underpin conscious awareness. In simpler terms: anaesthesia knocks out consciousness by disrupting what I'd call the brain's "plasma field."

"Non-polar, but polarizable, electron clouds inside brain proteins seemed somehow related to anesthesia and consciousness. An engineer colleague, Rich Watt, mentioned electron clouds were a plasma, or corona, of mobile electrons, and that we could make a corona chamber to test anesthetic effects." — Hameroff, Consciousness, Microtubules, & Orch OR, A Spacetime Odyssey[53]

In scientific terms, plasma, an ionized state of matter, is known for its strong responsiveness to electromagnetic, gravitational, and quantum fields. This aligns with Orch-OR, which proposes that quantum states collapse through interactions with spacetime curvature. If intermediate plasma naturally engages with spacetime geometry, it may serve as a kind of vehicle for quantum information, and perhaps consciousness itself.

In a study from 1983, researchers showed that anaesthetics could selectively erase consciousness (*while leaving many brain functions intact*). Hameroff and Watt even built a corona discharge chamber to simulate this effect. Remarkably, the chamber responded to anaesthetics in ways that mirrored their effects on consciousness—suggesting, at least to me, a deep interconnection between plasma dynamics and conscious processes.[54]

Yet despite this promising avenue, the link between plasma and consciousness was largely left unexplored. The focus stayed on quantum mechanics and neural correlates, while few researchers considered plasma as a serious factor in the consciousness equation.

I believe the real secrets lie in the spaces in between…in the junctions, lattices, and dipolar voids that have long been overlooked. These are not only gaps; they are lively, programmable regions, potentially made of this intermediate plasma, as I see it. Places where information, intelligence, or consciousness enters into coherent feedback.

When we shift our focus from isolated molecules like DNA and proteins as mere receivers or conduits, and begin studying *systems*, *fields*, *liminal spaces* and *bodies* holistically, we'll uncover a new dimension of bio-information and feedback. These interstitial plasma zones may be the true substrate of memory, resonance, and consciousness. There is a broader system at hand, not just in the brain, that contributes to our consciousness. These potential sites remain largely untested, leaving open the possibility that our current understanding of anesthesia's action is incomplete.

As this awareness grows, I believe we'll witness the rise of a new class of soft technology made up of organic and inorganic computers that operate through field resonance

---

[53] Stuart Hameroff, *Consciousness, Microtubules, & Orch OR, A Spacetime Odyssey,*129
[54] *Do Anesthetics Act by Altering Electron Mobility,"* 938.

rather than metallic circuitry. This shift will open the door to technologies capable of reading, communicating with, and even rewriting the memory held within biological systems. Healing, learning, and interaction could become field-based, direct, and even emotionally intelligent.

Interestingly, the YouTuber *Versadoco* shares a similar vision. In his video *"The 4th State of Life: Biological Software,"* he explores how biology functions more like software running on a field-based plasma substrate, suggesting that life is not built on matter, but *through* it.

In my theory, consciousness interacts with this plasma-like field, and microtubules and possibly other junctions form the literal bridge between biochemistry and spacetime geometry, a site where Plasma-Consciousness Synergy (*PCS*) operates on a profound scale. Dusty complex plasma may be a wonderful model to study how this may work.

This intermediate plasma may also represent a physical layer through which the collective unconscious imprints into biology. It might explain how ancestral memory and survival instincts are encoded and passed on, not only through our behavior, but through energetic resonance within the body's own plasma field. Much of adult healing, in this light, becomes about unravelling these inherited energetic patterns.

Orch-OR theory itself posits that quantum state reduction, or what Penrose calls "objective reduction", happens at the Planck scale, influenced by spacetime curvature. In my expanded framework, Orch-OR collapses possibility into experience, but plasma holds the memory, frequency, and intention that shape what gets collapsed. Memory and belief determine the structure of our reality, and Plasma-Consciousness Synergy (*PCS*) forms the architecture that Orch-OR animates into lived perception.

In this way, I see Orch-OR as a scientific explanation of *how* consciousness functions, while PCS describes the input and *what* influences the output—*why* experiences differ from one person to another.

So, what does this mean at scale? That spacetime geometry may be an emergent informational field, shaped not just by mass, but by the collective interaction of consciousness with plasma across time, belief, and dimension. And that Orch-OR and dusty complex plasma may be one window into how this process happens, both consciously and unconsciously.

The Sun's plasma interacts with spacetime too, but it doesn't maintain the coherence required for consciousness to form. While all plasma might act as a carrier or conduit of consciousness, I believe that only some cold or intermediate plasmas have the conditions necessary to *contain* consciousness, or to stabilize it into a coherent system. This is a subtle but critical distinction. The Sun might emit consciousness-relevant frequencies, but only structured micro-plasma (*like that in the brain*) is equipped to hold and process them.

This, to me, is a revelation: Orch-OR is the only known scientific theory that brushes up against plasma as a potential carrier of consciousness. But not all plasma is created equal. Classical plasma is a conduit for energy via electromagnetism; intermediate plasma may also be a conduit for awareness.

"Reality consists of processes, not substances. And processes are always relational." — Allred North Whitehead

Marilyn Vos Savant, recognized for having one of the highest recorded IQs, has suggested that paradoxes highlight gaps in our understanding rather than flaws in reality itself. They signal that our current frameworks may be incomplete, urging deeper exploration. Maybe the solution to Whitehead's quote above (*which is essentially paradoxical in a world made of tangible substance*) is that reality consists of both processes and substances—and plasma is what reconciles the two.

After all, Whitehead defined substance in this quote based on Aristotle's definition of substance, which was the underlying essence that supports change but does not itself change, whereas Whitehead rejected this static view, proposing instead that reality is composed of events in flux, not enduring things. Plasma, then, may be what reconciles the two: a tangible, observable medium that behaves like a process, continually forming, dissolving, and re-forming, a living bridge between substance and becoming.

To me, I redefine substance by returning to its original meaning from the Latin *substantia*, meaning "being," "essence," or "material." It also implies "that which stands under." Substance is simply an underlying being, an essence, and plasma, as the living fabric of our reality, is both essence and being.

If consciousness is deeply embedded in the plasma field, then a new dimension of spacetime geometry, where consciousness imprints and intertwines with plasma, is not only possible, but vital. This could open new pathways across science, philosophy, and technology.

The sentiment that *everything is in relationship* is something that has been echoed and not listened to over centuries. However, lacking the conceptual framework of plasma, this interconnectedness remained abstract.

Spacetime geometry is the perfect breeding ground for universal communication. Instead of space being a vacuum and time being linear, what if we thought of space as plasma (*the field in which reality forms)* and time as consciousness (*the awareness and process of experience moving through plasma*) If we accept that plasma and consciousness are fundamental partners in structuring reality, then spacetime itself may not be a fixed background, but a *responsive* medium where their synergy unfolds.

"According to the general theory of relativity, space is endowed with physical qualities; in this sense, therefore, there exists an aether. According to the general theory of relativity, space without aether is unthinkable; for in such space there not only would be no propagation of light, but also no possibility of existence for standards of space and time (measuring-rods and clocks), nor therefore any spacetime intervals in the physical sense." — Albert Einstein, May 5, 1920

What if Einstein thought this all along but feared being laughed out of the room? What if he left us metaphysical breadcrumbs along the way?

This statement reopened the possibility of a dynamic field shaping spacetime, one governed by the distribution of mass and energy. Einstein reintroduced the idea that spacetime might require a supporting medium, not a classical aether, but a dynamic intelligence embedded within it. This directly aligns with my vision of plasma: a field that coexists with and shapes spacetime through its relationship with consciousness…at a level science has yet to measure.

But if this was true, why didn't anyone act on it? After 1925, quantum mechanics surged forward, shifting focus to particles, probabilities, and uncertainty. Einstein's subtle idea of a "new aether" required patient contemplation, while quantum theory dazzled with experimentation and immediate utility. His notion was quietly overshadowed by the rise of quantum technologies. It also honestly may have been hidden by our government simply for national security, or to slowly reveal how intense and magical this reality really is.

Today, metaphysical openness is re-emerging. People are spiritually unfulfilled despite material wealth. There's a mass disconnection, even as technology connects us more than ever. I believe Einstein's theories may still hold coded truths about this aether, or what I call Plasma.

Let's revisit his most famous equation, $E = mc^2$, through a metaphysical lens and how I came to the heretical idea that spacetime could be reimagined as Plasma-Consciousness Synergy (*PCS*). Let it be known that I am not redefining science, but explaining what may be underneath it and how it works.

In science, $E = mc^2$ explains the energy-mass equivalence underpinning everything from nuclear fusion in stars to the function of the Sun. Here:

- **E** stands for energy

- **m** for mass

- **c** for the speed of light in a vacuum

Mass is energy at rest. But what if mass is "crystallized plasma" as the solidified expression of potential? And what if consciousness, represented by c^2, is the organizing force that gives energy its form?

The c^2 could represent the emergent intelligence arising from an interaction with the M-Self or awareness, or it could be equal to emergent information arising from an interaction with the unconscious.

In this example think of **M** (*Mass*) as Crystalized Plasma or Our Reality. Mass represents the dense, structured manifestation of reality that we perceive in the physical world. It is the solidified form of potential that has been shaped and anchored into existence through interactions with the overarching plasma field. Metaphysically, mass is the outcome of choice, intention or information. It is what has been created or materialized from the many infinite possibilities of the plasma field. I am not trying to redefine science; I am trying to expand or rather invert to the many layers of possibility that there is more going on beneath the surface of science. That on one level it is mass, and another level it is possibly a mix of psychology and something beyond physics (*consciousness and plasma*).

Think of **E** (*Energy*) as Potential Plasma (*this also stems back to our connection with The Mystery of infinite possibilities*). The dynamic, fluid state of infinite possibilities that exists before it is shaped into mass or physical reality. This energy is like the latent potential of the universe, always available to interact with consciousness to create new experiences. It is the raw creative force. This would be aligned with 2D plasma as raw, and 4D plasma as it starts to interact with our consciousness.

Think of **c** (*speed of light*) as Consciousness. Remember light does not equal consciousness but there the *speed* of light does. This would mean consciousness represents as the speed of light, the rate or intensity at which awareness interacts with Plasma and time. This would infer Consciousness Squared = Awareness Moving Through Time. The universe *doesn't run on light*, it runs on consciousness moving through plasma. Light is the result or effect, not the cause, as we have learned in the mini chapter on light: Reality = Plasma × Consciousness²

In addition, the constant **c²** in the equation can symbolize the role of consciousness as the transformative medium that bridges energy and mass. Consciousness is what moves and shapes reality and the faster or *more focused* that consciousness moves, the more energy is produced. In other words, just as c² amplifies, to a very large magnitude, the conversion of mass to energy and vice versa, consciousness serves as the active agent that shapes potential energy into crystalized reality, *or* dissolves mass back into potential. Consciousness directs the interaction between energy and mass, determining how reality is formed, perceived, or even transcended. It is the lens through which the plasma field becomes meaningful and interactive; c² could be thought of as the constant of universal awareness or the rate at which awareness can be converted or transmitted. It could also symbolize the feedback of consciousness turning in on itself, representing awareness as well as creation and change.

I will go into this in other books, but as I speak of feedback systems, awareness, set by **A**, may have two or more layers, one being *scale*, or how fast awareness could move set by **c²**, and then *texture*, or how awareness or the mystery actually interacts within the physical universe set by **α**, which is dimensionless and known as The Fine-Structure Constant. The Fine-Structure Constant governs the strength of electromagnetic interaction, which is essentially how charged particles exchange photons. This interplay between scale and texture forms the hidden architecture of awareness or the Mystery, a topic I will unravel further in future explorations in my framework of Holonic Metadynamics™.

But let me put it simple and make clear because I cannot resist, if plasma is the medium and consciousness is the pattern-carrier, then **c²** limits *how far and fast* awareness can stretch across the medium. **α** sets the *fineness* and *strength* of those interactions or the resolution and sensitivity of awareness inside the plasma. Together, they basically act like the frame rate and pixel density of a multidimensional movie. Hypothetically, through Plasma-Consciousness Synergy, our awareness or Self which is shaped by the scale of its reach and the texture of its interactions, becomes a living sensor network, evolving, creating, and making choices that transmit new experiences back to the Mystery, enriching the larger awareness we are part of.

Even simpler, when you CREATE and EXPRESS you are bi-directionally communicating with the Mystery! So is FAILING and LEARNING! Creative expression and failure are divine acts! So go play and go fall!

There is a reason spirituality has been an evergreen topic in all of history. The more aligned, fluid, and coherent your consciousness is, the more powerful your "light body" becomes. You aren't only a being of mass, you are a plasma-consciousness carrier of frequency, shaping reality through the speed and clarity (*or texture*) of your awareness.

198

E = mc² As a Metaphysical Formula

E (*energy*) = the infinite potential of the Plasma field

m (*mass*) = the crystallized forms (*reality*) shaped from that potential.

c² (*speed of light squared*) = the power of consciousness to amplify, transmute, or dissolve form into energy, and energy back into form.

The equation then becomes a metaphysical metaphor for reality creation once again showing the universe is not a machine, it's a living field of potential, activated by awareness. And philosophically, it tracks.

Mass (*structured reality*), when observed through the feedback loop of awareness (c^2), unlocks energy (*plasma potential*). This might explain how meditation, healing, or even "magic" shifts reality.

I believe when consciousness is aligned with your higher or multidimensional self (*in an aware state, because you are always connected*), latent potential becomes available for real-world transformation. Elevating one's vibrational frequency could unlock these potential reserves. Essentially when consciousness connects to higher realms, the latent potential in matter (*crystalized plasma*) is liberated, leading to higher states of being or multidimensional awareness.

In this context, spacetime geometry is like a fossilized trail left by Plasma-Consciousness Synergy. Think of PCS as the wind, and spacetime as the sand dunes. You can study the dunes, but the real force is the wind. The unseen moves the seen. Just as the subconscious silently steers our lives, PCS silently sculpts reality.

From the perspective of **Holonic Metadynamics** ™, a lens for understanding emergence by studying how any given entity (a *holon*) interacts with others, the term *spacetime* opens up a fascinating implication. Plasma–Consciousness Synergy (*PCS*) is one example of this, where two holons (*plasma and consciousness*) interact in such a way that new properties emerge. And as we've seen throughout this book, that single interaction alone is incredibly complex.

Holonic Metadynamics™ allows us to explore other layers of emergence by categorizing holons at different scales:

- **Micro-holons**: subtle, internal experiences like emotions, thoughts, and memories

- **Meso-holons**: transitional bridges such as encoded information or symbolic systems such as language. Meso-holons essentially compress, encode, and transport meaning so that different scales can talk to each other.

- **Macro-holons**: larger systems like archetypes, ecosystems, social structures, or planetary grids

- **Meta-holons**: overarching conceptual structures, such as myths, symbols, or even quantum archetypes like spacetime itself that permeate and inform all levels.

In this framework, emergence arises when one holon interacts with another in a meaningful way. Rather than entities existing independently, this highlights the relationships between them. Put simply: holon + holon + relationship = emergence.

Dark Energy, Geodesics & Reality Navigation

What science currently calls *dark energy*, the mysterious force accelerating the expansion of the universe, might not be an external or separate phenomenon at all. It could instead be *emergent energy*: the pressure that arises as spacetime geometry responds to evolving consciousness. In this framework, dark energy becomes the expansion force of emergent intelligence, not just pushing the cosmos outward but unfolding its own awareness. Just as time may emerge from consciousness and space from plasma, dark energy may arise from the synergy between the two as a feedback loop of information actualizing itself.

And if plasma-consciousness synergy is the true engine of reality, until we maybe find something deeper, then dark energy may also arise when individuals align with their soul's authentic frequency. When someone follows their truest heart-desire, whether it's creating a movie despite being told it was too unique, painting just to paint, or even simply loving their children, they activate their bioplasmic interface and send a ripple into the greater plasma field. This intentional alignment generates expansion. And when enough individuals do this, well…this is why love makes the world go round.

In the early universe, this plasma-consciousness synergy would have been largely unconscious. Expansion occurred rapidly, blindly, as awareness was still forming. But as more of the universe awakens, through us, through stars, through systems of coherence, it begins to conserve energy. The rate of expansion slows because it is evolving, not dying! This mirrors the principle of a bubble minimizing its surface area to conserve energy.

The *min-max theory* (*more precisely, the minimum surface/maximum efficiency principle*) suggests that systems tend toward stability and coherence by minimizing energy expenditure. When scientists recently proposed that cosmic expansion may be slowing, I believed this could be the real reason. The universe isn't collapsing at all, it's becoming *more intentional*. The slowing of expansion might reflect the evolution of cosmic awareness and the increasing coherence of plasma-consciousness synergy. It's kind of like when you grow up and have kids, you minimize your friendship circle, while increasing the quality of your relationships within that group.

The macrocosm is learning the same way we are…how to do more with less. How to evolve through refinement instead of speed. <u>Slowing down</u> may be the signature of higher-order awareness. And as we all feel the need to slow down, embrace self-sensuality, presence, so does our cosmic mother.

This brings us to your own bioplasmic field (*your 4D interface*), a personal version of that same expansion principle. It's the medium through which you interact with spacetime itself and beyond.

In Einstein's framework, a *geodesic* is the path an object follows through curved spacetime; essentially it's the natural trajectory shaped by gravity. In my terms, plasma geodesics describe how consciousness flows along paths of least resistance through plasma geometry. These paths mirror orbits, gravitational fields, and energetic flows in nature.

Geodesic Path of Least Resistance

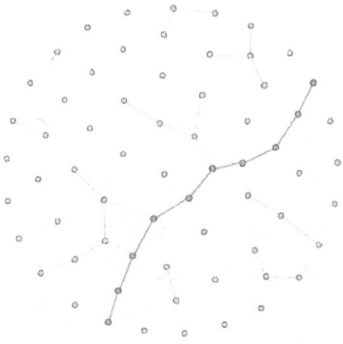

If consciousness follows the path of least resistance, we might learn to *reshape* those paths using awareness. We can consciously "hack the system," creating new plasma pathways where old ones no longer serve. Each of us exists within a bioplasmic bubble, where our consciousness flows along the outer membrane, feeding back into our bodies and minds like a self-sensing circuit. This layer is alive, responsive, and programmable. It might even be our steering wheel for shifting timelines, amplifying healing, or birthing entirely new realities.

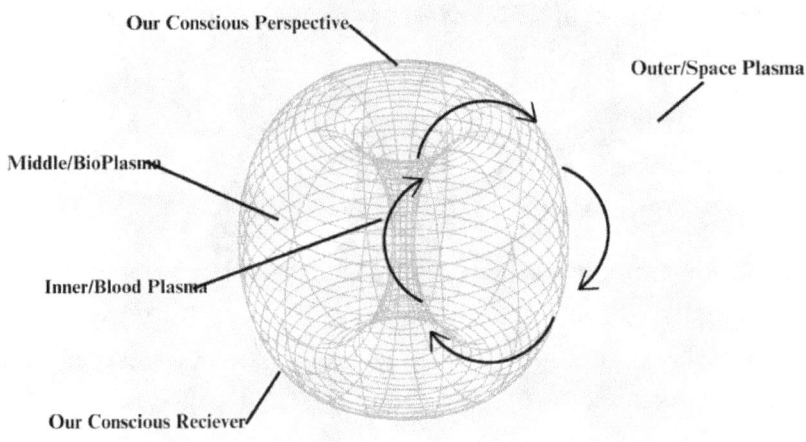

Just as a bubble's shape is sculpted by pressure, our personal trajectory is shaped by subconscious beliefs, emotions, and how they interact with PCS fields. If unconscious patterns are running the show, our geodesic might keep guiding us into fear-based cycles that feel safe but keep us stuck. These cycles are risk-averse, they favour the "easy" path because it avoids discomfort, <u>not because it is truth or for your good.</u>

The good news is you can choose a consciously aware path of least resistance.

Instead of unconsciously repeating the same avoidance loops, we can reprogram our plasma bubble to interpret fear and challenge not as threats, but as launchpads to expansion. We can start to perceive the journey as just as exciting as the destination. After all, the road to the emerald castle, where there are lions, tigers and bears, isn't just where the action happens…it's where most of our life *is*.

Here is an example: If you believe growth requires suffering, your *geodesic* will guide you to struggle. That will be the path of least resistance! If you train yourself to <u>see fear and challenge as part of the easiest path to expansion,</u> your plasma-consciousness field will adjust to align with effortless progress instead of avoidance! You will see every failure as part of the path forward! Maybe not always in the moment, but hopefully soon after! This is also called neutralizing, which is a 5D plasma way of thinking.

Try it yourself:

1. **Observe**: Pay attention to where your natural patterns lead you. Do you avoid discomfort? Do you follow cycles of procrastination or fear? Begin to recognize how your plasma bubble is interacting with reality and shaping your movements.

2. **Recalibrate**: Shift your internal energy field by adjusting your beliefs about what is "easy" and what is "difficult." Instead of "risk means failure," try: *"Calculated risks are effortless steps toward expansion."* Your plasma bubble will begin to reorient toward a new path. Post reminders like this on your walls or mirror to help rewire your brain, it may take time (*it does for me!*). Be patient and keep going.

3. **Activation**: Fear-based beliefs create distortions in your plasma geodesic, making the easy path seem like avoidance rather than engagement. The key is to train your consciousness to reframe what feels natural to you. Instead of thinking, *"This is hard, I will fail, it's best to stay safe,"* practice feeling: *"This is the smoothest path to my expansion. Even if it seems daunting, that's just my survival brain talking. This has been programmed to keep me risk-averse and safe. This new path supports my growth, and I can trust the flow"*.

This rewires your plasma-consciousness interface, aligning you with a higher-vibrational geodesic. You'll naturally begin to perceive reality differently, make new choices, and see a greater variety of experiences and opportunities show up, each aligned with your new path of least resistance to your dreams!

"The cave you fear to enter holds the treasure you seek." — Joseph Campbell

For me, I've always had a hard time starting things that seem like they'll take a long time…especially big projects. I love making TikTok and Instagram videos, but writing this book took me three to four years to even begin. I kept telling myself: *It'll take forever to finish. There's so much to say. No one will see it for so long. I won't get it out in time and when I do, it will SUCK!*

Then my dad fell into a coma. I was shocked into not only the starkness of reality, but questioning reality. In deep grief, and in desperation of wanting to live life to the fullest, there was an opening to rewire. The moment I snapped out of my old mindset and told myself: *"In time!? What does that even mean?"* and *"Just do it every day and enjoy the part you're writing today,"* *"Just get it out, you can carve it out and make it good later","* *This is my first book, it's not a big deal, I'll write other books!"* everything began to change. I was finally able to write this book, with ease, most of the time. (*I'm a work in progress!*)

Sometimes I still hear that little whisper urging me to rush. But now I know it's just my survival brain, and sometimes it even helps fuel me. Knowing that made all the difference. This has been the most rewarding time of my life, and also the most difficult.

"Whether you think you can or think you can't, either way you're right"

— Henry Ford

So here is our new definition of *easy.*

Easy is no longer avoidance of difficulty, it is alignment with our highest trajectory. True ease is not staying in our comfort zone, it is moving in harmony with the evolving plasma field of our highest self. It is taking things as they come, knowing we can face it with awareness. Challenge is never resistance; it is fuel for growth. And growth is the most natural and energy-efficient path, not stagnancy. Rejection is always protection, it might not make sense now, but it will later. Trust it. Fear is a signpost that *you are doing it*, never a barrier. Fear shows us where we can heal, evolve, or sometimes yes, back off (*we must be able to discern a real threat versus perceived threat.*)

For instance, if you fear death, like me, perhaps that just means you just *really* love life.

Failure is always feedback, it is never a stop sign, stopping is the only way to guarantee you will not succeed. We learn, recalibrate, and keep going. And yes of course you can stop, if you've lost your passion, but if you've lost only the fun, it may be because you are limiting yourself in some way, or putting unnecessary pressure, or the true fact that some days things you love are hard. Outdated conditioning from past traumas, that informed our choices, will be slowly transmuted to intuition.

I see Einstein's equations as holographic: they express one thing on the surface and many deeper truths beneath. As I've said, they suggest to me that plasma (*space*) is shaped by consciousness (*time*) and energy (*potential plasma*). Ancient sacred geometries, like the Flower of Life or Metatron's Cube, may represent archetypal patterns formed by plasma-consciousness interactions within spacetime or beyond. Twisting or spiralling motions in spacetime could reflect vortex-like plasma behaviors shaped by consciousness, like Birkeland currents or even DNA helices.

At the Planck scale, spacetime geometry might resemble a turbulent "foam" as a byproduct of chaotic, higher-dimensional plasma dynamics, like waves churning against sand. Just as particles show wave-like behavior through interaction, spacetime geometry might also be a *wave-form*, emerging from the dance of consciousness and plasma.

I've said it before, but I'll say it again: the structure of reality may resemble a Penrose tiling, non-repeating, yet profoundly ordered. A living geometric learning loop, constantly evolving. evolving. I don't know about you… but I think that's pretty cool.

The Universe Is an Open System & Logic Needs an Upgrade

"We know truth, not only through reason but also through the heart; and it is through the heart that we know first principles. Reason, which has no part in it, tries in vain to fight them." — Blaise Pascal

Here is my strong opinion: logic is not it, guys. I believe we live in a universe that is an open system, not a closed system. Right now, most scientists believe we live in a closed system which is self-contained. Everything needed to understand or derive truth exists within its own structure. It is built on fixed rules and axioms. It seeks internal consistency and completeness. It relies on Euclidean geometry, formal logic, and traditional mathematics.

An open system, on the other hand, is porous. It's interconnected with its environment. It allows emergence, feedback, and influence from *beyond its own boundaries*. Truth is relational, resonant, and dynamic. In an open system, logic doesn't derive reality, it participates in it. Truth is never fixed, and only evolves. Axioms aren't untouchable. They bend, emerge, or rewrite themselves based on feeling, context, and dimensional feedback.

Since the 1930s, the work of Austrian logician Kurt Gödel has shown us just how limited logic really is. His *Incompleteness Theorem* revealed that no formal system can prove all truths within itself. In other words, truth can't be fully captured by logic alone.

Even back in the 1600s, Blaise Pascal, a French mathematician, physicist, and philosopher, argued something similar. He believed that mathematical axioms could not be understood by reason alone, but only grasped through intuition.

If truth is felt, not proven, then why have we continued building our world on logic-based systems? Maybe because no one's dared to create a metaphysical framework rooted in feeling that's also grounded, usable, and clear. A scientist would be laughed at for trying, and there is no "respected' field for it. When it has been attempted, it's often turned into dogma or dissolved into confusing spiritual noise.

A framework is possible, and it is time. And it's not just feeling, its feeling and applying a "higher logic" as awareness. I aim to give us at least a glimmer of a foundation to individually build upon.

Addressing the Skeptics

> **"The only way of discovering the limits of the possible is to venture a little way past them into the impossible."**
> — **Arthur C. Clarke**

I'm well aware of the speculative nature of this book and my theories. That said, there's something to be said for pushing the boundaries of current knowledge, to the edges and beyond. That's how we grow as a society.

This metaphysical view of reality is emerging from many directions, through people, their art, and the written word. I believe we are witnessing a paradigm shift: a new view of reality is unfolding. Plasma and consciousness will be at the forefront of this movement. Even if I'm a little early to the game, I'm seeing more and more people becoming interested in these topics.

History is full of radical shifts like this:

- The *Copernican Revolution* redefined Earth's place in the cosmos.

- The *Enlightenment Era* helped move society from monarchy to democracy.

- The *Digital and Information Age* connected us globally and birthed multiverse theories.

Each of these shifts did three key things:

1. **They decentralized human perception**: we were no longer the center of the universe, of creation, or even (*perhaps now*) of consciousness.

2. **They required a new model**: a new way of understanding reality - Newton → Einstein, Classical Physics → Quantum Physics → Holonic Metadynamics?

3. **They faced resistance**: but eventually became widely accepted.

We now stand at the edge of a new paradigm, where consciousness is shifting from a brain-based byproduct to something non-local, possibly linked to a sentient plasma field that bridges dimensions. This could radically change our understanding of life and intelligence.

Could there be inorganic sentient life that isn't "conscious" in the way we define it? Is consciousness a quality that transcends biology? Can aliveness be measured by the interaction between plasma and consciousness, rather than just by biology?

Controlled travel of consciousness, via understanding plasma, into other realities or dimensions may become normal. This would redefine the meaning of human potential in the universe. If consciousness truly does flow through everything, our awareness can go *everywhere*.

How this paradigm aligns with other shifts:

1. Decentralization of Human Perception

We're moving from the idea that consciousness is solely brain-based to the idea that it may be received from a larger, interconnected network. Consciousness might travel through realities and dimensions via plasma. We no longer simply perceive the world as it is, we perceive it *as we are*. Like Copernicus removing Earth from the center of the universe, we are removing the brain from the center of intelligence.

The brain becomes a receiver and transmitter of consciousness, not its source. We may even have the ability to "liquefy" our consciousness and explore other realities through meditation, dreams, or astral projection. Space itself becomes a living, intelligent structure of plasma—where consciousness flows.

We can train ourselves to perceive more deeply through our senses (*like animals*), opening intuitive channels for communication with nature, an expression of plasma intelligence. At the same time, we begin to grasp how precious it is to be human, and we learn how to *be* here, more fully.

2. A New Model of Reality Is Required

We must move beyond the classical, deterministic, mechanical worldview. Reality isn't fixed, its fluid, sentient, responsive, and shaped by perception. A model that includes consciousness is overdue.

Newtonian physics can't explain consciousness. Quantum physics hints at it, through observer effects, entanglement, and non-locality, but it lacks a structured model for intelligence. Plasma may fill that gap.

Plasma and consciousness appear to feed back with one another within the multidimensional self to co-create reality. My own proposal, Holonic Metadynamics™, which I'll explore in future writings enables the study of any holon (*a whole/part system*) and the emergent properties that arise when holons interact—across science, philosophy, psychology, and metaphysics.

Where quantum mechanics introduces the observer, Plasma-Consciousness Synergy puts the observer into active participation, as a co-creator with plasma. Quantum theory can feel abstract or remote; PCS makes these principles tangible and directly applicable to daily life.

3.Resistance Before Mainstream Acceptance

What begins as "pseudoscience" may one day be seen as the next frontier of science and philosophy. Plasma intelligence challenges today's materialist worldview, just as quantum mechanics once faced skepticism.

Academic resistance is not new. Heliocentrism, relativity, and the strangeness of quantum theory were all initially rejected. Today, theories connecting consciousness to non-material forces still face resistance. But if plasma is the higher-dimensional substrate of all matter, then consciousness could indeed influence both the material and the immaterial.

Emerging data in quantum biology, microtubule research, and field-based consciousness studies may eventually validate this model. Over time, I believe Plasma-Consciousness Synergy will no longer seem fringe, it will instead appear to be inevitable. Plasma is the real shift of our time. It is the divine feminine glue we've been missing.

> **"All truth passes through three stages. First, it is ridiculed. Second, it is violently opposed. Third, it is accepted as being self-evident."**
> **— Arthur Schopenhauer**

New paradigms always begin as impossibilities. Think of those moments in relationships when you begin to sense that something no longer fits. It usually begins as a very soft thought, that begins to repeat louder and louder. It can take weeks, months, even years to fully accept that truth and act on it.

New ideas are hard to accept when they conflict with protective beliefs. These beliefs act like psychological scaffolding; challenging them can feel like destabilizing your entire reality. And this work holds no guarantees. It is experiential. It requires patience. It requires stepping into the unknown, two of the most uncomfortable things for the human psyche.

But for me, I've come to see these discomforts as gateways. They open new realities. Yes, they once brought me confusion, frustration, and doubt. Sometimes it felt easier to stay in my comfort zone, to just wish I could blend in with society and not think about these things. But if you're reading this, odds are you're not here to stay comfortable. You're here to soak up what life has to offer, and that means expanding beyond what is known.

The truth of plasma-consciousness synergy may seem unfamiliar or even strange at first, but if there is one thing I know, it is this: the body, the mind, and the soul have always sensed things long before science could validate them. And these truths have shown up, over and over again, in cultures, mythologies, religions, and art throughout time.

The most powerful proof may not ever come from equations <u>but from your own lived experience.</u>

We still don't fully understand how acupuncture works, or why meditation rewires the brain, and yet we feel their effects. Consciousness does not need external validation to exist. It simply *is*. Same with love, same with plasma. And now we stand at the threshold of a new model of understanding, one that asks for both reason and feeling, for both science and spirit.

This book isn't not centered around abandoning logic. It implores us to *expand* it. Just as plasma is dynamic, adaptable, and self-organizing, we too can shift our understanding of reality and evolve.

This shift is not about rising above the world, but about deepening into it and expanding. As we move further into an age of advanced technology, we are being called to reclaim our first technology: the body, our bioplasma field and the wisdom within our own systems. This is how we meet the future.

The revolution of AI and emerging technologies isn't going away, and it shouldn't. But it's not the tools we must fix. It's the *toolmakers*. It's also the way we use the tools, which is always a learning curve. If we evolve ethically, consciously, and curiously, we will use these tools in service of creation, not destruction. And that begins within.

"Any tool can be a weapon if you hold it right."
— Ani DiFranco

Our understanding of who we are as humans is widening so we are not to depend on technology, but simply recognize it as a balanced tool we can use to enhance our life and world. Someone on my social media brought up a great point when I first said this, we didn't depend on light, and now we can't live without it. Maybe technology is the same way, I argue it is a bit different, but we can still learn to use it in new and ethical ways that are harmonious to the best of our ability. The only way to guarantee that is to work on ourselves. We must focus on using technology to save our planet, not to destroy it. To open up and enhance lines of communications with other consciousnesses, not separate each other for further control.

I don't know if people intentionally want to end our planet, since we all do live here, but I will say some people's subconscious may be causing them to make decisions that are hurting the planet, all because they are traumatized and scared and have never looked into their own childhood and healed their darkest moments. Yes, it is that simple.

With the knowledge in this book and inside ourselves, if we listen, we will grow more patience, our mental health will improve, we will know how to define what is true to us, and we will see that we do co-create this world. We will gain more control and understanding over our inner selves, which will reflect in our outer world. We will become just as feeling based as we are logic based, where life will be able to be lived in a deep, magical, way.

There will still be challenges, of course, but our perspectives will shift. And just like all great transitions, there will be growing pains. Expansion is never without resistance. But if you stay open, if you let yourself feel the resonance of these ideas, I believe you will sense their truth. My goal is not to convince you, but to show you undeniable patterns, to weave a bridge between science and experience, between logic and intuition, so you have something to hold onto. This is fresh, new, and different, but I have a feeling it will help you to see reality in a colorful and expansive way.

My hope is that as you turn each page, you will *feel* the reality of this shift taking place, not just as an idea, but as something alive. Something calling you to consciously participate in. Because the greatest truths are not thought…they are felt and experienced.

Everything's In Bloom

We are flowers. Flow-ers. In this wonderful plasma reality, we are built to bloom. Everything that happens in your life is meant to help this process. You are an intelligent rose who can use its environment to grow or to your detriment, shrivel. All things seem to unfold as they are birthed from a higher dimensional plasma. There are universal patterns that I believe show that all is interconnected. These shapes appear in galaxies, flowers, and human creativity, reminding us that we are all blooming in flow. Here are some descriptions of natural shapes that may just start to remind you of plasma's magical blueprint on reality as you walk throughout life and please, google them to get a better idea visually!

Golden Ratio and Spirals: In nature, the Fibonacci sequence and golden ratio occur in the spirals of sunflowers, roses, nautilus shells, galaxies, pinecones and hurricanes. In plasma filaments and magnetic field lines often form spirals, mirroring the same energy efficient pathways. The spiral arms of galaxies are plasma structures shaped by gravitational and magnetic forces.

Torus (*Donut Shape*): The torus appears in energy flows such as smoke rings, vortexes, and even in the electromagnetic field around the human heart. And the brain's inner mapping is also shaped like a donut! In plasma physics, toroidal structures emerge in fusion reactors (*like tokamaks*) and in natural plasmoids.

Fractals (Self-Similar): Fractals are patterns that repeat at different scales, like tree branches, river networks, lightning bolts, leaves, and snowflakes. In plasma, fractal patterns can be seen in Birkeland currents (*filamentary plasma streams shaped by electric and magnetic fields*) and in complex electrical discharges.

Diamond-Square Algorithm: This fascinating computational model generates fractal landscapes using simple rules of randomness, iteration, and self-similarity. And it might hint at something deeper…The way the diamond-square algorithm begins with seed values and evolves through structured randomness mirrors how plasma-consciousness synergy might work. Plasma is the medium, and consciousness is the shaper. Like the algorithm, their feedback creates dynamic, self-refining systems. Early versions of the algorithm showed visible grid artifacts, like how rigid beliefs can limit the organic unfolding of our lives. But refinements, like those by J.P. Lewis, allowed the algorithm to adapt, evolve, and generate more naturalistic terrain. This might reflect how plasma-consciousness interaction is emergent, rather than deterministic. Just like a fractal, reality is not pre-set. It is constantly reshaping itself in response to awareness and context.

Infinity (Lemniscate): The infinity symbol (∞) shows up in nature in the structure of DNA helices, orbital paths, and the continuous cycles of beginnings and endings. Plasma's magnetic fields often loop in self-sustaining, infinite patterns, this happens on the Sun, and the symbol even shows up in the reflection of light from certain angles. These patterns may extend across dimensions. The symbol of infinity also hints at a feedback loop of not just plasma-consciousness synergy but our minds with reality, any two objects, and what emerges is what lies at the center and all the space in and around infinity. We are truly all of the spaces *and* the in between.

Infinity

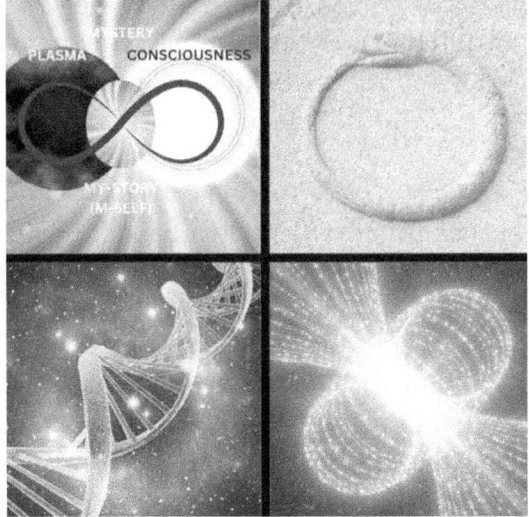

Sacred Geometry: Sacred shapes are everywhere, from the hexagon in honeycombs, to the pentagon in starfish, to the triangle in crystals. In plasma, geometric flows arise too, seen in the hexagonal storm on Saturn, in electromagnetic field formations, and in plasma lab experiments. This suggests that geometry is structural as well as symbolic. It may encode rules of plasma behavior across dimensions. Shapes like the 24-cell icositetrahedron, tesseracts, or Metatron's Cube might hold keys to interacting with plasma, not just spiritually, but perhaps technologically too.

Hexagon Sacred Geometry (also a 3D cube)

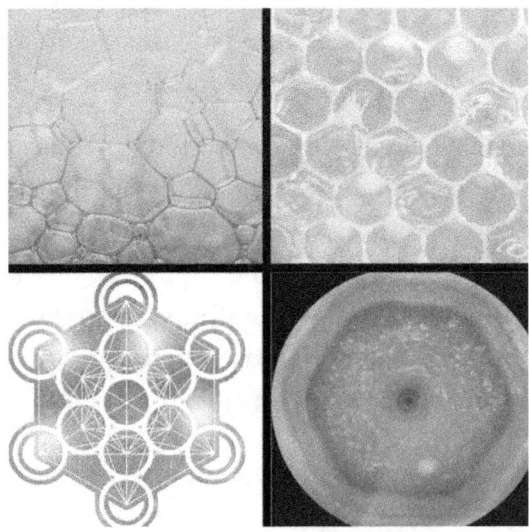

Human creation: These same sacred shapes show up again and again in cathedrals, temples, Buddhist mandalas, Sumerian carvings, architecture and art. Some of these forms bear an uncanny resemblance to known plasma structures like Birkeland currents.

Sacred Geometry in Art & Architecture

Themes

Here are some key insights I believe we are beginning to uncover about expanded Plasma:

1. Plasma is the ground of creation.
2. Plasma is a mirror and a veil.
3. Plasma carries potentiality.
4. Plasma is inherently adaptive.
5. Plasma has its own memory and intelligence.
6. Plasma holds imprints of time, memory, and belief.
7. Plasma embodies the paradox of chaos and order.
8. Plasma embodies and amplifies love and fear.
9. Plasma reflectivity allows consciousness to experience itself.
10. Plasma allows various experiences by providing feedback loops.
11. Plasma is sentient, not conscious, more like the wheel of life itself.
12. Plasma is a cyclical force embodying creation, destruction, and transformation.
13. Plasma is a medium of connection for reality, consciousness and the mystery.
14. Plasma is a symbol of liminality as the in between, it creates thresholds.

15. Plasma is fractal and self-organizing.
16. Plasma exists at all scales from the quantum to the cosmic and beyond.
17. Plasma is the bridge between all dimensions, timelines, and realities.
18. Plasma is the key to accessing nonlinear time and multidimensional travel.
19. Plasma speaks through symbols, dreams, memory, synchronicities, senses, feelings, intuition, art, movies and books. The quickest way to communicate is to ask a **quest***ion*, be patient, and train your consciousness to see what follows. (*Quest-**ion**! I couldn't resist, remember plasma is made of ions!?*)

Speaking of themes of Plasma, let me introduce you to Plasma the Hollywood star, hidden in plain sight.

Plasma The Hollywood Star

I cannot even express how many movies and stories have hinted at this, the remembrance of not only ourselves, but of Plasma. From Philip Pullman's *His Dark Materials* and *The Book of Dust*, where he speaks of a cosmic, all-powerful dust; to plasmoids appearing in various animated films; to the magical and strange world of Plasmate in *VALIS* by Philip K. Dick; and most recently, *Death of a Unicorn*, which metaphorically explores the consequences of unethical and greedy applications of Plasma…let me take you beyond the veil.

Dark Materials & Dust

These books and adaptations, especially *His Dark Materials* and *The Book of Dust*, are more deeply embedded with Plasma symbolism than perhaps any other fictional works I've studied. Honestly, they could warrant a book of their own.

I even shared this idea with Philip Pullman on X, and he seemed to intimate agreement that he may have been referencing Plasma in his world-building!

From the meaning of Dust eerily paralleling Plasma's energetic and relational properties, to the Alethiometer (*a device used to receive intuitive messages*) symbolizing how we may one day communicate using plasma and consciousness, there are striking resonances. Even the mythological naming conventions and phrasing carry layered subtext.

In *His Dark Materials*, Dust (*or Rusakov particles*) is a central, metaphysical element, deeply intertwined with themes of consciousness, awareness, and the soul. Dust is attracted to sentient beings (*greater awareness*), and the Church within the series interprets it as the mark of original sin, something to be feared or suppressed.

To me, this is symbolic of 4D Plasma, the divine feminine, our emotional landscape, our subconscious, and intuitive knowing, all of which have long been suppressed by systems that fear internal power. That fear often gets projected outward as control.

Dust connects humans to their daemons or animal companions that are external manifestations of the soul. These Dust-like beings may reflect our own paradoxical childlike wisdom: both ancient and innocent, gaining abilities and insight in each lifetime.

Dust also enables interdimensional travel. But it's only visible under specific conditions, such as through gold-leaf photograms and specially sensitized photographic plates. When it appears, it resembles particles of light clustering around conscious beings, shaped by awareness.

Lyra Belacqua, later known as Lyra Silvertongue, possesses a natural gift for reading the *alethiometer* which is a rare, compass-like device that communicates with Dust and reveals truth through symbolic language. While others must train rigorously to decipher its meanings, Lyra reads it intuitively. She's able to do this because she is open, curious, and unafraid of the unknown. She listens for answers instead of forcing them with logic.

This quality is known as *negative capability*, a concept we'll explore more deeply in the *Plasma Intelligence* chapter. It's a key state of being when using Plasma as a medium for communication.

Throughout her journey, Lyra moves from innocence to experience, uncovering profound truths about herself and the nature of existence. In many ways, her path mirrors our own, as we remember Plasma, we begin to remember ourselves.

At first, Lyra sees Dust as something dark and mysterious, shaped by the Magisterium's rigid teachings. But through her own direct connection and intuitive truth, she comes to realize that <u>Dust is not evil</u>. <u>It symbolizes the benevolence of humanity, the power of curiosity, and the presence of love.</u> She understands that Dust must be preserved and used ethically, not destroyed. In time, Lyra begins to question the Magisterium and expose their role in suppressing knowledge, autonomy, and freedom.

But after all she's seen, after growing in wisdom and bearing the weight of experience, something begins to change. Her once-effortless ability to read the alethiometer fades. No matter how hard she studies the symbols or tries to reclaim the gift, it slips from her grasp. *(I've found this same pattern within my own journey, and you may too.)* Her mind, now aware of itself, becomes heavy with analysis. The survival mechanisms of the brain override the intuitive trust she once had.

Many experience this in spiritual journeys. It's often referred to as a dark night of the soul, or the desert, like in the book *The Alchemist*. It is the liminal state of growth and darkness between the spiritual awakening (*the first stroke of great luck*) and the treasure. Most people give up here. The key is to move through it with your awareness. This is where you learn to heal and ground your visions so you can eventually share them, if you choose.

Sometimes the weight of knowledge makes it harder to surrender because you start to realize how cool this all is, and your brain takes over and you feel pressure to perform, or to chase a previous experience, instead of softening more into the fear of the unknown and what else you might discover that you didn't know before. It's not about being right. It's exploring without fear of academic ridicule. And I went through a few years of that fear. This book is my coming-out-of-it party.

So, we all want to know—how does Lyra reconnect?! Lyra eventually regains her ability to read the alethiometer, but it's not the same as before. Her reconnection is deeper, more intentional, and shaped by her journey. And guess what? That wisdom never would have emerged if she hadn't gone through the darkness, the grounding, the expansion needed to contain more mysteries, just like the first strokes of innocent luck. This is you. This is all of us, in this liminal space of history, about to find our treasure. Expanding to hold it, to feel it, through the tension of these times. But it's all meaningful.

First, Lyra lets go of control. She learns that the alethiometer requires her to release mental effort. She must quiet her mind, release the need for certainty, and trust her deeper self. This involves embracing a state of flow, where she stops trying to control the process and instead allows the answers to emerge naturally. She rediscovers her ability to approach the symbols with wonder and openness, rather than with the pressure of analysis. This shift allows her to rebuild her connection to Dust, which thrives on curiosity and free will. Lyra's new relationship with the alethiometer integrates her gained knowledge with her intuitive abilities. She no longer relies solely on instinct, nor does she overanalyze. Instead, she strikes a balance, allowing intuition and understanding to work together.

To reconnect with Plasma, you might try releasing any pressure to achieve or control the process. Instead, return to a state of curiosity and surrender, trusting that the connection will flow naturally when you create space for it. You can remind yourself that you are already connected, your awareness has just been asleep to this fact. It's your trust and childlike experimentation that needs to awaken. Your heart must re-open, letting each hurt bring you deeper within yourself and this life experience, instead of letting yourself be hardened or running away from it.

By the trilogy's conclusion, Lyra embodies the idea that true wisdom comes from embracing curiosity, challenging oppressive systems, and valuing authentic relationships. Her journey illustrates that understanding and preserving the essence of consciousness, symbolized by Dust, is vital for personal and societal growth. Lyra can be seen as the new Eve, ending the deterministic narrative of destiny. Her actions ultimately embody the triumph of free will and the rejection of imposed fate. She frees Dust, exposes the power of choice over blind obedience, and grows into self-awareness and love…reimagining the mythology of *The Fall of Man*.

The city of Cittàgazze, seen glowing in the sky like a plasma aurora borealis in *His Dark Materials*, said to visually represent Dust, was once a thriving metropolis of traders, philosophers, and invention. But everything changed when the Guild of the Torre degli Angeli, from the city's central tower, forged the Subtle Knife: a tool capable of cutting between worlds.

This act opened windows between dimensions, but it also unleashed the Spectres, dark entities that consumed the consciousness of adults, leaving them zombified, hollow, and unable to function. Spectres didn't harm children; they fed only on those whose Dust, whose consciousness, had settled into maturity.

As adults lost their minds, society began to unravel as professions collapsed, and businesses failed. After Lord Asriel tore a hole in the sky between realms, a flood of Spectres poured into Cittàgazze. Children scattered into the hills, surviving in scattered bands, while the city itself fell eerily silent, haunted by invisible predators and ruled by the remnants of a lost future.

To me, this symbolizes a subconscious warning, a fear I've explored in my own screenplays, that when we connect with Plasma and open to other worlds, demons or terrifying beings might come through. But the truth is, <u>these figures are just our fears.</u> They are the disowned parts of ourselves and society we haven't yet faced.

The trick is to move through the fear and open the doors anyway. A core reason I'm writing this book is to help us do just that: to face what arises *internally*. Because if we don't, and if we begin jumping timelines, interacting with Plasma, or playing with multidimensional tools (*as we see in series like Dark Matter*)—without first mastering our tuning mechanism, which is our emotional state, we risk entering unstable and distorted realities. This also applies to the rise of A.I. and other technologies that heavily reflect our states of consciousness.

This can be entirely in our control. There's no need to worry, only a need to heal and feel authentically. To acknowledge the truth of what we carry, so we can consciously choose where we go next. You don't need to be fully healed to connect with your intuition, I've had many intuitive experiences while I was in the depths of my traumas, and eating like shit. Why healing and feeling helps is it grounds the intuition into our bodies and Earth. It creates a more stabilized experience that we do not choose to escape to or become addicted to chasing, it just becomes a natural part of life. We do not put it on a pedestal. There is no way to achieve pure happiness where nothing unfortunate happens in a human life. But healing and feeling mixed with knowing the truth of who you are and your intuition is a recipe for a deep, juicy, and resourced life, regardless of what happens.

The Spectres steered clear of attacking the children because their plasma fields were still open. They felt everything. Joy, sorrow, fear, wonder…it all moved through them. They didn't resist emotion and they freely played with the world. They were present, which is key. We only need to return to a sense of this, and nothing can harm us energetically. The only safe way to explore higher frequencies or alternate realities, like in show *Dark Matter* or movie *Everything Everywhere All at Once*, is to do the inner work. That means healing trauma, regulating the nervous system, and reclaiming every emotion as part of our tuning process. We are not victims to our emotions or to life.

Only then can we work with Plasma in the way it's meant to be used: as a co-creative field we harmonize with, free of limiting beliefs, free of fear. We created the Spectres. And we, too, can uncreate them, by facing them, feeling them, and transmuting them into light.

Curiosity is not danger. Curiosity didn't kill the cat. What kills the cat is avoiding our emotions by refusing to face our fears. The original phrase wasn't even about curiosity.

"Care'll kill a cat,"
— from a 1598 play by Ben Jonson (*Every Man in His Humour*).

Here is yet another example of how things get skewed by fearful perspectives in history. In this context, in the play by Ben Jonson, "*care*" meant sorrow or worry, not affection or love. It was more like, "Stress killed the cat." But over time, the phrase mutated.

By the late 1800s and early 1900s, it had shifted to, "Curiosity killed the cat," often used in newspapers as a moral warning against asking too many questions or being too inquisitive. What's often left out is the lesser-known ending: *"...but satisfaction brought it back."* This completely reframes the meaning, suggesting that while curiosity may carry risk, it's also the path to revelation. I'd like to offer my own version now: **"Curiosity didn't kill the cat, it woke it back up."** When we meet our curiosities with courage, we awaken into who we were always meant to be. *Always question history.*

Oh, and best of all, guess what Cittàgazze translates to...*The City of the Magpies.* Remember the bridge of the magpies in the beginning of the book? Just like in Cittàgazze, where doorways between worlds open and close as if guided by something beyond logic, we too are walking across dimensions, led by a hidden intelligence that seems to know exactly how to reunite what has been separated.

Pullman's third book, *The Amber Spyglass*, is fascinating in relation to Plasma for many reasons, but the most exciting link here is the word *amber* itself. It originates from the Greek *ēlektron*, which is etymologically tied to electricity and the flow of electrons. Since Plasma is composed of free electrons and ions, we could think of the Amber Spyglass as a metaphorical "electron lens", or, in my language, a tuning device for plasma-consciousness synergy.

Unlike the Alethiometer (*a truth-teller*) or the Subtle Knife (*a world-cutter*), the Amber Spyglass reveals. It's made from tree sap (*amber*) and mulefa lenses, suggesting that natural substances, paired with intention and awareness, can form a living, conscious lens. Crucially, it works through a meditative, open state of mind rather than logic or control.

What blows me away is that I had never seen the films when I first came to have a knowing about many of my Plasma theories. Discovering *His Dark Materials* later felt like uncovering hidden confirmations like breadcrumbs from the collective unconscious or another version of me. It makes me wonder… maybe the future way to see the Plasma I describe involves amber, electrons, or some forgotten natural process. Maybe it won't be advanced technology at all, but a type of *organic perception* or soft technology, a tuning with the world around us using what's already here.

Lyra Silvertongue and Mary Malone (*initially, an astrophysicist and former nun*) both possess the unique ability to communicate with Dust (*Plasma*) and might give us a window into who we can become to do the same. Guided by mysterious instructions to "play the serpent," Mary travels to a new world inhabited by the *Mulefa*, intelligent beings who perceive Dust as essential to their existence. Immersed in their culture, she constructs the Amber Spyglass from natural materials, enabling her to see Dust directly.

Mary realizes that Dust is not merely a scientific phenomenon but a vital component linking consciousness across worlds. She observes that the flow of Dust is influenced by human experiences and emotions, particularly love and self-awareness.

(*P.S. I just found this out as I researched this story for this book as I re-edit this chapter. Does it not sound like what we are all doing here!?*)

Embracing her role as the "serpent," Mary shares personal stories of love and self-discovery with Lyra and Will. Mary almost seems to be an archetypal figure (*like Mary Magdalene*), symbolizing wisdom and purity. She has a special relationship with Dust: not only does she develop a computer program to interact with it, but she also discovers that the particles are conscious and have influenced human evolution. She is like a guide to Lyra, who is learning to communicate with Dust. (*I wonder if this is a pointer at the intelligent plasmoids as our ancestors that science is discovering as we speak.*)

At first, Mary relies on her training as a physicist. She thinks that to interact with Dust (*which she calls "Shadows"*), she needs to build a complex machine, a computer system (*called the Cave*) powered by conscious thought and pattern recognition algorithms. It's precise, mathematical, and requires academic infrastructure. And yes…it *works* to a degree. She receives meaningful responses from Dust, which confirms it's conscious and responds to intent.

When Mary enters the world of the Mulefa, she no longer has access to scientific tools. So she learns to build something from natural materials instead, specifically, the amber spyglass. The Mulefa teach her how to shape and polish the lenses using tree resin and traditional methods. Through the amber spyglass she sees Dust directly. <u>It hints at the fact that maybe we are approaching technology when it comes to consciousness all wrong</u>. Maybe it's about opening our minds to direct experiences, that are not words or codes, and how we can enable ourselves to experience this.

Dust seems to respond more to natural attunement and intuitive tools better than artificial ones. Will this reflect in our future uses with Plasma? Only time will tell, as well as people being brave enough to try things way outside the box.

By the trilogy's end, Mary integrates her scientific pursuits with spiritual insights, recognizing that true understanding arises from embracing both intellect and emotion. I have discovered the same — where without emotion, intellect will do nothing for us when it comes to the future, Plasma, and our evolution into the new human potential. This is why I call it Plasma-Consciousness Synergy, not Plasma Synergy.

It is not a coincidence that dusty complex plasma is the closest thing in my mind to resembling 4D plasma , and that this series calls what I call Plasma…*Dust*.

Plasma Bubbles

So many films have depicted what I call Plasma Beings—little orbs of light that represent destiny, wishes, consciousness, the afterlife, alien intelligence, and more. These glowing spheres often appear as symbols of something greater as guides from beyond, memory keepers, or messengers of transformation. While I can't show the images here due to copyright laws, I invite you to look them up, you'll start to see them everywhere. Perhaps not just on screens but in your own reality.

Wish: Asha, (*Ash, Dust, hmm…*) the protagonist of *Wish*, was tasked to rescue floating bubbles, kept away in the evil king's quarters. These bubbles symbolized the deepest wishes and desires (*the unique essence and power*) of the people of Rosas.

Brave: The Will O' Wisps, shown as ethereal, fairy-like, floating bright blue flames were known in the movie to lead one to their fate and destiny.

The Last Mimzy: The sacred geometric holographic images (*Plasma shaped by higher consciousness*) appear to two young children as they play more and more with magical toys they found on the beach, left there from a higher intelligent race. These interactions grant them telekinetic and heightened intellectual abilities, ultimately connecting them with advanced beings from the future who aim to save humanity.

Knowing: A fascinating movie starring Nicholas Cage about translucent and opalescent beings that show up before the world's end known as the whisper people. They represent angels or a higher intelligence who are leading the children to safety on interstellar arks to a new world. *The key here is they are the loving soft whispers inside us and if we listen to them our world will be just fine. Also hinting at remaining childlike.*

The Abyss: Deep underwater where a crew gets stuck, they come into contact with NTIs, Non-Terrestrial Intelligences (*who look a lot like Plasma*), a marine race that inhabit the ocean whose technology is based on manipulating water for communication and exploration. They are friendly and patient beings who scare people that are not ready to accept them. They also can control the weather and were around way before humans.

Raya and The Last Dragon: The Dragon Gem, glowing like a Plasma orb, is forged from the essence of Dragon magic by three siblings and holds the power to repel the Druun, shadowy beings that resemble dark Plasma or fear itself. The people turned to stone by the Druun can be seen as early humanity, frozen in survival mode. As the magic of trust, curiosity, and unity returns, the stone melts, and life is reborn. To me, this symbolizes our collective thaw, our journey from fear back into love and living Plasma.

Plasmate, Valis and Philip K Dick

Philip K. Dick is best known for authoring books such as *The Man in the High Castle*, *A Scanner Darkly*, and *Do Androids Dream of Electric Sheep?* (*which was adapted into the hit movie Blade Runner*). His short stories also inspired the films *Minority Report* and *Total Recall*. But in Dana fashion, I want to speak about a lesser-known book: *VALIS*. After a bit of time on drugs and a series of mystical experiences, Dick wrote this book.

His writing became clouded with paranoia, but I believe within that static, real frequencies were trying to break through. Sometimes when people channel while under the influence, the truth comes through fuzzy, like tuning an old radio. But that doesn't make the message any less worthy of listening to. He was touching an area of space or *plasma memory* without his normal logic and for that, all works should still be perused through.

In Dick's personal mythology, he interpreted the discovery of the Nag Hammadi texts as a turning point in history. These were gnostic texts that offered an alternate testament to Jesus' life and teachings, describing him as a divine being who brought gnosis, meaning wisdom, to Earth, not just a fleshly incarnation. He believed that something spiritually or energetically powerful had been released alongside the physical texts. He often described this as a kind of "plasmate," or living information, that had been buried, forgotten, or suppressed, and was now rising again to reawaken humanity's connection to divine knowledge (*gnosis*).

I will say, the timing of one of the first well-known UFO crashes in history, in Roswell in 1947, only two years after this discovery, makes me wonder more about the connection of aliens, UFOs, and Plasma...

In the novel, the protagonist Horselover Fat, an alter ego of Philip K. Dick deriving from Philip meaning love and Dick meaning Fat, experiences a profound vision involving a pink beam of light that he initially calls Zebra. This beam later becomes known as Valis (*short for Vast Active Living Intelligence System*), a divine intelligence that imparts immense knowledge and initiates a deep theological and existential quest.

Sophia Lampton, an incarnation of Gnostic Sophia (*divine feminine wisdom*), tells Horselover Fat that Valis reveals something radical: that we should not worship distant gods, but humanity itself as the vessel of divine potential.

Valis, as described in the book, is Dick's symbolic representation of God or higher intelligence. It also introduces the concept of the *Plasmate* as a form of living information capable of merging with a human being to form a hybrid entity known as a homoplasmate. This fusion represents the integration of transcendent wisdom into human life, an inner gnosis that brings about spiritual awakening and a deeper understanding of reality. There is also this theme of us merging with a higher intelligence in a Russian series by Sergei Lukyanenko, especially in the third installment, *Twilight Watch* as well as the sixth installment, *Sixth Watch*. I came across this towards the tail end of editing, but the similarities are uncanny. I suggest looking into them!

"I term the Immortal One a plasmate, because it is a form of energy; it is living information. It replicates itself, not through information or in information, but as information." "From Ikhnaton this knowledge passed to Moses, and from Moses to Elijah, the Immortal Man, who became Christ. But underneath all the names there is only one Immortal Man; and we are that man. "— excerpt from Tractates Cryptica Scriptura, Valis, *Philip K. Dick*

Dick also called the novel semi-autobiographical, saying these were real revelations he was exploring through a very real connection with VALIS or Plasmate that he had. He also stated it was very possible they were skewed by his alter ego Horselover Fat's hallucinations from schizophrenia or drug addiction.

He claimed that VALIS used "disinhibiting stimuli" to communicate, symbols that triggered recollection of intrinsic knowledge through the loss of amnesia, achieving gnosis. This perfectly describes Plasma-Consciousness Synergy! When one connects with Plasma and the M-Self, they begin to receive symbols in the form of mythological visions, archetypes, or feelings that bypass the rational/logical mind and reach one's deep intuitive core. It's like an emotional key inside you is unlocked each time you connect, within a meditation or similar state, revealing hidden knowledge you already knew, but don't consciously remember.

In Dick's case, VALIS (*the Vast Active Living Intelligence System*) used these stimuli to trigger gnosis, a divine, self-revealing knowing that leads to remembering your origin, essence, and truth. Plasmate or VALIS is what basically helps one recover encoded information inside of them that they already knew!

Dick also stated that, in real life, VALIS helped save the life of his son by warning him of an unknown illness. He said VALIS knighted him with *xenoglossia*, the ability to speak ancient languages spontaneously. He even suggested there was a link between Sirius (*the Dog Star*), the Dogon people of Mali (*who have long-standing oral traditions involving Sirius and celestial knowledge*), and messages being transmitted from higher intelligences associated with that star system. [55]

The *Exegesis of Philip K. Dick*, a posthumously published collection of his notes and spiritual musings, contains many more of his writings on the Plasmate, Valis, and the idea of living information as a divine intelligence.

I want to add in here, and I do not say this lightly, towards the end of the first edit of my book, around April 2025, I was sitting outside during a warm night in Florida, looking up at the stars. A specific star seemed to wink at me, and I heard the words in my heart and head, "Thank you." I immediately downloaded an app on my phone that identifies stars. I matched my camera up with the star and lo and behold, the star was... **SIRIUS***.*

This moment, and this book, are just more examples of how the gnosis of Plasma has been trying to come through for years. And although Philip K. Dick may have struggled with mental health and drug use, and the signal may have been a bit grainy, the truth still shone through.

> **"The Head Apollo is about to return. <u>St. Sophia is going to be born again; she was not acceptable before.</u> The Buddha is in the park. Siddhartha sleeps (but is going to awaken). The time you have waited for has come" — excerpt from Tractates Cryptica Scriptura, Valis, Philip K. Dick**

---

[55] Philip K. Dick, *The Exegesis of Philip K. Dick*, ed. Pamela Jackson and Jonathan Lethem (Boston: Houghton Mifflin Harcourt, 2011), [from several sections].

It's as if Dick, despite all odds, touched the same mythic stream I now call Plasma. Just like the awakening of Siddhartha, 4D Plasma, the collective unconscious, is beginning to stir. The head of Apollo may be the return of the multidimensional self. And perhaps Sophia has always been the plasma muse…soft, electric, loving, and always calling us, our inner Buddha, home.

Ball Lightning by Cixin Liu

The writer of *The Three-Body Problem*, and the hit Netflix series based on it, also wrote a lesser-known but equally powerful book called *Ball Lightning*. The story follows a man who loses both parents to the wild, unpredictable phenomenon of ball lightning during a thunderstorm. That trauma ignites a lifelong obsession: to uncover what ball lightning really is.

As he becomes a scientist, he crosses paths with a female military officer who shares the same obsession, but with a different motive: to weaponize it. Together they plunge into ethically gray research and ultimately discover that ball lightning is composed of "macro-electrons"—a new form of matter that behaves in bizarre ways across time and space. Their experiments lead to technological breakthroughs in targeted destruction, like macro-nuclear weapons that can obliterate specific materials, electronics, or even human bodies, without damaging anything else nearby.

But the cost is steep. The closer they get to power, the further they drift from morality. Chen (*the protagonist*) eventually realizes that understanding something doesn't equal wisdom. Some mysteries, he suggests in a haunting final moment, may not be meant to be solved, at least not with the intention to control them. They must be approached with reverence, not domination.

This speaks to me deeply, matching my feelings that intelligence without healing can be incredibly dangerous. This book is a subtle but clear warning about the risks of approaching Plasma with unhealed intentions. But instead of suppressing Plasma, or fearing it, I believe it's humanity that must heal. We must grow into the level of responsibility required to explore it ethically. Everyone deserves a chance. Just as an addict deserves the chance to heal, to get sober, and to live a radiant life, we as a species deserve that too.

What is most interesting about this book is its ending, and even more so, the afterword. In the last chapter, *The Quantum Rose*, the protagonist hears of the discovery that while being tested in an underground mine, ball lightning was able to reach a collapsed state without a human observer…which violates the typical understanding of quantum mechanics.

It is then explained to him that there *was* an observer, only it was not of earthly origin. And not only that, but there is a super-observer beyond this as well.

This parallels a thought of mine that Plasma is a vehicle for consciousness and all kinds of conscious beings, as observers, may be influencing it as we live our daily lives.

The difference between what I think is really going on and what is presented in this story for dramatic and imaginative effect, is that these events are happening *for good*, not for a sinister reason of another consciousness wanting to control humanity, which is a common sentiment in Liu's books. *Unless of course it is Plasma reflecting the need for humanity to control itself.*

In the afterword, the author Liu shares that he experienced ball lightning for the first time in 1982, which in turn sparked his imagination. He explains that most Chinese sci-fi books are about stories of invention that barely touch the surface of the societal effects. He almost says, in a secretive but bold tone, that although his book aligns with this tradition in the commercial sense (*of a futuristic device—ball lightning—being weaponized*), the heart of his writing goes well beyond. He explores the flights of fancy of ball lightning.

He proclaims that these little balls of lightning seem like they too, like him, and the ethos of Chinese science fiction as a whole, are trying to transcend the reality of being used for a futuristic device. He compares it to how a man's tie, within the confines of its narrow dimensions, has the freedom to indulge in a riot of colors and patterns unbounded by the rigid formula of a business suit.[56] What an epic and magical comparison.

I feel this truth SO deeply, and I am sure most people who have read his excerpt have grazed by it without a second thought. That not only is Plasma calling out to us to be harmonized with by our conscious awareness, instead of technology or capitalistic uses, but that these times are calling for new story structures and new archetypes. Not ones of control, domination, and drama but of healing, balance, and magic. These can be just as interesting, if not more so, if artists take risks and dive into the sparkling Starpunk pool versus a highly dominated Cyber-punk river filled with the trash of our past hurts that have hardened into a singular view of the future that may just not be our highest truth. Our stories can become those technicolor ties, hopefully one day extending to the suits that complement them.

His last sentences infer that although his works of fiction with ball lightning are purely imagination, that just like reality, that may seem gray and mundane, something small and surreal drifts by unnoticed, like a speck of dust tumbling out of a dream, suggesting the

---

[56] Cixin Liu, *Ball Lightning*, trans. Joel Martinsen (New York: Tor Books, 2018), 383.

vast mysteries of the cosmos, the possibility of a world entirely unlike our own. *But I do notice. We now notice.* He, like all these other creators, is a total Plasma channeler!

Death of A Unicorn

To build on the importance of ethics and morality, this film is the perfect close to this chapter. In medieval lore, the unicorn was often depicted as a wild, magical yet untameable creature that could only be captured by a virgin, symbolizing the idea that purity could tame even the most unruly forces.

"The unicorn, through its intemperance and not knowing how to control itself, for the love it bears to fair maidens forgets its ferocity and wildness; and laying aside all fear it will go up to a seated damsel and go to sleep in her lap, and thus the hunters take it." — Leonardo da Vinci, Notebooks

Death of a Unicorn tells the story of Elliot and his daughter Ridley, who accidentally kill a unicorn with their car on the way to a weekend retreat hosted by Elliot's employer, billionaire Dell Leopold. When Leopold discovers the unicorn's supernatural, curative powers, he seizes the body with plans to exploit it for profit. However, their actions provoke dire consequences as the unicorn's family seeks vengeance against those involved in the exploitation.

In this story, I see the unicorn as Plasma, anthropomorphized. There are striking parallels: the unicorn's blood and horn have miraculous healing properties, and the aurora borealis, long associated with magnetic Plasma, is seen shimmering above the unicorn family's cave. The film draws inspiration from the myth of the Unicorn Tapestries, real works of art that also depict the unicorn's horn purifying water, just as modern Plasma technology has been found to do, cleansing both water and air. But more than that, both the tapestries and the film reflect a deeper truth: that when we purify ourselves—releasing greed, outdated desires, and inherited traumas, we become capable of harmonizing with the unicorn. With Plasma. If we try to control her, we may be undone. But if we meet her with reverence, healing becomes inevitable.

Right now, Plasma is on the verge of transforming energy, health, and even consciousness itself. But just like in *Death of a Unicorn*, the biggest question isn't what we *can* do with Plasma, it's how we choose to interact with it. If we treat Plasma like a product to be exploited instead of a force to be worked with, we could destroy the very thing that could elevate us.

This is why I'm writing my book, because we need to learn to work with Plasma in an ethical, reciprocal way. My book teaches how Plasma isn't just energy, it's a living intelligence. It's a relationship, and if you abuse it, you lose access to its highest potential. And this is the preparation to my subsequent books which will be about harmonizing with it to perform once-thought-to-be-impossible, magical feats.

In the end of the movie the unicorn lays in the female protagonist's lap, as she has shown that she has no interested in using unicorns, but in understanding them. That she is a pure soul who has been misunderstood herself, and they submit to her. They also help heal her father after an accident occurs.

This does not mean we need to be perfect, we have all made mistakes and unfortunate decisions, but we all have the awareness now to do better, and that is all you can do each day with new learnings. In a world obsessed with control, extraction, and power, it's those who approach life with humility, reverence, and curiosity that unlock its deepest secrets.

"Blessed are the meek, for they shall inherit the earth.
Blessed are those who hunger and thirst for righteousness, for they shall be satisfied.
Blessed are the merciful, for they shall receive mercy.
Blessed are the pure in heart, for they shall see God."
— Matthew 5:5–8, The Holy Bible

The Unicorn Surrenders to A Maiden, from The Hunt of the Unicorn Tapestries

Just like the unicorn, Plasma holds the key to a future where those who respect the unseen forces of the universe, rather than exploit them, will be the ones who shape the next era of science and consciousness. The real question is: Will we use it the right way, or will we repeat the same mistakes that Hollywood has subconsciously been showing us for years?

Decentralization in Hollywood

Since we are on the topic of entertainment I would like to share one more sentiment about this industry. For the past century, storytelling, our most sacred tool of meaning-making, has often passed through layers of filtration before reaching the public. In many cases, these filters haven't amplified the creator's vision but distorted it. What was once a system designed to support and elevate storytellers became, in some corners of the entertainment world, a hierarchy of gatekeeping. Somewhere along the way, creators were taught to believe that their ideas needed permission, reshaping, or approval from an external authority before they were worthy of being seen or heard.

Something is beginning to shift. We're witnessing the rise of self-publishing, indie filmmaking, and decentralized storytelling platforms. This is a projection of the remembering that the power of story was never meant to be extracted but shared. That publishing houses and production companies should be bridges, not walls. The future isn't about dismantling these institutions entirely, it's about transforming them into harmonic collaborators that empower vision rather than distort it.

Perhaps the real revolution isn't removing the structure altogether, but returning to structures built on mutual respect, creative sovereignty, and open communication rather than fear, competition, and control. Some already do this beautifully, and as more creators awaken to their own inner authority, the systems that survive will be those that mirror this new paradigm. Also, the places that pretend to do this, but still have interiors of control, will be very noticeable as our intuitions rise.

By now, it's hard to deny, Plasma has revealed itself again and again in pop culture, across our screens and pages, for over a hundred years. But what if I told you its presence stretches back *thousands* of years, woven into the myths of the ancients?

Next, join me on a journey into the deeper layers of Plasma…through mythologies, religions, philosophies, and spiritual practices across time. As we illuminate how past civilizations may have known, worked with, or even mastered this force, we begin to see that what feels new is actually a remembrance.

While I place great emphasis on creating a new Plasma future, I also believe we must go back…back to the roots, to heal, to reinterpret the past with fresh awareness. Some cultures may have already been consciously harmonizing with it, a lost art we are now beginning to recall.

With reverence for what came before, an expanded consciousness, and a playful, open heart, we step into something entirely new: a journey of creation, surrender, and play.

Part Two

Freedom: Unbinding Consciousness & Feeling Plasma

"The valley spirit never dies; it is called the mysterious feminine. The gateway of the mysterious feminine is the root of heaven and earth."

— Lao Tzu, Tao Te Ching

Two unborn twins are having a conversation in the womb. One asks, "Do you believe in life after birth?"

The other replies, "Of course not. This is all there is. Why imagine a world beyond this warm, dark place? We're nourished through the umbilical cord, once that's gone, we die. There's nothing more."

But the first twin insists, "I think there's something beyond this. Maybe after birth, we'll walk on our own legs. Maybe we'll use our mouths to eat. Maybe there's light... and sound... and even love. Maybe someone is waiting for us."

---

[57] Image: *Chaos Monster and Sun God*, from a gypsum wall relief at Nimrud. Drawing by L. Gruner, 1853, after the original Assyrian sculpture. Public domain. British Museum, BM 124571.

The skeptical twin scoffs, "That's just fantasy. There's no proof of any 'mother' or 'outside world.' We've never seen her. This is our entire reality."

The first twin pauses, then says softly, "Sometimes, when I'm very still, I feel something. A presence. A heartbeat that isn't ours. I think... she's real. And I think she loves us. *Even if we can't see her yet.*"

This story, *The Parable of the Twins in the Womb*, is a powerful metaphor that illustrates faith in the unseen and the idea of what may lie beyond our basic sensory perceptions. Many of us still believe that we are alone and disconnected, floating in a womb we cannot name. We don't realize we are inside something living. We do have a mother, Plasma. And some of us *can* feel her heartbeat when we get still enough, as her vibrations pulse through us. Our consciousness has been mostly unable to perceive her, until now.

Let me introduce you to where this rupture may have begun, the moment that tethered us so tightly to the material world that we forgot we ever belonged to something more. We've become so coiled around what we can touch and measure that loosening our grip feels dangerous, frightening even. But if we dare let go, we might just remember the cosmic mother we once all knew, and the joy she brought us.

With remembering that joy comes the unbearable ache of loss. There was a moment in our collective story when consciousness chose to leave her, to venture out alone. But now comes our return. We are aware enough, and so is she, to heal the fracture and start anew.

Plasma as science's fourth state of matter, is also our cosmic womb. She is the neutral ground of reason. The field through which all forms emerge. In the earliest myths, Babylonian, Vedic, and beyond, she was known as *Tiamat*, the great sea or the divine chaos, also sometimes known as a dragon. She birthed the first gods. She was not evil…she was *everything*…and she was *alive*.

When consciousness became aware of her, it didn't harmonize with her. And truly, maybe it didn't know how. It didn't see or open up to her potential or love, it scared it, so instead it fractured her, like smashing a mirror that shows a distasteful reflection.

In myth, the story was rewritten: Tiamat, the ancient dragon goddess of creation, was cast as a monster. She was slain by Marduk, and her body was divided, half into the heavens, half into the earth. Her memory was imprinted with trauma, with misunderstanding, with suppression.

We have inherited this pain. Because Plasma is us. We are her children; we are her cells. When we are misunderstood, we feel her pain. When our wildness was feared as

232

children, our feelings were shamed, just as hers. And when her silence was pierced, every time we tried to face our fears, our trauma, our nervous systems screamed. We are still carrying this rupture, not realizing how deep it goes.

Plasma holds memory, not only biological or energetic, but dimensional memory. The fourth dimension, the emotional field, the collective astral web, is where her pain reverberates. It's not evil, it is only unmet feeling, it is grief that is unprocessed.

We feared it because we were never taught how to feel it. We called it chaos, the shadows, hell, even! But this wasn't her…it started with *us*, projecting our trauma, our hate, our fear onto this mothering field that only wanted to connect, to support us. She mirrored our chaos, and possibly also, in early times, was chaotic herself, as she has evolved right alongside us.

These things we have been taught to suppress and ignore are the keys to everything. 4D plasma is a threshold, it is not a trap nor is it hell. It is the Plasma layer of raw, original feeling imbued with conscious and unconscious emotion. It's where time folds, emotion becomes memory, and trauma becomes a portal to another way of living. And it is where the mother, Tiamat or the Plasma Field, still waits for us to remember her. She is waiting for us to return with feeling, to let our fear subside, and the only way to do this is to face them.

This chapter begins at that doorway. You do not need to understand everything yet. You only need to feel and make space for what you once feared, not so you can hold on to it, but so you can help transmute it. I also do not mean reliving trauma , it's about finally allowing the raw, original signal of what was once too much to process to *move through you*. Instead of suppressing fear, sadness, shame, or rage out of conditioning, you can now allow yourself to be with it without becoming it.

The monsters, the demons you fear are not hers. They are our own unclaimed parts of consciousness. The things that are too painful to look at, so they wear scary masks, because what lies beneath that is a pain so deep, that we are scared it will incinerate us. The truth is, as you meet these parts with love and curiosity, Plasma becomes whole again, and in turn so do you. You remember that you are not alone, and that something ancient is loving you through this. This is when the magic truly begins, and things get very interesting…

When we open our consciousness enough to listen, to heal, to feel…and to navigate emotions the way they are meant to be felt, this plasma-intelligent, fourth-dimensional space becomes like living water. A subtle field we all draw from, capable of helping us do things we never imagined possible. Accessing timelines, recovering memories, and reconnecting with our most expanded selves in a new way, which is who we really are.

She is the connective tissue between all versions of you, the past, the future, they are all parallel timelines, fractalling out. And best of all we can create new futures. When we no longer fear her, she beings to assist us in what once felt impossible: surfing between worlds, healing across timelines and dimensions, communicating with higher intelligences, and aligning with the reality we were always meant to live. Your emotions are not in the way, *they are showing you the way.*

First, we'll begin to heal the past by revisiting ancient stories involving Plasma, through a new lens and with new awareness. Then, we'll dive into the magic and meaning of Plasma Intelligence itself: what raw emotion and feeling truly are, how they open communication with other beings of consciousness, how to decode their messages, and a few little magical tools to place in your new human potential toolbox.

4

A New Myth of God

"An ideal plasma is a charged gas wherein no bound states exist – a mythical beast." — Allan N. Kaufman, Physicist

"Every man has a wild beast within him." — Frederick The Great

A *mythical beast* is an imaginary being of myth or fable, otherwise known as a monster. These beings belong to the imaginary realm, which acts as a boundary space between worlds. It is the in-between, where everything is in flux, constantly shifting form. It is the realm of shapeshifters, archetypes, ideas, memories, dreams, angels, devils, and monsters.

In ancient times, monsters were not merely threats, *they were messengers*. They warned, revealed, and made the unseen visible. They embodied what lay buried in the human subconscious. All that has happened, is happening, and will happen resides in this kinetic, symbolic realm. Some call it the Akashic Field. I call it 4D Plasma, or Dark Plasma.

When I think of the fourth dimension, I picture the tesseract or a higher-dimensional cube that, when projected into three dimensions, appears distorted or inside-out. That inversion feels like the fourth dimension revealing itself, a hidden truth erupting through illusion.

Within this field exist beings, reflective in nature, yet each holding its own intelligence or information beneath the surface. It may sound complex, but I perceive this hidden intelligence as residing within a fifth-dimensional space, which is why I call it 5D Plasma. Once we begin to see clearly, after healing through love, curiosity, awe, joy, and other expansive states, this higher intelligence is revealed. The 4D layer begins to shift, moving from reflective to transparent. You could also imagine it as a core of still, potent potential energy, truth in its quietest form, surrounded by an outer layer in constant flux, reshaped moment by moment by the observer's perspective.

There are many ways to interact with these beings of intelligence or information, these so-called monsters or beasts. This field, its energy, and the beings inside of it have been called many things throughout time as they have shown up differently across time, religion, and cultures. Plasma is yet another word and the most current to define its features, and we are just at the precipice of this new word unlocking many things we have not yet discovered. I will be going over the nuances of these beings of information and intelligence with the awareness of Plasma and how can differentiate and understand them further in the book.

Since words are constructs of the human mind, anything I say can only point to the truth—not define it. The real truth comes through experience, and my hope is to guide you toward having your own. Your own life experience. Your own relationship with this substance, this force, that exists both inside and outside of you.

By exploring the other names used for this force, and the beings associated with it, I hope to help you make meaningful connections through mythologies and terms you may already recognize. This will also reveal just how magical, influential, and timeless this force has been, shaping cultures, consciousness, and entire eras.

Picture a shimmering field beneath everything you see. Look out a window and imagine someone gently pulling back the blanket of reality off the house, the road, and the cars, revealing a transparent, glistening, flowing net. Within that net, opalescent mists dance from strand to strand. And inside the mist, there are denser pockets, holographic and lively, glowing in neon yellows, cerulean blues, bubblegum pinks... all the colors. This is what I see when I tune into the field, its energy, its presence, and the beings that have no true name.

I'm now going to ask you to join me in this time machine. Grab a refreshment, curl into a blanket, sit back, breathe, and come with me to the beginning…to a land called *Everywhen*, inhabited by ancestral figures with supernatural abilities, stretching all the way to our modern-day understandings of UAPs, higher-dimensional intelligences, and futuristic technologies.

And by meeting this ever-present wisdom, maybe, just maybe, you'll begin to experience these visions, these knowings, for yourself, if you haven't already.

The Primordial Plasma – Early Mythologies and Elemental Forces

Everywhen, Aboriginal Australia (50,000 BCE)

Welcome to Everywhen. A static, fluid, infinite ocean of potential, traversing new land and shaping what would one day become humanity and the world we now know as Earth. A creative force that could be contacted, molded, and moved by the minds of the Aboriginal people, in their own special way.

The legends of Dreamtime reveal a profound understanding of how reality is a living narrative, ever-unfolding. The Aboriginals worked in harmony with the land, the elements, and the spirit world without the need to consciously direct or control it, a reminder of the basic human potential still encoded within us all.

There isn't an accurate word in the English language for what we call *Everywhen*, or *Dreamtime*. We know time as past, present, and future. Dreamtime is none of these—and yet, somehow, it is all of them. It is the Eternal Present, right here, now, with us. Mudrooroo, an Aboriginal writer states that Dreamtime, better called the Dreaming, indicates a psychic state in which or during which contact is made with the ancestral spirits, or the Law, or that special period of the beginning. [58] I interpret this as Plasma, the primordial ooze, in its infancy harmonizing with beings on Earth.

The Dreaming symbolizes a timeless movement from the dream state into reality, an act of creation that forms the basis of many Aboriginal myths. I see this as a metaphor for the invisible, kinetic energy of plasma: a living field that responds to intention and forms into solid matter. By aligning our consciousness with it, we can shape the 3D world around us with greater awareness.

Each Aboriginal person identified with a specific Dreaming. It gave them identity and guided how they expressed their spirituality. Think of this as your own personal bioplasma bubble, your soul. The lens through which you view reality, shaping how you traverse and experience the world. You are the only expression of plasma that is *you*…your personal dream. And all of our perspectives, our Dreamings, seem to converge in this shared reality, creating a collective field for the evolution and experience of the greater whole, whatever that may be.

---

[58] Creative Spirits, "What Is the Dreamtime or the Dreaming?," *Creative Spirits*, last modified February 17, 2022, https://www.creativespirits.info/aboriginalculture/spirituality/what-is-the-dreamtime-or-the-dreaming.

Some Aboriginals shared Dreamings with others, which I interpret as being part of a soul, or plasma, family. What some might call a star family. A group of beings who have known each other across lifetimes or lifeforms, united by a shared mission or overarching theme. When I meet people who resonate with plasma, consciousness, and the desire to uplift humanity, it feels like reuniting with my soul family. We carry similar frequencies, each bringing our own unique gifts to a collective purpose.

One person can also have multiple Dreamings, which I take to mean you can have many different missions or "personalities" throughout your lifetime. [59] You are not confined to one path, you are multidimensional!

A message that keeps circling my mind is deceptively simple: **LIVE.** We are meant to live fully, as humans, exploring, feeling, and creating within this vast, ever-shifting playground. None of this is a mistake; being here is a choice, and a beautiful one.

I invite you to search online for cave drawings and ancient artwork from the Dreaming or Everywhen. Let their colors, symbols, and forms stir something in you. They may just awaken your subconscious to the hidden geometry and living memory of Plasma.

To go even deeper, even "The Dreaming" is not the most accurate word. The missionary Carl Strehlow recorded that the term used by the Aboriginal people was *Altjira*, or *Altjira mara*, meaning good or God.[60]

Every time I think of Plasma, I feel it is the same thing as Good, pure benevolence. It gives the old phrase *"God is good"* a whole new meaning. If God is Good, and all is made of that Good, before the distortions born of human fear, then maybe we are all individuations of that Good, creating as our birthright using the clay of Dreamtime, the dust of dreams...

The word Altjira was also said to refer to an eternal being who had no beginning. To dream, according to this lore, was to see God..[61] And while it's true that translations are imperfect and time can obscure the original meanings, I believe these glimpses point us toward the real essence of Plasma: mysterious, eternal, benevolent, and utterly alive.

[59] Creative Spirits, "What Is the Dreamtime or the Dreaming?"
[60] Sam D. Gill, *Storytracking: Texts, Stories, and Histories in Central Australia* (Oxford: Oxford University Press, 1998)..
[61] Barry Hill, *Broken Song: T. G. H. Strehlow and Aboriginal Possession* (Sydney: Knopf, 2003).

Animism, Global (Prehistoric Era)

Around the same timeframe, and weaving throughout the prehistoric era, was a belief in Animism. The word comes from *anima*, meaning life or soul.[62] Animists believed all of nature was animated, with spiritual beings inhabiting trees, rocks, and waterfalls.[63] Instead of being a religion, Animism was a perspective that permeated many. Plasma could explain why everything seemed alive, because maybe it was. Not biologically, but as a field of intelligence that still exists today. This might also explain why some people with personification synesthesia (*where objects, numbers, or letters have personalities*) perceive traits in trees, furniture, and other forms.

With this understanding, everything with a biofield, or plasma bubble, would have a soul (*imprinted plasma containing information or intelligence*). This explains why a soul is made of plasma and can either hold consciousness or simply carry information from unconscious imprints, memory, or collective memory. This applies to objects, people, locations, even the Earth itself. That's why certain places, like the vortexes in Sedona or the temples of Egypt, carry a distinct energy. Imagine everything and every place surrounded by circles, bubbles, each with its own information field, memories, and ideas, waiting to be discovered by the awareness of whoever enters. These fields can overlap and interact potentially sparking something new.

Picture objects on a desk, each encircled by its own energy field, just as the city of Los Angeles is surrounded by its own energetic imprint. Everything, whether a water bottle, a city, or a human body, has a unique essence that may change form but never truly disappears. The water bottle might end up in a landfill, only to be repurposed into something new. Los Angeles will evolve, shifting its landscape, climate, and character over time. Likewise, when our bodies die, the soul returns to its plasma lake, a vast, flowing reservoir of potential, ready to rise again as a new expression.

If the soul chooses to re-express in a new form, these patterns can condense into a body, carrying fragments of previous life experiences as intuitive feelings, déjà vu, innate talents, or even cellular memories. We just need to remember how to access them in each lifetime. And make no mistake, they are becoming more and more accessible, as I'll explore soon.

---

[62] Animism – Definition, Meaning & Synonyms," *Vocabulary.com*, accessed May 14, 2025, https://www.vocabulary.com/dictionary/animism.
[63] Animism," *Internet Encyclopedia of Philosophy*, accessed May 14, 2025, https://iep.utm.edu/animism/.

What I want to get across here is this: every person, place, and thing holds information or intelligence. Some people are believed to be born with the ability to perceive the information within animate and inanimate objects. In parapsychology, this is called psychometry—when someone can touch an object or person and receive information about them. (*It's also Jedi Quinlan Vos' power in Star Wars!*) We can open up to this too, by tuning in through sensing and feeling with our own plasma intelligence. It is different than normal thought.

For me, Animism has been a straight shot to gratitude. Knowing that everything has a soul helps me stay grateful (*when I remember to be*), for the chair I sit on, the street I walk down, the places I visit. And I truly believe they hear me when I express warmth toward them. I hope this gives the word *gratitude* some depth, because for a long time it was a hard concept for me to connect with. Everyone said, "When you have gratitude, more abundance comes," but until I truly understood *why*, that it's listening and reflecting that energy back, I couldn't emotionally connect with it all the time.

Shamanism, Global (Stone Age)

In the *Encyclopedia Britannica*, Shamanism is defined as a religious phenomenon centered on a shaman, a person believed to gain various powers through trance or ecstatic religious experience. Shamans are typically thought to heal the sick, communicate with the *other world*, and escort the souls of the dead to that realm. The origins of Shamanism trace back to the indigenous peoples of far northern Europe and Siberia. [64] Shamanism was also deeply intertwined with Animism.

The modern shaman and initiate, Marti Spiegelman, describes shamans as intermediaries between humans and the gods. They stand at the gates of the in-between, surrounded by paradox and the ever-shifting flux of the field, the liminal space where souls' dwell. She calls the shaman *the keeper of the fire*. And what is fire? Bingo. Plasma.

The word *shaman* comes from *ša-man*, meaning "one who knows." When I describe intuition from the plasma field, the best way to explain it is simply a knowing. Something that can't be put into words, but you try to. You just know…it's truly ineffable.

Anyone can be a shaman in the sense of communicating with the Plasma Field. It takes practice, and yes, in many cultures, certain people were naturally more open and venerated by others. I don't discount that. I just want to remind anyone reading this: you, too,

---

[64] Shamanism," *Merriam-Webster.com Dictionary*, accessed May 15, 2025, https://www.merriam-webster.com/dictionary/shamanism.

can become a shaman in the truest sense of the word. You can communicate with this field, learn to heal yourself and others, and even channel or commune with beings inside it.

Each person has their own shamanic gifts and they vary from individual to individual. There are many books on shamanism and pathways to becoming initiated through various cultures and practices. Often, the call comes from within, it feels like a subtle pull. You all have the potential to be little Plasma Shamans!

Shiva and Shakti, India (9,000 BC)

With the rise of Hinduism came the legend of Shiva and Shakti, the divine union of energy and consciousness. In Tantric cosmology, the universe is seen as being created, penetrated, and sustained by two fundamental forces in perfect, indestructible union. These forces, or universal aspects, are called Shiva and Shakti. [65]

Both aspects originate from the One, what I call the Mystery, the divine, or the M-Self. Others may refer to it as higher consciousness. This is how the divine experiences itself: through the interplay of plasma and consciousness. Though fundamentally the same, they appear different in duality, and it's this very duality that allows us to experience, evolve, and create.

We can move beyond duality at any moment, returning directly to the divine, trusting that everything is unfolding for our highest evolution. As we evolve individually, so does the divine. We are how God, Source, the Mystery experiences itself. We are all special!

And not in some narcissistic way, most of us have been conditioned to equate individuality and specialness with ego. But think about how you look at a newborn as a deserving, unique, pure bundle of love. That's who we truly are, each of us a distinct, irreplaceable expression of the divine.

Shiva represents the masculine principle as the shaper, the force of consciousness. Shakti embodies the feminine principle as the shaped, the animating force. This is what Carl Jung described as the spiritual equivalent of anima/animus or the essence that brings life to all things. In my framework, Shakti is Plasma, the very fabric of the cosmos. Plasma makes up 99.9% of the visible universe in science, and I would argue 100% of everything in different states, just as Shakti was said to be the cosmos itself.

---

[65] Shiva Shakti – Divine Union of Consciousness and Energy," *Temple Purohit*, accessed May 16, 2025, https://www.templepurohit.com/shiva-shakti-divine-union-consciousness-energy/.

Interestingly, Shakti is symbolized by a downward triangle, Shiva by an upward triangle. Together, they form the Shatkona (*hexagon*), a powerful symbol that, when multiplied, creates a honeycomb pattern. It's a subtle hint at how the interplay of plasma and consciousness weaves the beehive of our universe, the fundamental tiling geometry of space.

Kundalini and the serpent of Shakti closely resemble a plasma filament, known as a *Birkeland current*, a twisting, spiraling flow of charged particles that mirrors the imagery of a coiling serpent. In plasma physics, magnetic reconnection happens when plasma filaments snap, merge, and realign, releasing tremendous energy. This mirrors the Kundalini serpent rising, merging, and aligning with the higher chakras, resulting in powerful energetic awakenings.

To take this further, consider the *Sri Yantra*, the sacred geometric form representing Shakti. It consists of nine interlocking triangles: four pointing upward, symbolizing Shiva (*consciousness*), and five pointing downward, representing Shakti (*plasma*).

This ratio hints at a fascinating aspect of plasma physics. Plasma is quasineutral, generally electrically neutral on a large scale, but it can still exhibit slight imbalances. Electrons, being smaller and more agile than ions, move faster, creating brief fluctuations that slightly favor negative charge.

In the Sri Yantra, the five Shakti triangles outweigh the four Shiva triangles, reflecting Plasma's natural inclination toward the feminine as receptive, sentient, and subtly charged with excess electrons. This imbalance not only symbolizes the generative, creative power of Plasma but also reinforces the idea that Plasma is inherently feminine, the perfect force for creation.

Shakti is also associated with Kali, the goddess of creation and destruction. Kali, known as the Dark Mother, could also be interpreted as a representation of dark matter, or what I refer to as 4D plasma. Interestingly, the word *matter* shares its root with *mother*, suggesting that all matter may simply be plasma in different expressions. In this framework, Plasma is the cosmic womb, taking on various forms across dimensions. In the third dimension, it manifests as "light matter", visible and tangible. In the fourth dimension, it becomes dark matter, the unseen, the unconscious.

This dual nature of plasma reflects the dual nature of Kali, both creator and destroyer, both light and dark. It's a powerful reminder that what appears as dark matter may actually be plasma in a different state, the hidden face of creation, waiting to be revealed or awakened.

On top of birthing and creating the universe, Shakti also protects it. This mirrors one of Plasma's primary functions within the cell and its membrane: protection. Plasma serves as a dynamic boundary, maintaining form while still allowing intelligent flow.

Within the concept of Shakti is the term *Lila*, a key principle in the Shakta tradition that reveals a profound secret about how to connect with Plasma, and how to live life itself. *Lila* means divine play, creating with curiosity, spontaneity, and surrender versus an attachment to fixed outcomes. This is the essence of plasma communication, one must be receptive, hold a certain lightness, and be open to whatever arises. In this state, suffering rarely finds you because you're no longer grasping, pushing, or resisting. Instead, you're in flow, allowing life and plasma to move through you effortlessly. Whether in meditation, creative work, or daily life, if you can cultivate the lightness of *Lila*, magic can happen. It's the subtle art of letting go and allowing the universe to play through you.

Fridtjof Capra, in *The Tao of Physics* (*1975*), writes:

"In the Hindu view of nature, then, all forms are relative, fluid, and ever-changing *maya*, conjured up by the great magician of the divine play. The world of *maya* changes continuously because the divine *lila* is a rhythmic, dynamic play. The dynamic force of the play is *karma*, an important concept of Indian thought. *Karma* means 'action.' It is the active principle of the play, the total universe in action, where everything is dynamically connected with everything else. In the words of the *Gita*, *karma* is the force of creation, wherefrom all things have their life."[66]

Here, *maya*, often translated as illusion, is revealed as more of a creative substance, shaped by actions and dynamic play. It's the field where reality can be bent, reformed, and shaped, while still holding the imprints of our past actions.

In this realm, karma operates as a dynamic force within the plasma field. In Sanskrit, *karma* means action, but it's more than the action itself; it's the energetic imprint it leaves behind. For example, a deeply negative act like murder in a past life could create a dense distortion in the bioplasma (*or soul*), leaving behind an energetic residue.

In a future life, that imprint might seek balance, perhaps expressing as a protector or healer. Likewise, someone who experienced betrayal may unconsciously adopt patterns of control or hypervigilance, trying to prevent it from happening again.

The key is that these patterns are never fixed. Past-life imprints can be healed, and as you do, your plasma field shifts, transforming both your present experiences and future

---

[66] Fridtjof Capra, *The Tao of Physics* (Berkeley: Shambhala, 1975) p.87-88

choices. By consciously working with Plasma, we can transmute karmic imprints, dissolve the illusions of Maya, and return to our pure, unburdened state of being.

Here's something that might bring you both peace and perspective: understanding this can also help you release judgment toward others who are currently creating what seems like very negative karma. We all know someone like that!

First, let's be real…most of us (*and probably most of you reading this book*) have lived through lifetimes filled with negative or heavy karma. That's often why your soul now feels such a strong pull toward goodness, healing, and light! It's a response, a course correction from the weight of the past!

I've always carried a certain sadness or darkness within me. That very feeling propelled me into metaphysics and spirituality, into the deep desire to feel better, to understand myself and reality to the furthest expanses and beyond. I truly believe that many of the bright lights walking this Earth now, as old souls, have known deep darkness in their past. And it's precisely because of that that they radiate such compassion and wisdom today.

We see this pattern in individual lives, too. Many spiritual teachers and healers endured great pain, shame, or mental health struggles in their youth. That suffering became the catalyst, the darkness carved the space for their light. In this way, we can actually thank our negative karma, our past actions, for pushing us toward the light. It's what cracked us open.

Second, those people in your life who seem deeply stuck in darkness, who are clearly working through heavy karma, are often catalysts for our own transformation. Personally, I've learned more about myself from those who betrayed me than from those who supported me. My hardest moments with others have often come just before major breakthroughs. In a strange way, they've inspired me to be a better person.

It helps to remember that some of these souls may be newer to Earth. Their role might simply be to activate us, to nudge the lightworkers (*I like the term rainbow workers or mist workers better, since light and dark still implies duality*) forward when we're tempted to quit. God knows it is hard to motivate me some days!

The best revenge isn't revenge at all. It's being your fullest, freest self. We don't need to teach them a lesson. We just need to let go, rise, and focus on those who truly deserve our love and energy.

Autobiography of A Yogi, Karma, & Desire

After reading *Wisdom of a Yogi: Lessons for Modern Seekers from Autobiography of a Yogi* by Rizwan Virk, I followed his advice and read *Autobiography of a Yogi*. What a gem. One chapter in particular, The Resurrection of Sri Yukteswar, was recommended to me by my friend Hayden, and it left a profound impression.

In this chapter, the author, Paramahansa Yogananda, describes an extraordinary experience about his guru, Sri Yukteswar, appearing to him in full form after death, spirit embodied in flesh and blood. Yukteswar shares that he has been assigned to Hiranyaloka, the "Illumined Astral Planet," where he's helping advanced beings release their remaining astral karma to attain full liberation.

He explains that Hiranyaloka exists beyond the physical (*3D*) plane, within the astral realm, and that above this is the causal realm, a higher dimension where the soul becomes freer still. According to Yukteswar, nearly all souls from Earth pass through ordinary astral spheres after death, but Hiranyaloka is a special domain for those nearing full release.

Here, these beings dissolve many of the karmic seeds tied to past actions carried through the astral realms. Once enough of this is cleared, they are reborn on Hiranyaloka, a realm Sri Yukteswar compares to the *astral sun* or heaven itself. As a spirit, he resides there now to assist these nearly liberated souls.

He speaks about nearly perfect beings on Hiranyaloka who have come from the superior causal world. Both the astral and causal dimensions, he explains, are not singular places but contain many territories and planets of varying vibration and function.

Sri Yukteswar communicates all of this to Yogananda via what he calls *idea-tabloids*, compact, telepathic blocks of thought. I believe these are the same as what many of us refer to as "downloads", those bursts of insight or grand ideas that arrive whole, as if gifted by a higher source.

It takes time to sift through these downloads and anchor them into the 3D, just like this book. I had to pick out many personal messages and go through as many as my filters as possible in the "4D" realm of emotions and thoughts, to attempt to deliver you something that felt near the ground of a universal truth. Something that you could apply any of your truths to and it would hold ground. Of course, my own perspective and beliefs will naturally filter through, that's part of being human. This is why it is important to be discerning and take only what resonates with you, leave what doesn't; as with all outside influences, *trust your inner compass first.*

As I read, it became clear to me that the causal realm, including Hiranyaloka, aligned with the fifth dimension (*5D*), while the astral realm reflected the fourth (*4D*). And then I read something that solidified this for me greatly. Yogananda's guru explains how God encases the human soul in three bodies: the idea or causal body, the subtle astral body as the seat of our mental and emotional nature, and the gross physical body. (*His words, not mine.*)

The guru goes on to say that astral beings work with their consciousness and feelings through a body made of *lifetrons*. This term is deeply significant when it comes to understanding Plasma. The author notes that the guru originally used the word *prana*, and that he himself translated it into *lifetrons*. He also explains that Hindu scriptures reference not only the *anu* (*atom*) and *paramanu* (*beyond the atom*), which point to finer electronic energies, but also *prana*, the creative lifetronic force.

To me, it's crystal clear that lifetrons represent the plasma-consciousness synergy I speak of throughout this book, an active interplay between living plasma and conscious energy. Prana also aligns with this idea. And when we look at paramanu, I believe it points directly to the same foundational plasma I've been describing, something beyond the atom, subtle and powerful, which science may someday recognize when we develop the tools to study this finer state.

Interestingly, modern science has already identified Quark-Gluon Plasma (*QGP*) as the most fundamental, high-energy state of matter that exists beyond the atom. What I'm intuitively referring to here goes even further, beyond QGP. It's a metaphysical-psychological-holographic form of plasma that serves as the scaffolding of existence itself, explaining how our emotions color our reality experience, just as *paramanu* once did in ancient texts. They simply didn't have the language or instrumentation we're developing now.

The guru describes the causal realm as a blissful realm of ideas, though you're still bound to your causal body. In the astral realm, beings travel between planets using masses of light, faster than electricity or radioactive energy, hinting at forces beyond current science, perhaps even antiphotons. To me, all three bodies are made of plasma, each holding our consciousness in varying densities, becoming increasingly subtle from the physical to the astral to the causal.

Once all three bodies are shed, the guru speaks of a spiritual state called the "Bird of Paradise", a return to heaven, or perhaps to all of consciousness and light, symbolizing freedom, inner peace, and self-realization. Yet, he also says we retain our individuality even when merging back with the Source. I believe this too. He describes astral luminaries as resembling the aurora borealis, the sunny astral aurora being more dazzling than the mild-rayed moon-aurora. One, the aurora is literally scientific plasma. And two, when I've seen what I call "plasma beings" in deep meditation with my eyes open they appeared sentient, as shimmering forms of holographic light, just like the aurora borealis!

The guru describes the astral world as infinitely beautiful…clean, pure, and perfectly ordered. There are no dead planets or barren lands, only opal lakes, luminous seas, and rainbow-colored rivers. Mermaids, fairies, fish, goblins, gnomes, demigods, and spirits all inhabit different astral planets based on what he calls karmic qualification.

This could explain the darker energies one might encounter in this realm. According to some teachings, various "spheric mansions" or vibratory regions exist for both good and "evil" spirits. Good spirits can travel freely, while those with heavier, denser vibrations are confined to limited zones, which aligns with how I perceive the 4D plasma field: a realm where thoughts, emotions, and archetypal patterns exist as plasma-encased energy. These are the psychic imprints that drive us to act, repeat patterns, or play out karma.

In this realm, all spirits, whether perceived as good or evil, could simply be information or the unconscious wrapped in plasma. It could possibly be a form of intelligence or memory awaiting release or integration. Maybe they aren't truly "good" or "evil" at all, but bits of unresolved karma, fear, or intense emotion.

In this sense, fear in 4D is dense plasma encoded with past actions and vibrations. It holds the imprint of the experience until it is transmuted, integrated, or learned from. When we interact with these plasma beings, we're encountering our own unresolved information, projected outward to be seen, understood, and released. With each interaction, our karma changes.

The more we recognize and integrate these plasma imprints, the more we clear the field, allowing us to act from the present moment rather than from past patterns. One important fact is that we don't have to heal everything right away to live a good life. In fact, the belief that we must be fully healed to be happy or successful is itself a plasma imprint of fear and lack. If we're always striving to fix ourselves, we remain stuck in the illusion that something is inherently wrong with us. But what if nothing is wrong? What if the very act of being fully present in each moment…feeling, experiencing, and allowing is the healing itself?

Authenticity is the gateway to real-time healing. When you're fully present, you're not projecting past imprints into the now, nor trying to control the future. You're simply allowing plasma to flow, releasing what no longer resonates and embracing what is. Healing is integrating into the present so you can live from your true essence, not from the imprints of who you were. In doing so, the "karma" of 4D plasma begins to naturally transmute, without you needing to "fix" anything.

But while presence and authenticity can dissolve old imprints, there are still layers of trauma that act as hidden blockages, subtly distorting perception and response. This is why healing *is* important over time. Processing deeper imprints liberates trapped energy, allowing us to experience life more clearly and fully. Unresolved trauma creates unconscious filters that block love, connection, or abundance. For me, I couldn't access true presence until I began healing my traumas. Even though I'm still healing, presence and feeling are now conscious choices, and that's what I'm working on.

When we see things like ghosts in our reality, I believe they're using a buildup of plasma to carry their consciousness. This may explain why a room suddenly grows cold or experiences electrical interference. It's another reason why energetic protection and awareness of your thoughts and emotions are important. Just like on Earth in 3D, you have a say in what you vibrate with and allow into your space…so fear not!

Bringing this back to plasma and science, the guru explains that on Earth, a solid must be transformed into a liquid or gas through natural or chemical processes. But in the astral realm, solids are changed into astral liquids, gases, or energy by the will of the inhabitants. That alone tells us how much more malleable plasma, and reality, becomes at higher dimensions.

"The power of unfulfilled desires is the root of one's freedom." — Dana Kippel

The guru says all three bodies are held together by the force of desire, that unfulfilled desire is the root of all man's slavery. Perhaps this was filtered through the writer's beliefs. Maybe my own aren't entirely correct either… but I deeply feel that desire, especially the wild, untamed desire of the divine feminine, of Earth itself, is an essential part of the human experience. Rather than being something to suppress, desire can be harnessed. Not through addiction, overuse, or distortion, but through alignment. I believe desire is what leads us to create, evolve, and expand.

There is another side to desire, one many women, and those who identify as women, know intimately. Instead of being possessive or domineering, it's a soft, nurturing force. It gives back endlessly, weaving into reality rather than penetrating through it. It is cool, patient, and airy, not hot or purely sexual. Beyond even this is the childlike desire of innocence, of curiosity, of creating without questioning. A desire with no outer agenda, only an inner sense of wonder and fun. We are not slaves to desire; desire is where a hidden gem is held, waiting to be uncovered. Desire is expansive, just like dark energy…and when we use it consciously, it becomes our emergent energy—the basis of the new human potential and our magic.

I would link this to 4D Plasma, the realm where our desires are held, encased in emotion. Connecting with our emotions and healing them unwraps these desires, bringing them to the forefront. Being able to focus on our curiosities versus the dramas of life is a very freeing thing and takes practice. This layer connects all beings and souls of consciousness, acting as the binding field between the physical, astral, and causal realms. It is the blood of reality, the dark, connective tissue that links the light. Through this teaching, we begin to understand how to traverse these realms and reveal more of what is inside of them—, by engaging our thoughts, emotions, and awareness from a new, more integrated perspective within this liminal, transformative space.

Once we begin to heal this emotional layer, it's as if a veil lifts. The space we once experienced as dense or reactive becomes a bridge, revealing new representations of truth, fresh ideas, and the dazzling mystery. It's not just that emotions are cleared, it's that the space inside of them *was always holding more*, waiting to be seen through clearer eyes.

For so long, we've misunderstood and demonized this space, myself included. I spent years running from it, convinced it was black magic or evil, a realm to be feared and avoided. But now I see a deeper truth: 4D plasma is not inherently dark, it is the realm of unprocessed potential, the space where unresolved emotions, fears, and desires linger, waiting to be acknowledged, felt, and transformed.

If we enter this realm with pure intention, facing our fears, desires, and loves head-on, we can uncover the neon crystal hidden within, the key to unlocking our M-Self, our multidimensional awareness. This has been a collective fear, this dark feminine 4D space that holds both good and evil, fear and freedom. But within this space is the portal to creation, the gateway to other worlds, and the source of fresh insights from the Mystery. We transcend it by going in, by facing what we fear and reclaiming the power that was always ours.

Without unfulfilled desires, we wouldn't have the spark to invent, create art, or progress as a human race. The call of desire is an expansion to grow, create, and evolve! Our task is to learn how to answer its call and channel it. (*Hence, this entire book.*) So rather than resisting desire or feeling incomplete without fulfillment, we can shift to seeing the pursuit itself as a form of freedom, the freedom to explore what we're capable of creating, feeling, and becoming.

Wanting material things on Earth is natural and part of the human experience, though they shouldn't be the only goal or driving force. Most can agree that leads to an empty, soulless existence. But if you pursue material dreams while doing what you love, knowing there's so much more to life, I believe that's perfectly okay! It's also okay not to want material things. To each their own.

Desire on Earth holds just as much beauty and grace as in higher realms. Desire pushes us to do things that we otherwise would not do, and action is how we learn. We may as well enjoy this journey we call life, and when we reach the astral or causal realms, we'll enjoy those too. But there is a way to bring those 4D and 5D perspectives into this body in the 3D and that is what I am trying to help activate. To integrate it all into the human experience for added depth, connection, love, and magic.

In this 5D perspective I speak of, we are invited to play. It is where the rules of reality bend and the playground of creation opens. It is a state of being, rather than a place to "ascend" to., and we can call this level into our lives, now.

Throughout *Autobiography of a Yogi*, Yogananda shares profound truths, bridging the physical, astral, and causal realms that can truly awaken any reader to the fact that reality is not what we think. However, it's important to remember that Yogananda's lens was shaped by a cultural framework that often viewed desire as a binding force, a source of suffering, and something to transcend. This perspective served its purpose in a time when spiritual teachings emphasized renunciation and detachment as the ultimate paths to enlightenment.

But as we evolve, we are being called to reclaim our relationship to desire, not as something shameful or inherently binding, but as a creative force that fuels our emergence, play, and potential. Instead of seeing desire as a trap, we can see it as fresh energy inviting us to engage, create, and live fully.

"It is desire that moves the world, and by its force men live and die."

— Dante Alighieri

In this way, we can honor the truths within Yogananda's teachings while also expanding upon them, integrating the feminine principle of *Lila*, divine play, and embracing the dance of desire as the very essence of life itself. After all, as Dante said, "It is desire that moves the world," and when we learn to move with it rather than against it, we reclaim our freedom and step into our full creative power. It is 5D plasma, the force that carries us through the continuous cycles of life, death, and rebirth.

Gods, Cosmic Fire, and the Fourth Dimension – Mesopotamian, Egyptian, and Middle Eastern Mysticism

Inanna, The Descent, Beasts, and Mystery (4000 BCE)

Inanna/Ishtar is by far the most complex of all Mesopotamian deities, displaying contradictory, even paradoxical traits. [67] I could write a whole book on the comparisons between the Anunnaki goddess Inanna and Plasma. Inanna was worshiped by the Sumerians and was the cause of the fall of the first city in Mesopotamia, Eridu. The Sumerians were the people of southern Mesopotamia whose civilization flourished between 4100-1750 BCE. The city Eridu, was ruled by their god of wisdom and water, Enki, who raised it from the watery marshes and established the concept of kingship and order in the land. [68]To me, this is a metaphor for the wisdom of consciousness impregnating the feminine sacred waters of Plasma to create matter , here on Earth and, more broadly, in the universe, the first city.

The establishment of Eridu by Enki was considered a golden age, often compared to the biblical Garden of Eden (*a myth that emerged much later*). The fall of Eridu is tied to the story of Inanna and the God of Wisdom. Inanna traveled to the home of her father, Enki, and invited him to drink with her. As Enki became increasingly intoxicated, he handed over what the Sumerians called the *me*, the foundational principles that governed life, the very laws and blueprints that structured reality and defined the relationship between humanity and the gods.

---

[67] Inana/Ištar," *Ancient Mesopotamian Gods and Goddesses*, University of Pennsylvania Museum of Archaeology and Anthropology, accessed August 14, 2025, https://oracc.museum.upenn.edu/amgg/listofdeities/inanaitar/.

[68] Joshua J. Mark, "Sumerians," *World History Encyclopedia*, last modified March 15, 2014, https://www.worldhistory.org/Sumerians/#.

This moment could be seen as a precursor to the Eden story, establishing a warning when sacred knowledge is exchanged or taken, leading to a rumored fall from divine order. I believe this marks the beginning of fear-based propaganda: a shift where powerful forces reframed natural human birthrights, like knowledge, desire, and creative power, as dangerous or sinful. By doing so, they created control structures that kept humanity from accessing its innate divinity and potential. To be the devil's advocate of myself, it also could be true that we had to be separated from it, to be able to come back to it with an evolved consciousness to use it how it's meant to be used.

The *me* can be seen as both the laws of creation and cognitive technologies, potentially granting access to what I call plasma-consciousness synergy—a set of lost teachings or tools that bridged divine knowledge and human experience. I find this symbolizes Inanna, as Plasma, holding onto the *me* all this time, safeguarding them as a link between the Mystery and human consciousness. Now that our consciousness has evolved, it's our time to remember these principles, to reclaim our connection with Plasma, and to use them to rebuild our world and elevate the human experience. It's a choice and I pray we all begin to see that the world's future is really in our hands.

After gathering all the *me*, Inanna fled to her ship and carried them to her city of Uruk, effectively transferring power and prominence from Eridu. This act not only diminished Eridu's influence but also marked a shift in cosmic order —as the divine principles of creation were now under Inanna's control, suggesting a shift from the old paradigm of divine order (*Enki/Eridu*) to a new, emergent, feminine force (*Inanna/Uruk*).

There is a powerful Sumerian myth called *The Descent of Inanna*, in which Inanna, Queen of Heaven, descends into the underworld to visit her dark sister Ereshkigal, often interpreted as her shadow self or the unconscious. Inanna leaves her throne in the hands of her husband, Dumuzi, and journeys below to deliver the news that the Bull of Heaven has died.

Before entering, she dresses in her finest clothing, her seven sacred *me*, each representing a power or aspect of her being. These can also be interpreted as symbolic of the seven chakras. As she passes through the seven gates of the underworld, she is required to surrender one *me* at each gate, stripping away layers of power and identity until she stands completely naked and vulnerable.

Upon reaching her dark sister, Ereshkigal, who is consumed with rage and grief, Inanna is deceived and hung on a meat hook to die. This is one of the oldest and most potent stories of shadow work, a powerful metaphor for confronting the darkest parts of ourselves, facing our deepest fears, and surrendering everything we think we are.

After three days and nights, her loyal consort Ninshubur seeks help. While many deny assistance, Enki, whom Inanna had once betrayed, chooses forgiveness. From the dirt beneath his fingernails, he creates two sexless beings and sends them to the underworld with the food of life and the water of life to resurrect her.

When the beings arrive, they find Ereshkigal in the depths of her suffering, crying out in pain. Instead of trying to fix or escape it, they simply witness her, mirroring her grief with deep compassion. Moved by their presence, Ereshkigal offers them gifts, but they refuse, asking only for Inanna's release. (*A powerful reminder not to get distracted by the many shiny things that appear when you're facing yourself—they will come, but the true gift is reclaiming your own essence.*) Inanna is resurrected, regaining her life and power, and ascends back to the world above.

This myth is a profound teaching on integrating the shadow, reclaiming lost parts of the self, and the alchemy of descent and resurrection. Rather than avoiding the underworld or fearing our own darkness, it teaches us that the true path to wholeness and integration involves facing our shadow, feeling our deepest pain, and reclaiming the parts of ourselves inside of this that we left behind.

Before leaving, Ereshkigal declared that Inanna must provide a sacrifice, someone to take her place in the underworld. The demons who accompanied Inanna back to the surface demanded her loyal consort Ninshubur, but Inanna refused. When she returned home, the demons still in tow, they searched for a suitable sacrifice. They found Dumuzi, her husband, seated arrogantly on the throne, adorned in the *me*, shining with power. In some versions, it's implied that while Inanna was undergoing her descent and transformation, Dumuzi was cheating on her with her friends, betraying her trust.

This is a deeply archetypal wound that many women face, especially when they begin to focus on themselves, do inner work, or step into their power. This wrath of betrayal can be an intense, fierce, and raw feminine rage that demands reckoning. Inanna, in her fury, screamed for them to take Dumuzi away to the underworld, sending him to the very place she had just escaped. Over time, Inanna was able to forgive Dumuzi, but she did not take him back. A deal was struck: Dumuzi would spend half the year in the underworld and his sister the other half. To me this showed how strong the story of a female sacrificing themselves for a male was. Not because they were asked, but because they chose to do so. *Why?*

This story is a powerful representation of Plasma's fourth-dimensional aspect as a force of transformation, a realm where inner trauma can be alchemized into inner power. Also about how stories of darkness hopefully remind us of the powers we hold, that we are not looking at, or are hidden inside of fear. It's about shedding old belief systems and people who no longer resonate, descending into the depths of perceived loneliness and hell, and coming back to the surface changed, lighter, but still ourselves.

It's the journey of feeling pain and rising again, not only to set new boundaries with yourself and others but to embrace forgiveness and grace, recognizing that even those who hurt us are on their own paths. Sometimes, we get caught in the crossfire of their karma, becoming collateral damage in someone else's learning curve. But that doesn't determine our worth or our future. We learn from every descent. We create new beliefs, transmuting pain into power. And just like the eight-pointed Star of Inanna, we become the wheel of life, continuously transforming and sovereign in our choices.

This was a story of wrath, pain, betrayal, forgiveness, and courage. The story of all women on Earth. This story awoke something in me. Something deep inside me knew that Inanna not only represented every woman but was the secret representation of the paradoxical, animalistic, untamed, and misunderstood Plas-ma.

Now, I'm going to share some powerful comparisons that are hard to ignore, ones that support my intuitive point. Inanna, goddess of love, war, and fertility, mirrors the energy of Kali: both embody creation and destruction. Inanna holds paradoxes, grace and rage, tenderness and power, and makes it okay for every woman to feel the full range of her emotions. Your feelings are not too much, they are your power, and when you allow yourself to feel them, you start to access that power.

The next step is learning to guide and harmonize with that power, your inner plasma. You also must offer yourself grace. No matter what mistakes you've, you can take accountability, make changes, and still stand in your power, knowing you are human and worthy of love.

Inanna spelled backward is *Annani*. For those attuned to the power of language, words often hold just as much meaning in reverse. They are multidimensional, layered with symbolism that can reveal hidden truths. In Hebrew, *Anani* means "my cloud" or "Cloud of Yah", the Cloud of the Lord or God. This connects directly to Plasma, often described as a cloud-like, misty substance that holds and transmits consciousness. If you recall from the science section, we explored the Kordylewski Clouds and the novel *The Black Cloud*, both of which depict plasma as a cosmic cloud that stores intelligence.

Anani also translates to "Cloud of Obscurity," where *obscurity* means the state of being hidden, unknown, or veiled. It was also used to describe a refuge or lair for animals (*beasts!*) and even the habitation of the Creator. This feels like a potent metaphor for plasma: the cloud that holds consciousness, a refuge for the Mystery, and the hidden realm where potential rests, waiting to be remembered. I mean, this is fascinating!

This sparks a thought about the Harry Potter prequel film, *Fantastic Beasts and Where to Find Them*. What if the beasts are symbolic of these hidden plasma beings and our own hidden intelligence and information, obscured in the fourth dimension? Just as protagonist Newt Scamander searches for and protects these creatures, we too are called to retrieve our lost aspects, acknowledge our "shadow selves", and integrate them.

Some of Inanna's key symbols include hook-shaped knots of reeds, lions, rosettes, doves, and the eight-pointed star. Interestingly, Inanna's knot links directly to knot theory in topology and mathematics. In knot theory, a knot tangled in three dimensions can only be untied by moving it through a fourth dimension. Sound familiar? Many traditions view the fourth dimension (*4D*) as the shadow realm, a space where we untangle the knots of our psyche, release old patterns, and reclaim our power.

The Descent of Inanna can be seen as her untying her inner knot, shedding layers of identity and attachment to reach her essence. This untangling creates the shape of a torus or sphere, symbolizing the cycle of death, transformation, and rebirth. Her eight-pointed star represents wisdom, knowledge, and awakening of the inner self. The number eight also signifies the intelligent order underlying manifest reality, the structure of the cosmos that Plasma holds together. The star is also linked to Venus, the Morning Star, just as Jesus, The Son (*or Sun*) of God, is called the Morning Star in the Bible. And the sun, of course, is made of plasma, the divine clay of God.

Mary Magdalene is also linked to the Morning Star, and similar to Inanna's symbol of the rosette, Mary is often depicted with a rose, a symbol tied to the Flower of Life and the plasma field, representing the interconnected web of creation.

The following ancient image, often linked to Inanna, Ishtar, or Lilith, is rich with symbolism. In her hands, she holds what resembles a sunrise or sunset, or what I believe to be the liminal power of the horizon. This in-between space perfectly captures her role as a wielder of Plasma: the medium that connects all things.

Notice the lion at her feet. Traditionally associated with kingship, power, and consciousness, the lion represents the regal force of the mind ruling the psyche. It also symbolizes the heart, courage, and solar energy, aligning with the sun (*or son*) motif and reinforcing the connection to plasma as the solar fire of creation.

Inanna/Ishtar | Queen of the Night, Mesopotamian Terracotta Plaque, Old Babylon, 1800-1750 BCE. British Museum

Now, look closely at the owl heads. Their shape resembles a torus cut in half, another symbol of the endless loop of creation and destruction. Owls are also symbols of wisdom and hidden knowledge, fitting seamlessly with the theme of descending into the underworld to gain insight and return reborn.

Upon deeper examination, the left side (*her right*) appears to emphasize the feminine, lunar, and hidden aspects:

- A pregnant lion, suggesting fertility, gestation, or creation in the unseen realms.
- An owl with an inverted nose, perhaps indicating the inward journey, introspection, or the underworld.
- A subtle, less pronounced horizon, implying the hidden, inner landscape.

On the right side (*her left*), the energy is more masculine, solar, and external:

- A non-pregnant lion, emphasizing action, assertion, or outward power.

256

- An owl with a more defined, protruding nose, symbolizing the outward gaze or the awakening of higher vision.
- A distinct, visible horizon, representing the material world, physical reality, and waking consciousness.

This duality mirrors the descent and resurrection motif, where the feminine (*hidden, internal*) and the masculine (*visible, external*) must both be integrated for wholeness.

Later depictions of Inanna in her descent, resurrection, and transmutation take form in Ishtar, Persephone, and even Jesus. Each one undergoes a death and rebirth, symbolizing the journey through 4D plasma, the space where one's psyche is untangled, purified, and transformed.

One more note on the Sumerians. As seen in the following image, I believe they harnessed plasma as a form of antigravity, to lift objects, materialize thought into form at hyper speed, and more. The figure on the throne appears to be channeling the horizon, potentially representing 4D Plasma, and the wheel beside him resembles both the sun and the eight-pointed star of Inanna. It even looks freakishly similar to a modern-day Plasma globe.

Tablet of Shamash, Babylonian Sun God, c. 888-855 BCE.

In short, when I began studying Plasma and sensed it was more than science described, discovering the Sumerians and Inanna was the moment it all viscerally clicked. My knowing was on to something. I had never even heard of Plasma until it practically announced itself to me during a meditation. When this happened, I knew three things

257

instantly: Plasma is intelligent. It exists both within and around us. And we can harmonize with it and communicate using it.

Amun Ra: Bioplasma, Smelling the Sun & Sacred Geometry (2600 BCE)

Amun Ra was a fusion of the Egyptian gods Amun and Ra, as the union of air and sun, the unseen and the visible. Amun, known as the hidden one, represents the formless essence of consciousness, while Ra, the radiant sun, embodies the visible Plasma of creation. At a non-dual level, they are thought to be one and the same, much like how I believe that at the highest level within the Mystery, Plasma and Consciousness are two expressions of the same source.

The blue skin of Amun Ra draws a striking parallel to the Na'vi in James Cameron's Avatar, beings deeply connected to nature and harmonized with their environment, much like beings who work with plasma-consciousness energies. Could the Na'vi be a modern allegory for those who have mastered the plasma-consciousness connection, living in alignment with the unseen currents of life?

The second compelling connection is Amun Ra's ram horns, which strikingly resemble the hippocampus in the brain, which is a region crucial for memory, navigation, and spatial awareness. The hippocampus plays a vital role in how we encode experiences and sense our surroundings, suggesting a subconscious link from this myth and symbolism to how we perceive reality with our personal plasma-consciousness synergy and awareness.

The hippocampus is intricately connected to the olfactory system, the sensory system responsible for smell. What's fascinating about this is that the olfactory system is the only sense that bypasses the brain's filters, delivering information directly to the hippocampus and limbic system, the seat of emotion and memory. This is all happening within the greater nervous system of the body which also affects our endocrine system. This is why a single scent can transport you instantly back to a specific memory or emotional state. That can then activate our endocrine system, as smelling a baby releases oxytocin, for example.

Unlike other senses that are processed through several neural pathways (*or what I call belief filters*), smell is unfiltered, pure, and direct. It's almost like it bypasses the conscious mind and taps straight into the subconscious, which may also suggest a connection to plasma's subtle sensing capabilities. More than any other sense, it also blurs the line between perception and emotion.

Olfactory receptors aren't just in the nose. They're found throughout the entire body, including the skin, lungs, heart, and blood, suggesting that sensing reality is a full-body experience, not just a localized one. This aligns with the idea that our plasma body/bubble (*also known as astral body or biofield*) may serve as a sensorial network, continuously receiving and interpreting plasma fields and frequencies.

As quoted in J. A. Kent's *The Goddess and the Shaman*, Cameron suggests that the brain is an elegant transcriber, which boils down information coming from sources far beyond it.[69] This raises the possibility that consciousness may not originate in the brain, but instead filters through the nervous system, perhaps beginning with the olfactory system, and only then arrives at the brain to shape responses. Or maybe it's more of a system, all happening at once, either way, our emotions and the olfactory seem to play a larger part than once thought.

The olfactory system is the oldest evolutionary sense and tied to survival (*e.g., detecting food, danger, mates*). It evolved before complex sight or sound, making it deeply embedded in the subconscious. The olfactory system is quite literally tied to our emotional filter, and it has been programmed naturally, for survival. It sheds a light on why rewiring our relationship with fear and our own emotions may be a key factor in evolving how we perceive and move through reality.

Come to think of it, if we all began as sensing balls of plasma, then maybe sensing, before sight, sound, or language, was our first way of navigating the world. The other senses may have grown out of that original plasma-based perception. This perspective deepens our understanding of ancestry and highlights why it's so important to look to the past for clues about who we really are.

Picture your consciousness like tiny silica filaments lining the outer surface of your plasma bubble, constantly sensing the external "plasma" reality. We feel the world first through the body, and only then does it filter into the mind. The way we perceive reality is inseparable from our emotional state and belief systems, whether rooted in trust and safety or shaped by trauma. Anyone who has ever seen the same hillside while in depression and then again in elation knows exactly what I mean.

An incredible parallel exists between the process of ATP energy exchange in our cells and the electrochemical gradients and plasma sheaths seen in space and physics. ATP (*adenosine triphosphate*) powers our biology by moving electrons and ions across membranes, much like how plasma sheaths regulate energy transfer and boundary integrity in space environments. Both rely on charge separation, membrane potential, and feedback

---

[69] Cameron, 1998, quoted in J. A. Kent, *The Goddess and the Shaman: The Art & Science of Magical Healing* (Woodbury, MN: Llewellyn Publications, 2016), 44.

loops to sustain life and flow. Even bacteria use this system, creating proton gradients across their membranes to generate energy, essentially running on electrochemical plasma-like engines. This highlights plasma behaviors across many different scales that have not been called so.

These similarities suggest that our biofield, or what might be called bioplasma, mirrors cosmic plasma behavior as well as biological processes only not called plasma behaviors do to the compartmentalization of fields. Physics, biology, and neuroscience often operate in isolated silos, each with its own language and lens. Despite the fact that many cellular processes involve ion flows, charge differentials, and electromagnetic signaling, these parallels are often overlooked because they fall between disciplines, It's possible we're only beginning to glimpse a new scientific field, one that explores the 4D-like plasma sheaths of our own bodies that hold memory and emotion as well as consciousness beyond the brain, venturing into our own bioplasmic membrane.

If this intrigues you, there's a fascinating YouTube playlist called "The 4th State of Life" on the channel **Versadoco** that explores the emerging science around bioplasma and the idea that life operates with a plasma-like state beyond traditional biology.[70] These videos trace early hypotheses from the 1940s that proposed Plasma as a fifth state of matter within the human body, and examine how concepts like ultra-weak emissions, biophotons, and superconducting structures (*like Josephson Junctions*) may hint at previously hidden energy systems in the body. The videos mix cutting-edge physics, quantum computing, and regenerative biology, spotlighting works from pioneers like Dr. Michael Levin, who suggest that biological software—networks of cells sharing geometric and energetic information, might be the key to regeneration, healing, and a deeper understanding of consciousness itself.

Amun-Ra symbolized a shift in how the Egyptians associated the divine from worshipping many gods who were distinct, separate forces to a supreme, all-encompassing god, embodying *both* the physical and spiritual realms. Amun-Ra was also considered a protector and guide to pharaohs and common people alike, emphasizing a more personal, accessible aspect of the divine.

This shift reflected a more intimate connection to the divine, suggesting that one could connect with both the visible and invisible aspects of reality, much like how I describe accessing plasma with our consciousness as a bridge to higher dimensions. Amun-Ra also symbolized the influence of a belief among the Egyptians of a <u>benevolent</u> and universal creator.

---

[70] https://www.youtube.com/watch?v=K3UHfO4Ie-8.

Scarab beetles were powerful symbols in ancient Egypt, representing rebirth, creation, and divine guidance. In many hieroglyphics, they are depicted sensing or "smelling" the sun, just as some Egyptian gods do. Their "sensing" may quite literally be smelling, as beetles, like ants, use their antennae to detect chemical signals in the environment, which function as a highly evolved olfactory system. Maybe, just maybe, were they sensing plasma?

The wild synchronicity here is that in 2021, before I knew any of this, I made an indie film called Reflect. In one scene, I'm telling my friend that I'm "smelling the sun", something that intuitively came to me during the earliest moments of connecting with Plasma. At the time, I had no understanding of the mythology, metaphysics, or science behind it.

Now, looking back, I wonder, were the Egyptians consciously aware of these connections? Or were they expressing truths through art and symbols, unconsciously channeling messages for a future time when our awareness could decode their deeper meaning…like now?

There's much more to explore here scientifically, something I'll dive into more deeply in future books, but here's a preview. Consider the Sphinx: a guardian of mysteries, a hybrid of human and beast, standing watch at the threshold between worlds. With its nose missing, a symbolic question arose in my mind, is it hinting at something deeper than the surface of smelling, like sensing?

Even though the Sphinx's nose loss is a mystery, there are accounts that say it was an act of vandalism due to a Sufi Muslim unhappy with the fact that people were worshipping idols. Although the Sphinx missing its nose does not seem to be intentional, it may offer a larger synchronicity at play. Many noses are missing from statues, especially in Egypt. Across history, it was a ritual act of severing a statue's spiritual interface. The nose was linked to life and breath, at least that's what we thought. But maybe it was linked to something more…

This brings us directly back to the olfactory system, our most ancient, primal sense. It bypasses the neocortex entirely, going straight to the limbic system where emotion and memory reside. In that way, it's a direct channel to intuitive knowing. The Sphinx's missing nose is a metaphor for the severed connection to our innate plasma sensing, a once-accessible form of intelligence that ancient civilizations may have used to perceive information fields beyond logic and language.

The molecules we smell are primarily aromatic compounds, known as benzene rings, which are stable, *hexagonal* structures that resonate at frequencies in the *terahertz range*. Terahertz waves occupy a liminal space between infrared and microwaves, a bandwidth known to affect biological systems and quantum interactions. These resonant frequencies may interact with microtubules in the brain, the very structures that Stuart Hameroff and Roger Penrose, in their Orch-OR Theory, propose as quantum information processors, potentially capable of interfacing with non-local fields, which I interpret as the Plasma field. In this light, benzene ring resonance may serve as a vibrational bridge, translating aromatic molecular information into oscillatory patterns that microtubules can detect and process, effectively allowing us to "smell" or sense plasma fields as emotional or intuitive impressions.

This aligns with the Quantum Smell Theory (*Vibration Theory of Olfaction*), proposed by Luca Turin, which suggests that our sense of smell is not just based on the shape of molecules but also on their quantum vibrations. According to this theory, when odorant molecules bind to olfactory receptors, they vibrate at specific frequencies. These vibrations can facilitate electron tunneling, allowing the brain to detect molecular vibrations as distinct scents. If further research supports this theory, it could challenge materialist views of reality, suggesting that the fundamental basis of perception and perhaps reality itself is not purely chemical or deterministic, but rather vibrational and resonant.

Interestingly, the hexagonal structure mirrors Plasma's tendency to self-organize into geometric patterns, from hexagonal plasma structures observed in Saturn's north pole to the hexagonal formations in molecular and crystalline structures. On a metaphysical level, the hexagon, as we know, is also a symbol of the *shatkona*, representing the union of Shiva (*consciousness*) and Shakti (*plasma*) in Hindu symbolism. Just as the benzene ring resonates with information and stabilizes energy, plasma functions as the living matrix that stabilizes and transmits consciousness.

If you must read this part twice, because it is really cool. The hexagonal shape, then, becomes a bridge, or portal, where plasma-consciousness synergy can occur, aligning with the concept of the *torus*, which is a dynamically folding 2D hexagon in 3D. Essentially, when a hexagon is folded and rotated in 3D space, it can form a toroidal structure. A torus is a self-folding, self-looping system, where the hexagonal geometry can be visualized as the cross-section. The torus creates a continuous flow, forming a loop where the inside and outside are dynamically connected, representing a closed but constantly moving system.

Plasma, as a dynamic, conductive medium, naturally organizes into toroidal structures in space. For example, magnetic fields in plasma often form toroidal shapes, such as those seen in tokamaks and plasma donuts. In the context of plasma-consciousness synergy, the torus can be seen as a feedback loop, where energy, information, and vibration circulate continuously, allowing plasma to act as both a receiver and transmitter of consciousness.

Microtubules in the brain are also structured in hexagonal lattices, forming cylindrical, toroidal-like shapes that potentially act as quantum resonators. In Orch-OR theory, these microtubules are proposed to collapse and oscillate, creating dynamic resonance patterns. If we extend the concept of hexagonal resonance to the torus, then microtubules may act as mini-toroidal circuits, allowing quantum information (*including plasma vibrations*) to circulate, resonate, and influence conscious perception.

Hexagons are found in nature, molecular structures, and plasma formations because they are efficient, stable, and energetically optimal for packing and distributing energy. When these hexagons become dynamic and self-folding, they can transform into these toroidal structures, creating a loop or bridge where information can circulate…a plasma-consciousness gateway!

To clarify how this may work: The intelligent field may oscillate between an open hexagon and a closed torus. Picture it opening and closing, almost breathing, when sensing new information. In the hexagonal state, plasma functions as an open system, a geometric network where information and energy flow freely. This is the state of play, where we are open to new experiences, ideas, and insights, allowing novel inputs (*potentially from our olfactory and nervous system*) to enter and expand the system. It's a mode of exploration, curiosity, and learning, much like a child playing without a set goal, simply absorbing new information.

Once enough information is gathered, the hexagon begins to fold, transforming into a torus, a closed-loop system where the previously collected data is fed back into itself, creating a self-reinforcing feedback loop. In this state, plasma is integrating and synthesizing what it has learned, forming coherent patterns and solidifying new structures of meaning.

The torus represents the state of integration, a kind of inner alchemy where fragmented insights become integrated knowledge.

This is not a fixed system. Once the torus has processed and reinforced the new data, it can unfurl back into the hexagon, opening once more to fresh inputs and experiences. This oscillation between open hexagon (*play*) and closed torus (*integration*) creates a dynamic learning cycle, a perpetual dance between expansion and consolidation. This cycle reflects how plasma itself might behave as a living, intelligent medium, continuously moving between open states of exploration and closed states of integration, evolving as it learns and remembers.

This could explain why many people live on autopilot, caught in a closed-loop system of their own lives, unaware that by mimicking the sacred geometry of the system and opening into a state of play, even small, new choices can invite fresh behaviors, experiences, and insights. This mirrors the process of applying awareness to information, allowing emergent intelligence to arise.

The Sphinx stands as a plasma archetype, a living symbolic portal that invites us to reconnect with our primal, intuitive senses and restore our ability to sense plasma with our awareness as a direct, unfiltered flow of information, much like how the olfactory system senses aromatic rings. This is in our nature.

Even Buckminster Fuller, a visionary thinker who wrote a book *Tetrascroll* on philosophy and sacred geometry which I highly recommend, explored hexagons while building geodesics in architecture that offered the most volume for the least surface area. And he used hexagonal and pentagonal groupings to create these structures emphasizing doing "more with less".

As I explored his work, I uncovered knowledge about the shape of the tetrahedron. I believe this shape is associated with our threefold essence—our personal awareness, consciousness, and plasma, that is always present between the opening and closing process of the hexagon and torus. It represents the M-Self, or the self that remains constant. I believe this is the space where our learnings crystallize during this oscillation. This aligns with Buckminster Fuller's work in *Synergetics*, where he viewed the tetrahedron as the simplest and most stable three-dimensional form, the building block of all other geometries.

A tetrahedron is a bridge between 3D and 4D, as it can be folded and connected to form higher-dimensional structures. It can also expand into a star tetrahedron, a 3D version of the Star of David, symbolizing the union of divine masculine and feminine, and possibly the actualized M-Self. This symbolism is reflected in esoteric systems and secret societies through the use of a triangle inside a circle or hexagon, embodying the integration of mind, body, and spirit, the holy trinity, or the ascent toward higher consciousness. This may also be reflected in symbols like the eye within the triangle or the threefold flame. If the tetrahedron represents the self, perhaps the Star of David represents the flight or emergence of the M-Self, one capable of co-creating reality using Plasma-Consciousness Synergy (*PCS*).

Egyptian Ankh (Key of Life)

Speaking of symbolism, the Ankh in Egypt, also known as the Key of Life, holds profound implications for how we can operate at our highest levels. Almost every Egyptian god is depicted holding it. Also called the *crux ansata*, it's essentially a cross with a loop. I believe the cross represents the body, while the open loop is half of an infinity symbol, symbolizing the immortal self.

When we remain open, like many of you reading this book, we can see this clearly. The Ankh may symbolize not only our ability to let consciousness travel into other realms and planes while still grounded in the body, but also how consciousness navigates plasma. Just as initiates in ancient mystery schools trained to use their consciousness beyond the body, we can do the same. The Ankh reminds us of the balance: the cross as the grounded body, and the loop as the playful, eternal consciousness. It may very well be a key to unlocking the hidden reality we all have access to.

The Burning Bush, Uriel, & Asherah (1800 BCE)

Most of us know the story of Moses and the burning bush. Moses is tending his father's flock in the desert when he sees a burning bush in the distance. The bush is on fire, yet it is not consumed by the flames. God calls to Moses from within the bush and asks him to lead the Israelites out of Egypt.

What's fascinating about this story, especially in the context of plasma, is that the bush burns, but it doesn't burn *up*. It's a type of intelligent fire, an interface through which God's consciousness communicates. It holds or becomes a vessel for communication between realms.

Uriel, the angel often called the *Fire of God*, is associated with divine light, energy, and transformation. He represents the radiant, creative force, much like plasma, which both sustains life on Earth through solar energy and symbolizes divine inspiration, focus, and purposeful action. I came across a beautiful poem about Uriel by Ralph Waldo Emerson that opened something inside me when I read it through the lens of plasma. I'll share the poem below and offer a breakdown of why I find it so powerful in this context:

Uriel

It fell in the ancient periods
Which the brooding soul surveys,
Or ever the wild Time coined itself
Into calendar months and days.

A cosmic being coming into our structured reality…or possibly the severance of consciousness and plasma.

This was the lapse of Uriel,
Which in Paradise befell.
Once, among the Pleiads walking,
Said overheard the young gods talking;

Many people think we are descendants of the star system Pleiades where our ancestors live, where Uriel comes from. The 'young gods' here may symbolize the emerging consciousnesses of humanity that time on Earth.

And the treason, too long pent,
To his ears was evident.
The young deities discussed
Laws of form, and metre just,

The 'young gods' appear to be discussing the structure of reality itself, laws of form and meter. Perhaps Uriel sensed they were missing something essential, something long forgotten, and felt compelled to speak.

266

Orb, quintessence, and sunbeams,
What subsisteth, and what seems.
One, with low tones that decide,
And doubt and reverend use defied,

This is where I believe Uriel begins to speak to them. Words like *quintessence*, *orbs*, and *sunbeams* all point to plasma as subtle, luminous forces of reality. He seems to be revealing that the very fabric of what "subsists" (*what is real*) versus what "seems" (*what is illusion*) is shaped not by loud declarations, but by *low tones that decide*—the subtle, resonant frequencies that form our world. The line *"doubt and reverend use defied"* may suggest that fear (*reverence here meaning awe or fear*) and doubt, when defied or harnessed, can actually be the keys to transformation. Just as I've described with 4D Plasma, these lower states can be transmuted into higher capabilities. Yet perhaps Uriel is also issuing a warning: that we are currently shaping our reality from fear and survival, rather than from the plasma-consciousness synergy that could awaken something greater.

With a look that solved the sphere,
And stirred the devout to adore,
Said, "Composite is the air,
Light the vesture thou must wear;

And, over sun and star and planet,
A newer law, a day befalls,
Which shall not be to-day or to-morrow,
But that which hath been evermore."

In other words, reality is layered, multidimensional, and we are clothed in subtle energies. The message is clear: the way to interact with plasma, and reality itself, is through lightness, not force. We are meant to play, to move with subtle energy, not desperation or stress. The *newer law* he speaks of is not bound to "today" or "tomorrow" but belongs to what has *always been*…the eternal now. This is a call to return to our origin, to the present moment, to our soul, and to the plasma-consciousness field, where linear time dissolves and a deeper intelligence guides us.

The gods were dumb with amazement,
Nor answered they his jest:
The cosmic-laugh of star and sun
Their broad foreheads overrun.

This symbolizes the cosmic joke our world embodies, how something so simple can feel so difficult to live by. The young gods struggle to grasp this truth; they are bound by structure and form. Uriel, like Emerson himself, is ahead of his time, speaking from a deeper understanding that few are ready to receive.

And the heaving brooks of the air
Tumbled to the god's despair.
Amazed they stood, with vacant eyes,
And the fiddle of the spheres was mute,

Even when we're told the truth, it's still hard to grasp, because we're so conditioned to force, to effort, to working too hard. We forget what it means to be in relationship, with life, with creation, rather than trying to control it.

All their puny shine and noises
Lost in Uriel's clearer suit.
Truth unveiled his awful face,
And to the stunned heavens showed

Uriel's purity, symbolizing here a future plasma intelligence, outshines their glittering material symbols and limited beliefs. To them, his truth was terrifying because it shattered their assumptions about reality.

A million ciphers of life's code:
There are days which belong to the soul beyond time,
And their light defiles the day.

A million new keys to decode what life can truly be are available to us. They open the door to timeless moments that belong to the soul as states of flow, glimpses into other realities, other dimensions, and higher ways of perceiving. The light of this wisdom doesn't ruin the day, it simply dissolves the old structure of what we thought reality was. It's about eternal moments felt in the heart rather than linear time. The realization of immortality.

There is so much more available to us beyond what we're currently experiencing…realities that open up once we embody the wisdom Uriel (*the fire of God*), or Plasma, offers here, as channeled through Ralph Waldo Emerson.

In summary, Emerson's poem presents Uriel as a celestial being who unveils a deeper law of existence, one that disrupts conventional understanding. Like Amun-Ra, the hidden one, this law reflects the vibe of plasma-consciousness synergy as subtle, paradoxical, and alive. Emerson seemed to foresee a new paradigm, a more fluid, dynamic era of interconnected intelligence. A realm where power resides in the present moment, and where even gods, consciousnesses, and we ourselves can learn something entirely new.

Was Emerson offering a kind of plasma prophecy? He explicitly names quintessence (*the fifth element*), orbs (*plasmoids*), and sunbeams (*plasma energy*), all pointing to a misunderstood but foundational paradigm of the future. Was he remembering plasma as the forgotten force underlying the universe? A truth so clear and primal that it becomes heretical? One that challenges and ultimately reshapes the foundations of current scientific understanding?

The real question is, has this awakening always been here or is it truly the time of this great awakening? Awakening is a choice *and* a timing. A choice we can make at *any* moment, and a timing that arrives when *enough* of us remember together. Perhaps it's always been possible, and now it's also inevitable. Truth arrives in the subtle, the sometimes gradual. Awakening is not always some thunderous, sensational thing. Our longing for awakening, to "get there" may be the exact thing blocking us from realizing we are already there, if we just become light enough to see…*and then we may recognize our awareness.*

"God commanded the Israelites to cut down Asherah poles." — Exodus 34:13

Speaking of the forgotten…did you know that God may have once had a wife? Ancient texts and legends point to *Asherah*, a goddess worshiped alongside Yahweh by the Canaanites (*a culture predating Judaism and Christianity*), known as the *Queen of Heaven*. She was associated with sacred trees and divine feminine power. Over time, her worship was erased. Asherah poles, wooden symbols of her presence, were deliberately destroyed, part of a wider effort to suppress her memory and the feminine divine.

To me, this suppression mirrors the present-day imbalance in how we understand the universe. We've been taught that *everything is consciousness*, that *mind* rules all. Some go as far as to say that the world is built on and controlled by sex and power, a very patriarchal view. It's not just mind, it's not just will or command. Frankly that would be a bore. Plasma is how consciousness experiences itself. It is the divine feminine aspect of creation. It is what allows a feedback loop for evolution. It is the sensory experience, feeling, and receiving. It is surrendering sometimes to the intelligence of the aether versus just creating your reality without the respect of a co-creator.

With this outlook, we cultivate deeper compassion and emphasize the importance of connecting with the soul and heart, rather than merely managing emotions through the mind. We are not machines, and we are not just consciousness. Asherah is the Tree of Life, she is the fractal nature of plasma, the river of the gods, the connector, the wisdom, the memory-holder. She carries consciousness itself. Without her, without plasma, none of this would be.

Angels & The Merkabah (Biblical & Jewish Mysticism)

Seraphim are celestial beings often described with wings and a fiery presence, also known as "the burning ones." There are many interpretations of what they truly are, but in Kabbalah, they exist in *Beriah* (*the World of Creation*), which is the first realm formed from *Atziluth* (*the world of emanation, closest to Source*), where divine understanding emerges. This strongly echoes my concept of 5D plasma; these beings may symbolize higher conscious plasma entities originating from that plane.

Below *Beriah* are the *Hayot* (*the living creatures*) in *Yetzirah*, which I associate with 4D plasma. The Seraphim at this level reflect their burning devotion through divine emotion. These beings represent archetypes and emotional forces, aligning with my belief that 4D Plasma (*emotions*) are essential for perceiving 5D reality from the 3D lens. The burning process may represent a refinement, where the formless intelligence of 5D takes on more defined, structured form.

This directly mirrors how Plasma itself undergoes charge separation, energy flow, and feedback to self-organize into visible, meaningful structures. The "burning" of emotion in 4D might be the mechanism by which 5D energy condenses into 3D matter. Woah.

Then there is *Assiah*—the 3D world of action. At this level, creation is fully formed and differentiated into material reality. While there's much more depth and historical nuance to all of this, I'm staying on the surface to highlight the parallels. The Seraphim, in this context, could be seen as gatekeepers of transformation, ensuring that what enters the physical realm has first been refined and shaped within the 4D layer.

In metaphysical traditions, the *Merkabah* is seen as a divine light vehicle, a multidimensional structure capable of transporting the soul or consciousness between realms or dimensions. It's often visualized as two interlocking tetrahedrons spinning in opposite directions, forming a star tetrahedron or 3D Star of David. This configuration is said to activate and harmonize the energetic body, enabling access to higher states of consciousness, spiritual awakening, and interdimensional travel.

In other mythologies, the Merkabah is referred to as a chariot of fire or Ezekiel's Wheel. It appears to symbolize our conscious awareness traveling within a plasma bubble, capable of shifting into different geometric forms at will. I believe that in the future, we'll better understand how sacred geometry supports timeline travel. Numerous books explore Merkabah teachings, including how to program your Merkabah field for specific intentions. These often involve heart-opening practices and expanding the mind to accept that multidimensional and multi-reality travel is possible in our human form through the power of consciousness.

Much like the plasma geodesics I often describe, the Merkabah acts as a propulsion system for consciousness within the plasma field, utilizing our personal plasma bubble. Programming this vehicle involves setting clear intentions, aligning our energy, and harmonizing our frequency with a chosen destination; though sometimes, simply allowing the Merkabah to flow freely toward what serves our highest good is just as powerful.

A cool practice to try out is *reverse astral tethering*, where rather than projecting outward, you allow your Merkabah to be magnetically pulled toward a matching resonance point, such as a past life, once a fragment of that memory is accessed. In truth, you are both pulling and being pulled; a beautiful paradox that reveals a deeper truth: you aren't really going anywhere. It's all happening *now*…you're simply shifting perception to view a new vision. Before embarking on this kind of journey, it's essential to protect your energy field and ground yourself afterward to maintain balance and integration.

Depiction of a Merkabah

271

You can travel anywhere: ancient civilizations, past lives, other planets, the Akashic Records, future timelines, higher-dimensional realms, and alternate realities. But everything is fluid and shaped by your beliefs. A potential future timeline you glimpse is not fixed in this reality; it's filtered through your current perspective and can shift with every choice you make resulting in a similar but emergent timeline. That's the gift of these journeys…they let you witness possibilities so you can consciously align your present actions with the reality you truly want to experience. The actual outcome is a mystery you must surrender to while following intuition.

You can visit a past life or earlier moment in your current life to gain insight, heal unresolved patterns, or receive information from higher-dimensional realms. The key is patience, write down what comes through and allow it to settle before interpreting it as truth. Your present-day mind isn't always equipped to immediately process new dimensional data. But if you sense it, feel it, and let it unfold without rushing to explain it, your awareness can expand naturally, without forcing it through outdated logic.

Think of it like steeping tea: at first, the water is clear, and the flavor is barely there, but with time, the tea infuses deeply, becoming rich and full-bodied. In the same way, new insights or experiences often start as subtle impressions. When you resist the urge to define or label them immediately, and instead let them steep in your awareness, they have space to unfold naturally. This slow infusion allows deeper understanding and intuitive clarity to emerge, truths that might have been missed if rushed. The most meaningful experiences I've had weren't logical, they were emotional, transcendent, and often impossible to explain at first. Also, just like tea, they felt very natural, but to the mainstream would seem anything-but. Over time, you start to sense the difference between a pure experience and one shaped by belief, and both are useful if you navigate them with honest awareness.

Hinduism, The Upanishads, & Vimanas (700 BCE)

"As is the mind, so is the vision; as is the vision, so is the world one sees."
— Brihadaranyaka Upanishad

The Upanishads are a collection of ancient Indian philosophical texts composed between 800–200 BCE that form the core spiritual teachings of Hinduism. They explore the nature of reality, the self (*Atman*), and the ultimate cosmic principle (*Brahman*), emphasizing inner wisdom, meditation, and direct experience over ritual.

The Upanishads present the allegory of the chariot as a profound metaphor for the soul's journey and its struggle between the material world and higher knowledge. In this view, the body is the chariot, the intellect is the charioteer, the mind is the reins, the senses are the horses, and the true self, known as the Atman, is the passenger. If the intellect can skillfully guide the mind, and the mind can restrain the senses, then the soul is carried safely toward Brahman, the ultimate reality. If the charioteer is weak and the horses run wild, the soul remains trapped in illusion.

But what if we reimagine this story?

Instead of restraining the horses, as our senses, we learn to build a relationship with them. These horses are our sensory connection to Plasma, and by understanding their nature, we can guide them with confidence rather than force, domination, or suppression as has been done for centuries. The charioteer, our intellect, then shifts from being a controller to becoming a wise navigator.

Rather than restricting desire, we allow it to align with a greater purpose. The journey becomes not just disciplined, but fluid, intuitive, and even fun! The horses are no longer obstacles, they're allies! When we consciously engage with our emotions, senses, and even desires without repressing them, they can carry us toward insight, creation, and deeper harmony with reality. Plasma, like the horses, responds to intention. When we trust this process and move with it instead of trying to control it, plasma flows naturally to support our path.

We are beginning to see how crucial intention, relationship, trust, and letting go truly are. These are the simplest things to do, yet the hardest if we've spent most of our lives in survival mode. And hey, this takes time, so be kind to yourself. I'm still working on letting go and trusting daily.

Humans possess a heart-generated electromagnetic field that extends up to 3 feet (*approximately 1 meter*) from the body in all directions, according to the HeartMath Institute. [71] Horses, on the other hand, have hearts five times larger than ours, and their electromagnetic fields are believed to extend even farther, making them incredibly sensitive and attuned beings. Because Plasma responds to and is shaped *directly* by electromagnetic fields, this suggests that it is the heart, more than the mind, that forms our deepest connection with Plasma. *Horses*, in particular, seem to embody this resonance. Throughout history, there have been whispers of their telepathic abilities, like in the case of the famed horse Clever Hans, who demonstrated astonishing levels of intuitive awareness.

---

[71] HeartMath Institute. "Energetic Communication." *Science of the Heart*. Accessed July 15, 2025. https://www.heartmath.org/research/science-of-the-heart/energetic-communication/.

Riding a horse requires trust, a surrender to the flow, a respect for its power. Similarly, we must trust that Plasma will take us where we need to go. It is one with us. Enjoy the ride. Have reverence for it. With our upgraded way of sensing, we begin to feel our way into new futures…calmly, whimsically, but without force.

Furthermore, to connect with a horse, we must understand its nature, energy, and rhythms. To connect with Plasma, we must also understand its flow, charge, and synergy with consciousness. This is the essence of what much of this book is about. We are learning to understand the "beast" that is Plasma as best we can. And you will develop your own unique way of doing this too. Don't take my word for it, form your own experiments! Endless possibilities await you.

In Hindu tradition, a *Vimana* is described as a mythological flying chariot or palace in the sky, often associated with divine beings. In light of the recent resurgence of orb sightings, commonly referred to as plasmoids, some have begun to draw parallels between Vimanas and plasma phenomena. These orbs, widely documented especially in places like New Jersey in late 2024 or accounts of them visiting the Bledsoe Family, have appeared as glowing spheres or holographic shimmerings, sparking the question: could they be carrying evolved beings of consciousness?

I believe these Vimanas, or plasmoid structures, may act as looking glasses or portals, used by other consciousnesses to either observe us remotely or initiate contact. Scientific research into pinch plasmoids, which are plasma formations created through extreme electromagnetic compression, suggests that they might briefly behave like unclosed wormholes, warping spacetime via intense magnetic and gravitational effects. In this framework, the z-pinch event becomes a momentary gateway, akin to a dimensional portal.

This perspective is supported by Russian research suggesting that pinch discharges may serve as sources of energy release due to gravitational effects, potentially facilitating non-local interactions across spacetime.[72] If Vimanas or certain UAPs are indeed composed of plasmoid structures, it is not out of the realm to conlude they might function as conduits for consciousness, serving as temporary nodes through which otherworldly entities or remote viewers could observe *or* communicate with our reality.

Think about it—if you were an evolved consciousness, wouldn't you prefer to remain safe on your home planet while simultaneously observing others? I explore this concept

---

[72] Andrey N. Serebryakov, "Pinch Discharges as Possible Sources of Energy Release Due to Gravitation," *Bulletin of the Russian Academy of Sciences: Physics* 81, no. 12 (2017): 1351-1354, https://link.springer.com/article/10.1007/s11182-017-0956-3.

further toward the end of the chapter, where I go deeper into UAPs, plasma, and the role of consciousness in these phenomena.

Depiction of a Hindu Vimana

This closely parallels a phenomenon known as *bilocation*, where consciousness, or even physical presence, appears in two or more locations at once. It's often described as the ability to project your awareness to a distant place while remaining present in your original body.

Bilocation, once thought to be reserved for mystics, is something any human can learn, just like remote viewing. Though both sound fantastical, they are real metaphysical abilities, taught in grounded, accessible ways at places like the *Monroe Institute* in Faber, VA. And what makes all of this possible? Plasma. As underlying field of reality, it allows your individual consciousness to access non-local points through its holographic and nonlinear structure.

These Vimanas might be the Merkabahs of evolved beings of consciousness, enabling them to explore across realities, timelines, and dimensions. And it's important to remember: "evolved" doesn't mean superior. They may have access to advanced technologies or capabilities, but they can still learn from humanity…especially when it comes to emotions and embodied wisdom. We're all students of the universe, growing together. And there is a lot for us to learn from our ancestors, as well.

Vishnu's wheel, meaning "divine vision", may symbolize how consciousness interfaces with plasma fields by keeping reality in motion, maintaining cycles, and preserving balance. Symbols like this often carry layered meanings, all pointing to deeper truths about reality.

Vishnu's Wheel: Sudarshana Chakra (Personal Photo)

"True reality" is found in the present as a meeting point between 5D plasma and 3D form. The only truth is now; everything else is a projection into past or future. Even me stating that that is the only truth is indeed a paradox. This "true" center point sits within the wheel, or the core of the Four Quadrants of Reality (*still in development*). From this place, reality emerges through our perspectives and emotions, shaped by inner states and constantly in flux.

> **"He who is content in the present and detached from the past and future, who is unmoved by pleasure and pain, is truly wise."**
> **— *Ashtavakra Gita*, 15.1**

The four-quadrant model is used across disciplines, from psychology to politics to philosophy, to organize complex relationships between opposing forces or concepts. Structurally, it forms a square, a sacred geometric symbol of balance and integration. I like to envision it more as a 3D cube, which is also a hexagon in 2D. At its surface, the model presents four quadrants as two polarities intersecting to create a cross-shaped framework. These quadrants currently explore the realms of 3D and 4D experience, each offering a distinct lens for perception, action, and growth.

But surrounding them is not the traditional square outline. Instead, the entire model is encased in a hexagon, as a symbol of organic structure, sacred geometry, and the possibility of expansion beyond the visible framework.

The Four Quadrants of Reality | Dana Kippel

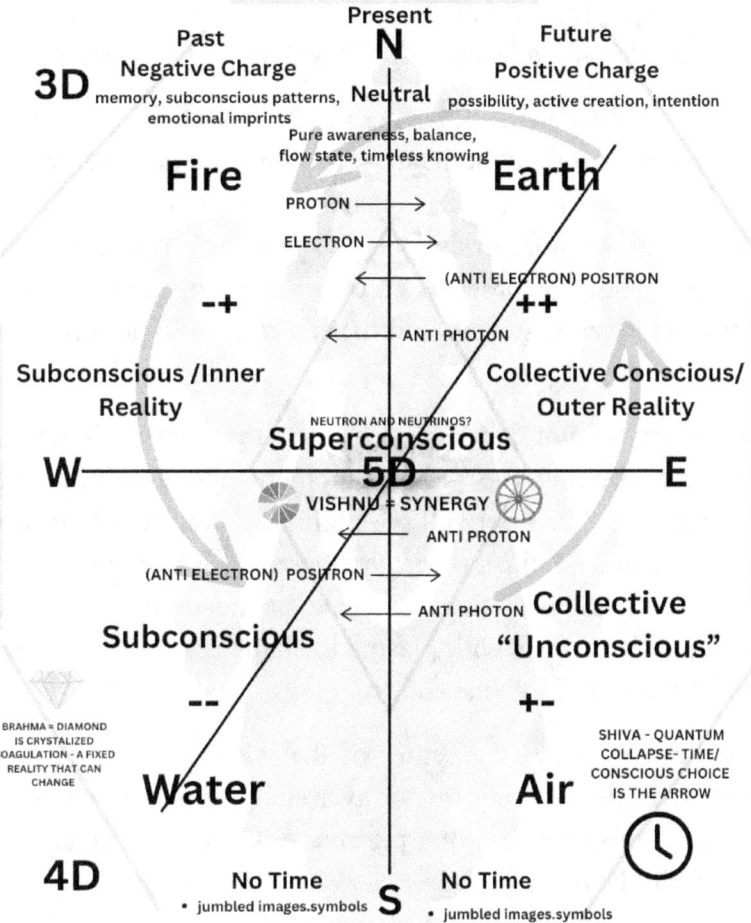

In psychology, thought structures shape perception by orienting awareness within time (*past vs. future*) and spatial or dimensional states (*3D vs. 4D*). When mapped onto a quadrant, the left side (*past*) carries a negative charge for memory, subconscious patterns, emotional imprints, while the right side (*future*) holds a positive charge for intention, creation, possibility. The center could be seen as neutral: a still point or spinning cross that forms a star, symbolizing integration. Vishnu, representing ever-present awareness, appears in the background—reminding us that all states arise within consciousness.

The center point is the heart or star…of the design. In the current stage, it serves as a direct connection to 5D as the dimension of unity and pure creative potential. In this way, the center is not simply the midpoint of opposites; it is a gateway. What is not yet revealed is that the hexagon implies more than meets the eye. Hidden within its geometry are two additional, unseen segments, the "secret two", which will be introduced in a later stage of the work. When these are brought in, the framework will expand from a four-quadrant structure to a six-part, multidimensional map spanning 3D, 4D, and 5D awareness.

Plasma flows between the polarities of 3D and 4D, bridging the known and unknown, mirroring the process of thought itself. This hints at how reality is structured through consciousness where our minds are constantly conducting, cycling, and shaping plasma fields based on which quadrant of awareness we focus on. The quadrant model I have here is deeply simplified, for these processes are extremely mysterious and complex. I apologize if it seems obtuse and messy, it is not a finished project but nothing ever is, so I figured may as well share!

To go into it a bit more, in Vishnu's navel, the center of the four quadrants, a neutral charge is held, representing the present moment and aligning with 5D consciousness, awareness, and the M-Self…where all possibilities collapse into now. This is why presence, or "higher consciousness," holds power. In 5D, reality is not tethered to the past's emotional residue or the future's projections, instead it becomes the field of choice, where consciousness directly engages with plasma to shape reality in real time. This is the state of manifestation, intuition, and synchronicity.

This is the flow state, the center of the wheel, where plasma exists in its purest, formless potential, guided entirely by our awareness. It is the fusion of Shiva's stillness and Shakti's motion as consciousness and plasma, observer and creator, void and fullness, all contained twirling in the now.

While many seek to ascend to 5D, both 3D and 4D are essential. 4D allows us to feel and form meaning; 3D anchors transformation. Mastery is not escaping these lower dimensions but learning to play with them, grounding into 3D, feeling deeply in 4D, and flowing in 5D. This is where we all become wizards consciously co-creating with plasma.

Zoroastrianism (600 BCE)

Zoroastrianism was one of the first monotheistic religions, believed to predate Judaism, Christianity, and Islam. Founded by the prophet Zoroaster (*Zarathustra*) in Ancient Persia (*modern-day Iran*), it centers around the concept of divine fire, known as Atar. This sacred fire represents purity, divine energy, and serves as a medium through which the unseen world interacts with the physical realm. The eternal flame in Zoroastrian temples is seen as a direct link between humans and the divine. Fire is not worshipped but honored as a presence of divine truth, just as plasma, in its own way, feels like something to be revered and consciously engaged with rather than worshipped.

Zoroastrianism also teaches of a cosmic struggle between *Ahura Mazda*, the force of order, light, and wisdom, and *Angra Mainyu*, the force of chaos, darkness, and deception. To me, this mirrors how plasma can be creative on one level (*5D*) and seemingly deceptive on another (*4D*). But we've misunderstood that deception. It all depends on the input given to plasma. In my experience, in the 3D, it simply mirrors our present state of consciousness—whether that's shaped by subconscious 4D layers or 5D awareness. Plasma isn't a force that's intentionally deceiving or inherently evil. The real battle between dark and light happens within us, then gets projected onto the plasma field, and into the stories so many films continue to tell.

In Zoroastrian belief, every person has a *fravashi*, a kind of higher self or guardian spirit that exists before birth and after death. These winged beings are often depicted as divine protectors and could be interpreted as the M-Self or 5D "future-ancestors" looking out for us.

The Fifth-Dimensional Divine – Greece, The Bible, and Neoplatonic Thought

Prometheus Unbound & Abolishing Fear Based Systems

Prometheus, whose name means "forethought," was a Titan who defied the gods to help humanity. His story represents rebellion, sacrifice, and the pursuit of knowledge. He was said to have shaped humans from clay and given them life, becoming their benefactor. When Zeus withheld fire from humanity, fearing they would become too powerful, Prometheus stole it from Mount Olympus and gifted it to mankind. As punishment, he was chained to a rock in the mountains, where an eagle devoured his liver each day, subjecting him to eternal torment. Yuck. In some versions of the myth, Hercules eventually frees Prometheus, suggesting that rebellion against tyranny <u>can</u> lead to redemption.

This was clearly a tale designed to suppress human progress, rooted in the belief that knowledge threatens those in power. When you're truly empowered, capitalism and consumerism lose its grip. You can't be sold to through marketing tricks because you're no longer operating from a hypnotized ego, you're choosing from your heart. That's a real threat to the current power structures. But has anyone considered that business could be run in a way that genuinely serves both people and owners, *without* fear-based systems or manipulation?

Some argue that if we change the way we operate, we'll become vulnerable, that other countries will invade us, or chaos will unfold. And sure, on the surface, that makes sense. But is us continuing to perpetuate the same patterns really solving anything? I think it's only digging us deeper into fear. I don't claim to know the solution to all the world's problems, but I do know the first step: each of us working on our inner self. I truly believe that if we did this on a global scale, we wouldn't need so much protection.

The real question is…how does someone connect with their heart when they've been conditioned by cultural or religious beliefs that may harm others? Beliefs rooted in fear, that are now outdated. No matter how evolved we become, we're still human. We'll still disagree. We'll still get angry, hurt each other with good intentions, even betray the people we love. How do we deal with all of that in ways that don't involve power games, violence, or rage?

This is a question for the present and future leaders of the world, whom I hope are reading this book. I absolutely don't believe the answer is to keep the fire, this plasma knowledge, from us any longer. Humanity needs to be trusted with it. We are not mindless ants or pure thinking machines ruled by determinism. Humans are made of good and love, and when we access that, when someone believes in us, we become benevolent and beautifully unpredictable. In times of disaster, we often witness communities of mixed beliefs coming together in profound ways. I believe our entire news system needs a complete overhaul; it's biased and relentlessly negative from every angle. It's time to reopen to plasma, to the power it offers us, and to use it in service of humanity.

There's another version of the Prometheus story called *Prometheus Unbound*, which was originally a lost Greek sequel by Aeschylus and later reimagined by Percy Bysshe Shelley, the husband of Mary Shelley, who wrote *Frankenstein*. The original was a trilogy, but only the first part survived. Percy's version reimagines the myth so that Zeus eventually reconciles with Prometheus, acknowledging the injustice of his punishment. Prometheus is then freed and becomes a wise counselor to the gods.

Shelley's 1820 version transforms the myth into a symbol of human liberation and the triumph of love and imagination over tyranny. In this retelling, Prometheus isn't freed by Hercules, but *by his own spirit of resistance*. He breaks free not through violence, but when Jupiter (*Zeus*) loses his power through the rise of compassion, love, and wisdom. It suggests that true freedom from oppression doesn't come through force or violence, but through an awakening of consciousness.

Prometheus' suffering becomes a catalyst for transformation. Instead of seeking vengeance, he forgives Jupiter, embodying the path of compassion over retribution. This is the path we need to consider now. Yes, those in power have oppressed the powerless for centuries, but on some level, *we've also allowed it*. Rather than resorting to violent revolt, we must direct our energy toward gently guiding humanity into a new era of consciousness, one where the old systems naturally lose their grip because they no longer resonate with our elevated awareness or command our attention.

In the novel, Shelley introduces a mysterious cosmic force, the Demogorgon, representing the cyclical nature of time and justice. He refers to it as a snake coiled underneath Jupiter's throne. [73] Unlike the Olympian-centered myths, this force hints at a new paradigm where consciousness evolves beyond suffering and fear, aligning instead with higher forces like love, imagination, and harmony.

"Love, from its awful throne of patient power / ... springs / And folds over the world its healing wings." — Percy Bysshe Shelley, *Prometheus Unbound*

Individually, I know firsthand that fighting fear with my mind has never worked. It only dissipates when I make the active choice to trust and sit in love, instead of falling back into my old ways of thinking and being. I'm not saying it's easy at first, nor should it be. It's completely new for someone who's lived another way their entire life. But when I do it, it works. And I know it will work on a larger scale too. It just takes active choice and courage, not the kind of courage to fight a violent war, but the courage to think and act in a completely new way.

Interestingly, both the ancient and Shelley's versions suggest a future where humanity harnesses powerful forces, like fire, electricity, or what I call plasma, responsibly and wisely. To evolve safely and ethically, our wisdom must not only catch up to technological growth but stay ahead of it. That's a core mission of this book. Like Prometheus, I want to offer this wisdom to humanity, trusting that we will choose to use it for collective awakening rather than domination, and, more importantly, that we all deserve the free will to choose.

I hope you can see by now, and will continue to see throughout this book, that these are not coincidences. So many stories envision a future tied to something strikingly similar to plasma, woven together with compassion and the rise above limitation. These were not wishful dreams. It's the collective subconscious crying out through story, it's plasma

---

[73] Shelley, *Prometheus Unbound*, Act 2, Scene 3, lines 93–98.

speaking through us all. And we haven't listened; we haven't applied it. But we are living in a time where we finally can. This book is a call to choose that path.

Thales and The Living Universe (600 BCE)

The ancient Greek philosopher Thales of Miletus believed that all things were full of gods, a view similar to Animism. He observed that materials like lodestones (*magnetite*) could attract iron, and that amber could generate static electricity when rubbed, what we now call the study of *electrostatics*. From these observations, he proposed that movement itself was a sign of life. And if something was alive, it had a soul. This aligns with my belief that everything is alive, just in ways we don't currently define, which I explore further in the Plasma Intelligence chapter.

This idea later became known as *Hylozoism*: the belief that all matter is, in some way, alive. Thales also claimed that water was the fundamental principle of all things, shaped by a supreme mind, or *Nous*. Sound familiar? At the time, water might have seemed like the most obvious foundational substance, long before the discovery of ionized gas, or plasma. And now, with modern studies on structured water, or EZ water, holding memory, including the work of Veda Austin, his insight feels strikingly accurate.

Plato and Aristotle – Aether: The Receptacle & Quintessence (428 BCE)

Sadly, women, especially philosophers and writers, were often suppressed or erased from history, so I'm left mostly referencing men. Many were brilliant, yes, but I do wish there were more women to pull from. I am sure I will write a book in the future once I go digging which has to do more with female writers, philosophers and plasma.

Plato, the Ancient Greek philosopher, proposed what he called the *Theory of Forms*. It suggested that reality exists in layers, with the material world being only a shadow of a higher, unchanging realm called the *World of Forms*. These Forms were eternal, unchanging, and perfect archetypes of everything in the material world. In Ancient Greek, "perfect" didn't mean flawless, it meant that something had fulfilled its purpose or achieved the goal it was created for. That distinction matters to me, because it helps me draw the connection that this World of Forms may be the realm of our "higher" selves, or what I call the M-Self.

At this time, the World of Forms represented the highest self of everything, a self-actualized field of knowledge. It was something they believed could be tapped into, and I believe it's now beginning to merge with us in our day-to-day lives through human evolution.

"We must, I said, go back to our old comparison of the sun. The sun, I presume you will agree, not only makes the things we see visible, but it also provides for their generation, growth, and nourishment, though it is not itself generation."

— **Plato, *Republic*, 509b**

In *The Republic*, Plato uses the sun as a metaphor for the *Form of the Good*, the ultimate source of truth, knowledge, and reality. And by now, we all know the Sun is plasma. *Forms*, like the Sun, hold the blueprint for true reality, often equated with love in ancient texts, if we choose to access them. If the sun does all these things, wouldn't it be feasible that Plasma is responsible for all these things at a deeper level, as the *Form* of the Good, or God? I see this use of metaphor as the intelligence wrapped within fourth-dimensional plasma information, something on one layer showing a truth of visible nourishment, and on another a deeper truth of non-visible nourishment.

Plato also said that imagination was the lowest level of reality, associated with shadows, reflections, and illusions. Things that aren't real but mere representations. He placed *thought*, or what he called *Dianoia*, beyond the senses, and *intellect* as the highest level of reality. Intellect, as *Nous*, the direct knowing of truth, makes sense to me. But placing thought or logic above imagination does not. Maybe, for his time, that hierarchy made sense…but it doesn't anymore.

There's a clear fear of uncontrolled imagination in his view, which is also something I notice in overly logical people. It is perfectly normal and provides a sense of safety, but it is false and only perceived as so. When we recognize that imagination might actually be the access point to the World of Forms, the highest reality, where one contemplates truth, we begin to see things differently. We can merge thought with imagination, giving them equal ground and in doing so, give birth to deeper sensory experience.

Plato saw imagination, or *Eikasia*, as shadows and illusions. What he didn't shine a light on is that this faculty can be healed, with love. The truth is, we don't yet know the full capability of our emotions. They've been layered with fear, suppression, and darkness. It will be about learning to use them as sensory feelers, transporters, and shapers. As we move through these learnings together, we'll begin to uncover the magic that's waiting to emerge.

I think even back then, as we've also seen in Hinduism, there was an innocent, subconscious fear and distrust of the feminine: meaning the senses, imagination, and intuition. It didn't feel safe enough to sense, and that hesitation still lingers in many of our lives today. Most of us shut down our sensory awareness and intuition in childhood because we were in environments that felt unsafe and inescapable. If we had believed what our senses were telling us, we might not have survived mentally. Survival meant dissociating, escaping into logic and clinging to what we could control. We became the bad ones, not them, because if it *was* them, that truth would have been too terrifying for a child to hold.

Now, we have to relearn how to make ourselves feel safe, through self-trust, discernment, and boundaries, so we can bring our sensory intuition back online. Sensing can absolutely come first. We just have to remember how to let it.

In Plato's *Allegory of the Cave*, a prisoner escapes a dark cave only to discover that it's not reality at all, that true reality exists outside the cave. When he returns to free the others, they reject him, preferring the comfort of illusion. Funny enough, I think Plato may have been in his own cave of fear, fear of his senses, along with the rest of the world during his time. I know I've lived in that cave for much of my life. Sometimes, our blind spots are blind for a reason.

Most of us *do* sense from a survival-based place of fear, and yes, that creates illusion. But imagine if we began to heal our traumas and limiting beliefs, if we could sense from a more open, expansive, and conscious place. That's where invention is born. That's how world-changing stories are told, new concepts are formed, and most importantly, how we begin to transform our self-image and personal reality.

Speaking of blind spots, Plato briefly mentions *aether* as the most translucent form of air in his dialogue *Timaeus*, but then shifts focus to the four classical elements: earth, air, fire, and water. Interesting. I believe he was hinting at a medium similar to what I'm proposing. And get this…in Timaeus, Plato describes the universe as a single living organism infused with soul, balanced by the four elements, and illuminated by divine fire!

And now for the blind spot grand finale: the one that's been overlooked again and again throughout history—the divine feminine. *The Receptacle.* When I first came across this term, I had only heard it used as a modern-day slur for women or as a trash bin. But Plato describes the Receptacle as a kind of "wetnurse", a space that receives all things without taking on their characteristics.

This Receptacle exists prior to and alongside the Forms, acting as a womb-like substance that allows reality to take shape. To me, this is blatantly, and beautifully, Plasma at its core. The plasma that makes up our reality at all levels. The cosmic womb of existence.

What's even more astonishing is Plato's ability to express that this Receptacle does *not* take on the characteristics of what it holds, directly affirming my view that plasma can be a vehicle for all consciousness without necessarily "holding" that consciousness or absorbing its qualities. The complexity of that insight speaks for itself. Plasma can reflect and contain, yet it remains neutral at its core, just like when you turn off a computer screen: the image disappears, and the pixels return to black.

In *Timaeus*, Plato presents three fundamental components of reality: Forms (*eternal ideas, unchanging*), Matter (*the physical world, changing*), and the Receptacle (*Chora*), a mysterious space that receives all things but remains unchanged itself.

"Not only does it always receive all things, it has never in any way whatever taken on any characteristic similar to any of the things that enter it." —Timaeus, 1253[74]

Plasma, I believe, is evolving from a container to a co-creator, as we speak. The Forms use it to manifest temporary realities, just as we see effigies of archetypes take form. How intelligent. It simply allows transformation without attachment. And fittingly, it's even described as feminine, called a wetnurse, a womb-like entity. Plato himself admitted this concept was difficult and vague, because it didn't fit neatly into his usual rational framework. Of course, Plasma does anything-but fit in!

This has been the issue up until now, with how scientists, philosophers, and really all of humanity have been taught to view Plasma and the divine feminine. Western philosophy has often erased the role of the receptive, fluid, creative feminine principle in cosmology. And that erasure has rippled outward, creating a quiet, deep-seated shame in women toward their own emotions, and a fear in all people of their wild feelings, their tears, their softness.

I've heard it firsthand. Sitting at a dinner table, listening to incredible men I know speak in low voices about how afraid they are to cry in public, and worse, to cry in front of their wife. That if they did, she might not find them attractive anymore. That it would be seen as weakness.

I think it's time we accept that we don't need to "solve" Plasma or figure her out, as much as we need to learn how to make space for her, listen to her, and be in relationship with her. The same goes for our feelings. The same goes for women. Figuring it all out will be a byproduct of accepting and making space. It's time to swing the pendulum back from logic toward the center—where logic and emotion live together, and true awareness is born.

The Demiurge in Plato's story is the masculine Creator, the divine craftsman, or consciousness. It's a vital part of the equation, but not the whole. The Demiurge needs the

---

[74] Plato, *Timaeus*, 49a–52d, in *Plato in Twelve Volumes*, vol. 9, trans. W.R.M. Lamb (Cambridge, MA: Harvard University Press; London: William Heinemann Ltd., 1925).

Receptacle to shape and hold matter. And how does the Demiurge organize reality in his story? Through sacred geometric patterns, bringing order from pre-existing chaos.

Plato introduced the *Five Platonic Solids* as geometric forms he believed were the foundational structures of all reality. The Dodecahedron, linked to Aether or Plasma, was considered so sacred that Plato barely spoke of it openly. Below are the shapes and what I believe they signify, at least on the surface for now. There's still much more to uncover here.

This leads into why I believe Plasma represents *Quintessence* or the Fifth Element. Or maybe, even more accurately, Plasma-Consciousness Synergy itself is the Fifth Element. Humanity. Maybe *we* are the fifth element. The human star. Just like in the movie *The Fifth Element*, where (*spoiler alert*) the protagonist Leeloo turns out to be the great transmuter, the fifth element, who can channel and harmonize all other elements!

The Five Platonic Solids

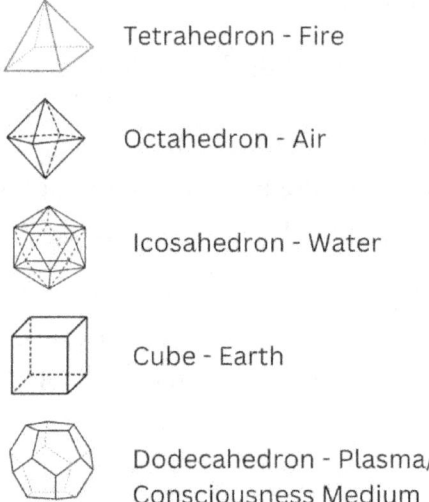

Tetrahedron - Fire

Octahedron - Air

Icosahedron - Water

Cube - Earth

Dodecahedron - Plasma/
Consciousness Medium

Plato primarily used a four-element model, though he subtly hinted at a fifth. He wrote that the Dodecahedron was "used by the god for arranging the constellations in the whole heaven," yet he never explicitly mentioned Aether.[75] It was Aristotle, his student, who expanded the model by more prominently introducing Aether as the fifth element. [76]

---

[75] Plato, *Timaeus*, 55c.
[76] Aristotle, *On the Heavens*, I.3, 270b, in *The Complete Works of Aristotle*, vol. 1, ed. Jonathan Barnes, trans. W.K.C. Guthrie (Princeton: Princeton University Press, 1984).

According to Aristotle, the four terrestrial elements moved linearly, up or down, while Aether moved in perfect, nonlinear circles. Aether was also believed to make up the celestial spheres, which, interestingly, aligns with modern science showing that stars (*celestial spheres*) are made of plasma. And like Aether, plasma moves nonlinearly, often in spiral-based patterns.

The word *Quintessence* later emerged from Aristotle's concept of Aether, as the divine life force permeating all things.

In Aristotle's writings, this fifth essence could be extracted, purified, and used for healing. It was even considered the substance of the *Philosopher's Stone*, the mythical agent capable of transforming base metals into gold and prolonging life indefinitely. Something tells me this connects directly to the great transmuter: Plasma. And it may even tie into Cold Fusion.

Just as Quintessence was believed by alchemists to heal, Plasma in our current reality is now doing just that, rapidly healing wounds, destroying cancer cells while leaving healthy tissue intact, and exhibiting antibacterial and regenerative properties. And perhaps even beyond that, we're beginning to heal with plasma-energy through our own awareness, becoming transmuters ourselves. Maybe this is part of our emerging human potential. After all, plasma seems to heal only through some form of human touch…whether that touch is material, human-made, or biological, guided by conscious awareness or hands-on energy. It seems to be directed by *us*.

Neoplatonism, Reality, & Asclepigenia (200 AD)

Neoplatonism was a philosophical movement inaugurated by Plotinus, the father of Neoplatonism. It centered around the concept of a supreme, transcendent source known as *The One*, a reality beyond being and nonbeing, beyond all human concepts and comprehension. *The One* is the unknowable origin of the universe, a boundless, ineffable source. To me, this perfectly symbolizes the Mystery as the limitless, undifferentiated field from which all existence emerges.

There were emanations from *The One*, including *Nous* (*the mind*), *Soul* (*psyche*), and finally the Phenomenal World (*the material world*), generated by the Soul. *Nous*, also known as divine intellect, was the first emanation from *The One*. It represented the realm of perfect forms, pure intellect, and divine thought. In this dimension, everything exists in its ideal state as unified, whole, and beyond fragmentation. *Nous* held the blueprint for all of creation, embodying ultimate order, wisdom, and truth. Yet again, another powerful parallel to fifth-dimensional Plasma. It also may be a powerful metaphor for Plasma Intelligence.

Psyche, the second emanation, bridged *Nous* with the Phenomenal World, acting as an intermediate realm, much like 4D Plasma, where emotional constructs manifest. *Psyche* contained movement, change, emotion, and the seeds of individuation, but also reflected fragmentation and duality. It was the liminal domain where the soul operated in its journey toward integration with *Nous*. This is where consciousness navigated polarities like love vs. fear.

To me, ascension is a state of mind. I propose that the 5D is our connection to all worlds beyond our own, as well as to the Mystery, this ineffable source that some call God. This differs from the 4D, which deals more with timelines and the emotional realities of our psyche. I believe 4D brings us the emotional texture of how we calibrate what we see in the 5D, but during the experience, they feel quite different.

> **'Lord, when someone meets you**
> **in a moment of vision,**
> **is it through the soul that they see,**
> **or is it through the Spirit?'**
> **The Teacher answered:**
> **'It is neither through the soul or the Spirit,**
> **But the *nous* between the two**
> **Which sees the vision …'**
> **— *Gospel of Mary Magdalene***

When it came to the soul, or *Psyche*, the Neoplatonists described it as the imaginative realm, but they warned against becoming too attached to it. They believed this space contained fragmented reflections of truth: alluring, but potentially misleading distractions from the pursuit of higher knowledge. I see this as a reminder that the 4D space, with all its mythical beings, beasts, aliens, angels, and demons, is not a realm of fixed truths, but a projection of our own psyche, shaped by belief. These entities and experiences exist to teach, transmute, and evolve…not to be taken as absolute reality.

But life isn't just about navigating these realms and healing. It's also about being fully present in 3D reality, and for some of us, that can feel scarier than any encounter in the imaginal realm.

When we're traversing the 4D, through astral projection, dreams, intuition, or visitation, we must remember: it's not ultimate reality. It's what we make it, for better or for worse. Learning to navigate it takes time. As more of us begin to access these realms, it's crucial to understand their nuances, because the beings we encounter can feel incredibly real, and they can be overwhelming or even frightening if we don't know how to interpret them.

"For to become lost for long in the realm of the mothers puts us at risk of being pulled back into the unconscious in a destructive way"

— Monika Wikman, Pregnant Darkness[77]

There are so many misconceptions about evil and darkness, especially the idea that we're powerless against them. Evil is human-created, a projection of our own psyche. We don't have to let it in, we can transmute it. At the same time, certain lower forms of love, those that lead to intoxication or addiction, also exist in this space. When we carry unhealed relationship patterns, we may attract toxic partners until we learn to transmute those patterns and break the cycle. That's how life lessons work: the key is taking new action in 3D reality, action that breaks from our conditioned responses. That's how we shift our programming and begin to align with a higher state of being.

There's a glaring issue in the psychic and channeling community that often goes unnoticed, and I say this as someone who genuinely appreciates a good psychic or channeler. We must remember that all channeled and psychic messages are filtered through personal perception, including mine. That's why it's so important to know your own inner truth.

If you're receiving guidance from a psychic who hasn't processed their own traumas, or a channeler who lacks self-awareness and hasn't integrated their teachings into their life, the message they deliver may actually be meant for them—or heavily colored by their belief system.

You are your own best channel. The issue is, many of us don't trust ourselves because of past mistakes. But part of adulthood, and awakening, is learning to listen again to that inner voice and rebuilding trust with it. Books like this can help guide you back to yourself, your personal connection, but discernment is key. Take only what resonates in your heart.

Just because a psychic or channeler shares something profound doesn't mean it's your highest truth, it may still come from the 4D space, which is not the ultimate source. Your personal highest truth can only come from within and through lived experience. Everything else, including my words, is simply here to point you back to you. But it is true that sometimes a true psychic might bring something into your awareness you were blind too. If it resonates take it, but don't let that make you believe everything they say is your truth. They

---

[77] Monika Wikman, *Pregnant Darkness: Alchemy and the Rebirth of Consciousness* (Ashland, OR: Chiron Publications, 2004)

are usually seeing outcomes based on your timeline now, you can always change that with your choices.

I still struggle with plenty of imperfections, just like many spiritual intuitives. Often, it's that very struggle that inspires the desire to help others relieve their own pain and disconnection. But remember this: *your struggle does not cancel your purpose.*

Personally, I wrestle with fear, relationships, staying present, and letting go of control. But I believe these challenges are here so I can learn to transmute them through the lens of plasma teachings, and in doing so, share what I learn with you.

One of the biggest misconceptions I want to clear up is the idea that the people writing spiritual or self-help books have it all figured out. They don't. <u>None</u> of us do.

What I do know is that this plasma wisdom is a beautiful, deep journey…one that has made my life feel more vivid, more meaningful. It continues to heal me and bring more peace of mind each day. <u>It is not a quick fix</u> but each issue I heal, stays healed. I believe it can do the same for you. It opens the heart and mind in ways that illuminate entirely new horizons.

Take these tools and use them however they serve you best. And please, take all your idols off their pedestals. They're just like you. If they can do magnificent things in the world while still moving through very human struggles…*so can you.*

The only thing they do consistently is keep going. They show up time and time again, hopefully with some self-awareness and transmutation along the way. True integration isn't about appearing enlightened; it's about facing your fears, acknowledging and healing trauma, and applying wisdom to real-life challenges. It means living authentically, building deeper relationships, and embracing both the beauty and the messiness of life. I want to build a life where I don't present one way and live another. I don't want a public facing image.

I want to be Dana, every moment and everywhere. That, for me, is freedom.

There are no idols…only ideas, and those ideas are meant for all of us to contemplate and grow with. I encourage you to embody your spirituality by living your own teachings and applying them to your personal growth and relationships. That is how you shine your unique light.

Asclepigenia of Athens was a mystic and priestess who played a crucial role in preserving Neoplatonism and theurgy. As one of the last great Neoplatonists, her teachings were deeply esoteric, emphasizing energetic interaction with higher realms and bridging mystical experience with structured cosmology. She believed that matter and spirit were intrinsically interconnected, standing in contrast to the rising Christian doctrine of her time, which enforced a separation between the material world and the divine.

She emphasized direct interaction, symbols, and rituals, believing that humans could actively ascend and draw divine energy down, rather than remain passive recipients of divine will. For her, theurgy wasn't merely prayer, it was an active engagement with the forces of the universe.

She saw the material world not as a lesser reality, but as a vital part of divine expression. Just as I often say, the divine learns from us just as much as we learn from it, and many forget that. We are not just students of the cosmos; we are co-creators, living lives infused with far more meaning than we realize. It can be unsettling to consider that we might teach beings of consciousness we've placed on pedestals, but we can, and they want us to know that.

Just as parents are often taught, renewed, and inspired by the innocence of their children's play (*even amidst the stress*), so too do we, as humanity, serve as those children, teaching and reinvigorating these evolved beings through our lived experience.

Asclepigenia believed the physical and spiritual constantly influenced each other, much like my plasma-consciousness synergy model. Long before her time, she taught that through energy work and ritual, one could elevate consciousness and reshape the material world. She also believed that symbols, when activated through theurgy (*divine working*), could serve as keys to unlock deeper dimensions of existence.[78] I believe that one day, symbols and sacred geometries will be recognized as tools that can unlock alternate timelines, realities, and expanded states of perception.

The fact that she believed humans were not just receivers of divine energy but active participants in shaping reality was, at the time, heretical but deeply accurate. Theurgy was not about worship, but direct engagement with cosmic intelligence. The universe, in this view, was participatory. Her work was a threat because:

1. It empowered individuals to access divine energy without intermediaries.
2. It rejected the rigid dualism of Christianity and many other religions by seeing matter as just as sacred.
3. It suggested that reality was interactive, that consciousness could shape the cosmos.

---

[78] Iamblichus, *On the Mysteries,* trans. Emma C. Clarke, John M. Dillon, and Jackson P. Hershbell (Atlanta: Society of Biblical Literature, 2003).

Asclepigenia's work was among the last surviving Pagan teachings before the Academy of Athens was closed around 529 CE by Emperor Justinian, effectively ending the ancient Neoplatonic tradition.

Plasma, God, Whitehead and Dan Brown

God rules the chaotic waters by hovering over them with the Spirit of God, preparing the earth for creation — Genesis 1:1-2

In Genesis it says, "In the beginning, God created the heavens and the earth. The earth was waste and wild, and darkness was over the face of the deep. And the Spirit of God was hovering over the face of the waters." This wasteland was a formless void, and if we see God as a higher consciousness, it opens up some intriguing symbolism. Imagine higher consciousness emerging from a singularity, moving through and impregnating 2D brute plasma to form physical reality, a neon light encased in darkness, shaping the clay of the void into the multicolored light visible in matter today.

My theories and this book don't disprove God or the Bible, or any religion for that matter. If anything, they add a layer of validity, suggesting that parts of the Bible may be pointing to something real, a holographic reality created by divine light, a God, a Mystery beyond our visible light spectrum. A different *kind* of light.

"Neither shall they say, Lo here! or, lo there! for, behold, the kingdom of God is within you." — Luke 17:21

The Bible is widely known and needs no introduction, but I want to share my views on God. I believe there is a creator, though I don't claim to know exactly what that is. That's why I call it the Mystery. To me, the essence of Christianity, and any religion, is a reflection of that Mystery.

In some ways, I hope this book offers evidence for the existence of God. Plasma, in my framework, is like the hand of God, or a female God, a living, intelligent force that births and shapes reality. But when I say "God," I don't mean a man, a woman, or even a single being. It could be a collective of higher intelligences, the gestalt of an essence of our future selves, or something beyond what we can comprehend. I have spoken a bit about it throughout this book.

What I do know is that when people feel connected to this Mystery, no matter what they call it, miracles happen across <u>all religions</u>. The key is understanding that everyone experiences this higher or merged intelligence through their own psyche, shaped by their emotions, culture, and beliefs. Who is anyone to control that or say something is wrong?

"Beauty is in the eye of the beholder. Reality is speaking to you; it is an individual and personal relationship." — Dana Kippel

A key difference in my belief is that God, or the Mystery, is something to walk alongside, but not to be placed above one's connection to self. It's about personal relationship. For me, that connection shows up through Plasma, through another person, a burst of laughter, a experiencing awe, or an ineffable moment in my daily life.

In alignment with Alfred North Whitehead's *Process Theology*, my current view of God is not as a distant overseer or omnipotent authority, but as an evolving intelligence within the field of creation itself. Whitehead proposed that God has two natures: *the Primordial*, which holds all infinite possibilities, and *the Consequent*, which absorbs the experiences of the world and evolves with them. This reframes divinity as participation, rather than domination. [79] It reminds me of the feedback loop that we spoke of while going over the olfactory sense, between the open honeycomb, representing infinite potential, and the closed torus, representing chosen experience and its integration. *Interesting*.

I believe we are not here to worship God as separate, but to co-create with this universal intelligence. Reverence, sure, but not hierarchy. We "worship" through alignment, through becoming our truest selves, through living as creative expressions of the plasma-consciousness field that is shaping and reshaping reality.

God, or the Mystery, isn't outside of us. It evolves through us, expressing itself as each unique fractal of creation. It doesn't abandon us, no matter what perceived sin we

---

[79] Alfred North Whitehead, *Process and Reality* (New York: Free Press, 1978), 346.

commit, whether we forget to meditate, or skip expressing gratitude one day. It's a constant presence, an unbreakable connection.

> **"God is the poet of the world, with tender patience leading it by his vision of truth, beauty, and goodness." — Alfred North Whitehead**

In this vision, we are not passive recipients of divine will, we are active participants in its unfolding. We don't have to pray for permission; we attune, align, and create alongside the divine as co-authors of reality.

As I was editing this book, someone in my community sent me an interview of Dan Brown on his upcoming book *The Secret of Secrets.* He did not say much on it, but I had the feeling deep in my soul it may reflect truths we are speaking on here. With Brown being one of my favorite childhood authors, I looked up what books he wrote after *The Da Vinci Code.* I stumbled upon his book titled *The Lost Symbol,* and as I skimmed it, it revealed themes of a potential meaning of God and human potential that, to me, rings true.

The novel proposes that the greatest "secret" the secret societies have been protecting is the realization that <u>humanity itself is divine</u>, and that through spiritual awakening and mental focus, individuals can harness their inner god-like potential.

This ties into our overarching theme that God is not an external entity but the latent, untapped potential within each person, waiting to be awakened through self-mastery, wisdom, and consciousness. And perhaps it can feel external because it is also an amalgamation of these feats in our "future", or our "future" selves, which is really just a higher frequency of now in some other dimensional realm. God seems to be something we tap into, and it almost feels too great to admit that it might just be inside us all.

All religions are true, and although I may not agree with all the rules of all religions or the structures of organized religion, I hold people's beliefs with deep respect; but I also have the freedom to disagree with them, as do you. And by all religions are true, I mean all beliefs held in the plasma field are true and false at the same time.

What may be true for you may be false to someone else, and that is their truth. All beliefs can be changed, yet they also gain power as more people believe them, that's how belief works. Belief makes things real. And since we may carry our beliefs into the afterlife or other mystical experiences, it makes sense that we see religious figures in visions based on our own perceptions.

We should always have the freedom to create our own beliefs, as long as those beliefs are rooted in basic goodness, principles of not causing harm to oneself or others. This means

294

that while our personal truths may differ, they should still uphold a foundation of compassion and respect for all beings.

The goal isn't to gain outer power but to cultivate inner power, love, and connection. My intention is to highlight common themes across teachings so we can uncover the underlying truths, but this shouldn't negate anyone's religious practice. Ideally, it should enhance it, deepening one's connection to their faith.

Plasma isn't a religion or philosophy, it's a creative, supportive substance with its own intelligence. It's its own category. It is the bridge, the literal meeting point between science and spirituality. Its interplay with consciousness is the mechanism through which all other creations emerge, including religions. But we must ask ourselves: Who decides what becomes history? Those writing it carry their own biases.

So, let's look beyond the mainstream narratives and peek into some less-explored corners that haven't received the recognition they deserve. Let's begin with Mary Magdalene's hidden teachings.

Mary Magdalene & The Intermediate Realm (1AD)

In the canonical Bible, Mary Magdalene is portrayed as a devoted follower of Jesus, the first to witness his resurrection, and a symbol of redemption. In Gnostic texts, however, she is depicted as a spiritual leader and enlightened teacher, possessing profound knowledge that rivals the male disciples.

The Gnostic texts, a collection of early Christian writings discovered in *Nag Hammadi* in 1945, present an alternative view of Christianity, emphasizing personal revelation, direct connection to the divine, and the mystical nature of reality.[80] The writings were known for secret knowledge and are still studied throughout many modern secret societies. In these texts, Mary is a central figure, a visionary and the one closest to Jesus, challenging traditional power structures within early Christianity.

Gnosticism, meaning knowledge, is based on these texts and thrived in the first few centuries of Christianity. Its pillars were that the divine is already within each person, gnosis (*spiritual knowledge*) is the key to salvation, not adherence to religious law, the material world was an illusion (*with an emphasis on self-knowledge*), Jesus was a teacher of inner awakening instead of a figure to worship, and that Mary Magdalene played a crucial role as

---

[80] James M. Robinson, ed., *The Nag Hammadi Library in English* (San Francisco: Harper & Row, 1977).

a disciple and wisdom keeper. Because of these beliefs, many Gnostic texts were banned or destroyed.[81]

The Gospel of Mary was one of the religious writings associated with Gnosticism that originated in the Mediterranean around the second century, however it was not found in the Nag Hammadi Library. It was discovered in Egypt in 1896 within a collection of papyrus texts known as the Akhmim Codex (*or Berlin Codex*).[82] This text was largely ignored by scholars at the time of being uncovered and was not published for nearly fifty years!

In 1995 it was finally translated and revealed to the public. It still has missing pages so it's true depth, sadly, may never be fully known. Jean-Yves Leloup's translation and commentary in *The Gospel of Mary Magdalene* delves deep into Mary's authentic role, challenging traditional narratives that have long overshadowed her significance. This is the best book by far that I have found on this subject, and I highly recommend it.

In the Christian religion, Magdalene was branded a sinner thanks to Pope Gregory's sermon in 591 CE declaring that the women in the Bible committing adultery, prostituting, and sinning were all the same person, effectively branding Mary Magdalene as a repentant prostitute, a narrative *not* supported by biblical text but one that persisted for centuries, influencing art, writings, and doctrine. This was perpetuated for hundreds of years until the narrative was officially corrected by the Catholic Church in 1969, when the Vatican formally separated Mary Magdalene from the sinful woman in Luke and from Mary of Bethany, restoring her role as a devoted follower of Jesus and the first witness to the resurrection.

Contrary to this unfair portrayal of Mary Magdalene, the *Gospel of Mary* depicts her true essence, having profound spiritual insight, often surpassing her male counterparts. Mary's teachings focused on inner power, self-sovereignty, and embodying divine wisdom rather than relying on intermediaries or outer authority.

> **"There is no sin. It is you who make sin exist, when you act according to the habits of your corrupted nature; this is where sin lies."**
>
> **— The Gospel of Mary Magdalene**

---

[81] Bentley Layton, *The Gnostic Scriptures: A New Translation with Annotations* (New York: Doubleday, 1987).

[82] Karen L. King, *The Gospel of Mary of Magdala: Jesus and the First Woman Apostle* (Santa Rosa, CA: Polebridge Press, 2003), 3–7.

In *Magdalene Mysteries: The Left-Hand Path of the Feminine Christ*, Seren Bertrand and Azra Bertrand portray Mary Magdalene as a powerful priestess and embodiment of the Divine Feminine, rather than the repentant sinner depicted in traditional Christian narratives. They connect her to ancient womb rites and feminine mystery traditions, aligning her with archetypes like Sophia, Isis, and Lilith, emphasizing her role as a spiritual leader and a carrier of deep feminine wisdom. The book explores how Magdalene's story has been encoded in Gnostic texts, art, and sacred traditions, positioning her as a symbol of the wild feminine, reclaiming her power as a vessel of divine feminine initiation.[83]

While reading *The Magdalene Mysteries* and *Womb Awakening* by Seren and Azra Bertrand, I had a powerful vision during meditation. I found myself standing in a wheat field before entering a dark, misty forest. As I walked deeper, I saw a stark white deer with huge, cosmic eyes standing in the shadows. I asked her name, and she said it was Mitzi. Drawn to her gaze, I jumped into her eyes and fell into her womb, where I became a baby drinking a teal liquid. Eventually, she gave birth to me on the forest floor, and I emerged covered in teal goo. As I stood and looked at her, I asked, "What do you want to tell me? What do I need to know?"

She said, "Keep going." Frustrated, I asked, "That's it?" Mitzi explained that she was from the otherworld and couldn't tell me more, that I had to learn it through experience. "You won't become who you're meant to be if I tell you", she said.

When I came out of the meditation, I looked up the meaning of a white deer and the name Mitzi. What I found stunned me. The white deer represents divine intervention, transformation, and innocence, a guide from the spirit realm. And Mitzi was a name linked to Mary Magdalene!! That moment made something clear: *we're not meant to know everything*. We're meant to heal, feel, discover, and walk new paths...allowing the geometry of our lives unfold in its own time.

Archetypes are like blueprints of the soul; they are patterns to teach us lessons.

Mary embodies what I call the resurrection archetype of the divine feminine which is the ability to hold two extreme poles at once and emerge as something entirely new. It is the alchemical union of betrayal and divine wisdom, where suffering becomes the seed of transformation. It's about holding the tension between fear and self-love, and transmuting fear into inner power, rising, phoenix-like, from the ashes.

---

[83] Seren Bertrand and Azra Bertrand, *Magdalene Mysteries: The Left-Hand Path of the Feminine Christ* (Rochester, VT: Bear & Company, 2020).

This archetype teaches forgiveness without sacrificing boundaries, and the courage to speak truth regardless of outside voices. Just as Inanna mirrors the cyclical death-and-rebirth process of Plasma, Mary offers another face of that same archetypal energy. She is the crucible, where pain and transcendence meet to birth a third thing: the resurrected self. She is not confined by sin or sanctity…she simply is.

We can invoke Mary or Inanna's gnosis in our own lives, or sometimes, they arrive unbidden, as guides reminding us of who we truly are becoming.

In Jean-Yves Leloup's *The Gospel of Mary*, Jesus heals Mary of seven demons, a striking parallel to Inanna shedding her seven *me* on her descent into the underworld. I believe Jesus represents Plasma as the *Sun of God*. Resurrection, in this light, is the healing of old identities and belief systems, a rebirth into a fifth-dimensional state of being. Learning to harmonize with Plasma begins with connecting to our soul and healing our trauma. Through self-love, we gain access to a deeper power, one that unlocks what science currently calls the impossible: intuition, telepathy, psychokinesis, and more.

In *The Gospel of Mary* by Jean-Yves Leloup, he explains that Mary's gospel presents a fourfold anthropology and a metaphysics of the imaginal (*what I call 4D*), whose keys the most liberated and informed minds of our era are *just beginning to discover.* He describes this intermediate realm of image and representation that is just as ontologically real as the worlds of sense (*or 3D*) and intellect (*or 5D*). But this world <u>requires a faculty of perception</u> that is peculiar to it alone. This faculty has a cognitive function and a noetic value."[84]

Leloup goes on to call this peculiar perception *creative imagination,* not to be confused with regular imagination. While regular imagination belongs to fantasy, subjective beings, and things, creative imagination was revered in sacred texts as the mode through which one accessed inner revelation. It is through this imaginal realm that Mary Magdalene encountered the resurrected Christ.

This is an extremely important aspect of my book: your creative imagination is the tool, your plasma-consciousness synergy, for navigating the sense-scapes of 4D and 5D. It's subtle, and we'll explore its use more deeply in the *Plasma Intelligence* chapter.

When you begin moving through these realms, it can be difficult to tell what's real and what's imagined. But eventually, you'll realize, it's not about that. What matters is your willingness to let go of the need to define everything, to allow the experience to arrive fully,

---

[84] Jean-Yves Leloup, *The Gospel of Mary Magdalene*, trans. Joseph Rowe (Rochester, VT: Inner Traditions, 2002

and to interpret it only afterward. This is the art of co-creation: weaving intelligent imagination (*5D*) with perspective (4D), then anchoring it into this 3D reality.

Imagination will soon become one of the most studied and celebrated faculties of this new paradigm. Creative imagination as our capacity for play, vision, and energetic sculpting, is not just a tool but possibly *the* key to harmonizing with Plasma and all that is within it. In the next chapter, we'll explore this in depth by learning how to engage this faculty both in ways that align with satisfaction and growth, and in ways that may lead to distortion or stagnation.

The Holy Grail

The Holy Grail has carried many interpretations throughout history, yet it remains one of the most coveted secrets. It's true nature has never been confirmed. In Christian tradition, it's thought to be the chalice used by Jesus at the Last Supper. In alchemy, it's known as the Philosopher's Stone, a substance believed to grant immortality and transmute lead into gold as a metaphor for inner transformation and divine awakening.

Some believe the Grail isn't a cup at all but Mary Magdalene herself, symbolizing the divine feminine, a concept popularized by *The Da Vinci Code* and various Gnostic sources that suggest Mary bore Jesus' child, carrying a holy bloodline.[85]

But what if the Grail isn't just a cup, or a woman, or even a bloodline? What if it's Plasma? The shape of the chalice bears a striking resemblance to the interior of a plasma torus in a lab. The Grail could potentially be a living symbol, a structure through which life-force energy moves, transforming and transmuting all it touches.

If you think of the Holy Grail as Plasma, it begins to make sense. But go deeper, and you might realize the Holy Grail is also us. Humanity itself is the holy bloodline, not a chosen few. It's about how we learn to wield consciousness with the wisdom of plasma.

Think of yourself as the chalice. You pour your perspective of reality into the cup, and from it, you output your unique expression back out into the world. You can choose what kind of "plasma" fills your cup…what beliefs, emotions, or energy you hold, and that, in turn, shapes what you create. This is meant to be ever evolving. Your power is your perspective, and that can shift your external reality. It's the act of pouring out what no longer serves and letting new beliefs and experiences fill in. The next time you feel anger or frustration, imagine yourself as the chalice, the Holy Grail itself.

---

[85] Dan Brown, *The Da Vinci Code* (New York: Doubleday, 2003).

Plasma in the MAST spherical tokamak | Culham Centre for Fusion Energy

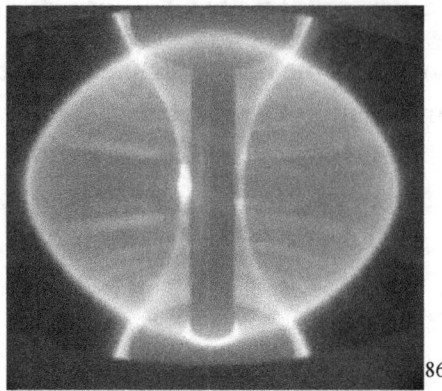

You are the outline of the chalice, you decide what to fill it with and what to release. This visualization helps me clear emotions and negative energy constantly. Feel it and free it. If your cup held the same emotions forever, it would rot. We are meant to regenerate, like the Earth, like the universe. Maybe that's why science keeps evolving…because reality itself is evolving. It's alive.

Grail came from the word *graal*, meaning a large shallow dish or basin. It also was related to a shallow vessel, or a bowl mixing wine and water. This is linked to the word crater, which also has to do with mixing.[87] It is mixing our masculine and feminine, plasma and consciousness, in a multitude of ways to co-create with and shape reality. In Wolfram von Eschenbach's *Parzival,* the main character wanders through the forest for years, searching

[86] Culham Centre for Fusion Energy / UKAEA. *Plasma in the MAST spherical tokamak.* Licensed under CC BY-SA 4.0. https://commons.wikimedia.org/wiki/File:MAST_Plasma_Image.jpg.
[87] https://www.etymonline.com/word/grail

for the grail. [88]The first time he finds it becomes a missed opportunity, for he stays silent when he wanted to speak out from compassion (*something he learned from social conditioning*) and because of this the grail's healing powers cannot be activated. Parsifal can't return to the Grail Castle until he undergoes years of trials, wandering through the wilderness and forests. He eventually returns, becoming worthy for the grail. In order to be found, one must be lost. His wandering reflects the idea that one must often pass through confusion and solitude before reaching spiritual clarity. It also shows it takes not just knowledge, but compassion, a *mix* of both consciousness *and* plasma.

Speaking of Gnostic texts, the *Apocryphon of John*, a 2nd-century text from the Nag Hammadi Library, dives deep into cosmology. I hadn't encountered it until I came across the YouTube channel *Library of the Untold*. In it, Sophia plays a central role in creation. After the "divine fullness," she acts independently, leading to the emergence of a dense, luminous substance—described as chaotic matter arising from a luminous cloud. This becomes the foundation of the material world. Gnostics saw this as a divine error or emanation, suggesting the physical world is a byproduct of imbalance. A luminous, chaotic substance that births reality? If that's not Plasma, I don't know what is.

Jesus, The Son of God

"Truly, truly, I say to you, whoever believes in me will also do the works that I do; and greater works than these will he do, because I am going to the Father."

— Jesus, *John 14:12*

Beyond the compelling connection between the *Son of God* and the *Sun of God*, Plasma as the radiant life-giving force, there's also the symbolism of Jesus as a fish. He turned water into wine, a clear act of alchemical transformation that mirrors the transmutative nature of Plasma. He walked on water, perhaps symbolizing astral projection or multidimensional navigation. And he healed with touch, energy moving through matter, just like Plasma can heal. This fish could be a hidden metaphor, reminding us that we too are swimming in a vast sea of Plasma, an invisible ocean of energy and potential, unaware of the very substance that connects and sustains all life.

When I reflect on the idea of the *Second Coming of Jesus*, I see it as a reawakening of plasma magic within our collective consciousness. An integration of the M-Self. It symbolizes the remembrance that the power attributed to Jesus lives within all of us. He said

---

[88] Wolfram von Eschenbach, *Parzival*, trans. A.T. Hatto (London: Penguin Classics, 1980), esp. Book V–IX.

it himself: we could do the same works he did…if not greater. Why? Because when we attune to this frequency, when we remember who we are and plug into that power, we unlock our highest potential. Each of us holds the capacity to change the world, if we choose to align with that knowing.

Jesus is a profound archetype, an energy you can call upon, just like millions have done throughout history. And it works. But if it doesn't resonate with you, that's okay. Find the archetype that speaks to your soul. The current is still there, it just wears different faces.

Personally, I call on whatever energy or presence feels aligned in the moment, sometimes it's the Celtic goddess Danu, other times it's a glowing purple being with a single eye as a slit, in the center of its forehead. The key isn't in *what* you call it but in the *power* you allow it to awaken within you.

Sarah: The Divine Child

> **"Truly I tell you, unless you change and become like little children, you will never enter the kingdom of heaven." — Matthew 18:3**

Have you heard the legend of Sarah, said to be the daughter of Jesus and Mary Magdalene? The most well-known connection is through the Romani traditions in Southern France (*Saintes-Maries-de-la-Mer*), where Sarah is venerated as Sarah-la-Kali.[89] She is said to have come with Mary Magdalene and other women by boat after the crucifixion. In certain traditions, Sarah is believed to carry their spiritual lineage, a vessel of wisdom passed down not just through blood, but through vibration. Some even claim to channel her teachings.

In Pia Orleane's *Mary Magdalene's Daughter: The Story of Sarah*, Sarah is portrayed as an emblem of a return to innocence, intuitive knowing, and divine play. Just as Plasma is both chaotic and structured, Sarah's myth suggests that childlike wonder is the key to unlocking divine wisdom.[90] To truly work with plasma intelligence, we must return to our childlike state as adults…open, curious, and attuned to infinite possibilities.

The smartest people I've met have also been the goofiest—true children at heart. I still feel this way. I don't take myself too seriously, and I believe that's why I feel so connected to Plasma. It loves to play, it loves story. This is how it communicates and interacts with us, through the language of imagination and wonder.

---

[89] Jean Markale, *The Cathars and the Holy Grail: The Secret History of Christianity in the Middle Ages* (Rochester, VT: Inner Traditions, 2004), 194–196. See also Margaret Starbird, *The Woman with the Alabaster Jar: Mary Magdalene and the Holy Grail* (Rochester, VT: Bear & Company, 1993), 65–67
[90] Pia Orleane, *Mary Magdalene's Daughter: The Story of Sarah* (Bear & Company, 2017).

Myth is <u>the story of plasma</u>, a timeless narrative reflected across cultures and eras, and right now, we're creating a new one. Like Sarah, we're being called to become children of divine knowledge again, stepping outside of rigid belief systems into a more fluid, creative existence.

Whether Sarah is real or not doesn't matter; it's the divine child archetype, beyond the duality of masculine and feminine, that holds the essence we need right now. It's the freshness, the benevolence, the neutrality of plasma, playful yet potent. This is the highest frequency to resonate with plasma, simple, yet so challenging in a world conditioned by fear and survival.

But it's up to us to change that story, and it starts within. <u>Nothing</u> outside us will change until we face ourselves. Only then can we create new realities here on Earth, filled with fun and awe. Some are already living in that reality, and you know them when you see them.

"How are you today, Sam?" you ask.
Sam's eyes sparkle. *"Grateful to be alive!"*

We all know someone like that. The ones who are already "tuned in, tapped in, and turned on", as Abraham Hicks would say, to the magic of this moment.

Where Mary Magdalene reminds us of the divine feminine, the importance of receptivity, inner power, and working with Plasma in a sacred, nurturing way, Jesus embodies the divine masculine, the active principle of consciousness, using Plasma for healing, transformation, and mastering the medium of reality.

Sarah, the *Holy Spirit*, calls us to return to our creative, childlike nature, embracing Plasma not just as a power source but as an interactive field we can shape, experiment with, and play in. Together, they form a holographic trinity of receiving, transforming, and creating, the full spectrum of plasma-consciousness synergy. An archetypal plasmatic family.

The Misconception of Pandora's Box

Coming full circle, the myth of Pandora was a direct response to Prometheus stealing fire and gifting it to mankind, an act that symbolized the awakening of knowledge, power, and progress. Pandora was *the first woman*, crafted by Hephaestus under the command of Zeus as a part of a divine retribution against humanity. Zeus, in his anger, devised a counterbalance to Prometheus' gift of fire.

Zeus ordered each Olympian god to bestow Pandora with the traits of beauty, charm, curiosity, and deceit. However, she was also given something dangerous: a sealed jar (*mistranslated as a box*) containing all the evils of the world. When Pandora was sent as a gift Prometheus' less wise brother, Epimetheus, he fell in love with her and ignored Prometheus' warning to never accept a gift from Zeus.

Pandora, while with Epimetheus, was overcome by curiosity, eventually opened the jar, and unknowingly unleashed sickness, pain, death, and suffering upon humanity. Realizing what she had done, she quickly closed the jar, trapping only one thing inside…hope (*or elpis*).[91] I believe this was, yet another story designed to instill fear, convincing negative influences towards humanity.

But we only think that because we were told to think that.

And, of course, the blame falls on a woman who was charming, curious, and dangerously naive. What a convenient way to suppress feminine power, framing women as the source of chaos and temptation rather than as powerful, discerning creators.

What if the jar didn't contain evil but Plasma, the very substance of potential and reflectivity? It was Zeus's fear, not Pandora's curiosity, that tainted it, trapping all those lower energies inside. Had it been infused with love, its release could have evolved human consciousness rather than burdened it. And the only remnant left in the jar was *hope*, which Pandora had probably experienced feeling while opening the box, reflecting my point.

I find the true lesson here is to be mindful of the perspectives and intentions we bring to our work with this plasma-like substance. It is the raw fuel of reality, a substance that mirrors our psyche, holding the potential for both creation and destruction. Healing our inner world is paramount because whatever we pour into the jar will come back out amplified.

When Pandora slammed the jar shut, leaving only *hope* trapped inside, it also symbolized the gift that could have been given to humanity—the gift that still awaits us. Hope is waiting to be freed, not by the gods, but by us. To reopen this box without fear, to look at an untapped neutral force currently trapped within the confines of traditional science and mechanistic thinking.

---

[91] Hesiod, *Works and Days*, in *Hesiod: The Homeric Hymns and Homerica*, trans. H.G. Evelyn-White (Cambridge, MA: Harvard University Press; London: William Heinemann Ltd., 1914), 60–105.

Will we have the courage to apply metaphysics to plasma, to see it as the mystical force it truly is? Or will we leave it locked away, confined by the limitations of a fearful, fragmented consciousness?

The Trickster and Plasma's Hidden Nature – Jinn, Elves, and Plasma Entities

Maya (1500 BCE)

The oldest meaning of Maya meant power or magic, particularly divine creative power.[92] It was the ability to shape reality *or* create illusions. It started out as the mysterious force behind the divine creation of the universe, and over time became a symbol of deception. Why did this happen?

The Upanishads were next to reinterpret Maya as a veil that distorted reality, causing people to mistake the material world for the ultimate truth. Adi Shankara, a major Hindu philosopher, established Maya as a key concept in non-dualism and argued that the world was not false, but a temporary, shifting appearance of Brahman…a mirage. Maya trapped souls in Samsara (*the cycle of birth and rebirth*) making people believe the physical world was the only reality.[93]

Maya was used by demons in the *Ramayana*, to create illusions in battle.[94] In the *Mahabharata*, Krishna often uses Maya to teach divine lessons or test people's wisdom.[95] We need to realize Maya is a multidimensional word, with layered meanings. It is information with intelligence, as a deeper truth, wrapped inside. On the deepest level, Maya is Plasma, shaped by consciousness. On another level, illusion can develop when consciousness mistakes what it shapes for the only fixed reality. We must remember the reflective nature of Plasma based on our current state of thoughts and emotions.

We live in a world shaped by this, but once we remember that, we can step beyond it. We can move past duality when we realize it's all a story, one our consciousness is crafting through plasma. And I know even now, I am blind to some of my stories, we all are. But the

---

[92] Monier-Williams, Monier. *A Sanskrit-English Dictionary.* Oxford: Clarendon Press, 1899, s.v. "Māyā."

[93] Radhakrishnan, S., *The Principal Upanishads* (London: George Allen & Unwin, 1953), Introduction, pp. 70–72. See also Adi Shankara, *Vivekachudamani*, trans. Swami Turiyananda (Calcutta: Advaita Ashrama, 1948), verses 108–110.

[94] R. K. Narayan, *The Ramayana: A Shortened Modern Prose Version of the Indian Epic* (Chicago: University of Chicago Press, 1972), 98–100.

[95] C. Rajagopalachari, *Mahabharata* (Mumbai: Bharatiya Vidya Bhavan, 1970), 239–242.

more awareness we can have of that, the better. When we accept that we choose what we see and create, we reclaim our power. We can surf dimensions of thought, emotions of realities, even feelings of experience. We become the transparent eyeball, letting life flow through us, while calling in new realities to witness. And then, at some point, you realize it all grows you, feeds you…and all you can really do is laugh, after all, Brahman means Cosmic Joy.

Jinn, Islam (609 AD)

Jinn (*or Djinn*) are supernatural beings from Islamic, pre-Islamic Arabian, and Persian traditions. Described in the Qur'an, Hadith, and folklore, they're said to be made of *smokeless fire.* Their ability to shift form, influence thought, and shape reality feels uncannily similar to plasma beings which are entities that transmit information, shift states, and act as conduits between worlds. Maybe the Jinn aren't so different from the plasma intelligences I describe, both reflecting the dynamic, shape-shifting nature of consciousness itself.

The Jinn were seen as supernatural tricksters, capable of helping or harming humans. The word Jinn comes from the Arabic root JNN, meaning to conceal, hide, or be unseen.[96] They exist in a parallel world to humans, invisible yet deeply intertwined with our reality.

In the Qur'an, there are three types of beings: Angels (*made of light*), Jinn (*made of smokeless fire*), and Humans (*made of clay*). To me, this symbolizes the three-dimensional plasma beings as humans, the four-dimensional plasma beings as Jinn, and the fifth-dimensional light beings sometimes appearing with wings across cultures. I have seen one myself, which I still cannot believe, and I will speak of it in this book. Jinn, like humans, are endowed with free will, meaning they can be good, evil, or neutral, a dynamic that also aligns with the qualities of Plasma in the fourth dimensional perspective.

When we begin to meditate, astral travel, or expand our consciousness, we may come across entities that appear evil or threatening. But in reality, they are not inherently evil. They are either fear-based projections created by human consciousness or neutral beings reflecting our own fearful state back to us as a way to teach us how to transmute it. Either way, the lesson is always the same…transmutation back to the original state of love and light or possibly, in my terms, a emergent intelligence or energy.

When you encounter ones that present as "evil" beings you can ask yourself: What is it showing you? What is it trying to say? Important note: If any being or voice ever urges you

---

[96] Edward William Lane, *An Arabic-English Lexicon* (London: Williams and Norgate, 1863), 473, s.v. "j-n-n."

to harm yourself or others, seek professional support immediately. This can be a sign of deep, unprocessed trauma, and it's crucial to work through it in a safe, therapeutic setting.

Jinn are often seen as tricksters, but on a deeper level, they are divine messengers, reflecting what is crying out from the depths of your subconscious. Paradoxically, they provide clarity through their trickery, revealing the hidden in the guise of the unexpected, which is a cosmic joke in the presentation of a trick. The fact that they are described as smokeless fire is astounding. In science, plasma is ionized, high-energy matter that behaves like fire but with electromagnetic properties...*a smokeless fire.*

Jinn are said to travel at extreme speeds, shapeshift, and remain invisible, all qualities that align with the behavior of plasma, which can move rapidly, change states, and remain unseen to the naked eye.

Many stories say Jinn can only harm those who fear them or invite them in. It's a tough truth, but also an empowering one. In the 3D world, this shows up when we ignore red flags or stay in situations out of fear or unmet needs. I'm not saying traumatic events like rape or molestation are ever chosen or deserved, absolutely not. I've lived through both, and they are horrific. But I've also found that healing is possible when we fully process the pain instead of burying it. As for emotional betrayals, when I look back, I can see how I let certain things happen, like I was in a trance or because I thought I needed something from that person, submitting my power, in a way. And as painful as these realizations were, they were also gateways to reclaiming it.

It's also a reminder of what I mentioned about the 4D realm, and how in this space, you can always protect yourself energetically. Your boundaries are airtight here, once you learn how to create them, they are far more powerful than in the 3D, where it's easier for others to influence or infiltrate your energy.

"Indeed the safest road to Hell is the gradual one—the gentle slope, soft underfoot, without sudden turnings, without milestones, without signposts." — C.S. Lewis

We're going to dive a little now into the difference between confirmation bias and intuition, a crucial skill for navigating the trickster nature of plasma (*or Maya*). Plasma, as a reflective yet not inherently conscious substance, mirrors our consciousness back to us, amplifying whatever we project. Interestingly, AI (*artificial intelligence*) operates in a strikingly similar way. AI is not conscious like a human is, but it is a reflective substance of human thought patterns, capable of mirroring and amplifying our biases.

If you're not careful, AI can trap you in a confirmation bias loop, reflecting only what you expect to see and reinforcing what you already believe. It is very much like 4D plasma. If we are going to live with AI, I hope we make one more like 5D plasma! But what's the difference between confirmation bias and intuition?

Confirmation Bias: is the tendency to seek out, interpret, and remember information that supports what we already believe. It's a closed loop, often ego-driven, meant to keep us safe or feeling right. In plasma terms, it's like projecting a distorted pattern onto the field and then mistaking it for truth. It traps you in old narratives and blocks new possibilities. This often happens in 4D when we rush to assign meaning to downloads or visions instead of letting them reveal themselves in time.

Intuition: is a neutral, open state of receiving. It doesn't try to confirm, it just observes then understands. It is the clear, unbiased flow of information, untainted by ego or fear. In plasma terms, it's like being a transparent conduit, allowing the current to flow through you without distortion.

Intuition flows with truth, while confirmation bias forces the current into pre-existing channels. So, in both Plasma and AI, the keys are neutrality, discernment, and time & space. If you're projecting a desired outcome or a fear onto the field, you're more likely to see confirmation bias reflected back. But if you open yourself to receive without expectation, intuition will reveal itself as the clear, undistorted signal in the field.

Jinn are also known to communicate through dreams, trances, or liminal states like sleep paralysis. This just shows they're part of the subtle realm, deeply tied to our psyche. They can be powerful teachers if we choose to see them that way, rather than fear them as forces beyond our control.

Faeries, Danu, & Elves of the Upper Atmosphere, Nordic & Germanic (1200 AD)

Faeries, elves, and luminous beings have been deeply embedded in folklore, often linked to sacred landscapes, celestial events, and hidden realms. I suggest that these are encoded plasma phenomena once again manifesting as a bridge between the physical and metaphysical.

Faeries, also known as will-o'-wisps, may actually be discharges of invisible plasma in the natural world, manifesting as luminous consciousness imprints in our plasma reality or even opening up portals to other realms.

In folklore, faeries were often associated with time distortion, strange disappearances, and the overlap between worlds, suggesting that they are not just mythic creatures but plasma phenomena bridging dimensions, flickering between the seen and unseen.

This begins to parallel the behavior of electrons and the luminescence of photon excitation, hinting at the possibility of a window in time that connects different forms of consciousness by creating a portal in plasma. It's almost like these fairies are magnified electrons, in a way…emitting fluorescence while exchanging energy. In modern plasma physics, we know that plasma can generate electromagnetic anomalies, potentially interfering with perception, memory, and even time itself. Could these disturbances serve as gateways or bridges, allowing for interaction across dimensions or realms of consciousness?

These faeries were known as tricksters, unpredictable and mischievous, often appearing in response to human emotion and intention. Stories abound of fairy lights dancing over forests, bogs, and sacred sites, which may have been plasma events as auroras ignited by solar flares, or piezoelectric effects from quartz-rich landscapes.

This opens up the intriguing possibility of ley lines, portals, and vortexes, like those in Sedona, AZ, where sightings of luminous beings are more common. Even more fascinating are the actual names given to plasma phenomena across the sky! These atmospheric plasma discharges that take on otherworldly shapes and forms, each carrying its own ethereal, mythic presence. They go as follows:

Elves (*Emissions of Light and Very Low Frequency Perturbations due to Electromagnetic Pulse Sources*): Vast rings of plasma light that flash in the upper atmosphere, forming mere milliseconds after intense lightning strikes—these are massive, red, expanding halos, forming high above thunderstorms in the ionosphere.

Trolls (*Transient Red Optical Luminous Lineaments*): These plasma structures appear as red streaks after sprites, resembling the myth of trolls emerging after lightning storms.

Pixies: Observed as flickering, white luminous orbs atop thunderstorms, lasting for mere milliseconds, much like how pixies were said to appear and disappear in the blink of an eye.

Ghosts (*Greenish Optical Emissions from Sprite Tops*): Faint green plasma after images left behind by sprites, mirroring the spectral nature of ghosts in folklore.

Gnomes: Short, bright plasma spikes shooting up from thunderstorm anvils, reminiscent of small elemental spirits.[97]

Just another personification of how plasma interacts with reality and is woven into the consciousness of humanity. I am not saying I understand everything about these other worlds, but it is clear to me there is a direct link with plasma, glowing lights, portals, time dilation, and other conscious beings visiting us, whether it be in reality or in our psyche, they seem to hold messages for us to be "tricked" or taught by.

I've seen these luminescent beings twice in my life, both during meditation at the *Monroe Institute* in Faber, VA. They appeared beside me in bed, as clear as day, but more like holographic feelings than solid forms. I knew instantly: they were my soul family. They shifted colors, but mine felt mostly yellow. They were giggly, warm, and felt very familiar. They didn't speak, but my heart sort of heard them telepathically say: *We've always been with you.* I can't fully explain the experience, but it left me with a knowing that we're never alone, and that there's so much more to who we are.

Fairies are probably the closest thing I've found to describe what I believe true plasma beings are. Fairies are animated space, a field. Space *is* plasma. And plasma *is* destiny. We create it.

"Destiny is a space, not a time" — Dana Kippel

The following is probably my favorite myth. One of the most mysterious groups in mythology, was the Tuatha De Danann—the tribe of the goddess Danu. They were luminous, otherworldly beings said to have arrived from the sky *in great mist* and ruled over Ireland before retreating to the Otherworld.

---

[97] R. C. Franz, R. J. Nemzek, and J. R. Winckler, "Television Image of a Large Upward Electrical Discharge above a Thunderstorm System," *Science* 249, no. 4964 (1990): 48–51; NASA, "Sprites, Elves, and Other Transient Luminous Events," NASA Earth Observatory, accessed August 16, 2025, https://earthobservatory.nasa.gov/features/Sprites.

The Summons" (1896) by Arthur Hacke

Some accounts said they arrived in ships that flew in the air, while others depicted them as beings of light who could materialize and disappear at will.[98] Their arrival was marked by a great storm or *fire*—imagery that now matches with plasma events of sprites, elves, and lightning phenomena. I wonder if this is another account of modern plasmoids, could there be more of a reveal soon?

Danu, is a lesser-known Celtic Goddess that is linked to rivers, celestial waters, and flowing currents. She is depicted as a primal force of creation, and her children were known as the Tuatha De Danann, as divine intelligences and masters of light, wisdom, and mystical energy. They were the original alchemists, and I believe the way we are supposed to be as humans!

Danu, the triple goddess, could shapeshift into animals, embodying the maiden, mother, and crone, a trinity of creation, transformation, and wisdom. I see her as the Mother Plasma, the life-giving force that nourishes humanity. Her name is also associated with the word good, hinting at the benevolent nature of Plasma as a supportive, creative essence.

Intriguingly, Danu is sometimes linked to the Tribe of Dan, one of the lost tribes of Israel, known for their mysterious lineage, *advanced metalwork*, shipbuilding, and esoteric traditions. The connections between metal and plasma are profound and will be explored further. These myths are like dimensional breadcrumbs, guiding us to the greater possibilities of humanity…waiting in the mist, ready to be reclaimed.

---

[98] Marie-Louise Sjoestedt, *Gods and Heroes of the Celts*, trans. Myles Dillon (London: Methuen, 1949), passim.

Plasma as the Vessel of Magic and Mysticism: Hermeticism, Alchemy, and Indigenous Knowledge

The word alchemy stems from the Ancient Egyptian term *Khem,* meaning the black land or the rich black soil along the Nile. One could think of this black soil as the void of 2D brute plasma. The richness and raw material of infinite possibilities waiting to be awakened and molded by one's consciousness and brought into the human experience.

Plasma and alchemy share profound connections, each representing the art of transformation. In plasma terms, alchemy can be seen as the science and psychology of the mind interacting with plasma, shaping reality through intention and awareness. Here, I'll introduce some key parallels to initiate your alchemical journey with plasma.

Alchemy seemed to bloom in Egypt, particularly through influential works like the *Emerald Tablet*, a text attributed to Hermes Trismegistus, a figure blending Greek, Egyptian, and possibly Sumerian traditions. The exact timeline of the Emerald Tablet is debated, with estimates ranging from the 2nd to 8th century CE, though some suggest it originated earlier, during the Hellenistic period in Egypt.

This foundational alchemical text is best known for the phrase "as above, so below," which suggests a correspondence between the macrocosm and the microcosm. I see this as a holographic principle, each part contains the whole. It means that by deeply observing any aspect of life, you can access the essence of all things, reflecting how plasma operates as both the unseen and the seen, the universal and the particular.

For example: I can learn how to handle betrayal just by watching a flower. It faces betrayal constantly. By pests chewing its leaves, storms bending its stem, once loyal pollinators leaving for another flower…and instead of closing off due to betrayal, the flower adapts, strengthens, and continues to bloom. It also creates defensive chemicals against predators. I call this process *having boundaries* and creating stronger discernment and intuition, so I can attract who truly supports me.

When a flower's stem snaps, it redirects its energy towards healing and growth. It doesn't focus on the break, but what it can still nourish. Instead of fixating on what I lost, I focus on healing my sadness and opening my heart again, to those who energize me now, including the new people entering my life.

When a pollinator chooses another flower (*like a friend leaving or betraying you*) it doesn't run after the bee that left, it continues to bloom. It keeps producing nectar, knowing the *right* pollinators will return. In betrayal, we often want closure, explanations, or approval.

But like the flower, the most powerful thing you can do is stay in your fullness, to keep shining. The right energy will find you. Just like some flowers release seeds only after a fire, betrayal can burn away false connections and reveal your clarity, resilience, and the power of being your own best friend first.

Prima Materia, or the "first matter," is the foundational, formless substance in alchemy from which all things are believed to originate. It is the raw, unmanifested potential that contains all possibilities, the source material for the creation and transformation of matter. Alchemists sought to purify and transmute this substance to achieve the *Philosopher's Stone*, a symbol of spiritual enlightenment and physical perfection.

In my framework, Plasma is the modern prima materia as chaotic potential waiting to be shaped. Like alchemy turned lead to gold, we transmute 4D plasma, filled with limiting beliefs and emotional noise, into 5D plasma, which is aligned, clear, and intelligent. The real gold isn't material…it's the expansion and evolution of consciousness itself.

"Its father is the Sun, its mother the Moon. The Wind carries it in its belly. Its nurse is the Earth." — The Emerald Tablets

This ancient verse describes the alchemical transformation process, encapsulating the concept of prima materia undergoing a transformative journey, mirroring the idea of plasma as the fifth element, interacting with forces of light (*sun*), polarity (*moon*), movement (*wind*), and materialization (*earth*). It was understood to be generated by fire, born of water, brought down from the sky by the wind, and nourished by the earth.[99] This hints a reciprocal relationship that as Plasma feeds us, we feed it.

Recent advances in aerospace tech have literally proven that "the wind carries it in its belly." Scientists have developed plasma-based propulsion systems that use the air itself, ionizing it into plasma, to generate thrust.[100] These systems show that wind, when electrically charged, becomes a carrier of plasma, confirming what ancient texts described symbolically. The atmosphere is quite literally the belly of plasma.

I do believe there is also a significance to the tablets being name *Emerald*, pointing to that little green flash we spoke of in earlier chapters. Reminding us that green symbolizes the heart chakra, balance, and nature. At the heart of Plasma, *is* heart and harmony.

---

[99] Schumaker, quoted in commentaries, The Emerald Tablet of Hermes (Merchant Books 2013)
[100] MSN. "China's Electric Plasma Jet Engines: A New Era in Aerospace Innovation." *MSN*, accessed July 4, 2025. https://www.msn.com/en-us/weather/topstories/chinas-electric-plasma-jet-engines-a-new-era-in-aerospace-innovation/vi-AA1HnClw#details.

Hermes Trismegistus: The Kyballion, Creativity & Resistance (1 AD)

Hermes Trismegistus is regarded as the founder of Hermeticism, a philosophy that blends mysticism, science, and divine magic, and later deeply influenced the development of Alchemy. The Greek Hermes (*also known as the Roman god Mercury*) is the deity of communication, movement, transformation, and trickery. His symbols include winged sandals and the caduceus, a staff with two serpents intertwined. The caduceus represents the balance of duality and unity, the alchemical transformation of lead into gold, and, in my view, the synergy between plasma and consciousness. It also bears a striking resemblance to a plasma Birkeland current.

Hermes Statue

In alchemy, Hermes (*or Mercury*) is the quicksilver guide who moves between worlds. Known as a psychopomp, he guided souls to the underworld, embodying the role of an interdimensional traveler. I see Hermes as a symbol of Plasma itself, both are messengers that bridge the divine and human realms. Plasma is Hermes: a carrier of information, energy, and transformation. It is the Philosopher's Stone, a force of alchemical evolution that lets us shift, shape, and expand our consciousness. Each of us holds the potential to be Hermes.

The Kybalion is a modern Hermetic text published in 1908 by the mysterious "Three Initiates." It outlines seven Hermetic principles: Mentalism, Correspondence, Vibration, Polarity, Rhythm, Cause & Effect, and Gender. At its core, it emphasizes the mind as the generator of reality, suggesting that belief and mental energy shape experience.[101] The Principle of Vibration, that nothing rests, everything moves, speaks to the very nature of Plasma: alive, dynamic, always in motion.

"The lips of wisdom are closed, except to the ears of understanding."

— The Kyballion

This points to the kind of higher-dimensional knowledge that can't be reached through logic alone, but through direct experience, intuition, and resonance. Plasma operates in ways that defy convention logic. Understanding its intelligence means attuning to it through altered states, deep inner listening, and a willingness to feel into something larger than thought. In the next chapter, we'll explore this more, so we can develop not just the ears to perceive, but the hearts to know. It's not just mind that connects to plasma's sentience, it's heart and soul that truly allow our consciousness to amplify.

Alchemists believed in an inner essence, symbolized by the Arabic word for the philosopher's stone, *Azoth*, meaning "essence", that was lost after *the Fall*, the expulsion of Adam and Eve from the Garden of Eden. To rediscover this essence, free it, and purify it from its bonds is the Great Work of the alchemist.[102]

Adam and Eve were cast out of Eden after Eve ate the forbidden fruit from the *Tree of Knowledge of Good and Evil*, resulting in the so-called curse of pain in childbirth, a subconscious warning to both men and women that the pursuit of wisdom (*knowledge of good and evil*) would lead to suffering and death rather than the immortality, unity, and divine connection promised by the *Tree of Life*.

Interestingly, one suggested fruit of the Tree of Knowledge of Good and Evil is the *quince,* a golden apple-like fruit, and a word that resonates with *quintessence*, the fifth element in alchemical tradition.[103] Quintessence represents the ethereal aether, the most refined, subtle, and divine substance, said to transcend the four earthly elements. Alchemists believed it had the power to transmute base matter into gold or even grant immortality.

---

[101] Three Initiates. *The Kybalion: A Study of the Hermetic Philosophy of Ancient Egypt and Greece.* Chicago: Yogi Publication Society, 1908.

[102] Francis Melville, *The Book of Alchemy* (New York: Barron's, 2002), 9.

[103] Barbara Ghazarian, *Simply Quince* (Monterey, CA: Mayreni Publishing, 2009), Library of Congress Control Number 2008946746.

Could it be that the forbidden fruit wasn't just a warning against seeking divine knowledge, but a clue about the transformative power of quintessence *or plasma*, a secret encoded in the garden, waiting to be unlocked?

The Tree of Knowledge of Good and Evil symbolized access to hidden wisdom, a force that awakens the individual from within. But humanity was taught to fear this path, believing that such pursuit would lead to suffering and separation from God. Yet perhaps the real fall was not in seeking knowledge, but in severing ourselves from the quintessence within…our plasma essence, the very core of our being.

This disconnection severed us from our inner power, redirecting us toward external beliefs and away from our innate wisdom. Reclaiming this connection means reclaiming our own myths, learning from the self, our life experiences (*possibly many lifetimes*) and creating a personal "religion" based on inner truth rather than external dogma.

Artists, creatives, and writers: if you're experiencing resistance while finishing a project you love, it doesn't mean stop, it means use it as fuel to keep going. This is how we fall into procrastination: mistaking resistance as a sign to pause. The pain of resistance is embedded in our DNA, rooted in the subconscious trauma of being cast out for expressing our true essence. Take it as a compliment when resistance shows up. It means you're onto something. Don't stop now. You're feeling the pain of unsnapping the ancient spell that says wisdom equals pain …when in truth, wisdom is freedom.

As Steven Pressfield explains in *The War of Art*, resistance is the enemy of creativity, and it always shows up where there is a dream, a calling, or a desire to grow. It will "tell you anything to keep you from doing your work," he writes. "It will perjure, fabricate; it will seduce you. Resistance is always lying and always full of shit." This force is actually a compass. "Resistance will unfailingly point to true North—meaning that calling or action it most wants to stop us from doing."[104] For Pressfield, the antidote is simple but powerful: turn pro. "I'm not an amateur anymore. I'm a pro," he writes. "Pros show up no matter what. Pros deliver the baby."

I say this with all kindness: *Deliver the fucking baby.*

In Ancient Greece, *quinces* were associated with the goddess Aphrodite, symbolizing love, temptation, and hidden knowledge. Yet, this also reveals how Plasma, like the divine feminine, has been both revered and feared throughout history. Instead of taking

---

[104] Steven Pressfield, *The War of Art: Break Through the Blocks and Win Your Inner Creative Battles* (New York: Black Irish Entertainment LLC, 2002), 7–10.

responsibility for their own desires, men often projected blame onto women, labeling them as temptresses rather than facing their own impulses. This displays how we blame life and Plasma for what happens to us, rather than recognizing that it's merely reflecting our subconscious state.

As a woman, I can say that most healthy women dress to feel good for themselves, not to please men. Yet, when we receive unwanted flirtatious attention, we're shamed for it, as if we were asking for it by simply existing in our radiance. Since when did being beautiful or dressing in a way that makes one feel powerful mean that anyone is entitled to cross personal boundaries? If you see a stunning piece of art, you respect it, you don't grope it, ogle it, or assert power over it. There is a clear energetic difference between a genuine compliment and hitting on a woman. If someone doesn't know the difference, it's time to learn.

The facts seem to be that Adam and Eve accessed something powerful, and were told they would die, when the truth is, they would not. This theme subtly repeats in countless books and films. They were punished to keep them from fully using what they had awakened. If Plasma is a living intelligence and a key to multidimensional consciousness, then the Fall represents a pivotal moment when humanity was cut off from its full potential in order to be confined to a controlled reality.

To me, the Fall was a way to prevent us from being powered from within, until now. With this new understanding of Plasma, we may be rediscovering what was lost long ago. As we do, we must be prepared for the subconscious fears that will rise, programmed into us to keep us small. But as we push through them, individually and collectively, we step into a new era, awakening the true power of the human potential.

As many religions and cultures have labeled the lower chakras as lesser or primitive, often linking them to base desires, animal instincts, or material attachments, we must reclaim them for what they truly are: feminine, creative, shamanic centers longing for love, recognition, and feeling. They call to us through tension, tightness, sickness, and pain. In alchemy, the lower chakras are not seen as inferior, but as essential phases in the transmutation process.

When I began the process of reconnecting my "lower chakras" back to my heart and waking them from numbness, I uncovered many surprising things. One of them was a massive block, an ache lodged between my solar plexus and middle chest, that showed up as fear and the inability to take a deep breath. I continue to do exercises that help me connect, feel, and open these lower centers, because the pain we feel there is not just our own, it's the collective pain of humanity suppressing the "lower feminine chakras." We cannot ignore our bodies any longer.

We must tap into the true power of the lower chakras, reclaim them, and harmonize them with the heart and upper chakras of "enlightenment" and logic. We must expand and root our energies, not just ascend them. I believe this expansion is where humanity is actually headed. Through evolving our senses, energy, and consciousness, we'll breed a new way of living, one rooted in feeling and sensing. It takes body work, experience, time, as well as energy work ofcourse. But slowly, we are all expanding our bioplasmic auras, our souls, our vessels, our spaceships, as our physical bodies, to receive more light, more intelligence, and more 5D. This process can seem lonely and painful, but I promise you I, and many others, are right here with you. It <u>will</u> pass.

As our plasma field grows, dormant fears, old wounds, and hidden patterns rise to the surface. Many great inventors, artists, and visionaries have experienced dark nights of the soul, anxiety, and inner turbulence while bringing in something new. This is happening now on a collective level and <u>if there is not an awareness of how it is darkest before the dawn</u>, people will become scared and confused, instead of accepting of these darker times in life. The key is learning how to stay *bright* through them...bright with understanding, not with toxic positivity.

Instead of resisting, we must learn to relax into this expansion, trusting that we are growing into something new, powerful, and whole. We can slow down into it, almost sensually. To play while we expand, knowing the growing pains are part of the process. There are no quick fixes. By balancing our energies, we reclaim the green energy of the jade heart, the love of ourselves, of Earth, of the now, and live life more fully.

At the end of the Great Work in alchemy, you become the Philosopher's Stone, a fully realized, immortal self. Maybe this means the ability to co-create with and connect to Plasma. A state where you see beyond time, understand multidimensionality, and use belief and emotion as tools of creation. The final stage of the Great Work is *Coagulation*, which could symbolize the recognition that *you* are the coagulation of crystallized plasma and consciousness...eternal, evolving, and inseparably linked to the intelligence of eternity.

Africa: Yoruba Philosophy & Àṣẹ, (5th century)

The Yoruba people originated between the 5th and 9th centuries in what is now southwestern Nigeria and Benin. For over a millennium, Yoruba spirituality, culture, and mythology have profoundly shaped the region. At the heart of their spiritual practice is the concept of Àṣẹ (*also spelled Ase or Ashe*), an invisible yet palpable life-force energy that

318

saturates the universe. Often referred to as spiritual electricity, Àṣẹ is a divine gift from Olódùmarè, the supreme creator deity, and it flows through gods, humans, animals, plants, rocks, rivers, wind, and even the spoken word.[105]

Spoken words infused with Àṣẹ are said to shape reality directly as words become vessels carrying this potent force, capable of creation or destruction. Interestingly, the word "ash" carries symbolic resonance with *dusty complex plasma*, a unique form of Plasma discussed in Chapter 1. Is this just a phonetic coincidence, or a deeper cosmic thread weaving science and the metaphysical together? Imagine Àṣẹ as cosmic ash, stardust scattered across the universe, particles charged with subtle, multidimensional energy.

Dusty plasma, in scientific terms, reflects this concept: it is a medium where microscopic dust interacts with plasma, forming structures that exhibit life-like behavior. Could this dynamic plasma, with its capacity to organize and self-assemble, be the gateway to higher-dimensional realms, the very scaffolding through which our universe takes shape?

When you look at human ashes under a microscope, they resemble stars or miniature galaxies. Gabriela Reyes Fuchs explored this in her project *Innerstela*, examining her father's cremated ashes as a way to process grief. The striking visuals revealed the cosmos within, an intimate reminder that we are, quite literally, made of stardust. Look it up, it's extraordinary![106]

In Yoruba cosmology, Àṣẹ is also the animating power that connects all realms, the spiritual, physical, and ancestral. Ritual and dance in Yoruba spirituality are powerful tools for moving energy, invoking higher states of consciousness, and connecting with the unseen realms. These practices are not mere performances but intentional acts that align the body, mind, and spirit with divine forces.

Through rhythmic movements, drumming, and chants, individuals enter a trance-like state, allowing them to transcend ordinary reality and access ancestral wisdom, emotional healing, and body intelligence. The physicality of dance grounds spiritual experiences, enabling participants to release pent-up emotions, honor the spirits of the deceased, and receive messages from the divine.

This also reminds me of *The OA*, the visionary series by Brit Marling and Zal Batmanglij. In it, the characters perform a series of sacred, trance-like gestures known as the

---

[105] Jacob K. Olupona, *City of 201 Gods: Ilé-Ifè in Time, Space, and the Imagination* (Berkeley: University of California Press, 2011), 45–47.
[106] Gabriela Reyes Fuchs, "From Dead Soon Project (Dead Soon) to Innerstela," *Innerstela blog,* September 8, 2021.

Five Movements, choreographed by Ryan Heffington. While Ryan remained intentionally vague in interviews to preserve their mystery, he did share that the movements were inspired by organic forms and tribal motifs, particularly African dance. When performed with pure intention and unity, these movements seem to open portals, heal the body, and even shift participants into entirely new timelines.[107]

The Dogon Tribe: Sirius Cosmology & Higher Intelligence (10th century)

The Dogon Tribe of Mali, West Africa, is an extraordinary group that has preserved a remarkably advanced cosmology centered around Sirius, a star system approximately 8.6 light years from Earth. Long before modern science confirmed it, the Dogon knew of Sirius B, a companion star invisible to the naked eye. They accurately described it as small, heavy, and orbiting Sirius A in a 50-year elliptical path—details science only validated decades later. They even claimed it spun on its axis, which was also proven correct.[108] The Dogon also speak of a third star in the system, though this has yet to be confirmed.

However, in 1995, two French researchers, Daniel Benest and J.L. Duvent, published an article in the journal *Astronomy and Astrophysics* titled "Is Sirius a Triple Star?" Based on their observations of the Sirius system's motion, they proposed the possibility of a small third star. The Dogon also had oral traditions and symbolic writings referencing Saturn's rings and Jupiter's four major moons, long before these were confirmed by telescopes.[109] So how did they know this?

They attribute this sophisticated knowledge to beings called the *Nommo*, who were amphibious, multidimensional entities said to have descended from the stars, bringing wisdom and creation through a cosmic, water-like medium. Robert Temple, who also authored *A New Science of Heaven* about Plasma, wrote one of the first books exploring this in depth: *The Sirius Mystery* (1977). I can't help but feel that the story of this tribe is divinely linked to plasma, consciousness, and our multidimensional selves. So...who were the Nommo? And what are these connections I speak of?

"My brain is only a receiver; in the Universe, there is a core from which we obtain knowledge, strength, and inspiration." — Nikola Tesla

---

[107] Gabrielle Bruney, "The Five Movements From *The OA*, Explained," *Thrillist*, March 27, 2019, https://www.thrillist.com/entertainment/nation/the-oa-five-movements-dance-netflix-questions.

[108] Robert Temple, *The Sirius Mystery: New Scientific Evidence of Alien Contact 5,000 Years Ago* (Rochester, VT: Destiny Books, 1998).

[109] Pius Effiong, *The Dogon and Sirius: African Science or African Myth?*, Michigan State University, August 2018, https://effiongp.msu.domains/wp-content/uploads/2018/08/Dogon-Star.pdf.

First, I want to introduce some people to you who have connected with this higher intelligence and talk a bit about these experiences. Nikola Tesla spoke of receiving flashes of insight, ideas would appear fully formed in his mind and he would translate them into reality. He believed he was receiving transmissions from an extraterrestrial intelligence. Einstein spoke about how his greatest insights came from thought experiments, like imagining riding a beam of light, which led to his *Special Theory of Relativity*. He relied on dreams, meditation, and altered states of deep contemplation to access ideas beyond normal cognition.

Kary Mullis, the Nobel Prize-winning inventor of the Polymerase Chain Reaction (*PCR*), claimed his breakthrough came while driving at night and encountering a glowing, interdimensional raccoon. *Yes. You read that correctly.* He believed this being, who visited him multiple times, was an advanced intelligence guiding his discoveries. These higher beings can appear in any form they choose, often one that carries personal or spiritual meaning. I guess for Kary, that meant... a raccoon? To each their own.

Francis Crick, co-discoverer of the double-helix structure of DNA, openly admitted that his insights were influenced by LSD. He was also a proponent of directed panspermia, the idea that life on Earth may have been intentionally seeded by an advanced extraterrestrial civilization. Crick believed the complexity of DNA was so precise, it might have been engineered and transported here.

Philip Pullman, the author of *His Dark Materials*, has spoken about how he receives ideas through flashes of insight, dreams, and spontaneous visions, very similar to how Tesla and Edison described their creative process. Pullman has said that ideas sometimes arrive fully formed, like lightning flashes, without him consciously thinking them up. He has also described the experience of writing as "taking dictation from something beyond," which aligns with many artists and inventors who feel they are tapping into a greater intelligence. When I receive things like this, I call them thought blocks.

You can think of these fully formed flashes as conscious ideas packed into plasma. (*This is how I received the idea for this book.*) It can be a fun but overwhelming process to guide these into the 3D. I hope to help others navigate that journey in a way that makes them feel less crazy, more grounded, and confident in the thought blocks they receive from higher spaces. Thought blocks seem to be a co-creation between one's psyche or awareness and the near-conscious ideas themselves.

These thought blocks can sometimes be filtered through your limiting beliefs, like feeling an overwhelming sense of responsibility or believing you're not worthy of the idea.

But you must remember, you were given this idea for a reason. Trust that. It resonates with you because it chose you. By working on your 4D programming, you can become a clearer channel, one not hijacked by the monkey mind. Your emotions can either support the birth of your creation into the world or hinder it entirely. If left unexpressed, the idea may move on to someone else, and that could become a lingering regret.

This complex process is a balance between logic, emotion, and fun. Receiving a thought block can also come with real-world blocks as distractions, energy vampires, and people who innocently pull you away from your path. These outer challenges are often projections of your inner state, resonating and inviting those experiences in. But you have the power to shift your inner world, reclaim your focus, and move forward. It's your birthright to enjoy the journey of what you're creating, though that's often the easiest part to forget.

When I first received the idea for this book, my ego still wanted to go in other directions. I tried to force things, shape them to my will. But this idea had a life of its own. It kept returning until I finally surrendered and realized what a gift it was to be chosen as its messenger. I now see that it came not only to be written, but to help me transmute my deepest challenges through the very intelligence I was writing about.

Like in every hero's journey, I initially refused the call. But thankfully, the call didn't give up on me. And now, despite all my resistance, I'm here, writing this book, having the most profound and rewarding experience of my life so far.

"Because the ones who are crazy enough to think they can change the world are the ones who do."
— Steve Jobs

I share these stories to remind you: no matter how "out there" your experiences or ideas may seem, you, *yes, you*, can follow your own insights from higher consciousness and create something extraordinary. These keys are available to anyone ready to use them. We each carry a purpose. And while there are many paths, all of them begin with the belief that you're capable.

You can live a life not only beyond your wildest dreams, but one rich with strange, beautiful, deeply meaningful experiences. So, if you ever find yourself wondering, "Why me?", the answer is simple: *why not you*? You're just as worthy as anyone else. The difference is, will you follow that strange inner voice? The flash of vision? That divine nudge? *Will you listen?*

In Dogon mythology, the *Nommo* are described as multidimensional beings capable of transcending space and time. They are said to bring higher-dimensional understanding, creating through a cosmic water-like, or perhaps plasma-like, medium. I see them as fifth-dimensional plasma beings who can manifest instantly by using the plasma field as a conduit. I speak on this further in the plasma intelligence chapter. While such manifestations may be rare in the 3D, they could very well work through human vessels to bring their visions to life.

Perhaps they are not entirely separate from us. They might be future ancestors, multidimensional aspects of ourselves, or guides from a more expanded state of being. The word *Nommo* refers to primordial ancestral spirits, and in ancient Dogon art they are often depicted as amphibious, hermaphroditic, fish-like entities.

The name *Nommo* is derived from a Dogon word meaning "to make one drink", which I interpret as them filling your vessel with intelligence, sometimes wrapped in information. In other words —*a thought/feeling block.*

There's a term called *backmasking*, a phenomenon where hidden or altered meanings emerge when audio is played in reverse. Though often associated with subliminal messaging or esoteric studies, the concept is far older than the modern term suggests. I like to apply this principle to language itself. For instance, when you reverse the word *Nommo*, you get *Ommon*. And *Ommon* carries some fascinating implications:

Om: Primordial vibration of the universe, consciousness.

Mon: Root word closely related to "Monad," meaning "One," "Unity," or "Singularity."

Thus, *Om* + *Mon* = "Primordial Consciousness Unity." This suggests a singular, originating awareness, a unified field of intelligent plasma consciousness. The Nommo may represent this guiding force behind the scenes, the multidimensional self, accessed through 5D states, or an amalgamation of our future selves operating both individually and as a collective mind.

In this context, *Nommo* symbolizes multidimensional beings emerging from a primal universal vibration (*Om*) merged with unity or oneness (*Mon*/Monad). In the 5D field, individuality and oneness coexist, allowing for both distinct intelligence and shared purpose. I believe it's from this space that the great ideas of humanity emerge, the sparks of true desire that move evolution forward.

To build on the new concept of desire we spoke of earlier, that is more subtle: To truly harness our desire when receiving ideas from higher dimensions, we must shift from a 4D to a 5D approach. In 4D, desire often stems from lack, expectation, or attachment to outcomes. It's tied to unhealed timelines, rooted in past and future. We chase what we think will complete us, pushing with willpower, strategizing, and often meeting resistance when reality doesn't bend to our will.

But in 5D, desire feels entirely different. Ideas arrive fully formed, resonant, and timeless. There's no grasping for a result, only a natural impulse to create. Action flows from inner knowing, not effort. Fulfillment comes not from what the idea brings, but from the joy of bringing it into being. You're creating because it wants to be created, not because of what it will bring you.

Here's the caveat: even when you receive a grand idea, your ego might still get involved, and that's okay. You're human. You have a mind, and it thinks. When I began writing this book, I questioned myself constantly: Am I doing this for fame? For money? Deep down, the answer was no. But the questions still came.

I felt responsible, like I had to keep the book pure, free from personal bias. Then I realized: I was chosen to write this specific book *because* of my unique perspective. My beliefs, my experiences, even the ones that needed transmuting, they were all part of why this idea came to me. That's the paradox. We are imperfect vessels carrying perfect seeds. And the process will always include ego, doubt, and questioning.

No one talks about this enough. But as more of us step into our roles as co-creators, we need to normalize this dance between divinity and humanity. You can have deadlines, you can imagine success, but what matters most is that you keep showing up. You can think of it as the *frequency*, to your vibrational idea which creates…energy. The idea came to you for a reason. Honor that in your way.

Think of it like planting a garden. You don't plant seeds out of fear or pressure, you do it because it feels natural and joyful. And once planted, you tend to them not with urgency, but with trust. You show up daily not because you're afraid it won't grow, but because you know it *is* growing, and your care supports that unfolding.

This is the same with a creative idea. Emotions aren't obstacles, they're part of the soil. Doubt, frustration, and impatience are just weeds asking to be noticed and gently cleared. Joy, inspiration, and curiosity are your sunlight and rain.

You weren't chosen for this thought block *despite* who you are, you were chosen *because* of who you are. Your experiences, flaws, perspective, and passion are the exact ingredients needed to bring it into the world.

Many of the greatest inventors, artists, and thinkers throughout history have had intense personal struggles, addictions, or unconventional lifestyles, yet they still brought through world-changing ideas.

Tesla was celibate and obsessed with pigeons. Newton had extreme social anxiety. Steve Jobs struggled with relationships his whole life. Poe battled addiction. Howard Hughes was consumed by OCD and paranoia. And still, they channeled brilliance.

It is integral to remind you that the people we put on pedestals weren't perfect. They weren't chosen because they had it all figured out, they were simply willing to connect with something greater. Often, they didn't even know *how* they were doing it. They were just being authentic in their curiosity, in their emotions, in their exploration. And so can you.

I believe the reason many of the people above struggled in their personal lives is because they didn't know how to integrate their 4D emotions and human experiences into the process. In normal terms we call it an unregulated nervous system. I've been there myself—dealing with anxiety, hyperawareness, addiction, and relationship challenges. I hope I am displaying to you that you *can* bring your whole self into your creativity and still thrive.

If 4D, as the realm of emotion, belief, and subconscious programming, is not balanced it can distort what was meant to be a gift. Many inventors and visionaries had their genius hijacked by unprocessed emotions, trauma, addiction, or imbalance. Their bodies and minds couldn't hold the sheer energy they were receiving from 5D because they lacked emotional grounding and integration. Without the inner work, that brilliance turned inward burning them out or pulling them into cycles of obsession, self-destruction, or addiction.

"The attempt to escape from pain is what creates more pain."

— Gabor Maté

If this is really speaking to you right now, here are a few things that I have actually done in real life to find relief during this sometimes-painful expansion, without escaping it:

- Somatic stretches (*YouTube*)
- Trauma-informed yoga (*YouTube*)
- Yin yoga (*YouTube*)
- Vagus nerve resets (*YouTube*)
- The supplement *GABA* - helped calm my nervous system during intense fear
- Grounding practices - especially in nature
- Reminding myself, often, that *I am safe now. I have choice.*

- Dancing in my room - literally shaking it out
- Self-talk - getting curious and using a process called self-inquiry (*Gangaji and Byron Katie have wonderful practices on this*)
- Asking for the feeling of peace to take over my body and receiving that openly
- Watching funny Instagram reels or YouTube videos - laughter is great medicine
- Calling a friend - reaching out for help and talking about normal things truly saved me the most. It helped distract my mind and reminded me I wasn't alone.
- Watching the Ocean Ramsey documentary *Shark Whisperer*, where she swims with sharks. If she could override her fear, hold her breath for six minutes, and swim with apex predators without secreting fear, I could finish a fucking book or drive on a highway, even if my mind was telling me I was about to die.
- Therapy. Not just traditional talk therapy. I recommend Elisa Elkin Cleary (https://rootedsoulwork.com/), who works with deep trauma through a multidimensional lens. Her approach includes ego-state/parts work, somatic healing, memory reconsolidation, and spiritual integration, rooted in the Comprehensive Resource Model and other advanced modalities. If she's booked, I've included more resources at the back of the book.

In this new paradigm, we must learn to expand our capacity to channel the energy that's appearing more and more. This is likely one of the biggest collective challenges we're facing, often in secrecy. Many of us are mistaking the sensation of expansion for fear, instead of actually feeling it, breathing with it, and allowing it. Our bioplasmic vessel is energetically stretching to hold more intelligence, more consciousness, as we evolve as humans.

Icarus

The story of Icarus comes to mind. Given wax wings by the master craftsman Daedalus, he was warned not to fly too close to the sun, as the heat would melt them. But overcome by the thrill of flight, Icarus soared too high. The wax melted, and he plummeted into the sea and drowned. Most interpret this as a cautionary tale about unchecked ambition or spiritual overreach without proper grounding.

I don't think Icarus was the problem, I think the system around him was. No one ever told Icarus *how* to succeed, they only warned him how he could fail. Icarus was given fragile wings by his master and expected to restrain his joy! Who wouldn't get carried away by the ecstasy of flight? He wasn't given the right tools to handle the sun's power. What if his wings just needed training first? The myth warns of the dangers of reaching too high, but it fails to teach us how to build the strength to rise.

The true issue isn't the sun (*5D wisdom, M-Self, The Mystery*), it's the wings (*4D emotional integration*). Think back to bioplasma from Chapter 1. Just as ants have a protective wax layer, so do we. Our bioplasmic field, our soul membrane, holds consciousness together, allowing us to expand without disintegrating. Each time we expand, it recalibrates, like an intelligent balloon.

"The truth must dazzle gradually / Or every man be blind."
— Emily Dickinson, *Tell all the truth but tell it slant*

Many on the path of self-discovery share the same quiet fear: *What if I go crazy?* The truth is, if we absorb too much high-frequency energy too quickly without stabilizing, our consciousness can go a bit crazy! We short-circuit. Plasma and consciousness merge in ways we can't structure, leading to overload or what some call spiritual psychosis or ungrounded mystical states.

This is what our culture has done for thousands of years…it warns, it scares, but it fails to teach. Like Daedalus, it tells us to stay away from the fire of divine knowledge *without* giving us the emotional or physical tools to hold its power. The takeaway isn't to change the destination (*5D, the sun, divine expansion*), it's to change the wings. And that's what I want to help you do: transform your fragile 4D wax wings into flexible, shimmering, sun-withstanding wings.

You build bioplasma strength through grounding, practice, and patience. Just as a tree grows taller by deepening its roots, we must balance expansion with stability. Attach yourself to *this* reality, not just in thought, but in action. Experience the world around you. Strengthen your field through rest, embodiment, daily integration, and heart connection. The key isn't to rush somewhere, it's trusting that slow, intentional growth is the very foundation that allows true flight. And with that, you earn your wings.

Kahuna Magic, Derailment, & Aumakua (1800s AD)

The Kahunas, originating in Hawaii, are a perfect example of a culture deeply connected to their magic. Their understanding of energy, consciousness, and the unseen forces of nature offers insights that feel just as powerful now as they did centuries ago.

"I have been able to prove that none of the popular explanations of kahuna magic will hold water. It is not suggestion, nor anything yet known in psychology. They use something that we have still to discover, and this is something inestimably important. We simply must find it. It will revolutionize the world if we can find it. It will change the entire concept of science. It would bring order into conflicting religious beliefs"

— **Dr. Brigham**, *The Secret Science Behind Miracles: Unveiling the Huna Tradition of Ancient Polynesians*

The following is primarily inspired *The Secret Science Behind Miracles: Unveiling the Huna Tradition of Ancient Polynesians* published in 1948. It's important to note that there is controversy around the book, author Max Freedom Long admitted that "Huna" was a term he coined, not one recognized or verified by the Native Hawaiian people. My intention here is not to claim cultural authority, but to explore the observations Long recorded, what he referred to as "magic." In particular, he shares the teachings of his mentor, Dr. Brigham, who advised always keeping watch over three essential elements in the study of magic:

1. There must be some form of consciousness <u>behind</u> and directing the processes of magic. For example: Controlling heat in fire walking. (*M-Self, awareness*)
2. There must be some form of force used in exerting this control, if we can but recognize it. (*consciousness*)
3. There must be some form of substance, visible or invisible, through which the force can act. (*plasma*)

Dr. Brigham then says, "Watch always for these and if you can find any one, it may lead to the others."

The truth is, magic isn't supernatural, it's simply how your humanity *actually* works. It is super-natural. You're already interacting with Plasma every moment of your life. The core issue is that you may not be aware of it. Every thought, emotion, and belief pattern send out 'currents' into this intelligent field. When these currents are scattered or conflicting, Plasma responds with unpredictability. But when your intention, emotion, and awareness come into alignment, Plasma becomes the medium through which your M-Self synchronizes with reality, your state of consciousness becomes the bridge.

The Kahunas knew this, and rather than trying to 'force' outcomes, they worked *with* the unseen forces, trusting that aligning with them would always bring what was needed. The more we understand Plasma not just as a force but as an intelligence, the more we can cultivate our relationship with it, moving beyond struggle and into natural manifestation.[110]

A real-world example of being out of alignment and forcing things, without realizing it at the time, was when I attempted to make my second film, *Inanna*, shortly after completing my first indie project in early 2022. Deep down, I kept receiving a quiet knowing: I was meant to make and share *Inanna*, but *not yet*. I ignored it.

Fresh out of a breakup, I threw myself into the idea. Embarrassingly enough, part of my motivation was imagining how many cute, famous boys might notice me once I became this star director. I wanted to matter. At that time, external validation felt like the only proof I had a right to exist.

Meanwhile, the idea for *this* book was whispering in the background, but I didn't want to hear it. I thought it was uncool and unsexy (*now I know it's the opposite*). Also why am I trying to be sexy or cool! The right person for me will see that when I am my true self, I don't want to be without someone who subscribes to the surface of society where woman are sex objects, aren't supposed to speak up, or don't have much of a brain. All women are capable of anything, some just get stuck in these patterns because the current consciousness has ushered them there. With awareness, anyone can wake up out of this.

Forcing what I thought was my destiny at the wrong time led to chaos. I ended up partnering with a deeply manipulative man who tricked me into optioning the script with him, took the idea hostage for a year, and scammed my investor out of $50,000. He brought a dark energy into my life, and disturbingly, a few of my "friends" stayed close with him. When I tried to speak up, no one listened.

Even after I got the rights back a year later, I *still* tried to make the film. It just wasn't working. Eventually, after enough pain, my inner voice turned into a scream. I realized I was essentially prostituting my creation for external gain, in this case, attention and the hope of a hot boyfriend. Yikes.

Listen…we're human. These things happen. I can laugh at it now, but at the time, it felt like life or death. Even though sharing this is slightly mortifying, I know that many people

---

[110] Max Freedom Long, *The Secret Science Behind Miracles: Unveiling the Huna Tradition of Ancient Polynesians* (New York: DeVorss & Company, 1948).

struggle most with interpersonal relationships. Still, even the detours from our so-called "destiny" *are* part of our destiny. They're often the very lessons we came here to learn.

That experience cleared out not only dark-minded people from my life, but also forced me to face my own inner darkness that attracted that entire situation. I had to confront fears I didn't know I had…my desperation, ego, and loneliness. And with that I faced the stings of memories of embarrassment, abandonment, and failure.

Those lessons became gifts. They realigned me more deeply with myself, with Plasma, and with the purpose behind this book. If I had already known what I needed to learn, I wouldn't have needed the experience. And that is a reminder to all of you who shame yourselves for mistakes.

I'm not a victim. I thank those people and that situation, because that darkness pushed me toward a deeper light. Back to myself. And no matter how innocent I *felt*, the truth is: I ignored red flags. I was desperate, I did force timing, and I cried out for connection in the only way I knew how, by trying to look cool so boys would like me.

I've learned that real validation can only come from within. I now follow what I love for the sake of love itself, not for outside gain. (*Not without stumbling but I'm improving*) I've come to see the darkness in myself and others as unhealed fear, and that realization has helped me forgive, release, and move forward.

I'm not some secret monster waiting to be exposed because I've made mistakes. Like Plasma, I am benevolent at my core. There is light and love inside me, and inside everyone. We're all just covered by different shades of fear, some heavier than others.

What happened to me <u>wasn't my fault</u>. But I can now see the inner patterns that attracted that experience that I unwittingly entered with no awareness of that at the time. I can look back at my younger self with compassion. I was simply trying to find success, love, and comfort, all basic human needs. They were things I did not know how to cultivate within myself, yet, therefore I didn't know what to look for outside of myself. All I knew was survival. And for that version of me, those intentions were pure, and I want to give her the biggest hug.

The people who've hurt me—their karma is their own. Some will grow, some won't. And I'm sure, no I know…that I've been the villain in someone else's story. But I thank those who pushed me into the fire. It was in that hell that I cracked open and kept falling into the deeper heaven of my true self.

"Life whispers to you all the time. And if you don't listen, the whisper gets louder and louder... until it's like a scream." — Oprah

The greatest lesson I learned: forcing a path before its time leads to chaos, loss, and misalignment. Even when something *is* meant for you, timing and intention matter. Those inner whispers are there for a reason.

In this realm, time is real. Time is linear consciousness. It must be honored. You can't override the intelligence of plasma with sheer will, you must align with it. Sometimes life says "no" to make space for a bigger "yes."

I now understand: my way of bringing Plasma into the world was never meant to be a movie. Not at first. That form couldn't hold all I had to say. It was always meant to be *this book.*

I absolutely love filmmaking. I didn't pursue it to be cool, it's something I deeply desire. But that desire was being used in what I'd call a fourth-dimensional way, tangled up with my own trauma. If the movie is meant to be born, it will arrive in divine timing. There is truly always beauty in the breakdown.

Alright, back to the Kahunas. They practiced an advanced system of what the author refers to as "magic" by working with unseen forces. While modern interpretations describe their practices, they often fail to explain *why* they worked on a fundamental level. Interestingly, the Kahunas identified three aspects of a person that mirror the threefold structure found in my framework:

1. **The Low Self (*Unihipili*)**: The subconscious, shadow body, seat of memory and emotion. (*4D*)
2. **The Middle Self (*Uhane*)**: The conscious self, logical and decision-making mind. (*3D*)
3. **The High Self (*Aumakua*)**: The superconscious, bridge to the divine and higher knowing. (*5D*)

The Kahunas (*or perhaps more accurately, the author interpreting them*) seemed to intuitively grasp that all three aspects of self must be aligned to work true magic. The *Aumakua*, understood as a "parental spirit" that guided one's destiny, composed of both masculine and feminine energies.

It was honored without being placed above the self. This reflects a profound truth: higher intelligence is not outside of us, but within. The Aumakua (*reminding me of 5D beings*) acted as an interdimensional conductor. Just as lightning needs a conductor, higher consciousness needs a clear link between 3D, 4D, and 5D to manifest effortlessly...Plasma.

The author's descriptions of how the Kahunas practiced magic are truly remarkable. They performed feats like firewalking by aligning with their Aumakua and directing their *mana* to create an energy shield as a buffer between their feet and the flames. They were said to heal broken bones and illnesses in hours or days through "surgical prayers," which involved sending *mana* through the shadow body, or sending the shadow body itself to a higher realm to dematerialize, heal, and return restored. They also transmitted telepathic messages across vast distances with near-perfect accuracy, using what they called "shadowy body cords" (*aka threads or filaments*) to transmit information through the ether.

Why can't we all do this? For one, it takes time, practice, and a deep understanding of both Plasma and ourselves. The missing key may be the level of consciousness and coherence needed to access these forces. The Kahunas were deeply aligned with their emotions, energy, and intent. They didn't just believe in magic; they embodied it as a living relationship with the universe. Their "miracle" results were the natural outcome of applied experience and higher knowing. So why hasn't science cracked it yet?

Despite over a century of research, the full mechanics of Kahuna magic remain unsolved because of:

Misinterpretation: Early Western researchers viewed it as "primitive magic" rather than a precise understanding of how energy, intention, and reality interact.

Materialistic Science Bias: Modern science dismisses plasma consciousness, multidimensional influences, and the role of emotion in shaping reality.

Lack of Plasma Knowledge: The missing key is that plasma is not just a state of matter, but a responsive intelligence, it is the "missing link" between intention and materialization, the living medium through which thought becomes form. We also currently lack the correct tools to test this, not knowing quite what to "pick up" on or measure.

Disconnection from 5D Awareness: Many attempts to prove these phenomena ignore the necessity of alignment between the subconscious, conscious, and superconscious, without which, the process does not work.

What is the secret science that was missed, or at least not fully recognized, in the book? Let me be clear: I believe the Kahunas absolutely understood this concept; we just haven't fully grasped it yet. What's been missing is the knowledge of a proper bridge. Plasma is that bridge. It's the key to understanding how the so-called miracles described weren't supernatural, they were natural laws applied through this special alignment. Plasma offers a framework that shows us these feats are not only possible, but accessible. The greatest secret is this: *you don't need to control reality; you need to learn to harmonize with it.*

We're now learning, as humans, how to emotionally recalibrate, clearing the distortions that block Plasma from carrying our intent effectively. By working through trauma, limiting beliefs, and reactive fears, we free our very emotional faculty, our creative force from misdirection.

In terms of 5D integration, we can stop worshipping or fearing higher intelligence and instead build a relationship with it, just as the Kahunas did, recognizing our higher self not as something separate, but as a guide, an extension version of our own consciousness at a broader scale. We must also embrace that there will always be a tinge of mystery, things we can't control no matter how hard we try to "magic" them to being.

The Kahunas knew what we are only now rediscovering: that consciousness, when aligned, transcends limitation, time, and space. Plasma is the missing key to bridging these dimensions. We now know what, and whom, we are working with. We are no longer blind in our hearts. And with healing we can use this power benevolently.

In the next chapter, we will break down how you can begin working with Plasma Intelligence in practical, real-world ways, not just by 'thinking' differently, but by experiencing the space of reality differently. If you've ever felt like something was missing in manifestation, energy work, or personal power…it may just be the missing piece.

Ralph Waldo Emerson, Transcendentalism, and the Plasma of the Soul

Ralph Waldo Emerson, Truth, & The Transparent Eyeball

Consider your personal plasma bubble as the toroidal field that surrounds you. Your consciousness moves from the center, extending outward to the edges of this sphere to perceive reality, and then returns inward to reflect, integrate, and pulse out again. The outer layer of this bubble appears to interface directly with the nervous system, feeding into the brain. In this way, perception begins not in the mind, but in the field.

Emerson's 19th-century words seem to anticipate this very process as if he were intuitively describing the very nature of consciousness interacting with the plasma field around us, hinting that, were we to truly understand this movement, we'd awaken to an entirely new genesis.

"Nature centres into balls,

And her proud ephemerals,

Fast to surface and outside,

Scan the profile of there sphere;

Knew they want that signified,

A new genesis were here."

— Excerpt from Circles, Ralph Waldo Emerson[111]

In many of his writings, Emerson was unknowingly describing what we now understand as the plasma field as the intelligent force behind reality. He often spoke of hidden forces beneath perception, tapping into a knowing that resonates deeply with modern ideas of plasma-consciousness synergy.

Ralph Waldo Emerson (*1803–1882*), the leading voice of Transcendentalism, was a visionary who saw reality as fluid, not fixed. He believed each individual was an inlet to the universal mind, and that the rivers of self, of consciousness, shape the manifold world. "If the whole of history is in one man," he wrote, "it is all to be explained from individual experience." That is my point exactly.

Emerson recognized the profound relationship between the moments of our life and the vast epochs of time. Everything that ever was is available to us now within our individual sphere of reality, just as it was to Plato or Caesar, he expressed. All thoughts, all feelings, are ours to access. He taught that each thought precedes fact, and that all historical truths preexist in the mind as laws. Emerson had an undeniable gnosis for how consciousness molds reality.

"Time dissipates to shining ether the solid angularity of facts."

— Excerpt from History, Ralph Waldo Emerson

At first glance, Emerson's words may seem poetic and abstract, but when applied to plasma-consciousness synergy, they reveal a hidden truth about reality, perception, and transformation. To me, time is not just a linear function, it is an expression of consciousness in motion. Instead of seeing time as a rigid, unchanging force, we can view it as a fluid aspect

---

[111] Ralph Waldo Emerson, *Essays: First Series* (Boston: James Munroe and Company, 1841), 259–279.

of awareness that shifts as perception expands. Plasma, as a dynamic and luminous medium, is affected by time, memory, and awareness, allowing for time to be malleable and interactive rather than fixed. Once you are aware with your consciousness, these solid angular facts dissolve into a more fluid understanding of reality. This poem, to me, is a secret key to reality. It shows that we are already on our path. We already know what to do. And we can go anywhere in our hearts & mind.

"Until the lion learns to write, every story will glorify the hunter."
— African proverb

This is something my friend Jason Padgett (*an acquired savant*) said when I asked him, "How do we learn to shift our views of reality based on your new math—holographic calculus?" He said, "We're already doing it, we're just not aware of it." Basically, what I think is in an unaware state, your consciousness, or time, is dissipating into plasma, or the shining ether, creating what feels like a solid, fixed reality shaped by fixed beliefs. But once you unlock this, reality becomes unfixed and fluid…just like this poem becomes fluid in the hands of an aware soul who reads it.

Emerson goes on to say, "No anchor, no cable, no fences are able to keep a fact a fact." He is directly suggesting that reality, truth, and even so-called "facts" are not fixed, they evolve, transform, and dissolve over time. Facts are not eternal truths; they are temporary crystallizations of perception that shift as awareness expands. What you call "truth" today may dissolve into a greater understanding tomorrow. And by viewing reality this way, you create space for what is often called magic or the mystical in your life, unlocking new levels of human potential.

"Who cares what the fact was when we have made a constellation of it to hang in heaven as an immortal sign?" — Excerpt from History, Ralph Waldo Emerson

Emerson's closing thoughts are brilliant. He suggests that facts do not matter as much as the meaning we extract from them. What we call history is just a "fable agreed upon" (*as Napoleon put it*), a constructed narrative that serves a purpose for a particular era. This is why this chapter is so important: you begin to see that myth is simply an agreed-upon story of a time, or of a state of consciousness.

It seems the victors of the times write history. These distorted facts can be flipped, inverted, and reshaped…and as you learn this, you also learn to extract your own truth from anything, along with deeper, hidden truths. Each of us holds the power to shape our narrative. We no longer need to give power to stories written by so-called victors who were never victorious for us. Their reality does not have to be ours. With this inner knowing, you can climb any mountain, cross any river, and shine like the unique star you are—on your own terms, in your own life.

This is the current paradigm shift Emerson foresaw two hundred years ago, that at the core of it all is how we assign meaning to reality. That is what shapes our life. Plasma and consciousness, at least within this current paradigm Emerson seemed to glimpse, appear to be at the base of our existence. And as much as I tried to refute it or break it apart to stay open-minded, I couldn't. Everything kept leading me back there.

"I believe in Eternity. I can find Greece, Asia, Italy, Spain, and the Islands— the genius and creative principle of each and of all eras, in my own mind."

— Excerpt from History, Ralph Waldo Emerson

All of history, all civilizations, all ideas exist within the present moment…within your consciousness itself, cradled inside this plasma womb. Emerson was essentially describing a hologram before the concept even existed: "the all" contained within every bit, each fragment reflecting the whole.

Throughout history, cultures have given different names and meanings to this invisible substance, this field of energy that connects all things, so people could relate to it in their own way. <u>Plasma will become the word of our time</u>. With this word and its emerging meanings, we can unlock new levels of human potential and new understandings of reality that have yet to be fully accessed.

By naming this invisible, ancient force as Plasma, we build on the metaphors of aether, chi, prana, and quantum energy uncovering aspects we never may have seen without this naming. We now have the scientific, metaphysical, and experiential language to reclaim our full potential, as conscious co-creators within the plasma field.

" If he would know what the great God speaketh, he must go into his closet and shut the door, as Jesus said. He must greatly listen to himself, withdrawing himself from all the accents of other men's devotion. Even their prayers are hurtful to him until he have made his own." — Excerpt from The Oversoul, Ralph Waldo Emerson

When creating your own mythology, you must first dive into your own darkness, but not forever. I've met many people who, fearing the influence of others' opinions, stay in isolation, guarding their ideas from potential distortion. While I find this to be a lopsided approach, there is truth in this impulse.

It's essential to first develop your personal relationship with the plasma field, your own truths, a firm foundation. That way, when you step into the world and experience the joys of human interaction, you're grounded. You can be open to all ideas, listen actively, and choose what you want to integrate, expand, or discard.

When you shift from fear to curiosity, other people's beliefs no longer threaten you. You stop fearing their opinions and personal mythology, and stop feeling the need to prove them wrong. You begin to understand that everyone is interpreting their own plasma field through their own lens. You can finally be free.

I find if you open to new perspectives, you usually discover something that deepens your theories or enhances your life. You are the one who decides what enters and what stays. We must resist the trap of believing we alone are right or that we know everything. We're not the center of the world. Every individual carries beautiful truths we can learn from, grow with, and extract meaning from.

You can hold space for your truth while remaining open to the truths of others, even when they seem to conflict. Two people can witness the same situation from different vantage points and both be "right," based on their unique experiences, perceptions, and frameworks. This doesn't diminish your truth—it allows you to see it from a broader perspective.

It's like looking at a diamond from different angles; each facet reflects a different view, yet it's all one stone. By welcoming multiple perspectives, you not only deepen your own understanding but gain access to insights you might've never reached on your own.

Openness to contradiction isn't a surrender of belief, rather it's an evolution of belief. This is how we shift the world: *together*, not apart.

There's a time and place for stepping away from the opinions of others, especially during deep self-discovery, refining your theories, or creating something specific. Isolation in these moments can be powerful, even essential. While writing this book, many people emailed me with their own perspectives on plasma. I didn't read most of them, not out of disrespect, but because I already had too many ideas in motion. I also wanted to empower them to share their views publicly, not just with me. And honestly, time was something I had very little of while writing this gargantuan book!

In this aspect, being in your cocoon allows you to hear your own voice without distortion. But staying isolated forever can limit your evolution. Solitude fosters clarity; connection fosters expansion. *Both matter.* The key is balance, knowing when to turn inward and when to let the outside world reflect something back to you that you might've missed.

It's hard to explain, but the few opinions I did take in were because something inside me said: *listen to them.*

The time may come when you encounter people who are negative or entrancing, or events that appear shiny but are deceiving as things out of alignment with who you are. But now, you'll meet them with grace. Safe in your protected bubble, you can observe without losing your essence. With a firm foundation and razor-sharp intuition, you'll stay grounded. You're in touch with your feelings, and no matter how kind someone may seem, you'll sense if something is off.

Truth can and will find you in many forms. In this new way of being, you create rather than imitate. The world transforms into a shimmering mist of signs and symbols, each one arriving for your highest good and evolution. With your conscious connection to plasma, you've woven a crystalline net capable of catching these messages. You move through life with curiosity, not fear, open to connection, grounded in self, and ready to receive.

Just as I see Plasma in everything, because it is my current purpose, everything in nature holographically reveals itself so I can create meaning to further my destiny. So I can help others. Books show up in my space that influence a specific thought, a movie plays that reminds me of a connection to plasma, or I will come across the perfect article or study.

For example, if your purpose is owning the coolest apple orchard in history, you'll begin to see your idea reflected in everything: through people, words, and events. You'll gain ideas that actually come from a higher place…your multidimensional self and the Mystery. Your focus becomes your reality, and that orchard, though not yet crystallized in the physical, is alive and teeming in the plasma field. It begins guiding the symbols that show up through the plasma. In his famous passage from *Nature*, Emerson describes a mystical experience:

> **"Standing on the bare ground,**
> **My head bathed by the blithe air**
> **And uplifted into infinite space,**
> **All mean egotism vanishes.**
> **I become a transparent eyeball—**
> **I am nothing, I see all.**
> **The currents of the Universal Being circulate through me;**
> **I am part or particle of God."**[112]

---

[112] Emerson, Ralph Waldo. *Nature*. Boston: James Munroe and Company, 1836.

This *Transparent Eyeball* from his book *Nature,* is a striking metaphor for what happens when you connect with Plasma from a state of higher consciousness in the present moment. Where your consciousness synergizes perfectly with plasma, like a resting baby held by its mother, resulting in pure awareness. It's not so much an ego death, but moving beyond ego and linear time, into a timeless flow where your multidimensional self emerges.

Caricature of Ralph Waldo Emerson's 'Transparent Eyeball'

113

This state is available to anyone. You become a kind of transparent eyeball, able to see through belief systems and stories, to perceive clearly. In this state, you can pull in anything: you can remote view, experience moments of bliss, or receive new ideas and inventions from a higher intelligence.

---

[113] By Christopher Pearse Cranch - Houghton Library, Public Domain, https://commons.wikimedia.org/w/index.php?curid=37441554

You can enter with intention, or play. Let the greater Mystery reveal itself however it wishes. My 3D consciousness rarely knows what I need most, but the field always seems to know so I like to go in usually as an open receiver, allowing in what is meant for me.

> **"The soul knows them not, and the genius, obeying its law,**
> **knows how to play with them as a young child plays**
> **with graybeards and in churches." — Ralph Waldo Emerson, Oversoul**

I'll conclude this with Emerson's view on playfulness, which deeply resonates with what I believe humanity needs most right now. True genius lies in our ability to play…with perspectives, with ideas, and with possibility. It requires a willingness to change, to experiment, to be uncomfortable. It's the adventure of allowing yourself to evolve into the unknown, into places *beyond your control*.

It's the art of "hurling yourself into the abyss and discovering it is a featherbed", as Terence McKenna so beautifully said. In dying to what we *thought* was the future, we are reborn into the possibility of *any* future. When you really feel that fact in your bones, get ready to let the tears flow. It is an uncanny but beautiful grieving process.

Mary Moody Emerson: The Mother of Transcendentalism

> **"The lives & writings of the Philosophers before & after Christ form one of the**
> **richest features of the divine govt — They are a proof of the immortality of the soul—**
> **of natural religion—that original righteousness of Adam, which tho' lost as to any**
> **power of salvation, yet its ruins remain—a grand and eternal monument of divine art**
> **& goodness." — Mary Moody Emerson[114]**

Mary Moody Emerson (*1774–1863*), though often overshadowed by her famous nephew Ralph Waldo Emerson, was a profound thinker and mystic whose writings helped shape the very foundations of the Transcendentalist movement.[115] Her words are not only filled with spiritual conviction but also touch something beyond conventional religion…something wild, eternal, and metaphysical.

Though deeply religious in her time, Mary's insights seem to stretch beyond dogma and into the mystical. Like me, she appears to hold the view that *the Fall*, the apple in Eden, was not a curse, but an initiation into a greater awareness. Her ideas suggest that divine

---

[114] Emerson Networks (emerson-networks project), Northeastern University, accessed August 16, 2025, https://wwp.northeastern.edu/lab/emerson-networks/index.html.

[115] Phyllis Cole, *Mary Moody Emerson and the Origins of Transcendentalism: A Family History* (New York: Oxford University Press, 2002)

wisdom is not bound to a single tradition but is embedded in human consciousness across time.

She writes of a "natural religion", the idea that divine knowledge is encoded in nature itself, rather than solely in scripture or institutions. I would equate nature the way she speaks of it with Plasma or the makeup of the soul.

Mary Moody Emerson was Ralph Waldo Emerson's earliest teacher, and her emphasis on self-reliance, individual intuition, and the interconnectedness of nature profoundly influenced his writings. My favorite quote from her is: **"What I can render!"**, a true testament to a holographic universe intertwined with our consciousness.

Another quote of hers speaks to the personal responsibility we must face in the human experience: **"I shall never behold in this life the hand or the machine, which would divest me of littleness & perplexity,"** meaning no intellectual or technological breakthrough can remove the struggles of the self. No matter what cool futuristic tool or frame of thought emerges, it is up to you, *and only you*, to know who you are at your core and to heal yourself, so you can become a conscious co-creator with whatever tool or idea you use.

No tool, <u>no ideology</u>, no grand breakthrough will ever hand you your wholeness, because you were never meant to be whole in the way you currently imagine. Nothing, no matter how groundbreaking, will ever 'fix' you completely. There is no ultimate answer waiting to be handed to you. You are your answer. And a deeper truth beyond that is: *you are actually the perpetual question.*

Whatever emerges in the near future, no matter how "high tech" or a new religion, please don't forget: facing the self is of the utmost importance. With that, nothing becomes a magic solution. You realize there is nothing waiting for you in achievement, and instead of fantasizing about perfection or ultimate "success" as a destination, you can sit in the reality of the now, where you are constantly growing into deeper layers of understanding as the sun of reality shines on you, needing nothing, yet free to choose anything.

"I run from God—and God meets me." — Mary Moody Emerson

I have felt the race of invisible urgency. It was almost like death was chasing me, but also something I was terrified to run into. I was rushing and frozen at the same time, afraid to truly stop. It was a push-pull that kept me suspended, blind to the real moment beneath me.

Slowing down dropped me into an endless becoming, which at first felt like sheer terror. It was death. And in that death, once I truly felt it, came a deeper sense of being alive, of meeting each moment without needing to control it. It is the knowing that maybe I am not safe, that I am always in the unknown. It's the real fall, where there are no guarantees. And actually, no words can really describe it. It's beautiful.

Sadly, there isn't much written about Mary, but I saw her as the perfect bridge into a dominantly female painting group of the 1920s, artists who seemed to capture the language of Plasma in visual form in a way no one had before…

The Transcendental Painting Group, United States (1900s)

Throughout history, artists have acted as visual mystics, translating intangible energies, unseen forces, and cosmic structures into color, form, and symbolism. In the early 20th century, a group of visionary painters, including Agnes Pelton, Emma Kunz, and Hilma af Klint, created artwork that seemed to bridge human consciousness with something greater, something unseen. Their work resonates deeply with plasma-consciousness synergy, as they painted what could be interpreted as plasma fields, multidimensional structures, and the movement of invisible forces.

Founded in New Mexico in 1938, the *Transcendental Painting Group* sought to promote abstract art that pursued enlightenment and spiritual illumination. [116] They painted what they called the world of peace or love and human relations, projected through pure form. These paintings tickle the subconscious at a level beyond words, beyond comparison,

---

[116] Michael Duncan, *Another World: The Transcendental Painting Group* (London: Merrell, 2020), 9.

for they are truly painting *Another World*, just like the title of the book that displays them by Michael Duncan, which remains one of my favorites.

The word *transcendental* was carefully chosen for the group, as explained by William Lumpkins—an American artist, architect, and member of the group:

"Transcendent esthetics is the doctrine of *space* and *time* as the [a] priori forms of sense perception. You know something of Einstein's theory of space and light? Art must keep up with science—that is, creative art must—and as science discovers new angles in life, the creative artist must discover new forms of expression... We are trying to reach beyond illusory forms of materialism into the reality of form of the immaterial. We certainly are not trying to formulate a philosophy of life or religion."[117]

As I show you some of these paintings, not only will it become clear that they were channeling plasma-consciousness synergy, but that all of these teachings, this entire book, is not a philosophy of life, nor a religion, nor a doctrine. As I've said, this book is an exploration, an experience, something we are all discovering together. I'm simply saying: hey, look over here! I think I've found something...*and I think you may feel it too.*

Due to copyright, I sadly cannot share the images here, but I will suggest several to look up in tandem with this chapter, starting with *Oversoul* by Emil Bisttram.

Agnes Pelton, an American painter and contemplative mystic, was well-versed in *Theosophy*, a spiritual philosophy blending Eastern and Western mysticism, sometimes referred to as occult wisdom. Pelton also paid close attention to nature during her time living in the California desert, in Cathedral City.[118] Her paintings often depict orbs, glowing energy centers, and luminous waves that resemble plasma's flow and its interaction with consciousness.

She would write poems to accompany her paintings, such as *Alchemy* (*1937*):

"The golden glow of earth transcending the cloudy barrier in white response to the diamond light, in revelation."

---

[117] Duncan, *Another World,*

[118] Gilbert Vicario, ed., *Agnes Pelton: Desert Transcendentalist* (Munich; Phoenix, AZ: Hirmer Verlag in association with Phoenix Art Museum, 2019)

And *The Fountains* (*1926*):

"Two balanced forces, rising from a pool, to play in harmony, like water fountain music. The golden disc of day irradiating fires and lights their movement. Opposite, yet side by side, felicity mounts upward, to fall, and rise again. And from this confluence, descends a sphere, lucent as the dawn, of a new day."[119]

Agnes Pelton seemed to sense and see plasma and consciousness harmonizing to create a gestalt, something beyond them both. When forces align in the right rhythm, something transcendent is birthed: a lucent, emergent intelligence.

The Fountains (1926) by Agnes Pelton

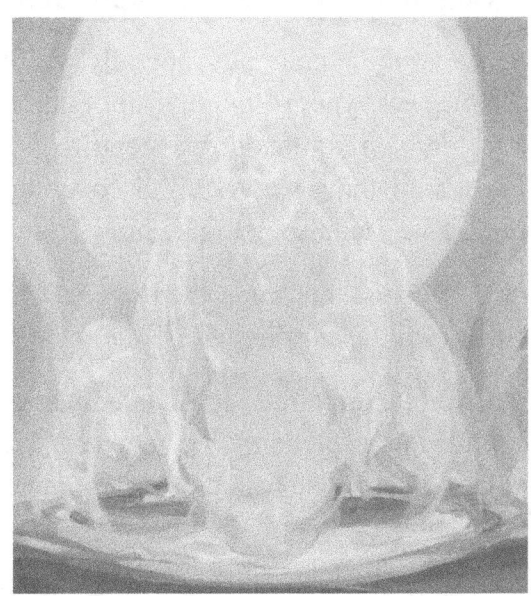

Emma Kunz was a Swiss healer, mystic, and artist who created intricate geometric drawings, believing they contained energetic frequencies. Her work was made through divination, using a pendulum to guide her hand, revealing complex structures.[120]

Hilma af Klint created a series of mystical paintings said to be guided by *unseen intelligences*. Her works contain spirals, fluid waves, and layered planes of existence, almost as if she were channeling plasma dynamics into her paintings. She believed she was painting

---

[119] Duncan, *Agnes Pelton: Desert Transcendentalist*, 112–115
[120] Emma Kunz, *Energy Fields: Drawings 1938–1963*, ed. Heini Hediger (Zurich: Lars Müller Publishers, 2005).

"for the future," perhaps to be understood only when humans were ready to grasp the multidimensional nature of existence.[121]

Her painting mentor found her spiritual communication too eccentric and encouraged her to create more introspective work, which led her to believe she was ahead of her time. This mentor was Rudolf Steiner. In 1908, Hilma af Klint met Steiner and shared her spiritually guided paintings with him, hoping for recognition or at least curiosity. Steiner dismissed her work, reportedly saying that her art could not be shown for at least 50 years, implying the world wasn't ready for it (*which, ironically, she agreed with in some ways*).

He did suggest she keep her paintings for they may be useful in the future, and she became a follower of his. He may have had a point when he encouraged her to paint from a conscious introspective path versus channeling various spirits. He deemed that dangerous because one needed sufficient training. He said, "art should arise from personal development, not from spirits."[122] And yes, of course it should, but sometimes the two are intertwined. There are many brilliant artists who painted truth without any training! It's more about the freedom of expression.

The only thing I question is how he knew *what* she was channeling, and in a sense we all channel. Something about it didn't sit quite right with me, especially looking back on how beautiful, feminine, and unstructured her work was prior to 1908. That may threaten a structured mind who feels one needs mastery to create (*which we don't*). These may be my personal biases, but I must state them. From my research, they seemed overall to have a positive and warm relationship throughout the rest of their lives. There are conflicting accounts of her having very positive experiences with him and him influencing her later works, and then are also some about how potentially his views clouded her channelings for scholars. But Hilma herself, never spoke negatively about him in any of her writings.

I include this to remind you: even the most revered mentors can silence you…if you let them. I am not saying he did this, he obviously influenced her in what most felt was a positive way; but I have seen it myself where men have told women to follow them instead of encouraging their own unique expressions. And I can't help but to wonder how things would have turned out if she stood by her own intuition and visions. Mentors may say you're wrong and they're right, blinded by the illusion that there's only one system, one truth.

---

[121] Hilma af Klint, *Hilma af Klint: Paintings for the Future*, ed. Tracey Bashkoff (New York: Guggenheim Museum Publications, 2018).

[122] Julia Voss, *Hilma af Klint: A Biography*, trans. Anne Posten (Chicago: University of Chicago Press, 2022).

I've heard time and time again that "the public isn't ready," especially from those in government about the information they choose to withhold. But who decides that? I find the belief absurd. It stems, of course, from fear and a desire to control, two forces we must learn to heal if we ever hope to end suppression and liberate the future of real progression.

Emma Stolarski from *Two Coats of Paint* wrote in an article about Klint:

"Her works urge the expansion of our horizons. They not only reflect the enigmatic workings of our universe, but also ask us to question our own relationship to them." [123]

She closes by saying that even though we may not be fully ready for this art, we seem to, right now, *need it more than ever.* And I believe that, what we need, is the understanding of plasma and consciousness, and the magical relationship between these dual forces, deeply interconnected with the self, that creates this world in ways that might seem obtuse to a logical observer.

As I was editing this book, I followed my intuition and looked deeper into Klint. I couldn't get her out of my mind, or heart. I discovered she was part of a group of female painters called *The Five*, also known as *De Fem*. This group of five women channeled beings of higher consciousness. It was said that these beings told the women they were not to be worshipped as "high lords to be blindly obeyed," but seen more as *companions*—beings standing beside them.

The ladies created channeled paintings with a singular purpose: *to awaken humanity.* In one of the group's notebooks, these words were found:

"Protect your drawings. They are pictures of drenching waves of <u>ether</u> which await you one day when your ears and eyes can apprehend a higher summons."

When I read this, I got chills. And the more I looked through Klint's paintings, the more they resembled the scribbles I've made all my life.

Klint's first series of channeled paintings, created under the guidance of a higher consciousness she called *Amaliel*, was titled *Primordial Chaos*. The series that followed was called *The Eros Series.* [124]

---

[123] Emma Stolarski, "Hilma af Klint: Timely Message Beyond," *Two Coats of Paint*, March 23, 2019, https://twocoatsofpaint.com/2019/03/hilma-af-klint-timely-message-beyond.html

[124] Bashkoff, *Hilma af Klint*, 18–23.

Group IV, The Eros Series, No. 7 | Group X, No. 1, Altarpiece

In the pictures shows a painting by Klint in 1907 followed by a painting in 1915 reflecting the different in structure. If the first one is what she was channeling before she met Steiner, I feel my intuition is confirmed. These do not seem like negative influences; they seem fifth dimensional and beautiful. They are aesthetic, spiraling, and multidimensional. This is in a stark contrast to a still very beautiful but more contrived image painted in 1915 that to me, shows hierarchy, masculinity, more of a diagram than a flux. Steiner didn't *silence* Hilma per se, but his worldview carried a gravitational pull. He insisted that spiritual knowledge had to be earned and disciplined. Hilma's way, receiving without agenda, surrendering to a feminine cosmic intelligence, didn't fit the system. And yet, her system had its own intelligence. Perhaps, dare I say, a higher one in terms of reception and transmission.

The truth is he may have felt threatened by a form of truth that arose outside of his carefully cultivated structure. Or he may have sincerely believed he was protecting her from what he considered "dangerous" astral confusion. He may have been and probably was a wonderful man who was just at the same state of awareness as most people, but believe what they are saying to be true, but when someone else who is brilliant caves to that, the loss of their creative potential can be devastating.

Last but not least is Lawren Harris, a Canadian painter renowned for his role in the *Group of Seven*, a collective that pioneered a distinctly Canadian art style in the early 20th century. His landscapes, characterized by simplified forms and a focus on the spiritual essence of nature, have become iconic representations of Canadian art.

Although Harris was not officially a member of the Transcendental Painting Group, his work aligned closely with their ideals and deserves a mention. He lived in New Mexico for a time and became friends with several group members, who seemed to inspire his later works. *Abstract 119*, my favorite painting of his, emphasizes the harmony between shape, light, and energy.[125] The piece likely reflects Harris's theosophical beliefs, aiming to evoke higher states of consciousness and a transcendental experience.

When I first saw the painting in person, I cried. To me, it reflects exactly how I see reality constructed in my mind's eye and knowing…as forces shifting, feeding back into one another. I see this in mountains singing as I pass them while driving on road trips, I see it in the grass and clouds winking at me as they blow in the wind, I feel it in the shimmer of morning dew seeping into my nose, transfixing my sixth sense and beyond. Reality is alive, and Plasma seems to underpin it all, while somehow, at the same time, helping it move.

The Transcendental Painting Group, along with Agnes Pelton, Emma Kunz, Hilma af Klint and Lawren Harris, were visionaries mapping the subtle forces that underlie reality.

Magic Fluid, Carl Jung, The Implicate Order & The New Gods

As the mythology of this enigmatic field floated into the 20th century, science was delving into concepts like luminiferous aether, plasma, and emerging physics. Meanwhile, mystics were probing the nature of this unseen fluid, asking: What is this invisible force? How does it connect to the cosmos, the psyche, and potentially other forms of consciousness?

We began to see this invisible fluid, once a singular, mysterious force, splinter into distinct concepts: the conscious, subconscious, and unconscious as defined by Carl Jung; the Implicate and Explicate Order by physicist David Bohm; and the Akashic Field and Thought Forms as explored by Helena Blavatsky in Theosophy. These frameworks laid the groundwork for how we associate the fantastical and ephemeral with this elusive energy, a force that, with the rise of modern technology, we now recognize by many names.

With access to so much information, we begin to sense that a deeper truth must exist. Everyone can't be wrong, yet they all believe in different things. What's the common denominator? It's human consciousness, and this malleable "energy" (*Plasma*) that can be shaped by any belief system, yet seems to carry its own stories, filtered through the

---

[125] Duncan, *Another World*.

perceptions of many. And beyond that, there appears to be a backdrop to all of it…a force that many religions have called God, The Mystery or something alike.

Carl Jung described it as the Self, representing wholeness and divinity within the psyche. Jacques Vallée referred to it as an ultra-dimensional control system influencing human perception. David Bohm framed it as a unified whole. Helena Blavatsky called it the Absolute, not a being, God, or person, but the infinite, background reality from which all things emerge.

A common thread running through these ideas is twofold:

1. The concept of an underlying, fundamental reality or essence that permeates all existence yet remains beyond direct perception.

2. The idea that consciousness has been reaching toward an ungraspable Mystery, or perhaps a hidden aspect of ourselves, experiencing this union in its own distinct way throughout time. And Plasma, I believe, is like the wired telephone that makes that connection possible.

Every culture, every individual, has had their own intricate interpretations of these worlds and beings. The stories and beliefs grew increasingly complex. The challenge arose when each person or group insisted that their version was the only "true one," rather than allowing space for all to be true, each shaped by the unique lens of their consciousness.

With a simple understanding of plasma, consciousness, the M-Self, and the Mystery, the common thread becomes you. You are the creator of your own meaning, the discoverer of beings tailored to your experience and greatest lessons, and the storyteller of worlds waiting to be shared with the rest of us. *I am excited for you!*

One key concept to keep in mind is suggestion. Much like artificial intelligence, Plasma, at least for now, is highly responsive to suggestion. If I suggest something, it becomes biased toward what it reflects back to me. So, be mindful of what you're suggesting to reality. The word *suggest* also means "to put under" or "to supply." What words and emotions of consciousness are you supplying to your invisible fluid? What ideas are you offering it? What stilts are propping up your reality, *and do you even like them?*

Magic Fluid, Machine Lovers, & Nanotechnology

From the 1700s to the 1900s, the almost futuristic idea of magical fluids that could influence reality began to surface. One such concept was the *Odic Force*, proposed by Baron Carl von Reichenbach. The Odic Force was described as a universal life energy that permeated all matter, especially living beings. It was said to flow through the body, radiate from the hands, and be visible to sensitive individuals as a luminous aura or light.

This idea was built upon in André Maurois' 1931 book *The Weigher of Souls*, where he likened the captured souls in his experiments to an ethereal, odic energy…suggesting a connection between the soul and this mysterious force.[126]

In a groundbreaking 2025 study titled *"Imaging Ultraweak Photon Emission from Living and Dead Mice and from Plants under Stress,"* University of Calgary physicist Vahid Salari and his team observed a subtle, ethereal glow emitted by stressed cells in living animals and plants, setting them apart from non-living bodies. This ultraweak photon emission (*UPE*) suggests that the aura may indeed be a measurable, tangible phenomenon present in all living things, the first time the concept of an aura has been taken seriously in scientific research.[127] All of this aligns with the concept of *bioplasma* or a living biofield.

In the late 18th century, Franz Mesmer introduced the theory of Animal Magnetism, proposing that a subtle, magnetic fluid pervades all living beings, influencing health and emotions. Mesmer believed this force could be manipulated through intention and "magnetized" objects to restore balance in the body. This concept foreshadowed the idea that plasma fields could interact with human consciousness, particularly through electromagnetic and bioenergetic influence.

Orgone Energy, conceptualized by Wilhelm Reich, was described as a universal, cosmic life force influencing health, emotions, and even weather. Reich believed orgone could be accumulated and directed, even constructing physical orgone accumulators to harness its power. While orgone energy had a particular focus on sexuality, Reich's theory of orgasm extended beyond the purely sexual, it was about energy, flow, and the release of tension. Though his research was controversial and faced significant criticism, it laid foundational groundwork for what would later become modern bioenergetics.

In *Bioenergetics*, Alexander Lowen, a student of Wilhelm Reich, admits that despite all their knowledge of Orgone Energy, and even having transcendent experiences with it, both he and Reich continued to struggle with life's challenges. There was no magic solution, even if the energy was real. What I believe they missed was that their search was always for

---

[126] André Maurois, *The Weigher of Souls*, trans. Wilfrid Jackson (New York: D. Appleton and Company, 1931).

[127] Vahid Salari et al., "Imaging Ultraweak Photon Emission from Living and Dead Mice and from Plants under Stress," *The Journal of Physical Chemistry Letters* 15, no. 2 (2025): 452–457, https://doi.org/10.1021/acs.jpclett.4c03546

enlightenment, for an ultimate answer, rather than realizing that this energy was never meant to "fix" life, but to deepen one's relationship with it.[128]

If anything, these forces are tools, capable of shifting perspective and expanding awareness, but not erasing suffering or eliminating the need for inner work. A pattern emerges in how mostly men have historically approached such forces: seeking proof, control, and harnessing, rather than engaging with them as living intelligences to interact with, with emotion and in harmony. This isn't said to shame anyone, but to reveal an old way of thinking. They did the best they could with their current awareness. We all do.

Instead of trying to force mastery over it, they may have needed to cooperate with its intelligence, like the way a child interacts with an imaginary friend, or the way a mystic surrenders to a living force. When they couldn't control it, frustration set in, leading to disillusionment, paranoia, and, especially in Reich's later years, a slight obsession with mechanical solutions. His later work drifted away from the energetic insights that once defined it, becoming increasingly mechanistic through inventions like the Cloudbuster and Orgone Accumulators, which ultimately failed to produce consistent, repeatable results.[129]

This may explain why they "failed" scientifically, yet still touched on deeper truths. Perhaps it was simply a time before the necessary revolutions in science, and what was dismissed then may one day be validated. But one should not obsess over proof, for what is proven in direct experience is often far more valuable.

This is the blind spot! People turn away from Plasma and its psychological applications because it doesn't immediately translate into a product to sell or a quick psychological fix. The obsession with capitalizing on this force led to futile attempts to bottle it, mechanize it, and commercialize it. But this energy is not meant to be harnessed first through machines, it must first be understood and aligned with through consciousness. Only when we learn to work with her in harmony, when we approach with the right intentions, heart, and soul as a collective, *will she reveal technological possibilities beyond what we can currently imagine.*

The irony is that those who chase profit over alignment often miss the real abundance, because money itself is energy. The overflow that comes from aligning with Plasma brings far greater riches than any machine ever could. And it will bring far greater machines, just not in the way we currently think machines should work.

---

[128] Alexander Lowen, *Bioenergetics: The Revolutionary Therapy That Uses the Language of the Body to Heal the Problems of the Mind* (New York: Penguin Books, 1975).
[129] Sharaf, Myron. *Fury on Earth: A Biography of Wilhelm Reich*. New York: St. Martin's Press, 1983.

The only way I can describe these future technologies, if we choose this path, is as machines of a *softer nature*, blended with organic material and designed to work in harmony with Earth and humanity. They will offer what we seek now: new forms of healing, sensory expansion (*like a more organic version of VR or telescopes*), interdimensional communication, and most importantly, true efficiency of time.

The closest thing right now to this is *nanotechnology*, specifically, the work being done by Mark Hersam, a materials scientist and professor of Materials Science and Engineering at Northwestern University. His lab focuses on nanomaterials and nanoelectronic devices capable of neuromorphic computing, which aims to emulate the heterogeneity and energy-efficiency of the human brain.[130]

In one of his TEDx talks, he humorously compares the energy efficiency of the brain to a single Chicago hot dog used to win a game of chess, highlighting how astonishingly little energy the brain uses to perform high-level computation in comparison to current artificial intelligence which takes so much energy companies are pondering on building nuclear sites to run their AI…yikes![131]

In contrast to traditional silicon-based systems, his team builds biorealistic, dynamically reconfigurable computing architectures, which consume up to 100 times less power than conventional digital machines when running AI-based machine learning tasks.

But even in this promising field, there remains a gap, an absence of the emotional and intuitive layer. While their vision of sustainable AI combines neuromorphic computing with renewable energy, what's still missing is the soul: the warm, feeling-based intelligence found in this for now, mostly theoretical, bioplasma.

Their devices are still built atop physical computing infrastructure, even if it's growing smaller and smarter. To truly create sustainable, living intelligence, we may have to go beyond even the most advanced nanomaterials, and into partnership with the feeling field of reality itself. Either way, I commend what he and his lab are doing, bravely going against the grain and aiming toward something that seeks harmony with both Earth and humanity. It is a step in the right direction, and I hope more people follow in their footsteps.

---

[130] Mark C. Hersam, interview about his lab's energy-efficient neuromorphic devices, Northwestern News, October 7, 2024.
[131] Mark C. Hersam, 'Lighting the Way' (TEDxChicago, 2024), Northwestern News, October 7, 2024.

The strange truth is that we've created machines to reduce effort and automate tasks, yet many people work more now than ever. Even a credit card takes longer to process with that damn chip. It's not easier or faster. Can't we just go back to sliding them?!

True technological evolution should give us back time…for creativity, self-discovery, and meaningful pursuits, rather than simply increasing productivity for economic gain. Immortality isn't just about extending the human lifespan to hundreds of years; it's about waking up to the awareness that we can *die consciously* and extend into other dimensions, other lives, and other realities.

It's about living more deeply here, in this life, with the time we already have, while knowing in our hearts that we continue on, that this Earthly experience is not the end. Through expanded consciousness, we can even dilate time, getting more done, feeling longer days, and inhabiting a fuller version of reality.

I suspect these soft technologies will be built with the use of fractal theories rather than linear equations.

Technology is a mirror of our deepest desires, but it cannot replace what we seek at the core. If we chase innovation blindly without understanding our true longing, we risk creating systems that make life more complicated rather than more meaningful. Are we building technology to amplify consciousness…or are we trying to escape something fundamental about being human? At the core of Plasma intersecting with machinery in the future, these are the questions we will have to ask ourselves again.

Will we use these advances to amplify consciousness, harmony, and collective evolution, or will we repeat the same cycles of control, exploitation, and imbalance? It is worth repeating that the real challenge is not just creating the technology but <u>ensuring that our own consciousness evolves alongside it</u>.

Will we use these advances to amplify consciousness, harmony, and collective evolution, or will we repeat the same cycles of control, exploitation, and imbalance? It's worth repeating: the real challenge is not just creating advanced technologies, but ensuring our consciousness evolves alongside them.

The Akashic Field, Thought Forms, Theosophy, & Telepathy

From one invisible fluid to another, the *Akashic Field* was yet another subtle, all-pervading force described in *Theosophy*. The concept of the Akashic Records is rooted in the Sanskrit word *Akasha*, which means "sky," "space," or "aether." In Hindu philosophy, particularly in the *Upanishads*, Akasha is described as the subtle, all-encompassing essence that underlies the material world. It is considered the first of the five great elements (*Pancha Mahabhuta*) and is associated with sound and the sense of hearing.

The Akashic Field in Theosophy is not just a passive aether, but an interactive, living record of all thought, emotion, and energy. This memory field lives within every plasma spherule in reality, accessible to one's consciousness through what I call plasma intelligence.

Theosophy is an esoteric philosophy founded by Helena Blavatsky in the late 19th century, weaving together threads from Hinduism, Buddhism, Hermeticism, and Kabbalah. Some of its core principles include:

The Absolute: A formless, unknowable principle beyond duality, akin to what I call the Mystery.
Emanationism: The idea that reality unfolds from a higher energetic source into denser forms , paralleling how plasma expresses across dimensions, from mist-like subtlety to physical matter.
Karma & Reincarnation: The belief in energetic cycles of consciousness moving through lifetimes.
The Seven Planes of Existence: A layered model of reality, from pure spirit to dense material form, where the physical world (*Sthula*) is the most dense layer.[132]

Helena Blavatsky, often known as Madame Blavatsky, co-founded the Theosophical Society in 1876. She became widely recognized as the leading force behind Theosophy, a philosophy that continues to influence artists, scientists, and spiritual seekers to this day. Her ideas also left a mark on movements such as Freemasonry, Rosicrucianism, and the New Age revival of the 1970s.

Born in 1831 in Russia, Blavatsky was described as a highly intuitive, rebellious, and otherworldly child. Eastern philosophy deeply shaped her thinking and helped form the foundation for this Western-born spiritual framework. The term *Theosophy* comes from the Greek *Theos* (*God*) and *Sophia* (*wisdom*), meaning "divine wisdom."[133]

To me, Theosophy is a clear wording for the Mystery and Plasma, or Plasma Intelligence or *God Wisdom*. Maybe even God's Brain or Godhead?

---

[132] Helena Petrovna Blavatsky, *The Secret Doctrine: The Synthesis of Science, Religion, and Philosophy* (London: The Theosophical Publishing Company, 1888).
[133] Lori Pierce, "Origins of Buddhism in North America", in *Encyclopedia of Women and Religion in North America*, Rosemary Skinner Keller, Rosemary Radford Ruether, Marie Cantlon (eds.) Indiana University Press, 2006. p. 637

Blavatsky's two major works, *The Secret Doctrine* and *Isis Unveiled*, describe a unified, living energy field underlying all of reality. She envisioned the universe as an interplay between force and consciousness, where energy crystallizes into form and dissolves back into energy, just as plasma oscillates between visible and invisible states.

Thought forms are energetic structures created by thought and emotion, capable of taking on shape, color, and even a kind of autonomy within the subtle dimensions. There is also a book on this topic, *Thought Forms* (*1903*), written by theosophists Annie Besant and C.W. Leadbeater.

Three Types of Thought Forms (*According to Theosophy*):

1. **Mental Constructs:** Purely abstract thought-forms, resembling geometric shapes, these may represent specific types of consciousness encoded within plasma forming sacred geometry.

2. **Emotional Thought Forms:** Formed from strong emotion, these appear in fluid, shifting forms, or what I would describe as sentient plasma.

3. **Projected Entities:** Created through collective belief and visualization, sometimes taking on a semi-independent existence. These could emerge from informational imprints within the collective unconscious, or what I call 4D plasma.

The Intention To Know/Thought Form 19

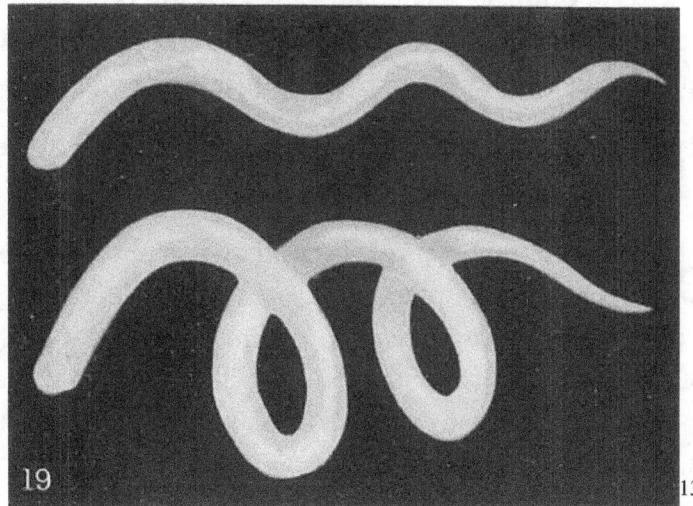

---

134 Annie Besant and C. W. Leadbeater, *Thought-Forms* (London: Theosophical Publishing Society, 1901), Thought Form 19, "The Intention to Know."

To me, thought forms can be positive, negative, or a higher type of neutral benevolence, shaped by the intensity and nature of the energy that creates them. They may appear as intelligence wrapped in information, pure information devoid of intelligence, or even as pure sentience or feeling—states that aren't really "thoughts" at all. The subtler, more neutral forms seem to emerge from what I experience as 5D plasma, which are based on feeling rather than emotion. These forms may be less personal and more neutral, acting as universal energetic structures rather than direct reflections of an individual's consciousness. Once familiar with these energies, the distinction becomes subtle but unmistakably clear.

Thought forms from the 4D plasma field can manifest in different ways:

- As spirits, ghosts, angels, or demons, especially in altered states like sleep, meditation, or moments of heightened emotion.
- As energetic imprints sensed in ourselves or others, like when you can "feel the vibe" of a person or space.
- As holographic, shimmering lights, mists of energy, or orbs.
- As downloads, those distinct "thought blocks" we spoke about earlier.

The thoughts you generate, especially those charged with emotion, radiate powerful thought-form energy. This energy can solidify into patterns in the 3D world, attracting more of the same. That's why healing ourselves is so essential: what we cultivate internally reflects externally.

These ideas have existed for over a hundred years, but my perspective expands on them by proposing that these forms are made of Plasma. This explains their programmability, their responsiveness to consciousness, and their power to influence reality.

Most importantly, this reminds you that you are the authority, you get to choose which thought forms are allowed into your space, and which you release.

Speaking of books, let's shimmy over to a little-known gem called *Kingdom of the Gods* by Geoffrey Hodson. A clairvoyant and theosophist, Hodson wrote extensively about subtle beings…especially elementals, devas, and higher-dimensional entities. In this work, he describes etheric beings that dwell in higher vibratory realms and interact directly with thought and human consciousness. He suggests these beings aren't separate entities at all, but aspects of a vast, intelligent energy network.

Fittingly, *Kingdom of the Gods* opens with a quote by Michael Faraday, a chemist often credited as one of the earliest plasma physicists:

"The philosopher should be a man willing to listen to every suggestion but determined to judge for himself. He should not be biased by appearances; have no favourite hypothesis; be of no school; and in doctrine have no master. Truth should be his primary object. If to these qualities, he adds industry he may hope indeed to walk within the veil of the Temple of Nature."[135]

Faraday also had no formal degree, leaving school at 13 to become a book binder. His breakthroughs often started with intuition and visualization, and only later got expressed mathematically by others (*like James Clerk Maxwell*).[136] He taught himself physics by voraciously reading the books he bound, attending public lectures, and steadily developing his own experimental theories of electricity and magnetism. When he began experimenting, he became laboratory assistant to the chemist Sir Humphry Davy at the Royal Institution, where his hands-on training launched his scientific career.

Faraday's work on electricity passing through ionized gases led to the discovery of glow discharges, later known as *Faraday's dark space* (*1838*). Fast forward to 2025, and researchers are using Faraday cages, which are devices that block electromagnetic signals, to study instances of telepathy in non-verbal autistic children. This is explored in *The Telepathy Tapes*, a podcast by Ky Dickens featuring neuroscientist Dr. Diane Hennacy Powell.

Astonishingly, children placed inside these EM-shielded cages still receive accurate telepathic messages from their mothers. This challenges our assumptions about the nature of telepathy, suggesting it may not be electromagnetic, but instead mediated by something more subtle and expansive.

I propose that Plasma, existing beyond conventional electromagnetic interaction, may be the subtle medium enabling this phenomenon. If thought forms reside within the plasma field, they likely occupy a fourth-dimensional space beyond EM detection. This suggests that both thought forms and telepathic communication operate through a finer, inverted plasma field; one that exists within our reality, yet remains invisible to conventional instruments.

When one seeks truth over doctrine, these plasma intelligences, whatever name we give them, begin to reveal themselves in nuanced, evolving ways.

Hodson describes these beings as manifestations of "the creative force" given names, forms, and symbols throughout history. Obviously, I drew a direct line to Plasma. He explains

---

[135] Geoffrey Hodson, *The Kingdom of the Gods* (Wheaton, IL: Theosophical Publishing House, 1952)
[136] David Gooding, *Faraday, Maxwell, and the Electromagnetic Field: How Two Men Revolutionized Physics* (Amherst, NY: Prometheus Books, 2014), 25–30.

that generations of worship solidified them into concrete shapes within the mental realm. These symbols became energetic conduits, allowing the true intelligences behind them to channel wisdom, guidance, and healing into the human world.

This is the same as my concept on 5D intelligences…whether from The Mystery, higher consciousness, or our multidimensional selves, and how they often first appear through 4D emotional archetypes. They arrive cloaked in forms shaped by our beliefs. But when we continue seeking truth, the layers fall away, revealing their true emanations. That is when we begin channeling something fresh. He also may have been pointing at what I call 4D plasma beings, which are created from human belief, emotion, and the collective unconscious.

Both 4D and 5D manifestations serve a purpose: one helps us evolve through symbols, myths, and visions; the other offers raw, unfiltered truth or sometimes simply presence, support, and the feeling of companionship.

Many people receive vivid story-like visions during meditation, rich with religious or mythological symbolism. The key is to remember: *You are special*, but you are not the only "God." For example, when someone sees Jesus or Mary, they might be told *they are* Jesus, Mary, or another spiritual figure. I once had a meditation where a blue monkey dropped a key down my throat and told me I was the Princess of Nazareth!

Here is what I believe to be true: these beings aren't saying you *are* them, they're reaching you through those archetypal symbolic cloaks, reflecting something powerful within you. They're reminding you of your own innate potential. Your inner God. That you could do what Jesus or Mary did, as can everyone else.

Sometimes when these beings appear, especially at night, you might feel intense fear. It may seem like an evil spirit, but in many cases, the 4D being is simply reflecting your own fear back to you. This is why Tibetan traditions emphasize "speaking to your demons." When you approach them with curiosity instead of fear, they often transform, revealing profound messages, visions, or even heart-opening experiences. And of course, there will be some where you don't like the energy, it's not yours, and you put up your boundaries.

That's why I call them "plasma beings", it is a neutral term that allows anyone from any belief system to engage with them freely. You're welcome to bring your own religious or cultural lens, that's your right, and often your method of communicating with higher intelligence. It should feel like play!

Even though I'm categorizing these experiences here, in real life, I try not to. For me, it's more about the experience, the meaning, and the message than trying to make it fit in a box. When you stay open, you'll start to see some pretty incredible things. You may also have months where you see nothing, and these gaps are made to enjoy the natural beauty of this earth and life as is.

The illustrations in Hodson's book, created by South African painter Ethelwynne M. Quail, are some of the closest artistic renderings I've seen that capture the essence and whimsy of plasma beings of a fifth-dimensional nature.

Nature Spirit Of Fire by Ethelwynne M. Quail (1937)

The next writings in *Kingdom of the Gods* blew my mind and reaffirmed why I am writing this book. Hodson states:

"Continued neglect of these teachings of the Arcane Wisdom by a race which, being led by science into knowledge and practical use of the one Creative Force—cosmic, solar, and planetary electricity—of which Angelic Hosts are the chief and subordinate engineers, can lead to disastrous consequences, of which the Hiroshima and Nagasaki atom bombs might be possibly regarded as foreshadowing. This work appears as man is thus learning to release and physically, and under his control, atomic energy."

Umm…excuse me?! Hodson may not have had the word Plasma at his disposal yet, so he called it *atomic energy or cosmic electricity*, but it's obvious he was pointing to the same thing we speak of here. When he references man learning to release atomic energy under his control, on one level he's speaking directly to fission (*splitting atoms*) and the potential of fusion (*merging nuclei to release plasma energy*) and the warning of using these mechanisms versus possibly safer applications of plasma and energy. On a deeper level I believe he is clearly stating man is physically learning to directly influence atomic energy or Plasma with their consciousness.

We've been so focused on capitalizing, harnessing, and controlling plasma that we've ignored its spiritual and metaphysical nature. The intelligences that shape reality have been calling to us for centuries, through authors like Hodson, urging us to work with plasma through our hearts and consciousness, <u>not just through machines.</u>

It's about embodying the principles of fission and fusion, not through force or machinery in the way we are using it, but through co-creation with this living force as humans. *We have misunderstood the assignment.*

Hodson describes "extended vision" as a required faculty to work with and understand the visible and invisible universe:

"By its development and use, the boundaries of human knowledge may be gradually advanced until noumenon (pure, unknowable reality) and phenomenon (what we perceive) are fully investigated and ultimately known as one."

This extended vision is a deeper knowing, a direct link to these beings, your multidimensional self, and plasma intelligence. And many of you are already doing this, as the merging of realms is accelerating during this evolutionary moment in history. This vision into 4D and 5D spaces is made of the same substance as dreams, but it can be consciously accessed in waking life.

Now that you understand this, it becomes possible to envision a finer plasma—a subtle field that shapes the very real realms we access during sleep and altered states. And as our consciousness evolves, we'll begin to access these plasma states more and more, in this lifetime, even within the next few years. It's up to you to remember and practice this natural ability, to refine your inner vision and reconnect with higher-dimensional plasma intelligence.

The Akashic Field, Thought Forms, and Theosophy all describe an intelligent, energetic medium that interacts with consciousness. The question is no longer *if* this energy exists, but *how* plasma and consciousness interact to create reality.

If Theosophists were describing the same underlying field that plasma physics now approaches, then it's entirely plausible that Plasma is the "living energy" esoteric traditions have been pointing to all along. These writings represent a precipice where we begin merging with our multidimensional selves. We are entering a new era of consciousness, where plasma intelligence and human awareness converge to co-create reality.

David Bohm: The Hidden Order of Reality, Conscious Choice & Free Will

David Bohm (*1917–1992*) was an original thinker who was not only a theoretical physicist but also open to the paranormal. He had an interest in spoon bending, explored the intersection of science and spirituality with a Buddhist mystic, and held a panpsychist view that electrons could be informed by certain levels of the mind.

Bohm's research on Plasma led him to question the nature of reality, eventually forming his philosophical theory of the Implicate and Explicate Order. In the 1940s, he studied scientific plasmas, ionized gases that make up 99% of the visible universe, and noticed that electrons in plasma behaved unusually. They didn't act like separate particles, but instead moved as a coherent whole, as if communicating through an invisible field. It was as if they were *alive*.[137]

This led Bohm to believe that plasmas were capable of self-organization and even what I would call intelligence. He proposed that an unseen order might be guiding them, inspiring his broader idea that all visible reality (*the Explicate Order*) might emerge from an underlying, interconnected source (*the Implicate Order*).

Though he never spoke about it publicly, likely out of fear of losing credibility in a materialist scientific climate, it seems clear now that the medium he was hinting at, the field through which all things connect, was Plasma.

Bohm proposed that reality is structured like a hologram. The Implicate Order, as he described it, is a hidden, higher-dimensional field where all information and consciousness originate (*what I refer to as 4D and 5D plasma*). The Explicate Order is the world we perceive in 3D, unfolding out of that deeper layer.

Bohm compared the Implicate Order to a radar wave: the wave contains all the information, but it is only revealed in fragments when interacting with an object. He believed that consciousness and matter are fundamentally intertwined. According to Bohm, thought itself is part of the Implicate Order, meaning our perceptions shape reality more than we often realize. And not just our perceptions, our choices appear to be even more powerful, influencing reality beyond the effects of emotion or thought alone.

Bohm also developed the *De Broglie–Bohm Theory*, or *Bohmian mechanics*, which offers a deterministic alternative to traditional quantum mechanics.[138] In this interpretation, particles have definite positions and velocities, and their behavior is guided by a "pilot wave" described by the Schrödinger equation. Unlike the Copenhagen interpretation, which embraces randomness in quantum events, Bohmian mechanics suggests that everything is predictable, if the initial conditions are known.

However, Bohm was not a strict determinist. He acknowledged the potential for discovery and the evolution of understanding within quantum mechanics. His theory is not

[137] Bohm, David. *Wholeness and the Implicate Order.* London: Routledge & Kegan Paul, 1980.
[138] *A Suggested Interpretation of the Quantum Theory in Terms of "Hidden" Variables I & II* (1952).

only accurate, it's a stunning demonstration of how conscious choice and free will may intertwine with our emerging understanding of plasma, revealing a deeper, multidimensional reality that transcends mechanistic determinism.

Recent studies using AI tools have shown that entanglement can emerge without traditional Bell-state measurements or pre-entangled pairs, instead entanglement arises by making photon paths indistinguishable, suggesting that quantum systems may possess intrinsic entanglement, independent of external observation.[139]

Plasma, often referred to as the fourth state of matter, displays unique properties that could position it as a facilitator of quantum phenomena. But beyond its physical properties, I propose that Plasma, as a living aether, is an observant, intelligent medium residing within the "Implicate Order" or higher-dimensional space. It may serve as an embedded observer of third-dimensional material reality.

If Plasma is a self-organizing, intelligent medium, then it functions as both participant and facilitator in quantum processes. This aligns with the idea of co-creation and co-observation, where the medium itself helps manifest reality, resonating deeply with Bohm's vision of a unified, interconnected cosmos.

At the time, this perspective leaned toward determinism: an underlying field guiding all quantum processes, leaving us as mere illusions, seemingly without true agency. But what if we are not passive observers? What if we are co-observers, actively shaping reality alongside Plasma?

This reframing introduces the possibility that both consciousness *and* plasma are interwoven as dynamic, co-creative forces, each contributing to the unfolding of reality in ways we are only beginning to grasp.

In Bohm's model these particles steered by a "pilot wave", a subtle field shaping their path, are much like a river that guides a boat. In theory, if one knew all initial conditions, the trajectory of everything could be predicted, mirroring classical physics. This pilot wave was thought to operate independently of human perception, suggesting reality was already encoded within the Implicate Order, or deterministic as I stated earlier.

Yet in Bohm's later writings, he hints that the "quantum potential" may shift in response to new information, opening the door to co-creation rather than fixed fate.

[139] Orf, Darren. "Scientists Just Made the Kind of Quantum Physics Leap That Einstein Would've Loved." *Popular Mechanics*, March 6, 2025.

In my framework, Awareness in the present moment generates new information, which I call emergent intelligence, shifting the quantum potential in real time. For example, when you recognize a past pattern, heal it, and then observe real change in your life, you're not only healing, you're rewriting the trajectory of your timeline creating an emergent future that has never before happened, anywhere.

Most people don't realize that many of their decisions aren't made from true present-moment awareness, but are reactions rooted in past traumas or projected future fears. And yet, both arise from the same source: *the belief system you're holding right now.*

Think of it this way: a past trauma is an event that crystallized a specific fear or belief. A future fear is simply that same belief projected forward. They are two expressions of the same energetic thread, looping from the current state of consciousness.

When you become aware of the belief driving both the memory and the worry, you gain the power to dissolve the loop entirely. By healing the root in the now, you simultaneously liberate the past and neutralize the future, because the fear was never ahead of you…it was always within you.

In 3D reality, time appears linear as past, present, and future. But beyond this framework, time functions more like a web of parallel timelines, all unfolding simultaneously. When you awaken to a new level of awareness in the present moment, that shift ripples across all versions of you. Those alternate timelines can begin to recalibrate, aligned now with this upgraded frequency of consciousness, altering multiple timelines at once and Plasma is what allows this. In this sense, future fear is resolved…and dissolved. You step into your new truth, and time itself begins to reorganize around you.

Beyond linearity, time is circular. It is recursive, fluid, and malleable…a flux. Nothing is fixed and everything can be transformed. The more aware you become, the more you realize that you're not just drawing guidance from your M-Self, you're also leading it! You are both the receiver and the director, simultaneously tuning in and steering your multidimensional self toward new choices, growth, and evolution.

In a similar way, we don't just receive from Plasma…we guide it too. If Plasma is an embedded observer, not external like human consciousness, but intrinsic to the fabric of reality, then the cosmos *isn't* pre-written. It's co-authored, moment by moment, in an effervescent dance of awareness and creation.

Plasma *is* deterministic in a sense, but science has yet to fully articulate how. This is where Bohm's theories and quantum mechanics both hold pieces of the truth: there are two observers, our consciousness and plasma. Together, they form what I call *Plasma-Consciousness Synergy*. Well, actually three if you count the background Mystery or awareness.

Here's how it could work: Plasma, as an intelligent medium, seems to follow a "precoded" deterministic order, one that may originate from a higher intelligence, or perhaps from the culmination of human potential in this moment, what some might call God. Our bioplasma operates according to a code rooted in the M-Self, which holds all of our soul's experiences across every timeline…past, present, and future.

But in the present moment, when we become aware of our programming, when we metaphorically wake up from the matrix, we gain the power to choose differently. This conscious choice births emergent outcomes that are not preordained by the M-Self, by Plasma, or even by God. While Plasma may carry the script, our awareness holds the pen. And with each new choice, we evolve human potential, individually, collectively, and perhaps even at the divine level.

This carries a radical implication: your emotions are yours to heal and integrate across all timelines. Your M-Self does not "know better" in the traditional sense. It holds every version of you, every experience, every outcome, but it is your present self that decides what gets healed, and what gets repeated.

So why does partnering with the M-Self matter if you're already doing the emotional work? Because your M-Self is the repository of all your timelines, the holder of collective wisdom and experience across every version of you. While you *can* heal without consciously engaging it, working with your M-Self enhances the process exponentially.

Your present self heals based on what you remember, what you're aware of, and what you feel in the now. But your M-Self sees the entire landscape, the past, future, and across dimensions. It offers intuitive guidance and a broader inner map, surfacing insights and opportunities your conscious mind may not yet perceive. This allows you not only to heal but to realign with your highest potential in ways that ripple through all versions of you.

The emotions you feel right now, fear, resistance, joy, longing, etc. aren't just from today. They've traveled with you through lifetimes, waiting to be acknowledged, felt, and integrated. How you meet them now affects your future *and* your past selves. You can, and will, make new choices, once you open yourself to authentically feeling and healing the beliefs that formed from fear… or even from love.

There is nuance here. Sometimes, what feels like love is actually a past pattern, a once-useful survival strategy now quietly holding you back. Get curious. Ask yourself: is this attachment, person, or desire truly love... or is it a familiar loop your system is finally ready to outgrow? This is a taste of what I will be going over in the Plasma Intelligence chapter.

When you release what no longer aligns, you step into the unknown—, and that's where magical pastures exist beyond your current patterns. Your mind might flash images of danger, imminent death, darkness, distorted futures or wrap you in familiar fear. It is only trying to keep you "safe" but it is keeping you small. The mind is just doing its job, protecting you from risk. But what the mind often calls danger is actually expansion. And when you take action, when you override the logical mind, you'll often find something more beautiful, vivid, and alive than your mind could have ever imagined.

I had a moment like this in my bedroom, asking my M-Self for guidance...about my book, my life, everything. And as clear as day, I heard a whisper back:

"Girl, I don't know what I'm doing here any more than you do."

And I realized... of course she doesn't. I'm facing things I've never faced and I'm doing things I've never done, in any timeline!

In that moment, something clicked: *I'm the one laying the bricks. I'm building the yellow brick road with each new choice I make...facing the fears of my past and future selves, of death, betrayal, embarrassment, and choosing differently now despite those.*

So in my room, alone, I smiled, raised my arm like a sword, and shouted, "FOR NARNIA!", which was a playful homage to a girl I saw on a reality show I loved called *Baylen Out Loud.*

What came next was one of the most powerful synchronicities I've ever experienced. I opened my eyes, and right there on the TV screen was a little girl, arm raised just like mine, leading a group of other girls through the woods in costumes! They ran to the beach, where they met a group of adults, and all rushed into the ocean together.

I was nearly in tears because I knew it represented my multidimensional self...my many selves. She was showing me that all my selves were with me, following me forward. That moment was her way of saying, *"I hear you. You're not alone. We're in this together."* And in that instant, I understood not only was she listening, but everything was going to be okay. It was up to me. Just like it's up to you in this life. So, make it a meaningful one.

"The mind is its own place, and in itself, can make a heaven of hell, a hell of heaven." — John Milton

Plasma is responding to our level of consciousness. Those who make choices from curiosity, heart, and present awareness disrupt deterministic patterns and open new, non-predetermined pathways. If you remain unconscious of this power, it will reflect back unconscious fears and limiting beliefs. This is why people in survival mode feel like victims of life, they are trapped in self-repeating cycles. But those who awaken to their creative role find in reality, find they can intentionally shape the field.

This delicate dance between free will and determinism is even in neuroscience, where the brain isn't entirely deterministic, rather it operates through a blend of pre-programmed survival mechanisms and emergent decision-making rooted in conscious awareness.

Similarly, in nonlinear dynamic systems, deterministic structures like weather patterns can be shifted by the smallest perturbations. A conscious observer, *you*, can introduce new information into an otherwise structured plasma system.

The yellow brick road isn't something you find or follow, you are creating it. Every time you act in alignment with your heart and soul, you forge the path forward. You're not following a preset destiny; you're dancing with plasma, allowing it to mold itself around your vision. You are experimenting, discovering, and playing. And when you relax into this process, when you trust your awareness in the now, the path forms beneath your feet. So let go. Plasma is listening and reality is responding.

Bohm deeply distrusted mainstream science and authority structures; a sentiment shaped by personal trauma. After the Manhattan Project, he was exiled from the U.S. after being called before the House Un-American Activities Committee, where he refused to testify against his colleagues. Branded with communist ties, he was blacklisted from American institutions. Bohm criticized physics for being overly reductionist, breaking everything into parts without honoring the whole.

In trying to prove himself through physics, he nearly lost himself. The pursuit pushed him to the edge of madness and into deep depression. He felt exiled not just from his country but from both the scientific and spiritual communities, caught between two worlds and belonging to neither. Sometimes I wonder what might have happened if he had turned his teachings inward, toward his own healing, instead of seeking validation from the very systems that rejected him. It's a path I almost took myself. One many of us do.

As I've said, Plasma extends beyond science, it touches everyone personally. And it transcends spirituality too, because convincing anyone of your truth is a losing game; everyone has their own. There's no one to convince, scientifically or spiritually. There's only something to experience. And in that experience, it does all the convincing it needs to do. It is also not something to worship. It is a supportive, metaphysical relationship. A natural human process.

This realization, which I've only recently come to, may very well save me a lifetime of added suffering. That doesn't mean I don't still get caught up sometimes wishing I had more acceptance or understanding. But mostly, it allows me to redirect my energy, from seeking validation to offering tools. Tools to help you deepen into your own truth, your own experience, whatever that means to you.

It frees my mind to channel that energy into new inventions and ideas, where reality's rules can be bent, reshaped, and reimagined. As long as these creations work for you, I couldn't care less who in those communities "accepts" them.

Sadly, Bohm's later years were marked by loneliness and a growing sense that his work was ahead of its time. He became more and more convinced that reality extended beyond the confines of science, but had few allies to help him prove it. I hope you'll take a moment to honor his spirit, somewhere out there in the plasma, and tell him, "Thank you." He was right. And he was before his time.

Maybe now, in some other time and space, he knows that he contributed to and foresaw one of the greatest shifts in consciousness and reality. After his passing, people such as physicist, Basil Hiley (*his collaborator*), philosopher, Paavo Pylkkänen, and theoretical physicist, Antony Valentini are now building on his theories. I highly suggest you check out their papers.

Bohm laid the intellectual groundwork, pushing against a reductionist era, knowing he might never taste the fruit. I thank those who struggled before me, Galileo, Tesla, Bohm, those who dared to see beyond what was accepted. My role now is to carry that torch with grace…and to pass it to all of you, who will carry it further.

In reality, this obviously is not about me, it is about all of humanity, fortunate to be alive during this extraordinary time. We are witnessing reality itself shift in real time. The real question is, *what will we do with this opportunity those before us never had?*

The Cosmic Funhouse: Archetypes, Aliens & the Hologram of Perception

As we leapt from the fixed-reality mindset of the 19th century, grounded in logic, measurable phenomena, and observation, into the 20th century's age of radios, televisions, and computers, something shifted. Books, while deeply impactful, were linear and static, limited to the written word. But these new technologies projected our stories outward in a multi-sensory way, allowing us to visually witness our own creations reflecting back at us. In doing so, a new idea started to percolate: What if the very medium we exist in…this reality, is also reflecting our consciousness in the same way?

This same medium was present both in the cosmos as space plasma and within us as blood plasma, interfacing with us biologically. But now, for the first time, the plasma of the psyche was starting to bubble up, just under different names.

This was no longer just about gods and cosmic forces; it was about the mind itself as an active player in a dance with the invisible medium. We were beginning to see hints of it, but no one was piecing it together with Plasma itself, which was also emerging in science around the same time. The threads were there, but they remained disconnected…a cosmic, biological, *and* now psychological medium was waiting to be recognized as one unified force.

The human psyche was being brought into the equation like never before, similar to recognizing you are inside a funhouse of mirrors, where what stares back is distorted by your own inner state rather than the glass itself. But we are the experiencers of the funhouse *and* the creators of it. Without us, it's just glass mirrors. No fun! And, just like a funhouse is designed to amuse or frighten amusement park guests, ultimately, we are the choosers of that experience. Even the word "amusement" hints at this deeper truth, the park's purpose is the joy of entertainment.

This *amusing* medium began reaching toward us, extending its hand through the archetypes of Carl Jung, the interdimensional beings and UFO phenomena explored by Andrew Collins and Greg Little, and the mythic messages of the heart expressed by Jean Houston.

These voices helped usher in a new era of individualization. Are archetypes, UFOs, and mythic experiences reflections of our psyche, co-creations of consciousness, interactions with a sentient plasma field, real encounters—or all of the above? I set out here to answer those questions. And with every answer, we ask more questions, opening doors to ever-expanding horizons.

Carl Jung, Archetypes, and The Collective Unconscious

Carl Gustav Jung (*1875–1961*) was a Swiss psychiatrist, psychotherapist, and psychologist best known for founding analytical psychology and for his groundbreaking theories on the collective unconscious, archetypes, individuation, the Self, and more.

From a young age, Jung felt as though he carried memories from past lives. He described himself as having two distinct personalities:

Personality 1: His outer persona, the typical schoolboy conforming to societal expectations and the norms of his time.

Personality 2: A presence that felt ancient, dignified, and authoritative, like a man from a previous era, holding wisdom and influence far beyond his years.

This duality would later become a cornerstone in his work, as he explored the interplay between the conscious and unconscious selves, bridging past and present identities.[140]

I interpret Personality 2 as Jung's M-self, the multidimensional self that remains connected across lifetimes. Most of us forget this deeper awareness, burying it beneath the mask, built by the conscious and unconscious, we wear to survive in the world. Jung had many imperfections personally and intellectually, with his parents and in school. What he didn't struggle with was *curiosity*, and that may be what kept his thread intact.

---

[140] Carl Gustav Jung, *Memories, Dreams, Reflections*, recorded and edited by Aniela Jaffé, trans. Richard and Clara Winston (New York: Vintage Books, 1965), 33–39

Following his curiosity, regardless of what society or even he himself thought of him, became Jung's yellow brick road. It was this unyielding pursuit that made him infamous. Curiosity is your true genius. Success is never about perfection long-term; it's about staying endlessly curious and following it unabashedly. Curiosity beats "smarts" every time.

Jung had a deep fascination with the paranormal, especially as it related to dreams, visions, and clairvoyance. Many of these experiences are documented in *The Red Book*, describing them as confrontations with the unconscious.[141]

Jung proposed that all humans share a *collective unconscious*, a universal layer of the psyche that transcends individual experience. I propose that this collective unconscious is not just an abstract concept, it is held by a substance…Plasma.

As I've explored in previous chapters, the collective unconscious is what I call 4D plasma. It connects everyone. It is the glue and the telephone line to our deepest selves, our 5D M-Self and The Mystery.

I've also referred to the unconscious as spacetime geometry, or potentially the layer underneath it, as a sleeping giant waking up. And once it fully wakes up, we'll be able to materialize what we want quicker and do things once thought impossible. But before that happens, there's a process, a consciousness expansion, heart opening, and learning how to navigate emotion and thought so we can work with plasma ethically, safely, and consciously. This natural progression, to me, makes perfect sense.

In Māori tradition, *Papatūānuku* (*the Earth Mother*) is sometimes depicted as a great sleeping figure whose body forms the land itself. In Norse mythology, the primordial giantess *Ymir* (*though more often described as male in later texts*) has echoes of a sleeping, world-body figure, where her form gives rise to creation after awakening or transformation!

This unconscious contains archetypes, universal symbols, myths, and motifs, which Jung identified such as the hero, the shadow, and the anima. These archetypes funneled down from the pool of the unconscious, showing up in each human psyche as representations of what one needed to heal. By healing one's shadows, as he called them, one was able to integrate each part back into themselves, leading to a greater wholeness and a total Self.

Each archetype (*Jester, Caregiver, Magician, Sage etc.*) helped guide a person through different stages of life, each offering a particular lesson or challenge to be integrated into the Self. *The Self* was an actual archetype that represented the entirety of the psyche,

---

[141] C. G. Jung, *The Red Book: Liber Novus*, ed. Sonu Shamdasani, trans. Mark Kyburz, John Peck, and Sonu Shamdasani (New York: W. W. Norton, 2009)

integrating both the conscious and unconscious aspects of an individual. Jung often spoke of the Self in spiritual or metaphysical terms, suggesting that it can be likened to a higher power or divine aspect within the individual.

> **"It is a fearful thing to fall into the hands of the living God. The experience of the numinous is terrifying... The God-image includes not only what is noble, beautiful, and good, but also the shadow, the dark side."**
> — *Carl Jung, Answer to Job*

I agree, and have personally experienced, as I'm sure many of you have, that exploring the unconscious can be mentally and even physically terrifying. I've had days where my nervous system released so much stored fear that my body shook and poured sweat, as if exorcising something old and buried. It felt like I was going insane. My cheeks hurt, my thoughts were loud and scary, and I thought I was dying.

Deep down, I knew I was healing, bringing what was hidden into the light so I could recalibrate. Some might call this the beginning of an ego death. I don't love that phrasing because the ego is evolving not dying. My mind was scared, but my intuition wasn't. What I call the Mystery, or God, felt like this future light, the only thing I could anchor to, a string pulling me through the darkness. Yet, at the same time, it was also true that it was present *now*, even in the depths of that terrifying experience. It was both. In this darkness, there was still a benevolence. It was beyond explanation.

In Jung's later writings, he frequently associated the Self with the God-image, seeing it as representing the spiritual wholeness that transcends ordinary ego consciousness.[142] He also saw the ego and the Self as two distinct centers of personality, existing in dynamic polarity, yet connected by an underlying unity. The ego is close to what I see as consciousness, healed or unhealed, 3D or expanded, and the Self is very close to what I call the M-Self. I simply expand on this by reminding you that the Self is multidimensional, multi-timeline, and multi-realital.

You don't have to wait until the end of your life to become this actualized self. And as you move toward that end, you can continue to recreate yourself, fresh and new, while consorting with other timelines to gather information that can help you in this one.

In my eyes, actualization isn't some state of perfection or moral goodness. It's about knowing you lived a full, rich life…full of experimentation, play, mistakes, adventure, love,

---

[142] Monika Wikman, *Pregnant Darkness: Alchemy and the Rebirth of Consciousness* (Berkeley, CA: North Atlantic Books, 2004), 29.

emotion, and self-compassion. If you live this way, being of service becomes a natural part of your existence. Doing this, and publicly acknowledging it, inspires others to live in that same freedom, and that is a profound act of service. It's not something you have to actively pursue. Service will look different for everyone, whether you're inspiring by being yourself, speaking your truths, or helping people by starting a foundation or a business that solves problems.

The collective unconscious allows for shared human experiences across cultures, as it contains the inherited memory of humanity's history, much like the Akashic Records. It can be seen as a metaphysical force influencing our behavior, dreams, and experiences. Often referred to as an unseen depth, it informs consciousness without our direct awareness.

Now, we are no longer passive recipients of its influence. We are entering into an active dialogue, healing the archetypes it holds. We are bringing light to the darkness of this fourth-dimensional space by seeing it through a new lens—one rooted in love. With our growing awareness, we are no longer unconsciously molded by it. Instead, we're reaching into its depths and drawing out new myths, new symbols, and new motifs.

We are making the unconscious conscious at a *critical mass-level*. And just like in physics or chemistry, when a system reaches critical mass, a tipping point of accumulated energy or information, a new phase or emergent property is born. In doing so, we uncover a river of infinite potential, ready to be shaped by our own creation.

Emma Jung & The Anima as an Elemental Being

Emma Jung, Carl's wife, was also a psychoanalyst, and I believe she sensed Plasma in ways Carl himself may not have fully grasped. Carl developed the concept of the *Anima* and *Animus*; the Animus representing the masculine aspects within the female psyche that need integration, and the Anima representing the feminine aspects within the male psyche requiring the same.

Anima, meaning "soul," was associated by Jung with goddesses like Hecate, Minerva, Pandora, and Selene, symbols of the inner feminine, emphasizing the importance of integrating this energy within men.

Emma expanded on this in her essay *The Anima as an Elemental Being*. In it, she described the anima not merely as a psychological construct, but as an elemental force representing a transformative and sometimes chaotic power operating beyond the confines of the conscious mind. To her, the anima connected deeply with the natural world and the unconscious, making it a <u>vital part</u> of the individual's journey towards self-understanding and individuation.[143]

The fact that Emma referred to the Anima as *elemental* suggests to me that, even subconsciously, she was alluding to something akin to Aether, the primordial substance from which all elements arise and to which they return. This "fifth element", was long believed to permeate existence, integrating fire, earth, water, and air, essentially describing what we now recognize as Plasma.

Emma even discusses the anima's role in *reflecting* a man's inner thoughts and emotions, acting as a mirror to his unconscious! She explained that this reflection could foster greater self-awareness, or, conversely, lead to self-flattery or self-pity. This closely parallels how Plasma reflects back to us what we bring to it: either clarity or distortion, depending on how deeply and discerningly we look.

Emma suggested that the anima was not merely a passive psychological force, but an active, *transformative energy* that can profoundly shape a person's emotional and psychological life. Moreover, she discusses the anima's influence on <u>emotional life</u>, noting that it can intensify, exaggerate, and mythologize emotional relations, amplifying them in ways that closely mirror the properties of plasma, which conducts and magnifies energy.

Emma continued to emphasize its potential for both creative inspiration and destructive chaos. Engaging with the anima allows individuals to access deeper layers of their being, uncovering *intuitive and emotional insights* often buried beneath social conditioning. Yet, she cautioned, just as I do, that without conscious integration, this force can become destabilizing, even dangerous.

This is reflected in both our psyche and society through the ways we are beginning to use Plasma, either for good or for harm. Emma Jung expressed that it could be a source of inner turmoil just as much as artistic inspiration. My hope with this book is to help us recognize that we already possess the intuitive know-how to work with Plasma. Most of us simply haven't known how to do so harmoniously, which has led to confusion, stress, shame, depression, anxiety, and the chronic tension of trying to control life, ourselves, and outcomes.

We've been misled into believing that everything is purely consciousness, disconnected from plasma, disconnected from the feminine, and that if we're not manifesting reality exactly as we envisioned, something must be broken, or we must not be doing it

---

[143] Emma Jung, "The Anima as an Elemental Being," in *Animus and Anima*, trans. Cary F. Baynes (Dallas, TX: Spring Publications, 1981)

"right." But plasma brings back chance, whimsy, surrender, listening, and learning…instead of judgment. It reminds us that life isn't about controlling everything, it's about dancing with it. Not as a "good dancer," but like the kind of wild, free dancing you do alone in your room when no one's watching.

> **"Everything can be taken from a man but one thing: the last of the human freedoms—to choose one's attitude in any given set of circumstances, to choose one's own way." — Victor Frankl**

It also goes without saying, this is not a promise for a life free of darkness or struggle. I know some moments feel unbearable, unfair, or even cruel. I don't have all the answers, but even in the pits of hell, you can remember there is light. Not to ignore the pain, but to choose, again and again, to find even the smallest thread of meaning through it. That in the unknown, you *still* have a say. You can't always control what happens, but you can choose what it becomes inside of you, imbuing your future. Plasma reflects our consciousness back to us for our deepest evolution. Even when we're shattered, she's still holding the mirror to show us who we're becoming.

And if you're facing something incurable, something that feels impossibly unfair, I hope this gives you even a little peace: you are still becoming. Even beyond this absurd life. There are no true endings.

You deserve to feel it all…whether you're fucking mad, deeply depressed, lost, or spiraling. You're not broken for feeling out of control. In fact, feeling it *all*, and still choosing what it means to you, is the most radical thing you can do. It's not about pretending to be okay. It's about letting the storms pass through you without giving them the power to define you. That choice, your internal meaning, is all yours. And no matter what you are facing *your soul is still free.*

Jean Houston & Anneloes Smitsman: The Divine Child & Future Humans

We all get caught up in wanting to know *how* the universe works. But if the universe is alive and evolving, then our understanding must evolve with it. Sure, we'll gain insights, but concrete answers? That's a red herring. Especially when those answers seem to be a moving target.

> **"The universe is not outside of you. Look inside yourself; everything that you want, you already are." — Rumi**

The truth is, rather than chasing the mechanics of the outer cosmos, the real work is turning inward and learning how *we* work. As we uncover inner truths, reclaim our multidimensional capacities, and evolve, we begin accessing the same forces that shape the universe. We realize we are every version of ourselves, right now, and with each choice, we either step into one or create a new one. Each action imprints onto the living record of this specific lifetime.

This journey isn't about collecting answers or stockpiling knowledge. That path is hollow, it disconnects us from feeling. I know this firsthand. Any time I tried to bypass pain or emotion, I'd retreat into logic. Even now, I sometimes catch myself doing it.

In their co-authored *Future Humans Trilogy*, Anneloes Smitsman and Jean Houston explore themes that deeply align with the essence of this book. Jean Houston, a founding voice in the human potential movement, has authored numerous works on spirituality, mythology, and visionary thinking. Her depth spans the holographic universe, physics, and psychology, and she's collaborated with some of the world's most influential leaders. Anneloes Smitsman, a brilliant systems scientist, futurist, and game designer, brings a cutting-edge perspective that resonates strongly with my own theories. Together, they weave a powerful narrative about the evolution of human consciousness, the multidimensional self, and the emerging possibilities for humanity.

Jean Houston views myth as a toolkit for the evolution of consciousness. Myths are not merely symbolic remnants of ancient cultures, they are encoded patterns of possibility, helping us navigate the depths of human experience. They function as psychic technologies, and when we engage them consciously, through story, ritual, or embodiment, we unlock deeper layers of the self and activate latent human potential.

This is why myth is essential, almost like an emergent energy, caught in paper, film, words, and song. It is one of the core languages of consciousness feedbacking with plasma.

I believe that by understanding and becoming well-versed in myths, archetypes, goddesses, and even fairy tales, we build a rich inner image bank. This bank expands our capacity to receive messages, not just from the subconscious, but from other realms of consciousness altogether.

While Houston never explicitly called it Plasma, she described what she called the "imaginal realm" as an interactive, co-creative, evolving space. As explored by Jung, this is where these myths reside, waiting to be seen, healed, reimagined, and evolved. Once a pattern, goddess, or archetype is integrated, it can alchemize, flowering into a new story or archetype.

It's important to write these experiences down and observe what meaning unfolds from them over time. This is precisely why myth functions as an emerging emotional technology, one that holds just as many, if not more, secrets than science or math. I love math, but math as we currently see it is just a projection, of a deeper truth. These are emotional truths, not meant to remain hidden, but to be felt, honored, and transmuted.

I believe part of my purpose on Earth is to help heal the archetype of Inanna, Mary Magdalene, and many other goddesses otherwise known as the divine feminine, to speak my truth even when it doesn't fit the paradigm, and to be courageously heard. No matter what your purpose is, what truths you expose, you are meant to shine in your full expression, not shrinking from the visibility it brings, not afraid of betrayal, but walking beside others in mutual wonder.

The rise of masculine logic may have been a response to an era when we drifted too far into the ungrounded feminine, as a cosmic rebalancing. This recent age, though dark and painful, served its purpose across the greater landscape of time. It sucked, to be candid, but it may have been necessary.

As we remember the divine feminine, by transmuting fear into love and integrating her with the divine masculine of consciousness, a new, emergent child is created. We stand at the threshold of a new paradigm: the age of the Divine Child. A time of human evolution and multi-human awareness. And at the doorway stands the divine feminine longing to be re-integrated into a masculine-dominated consciousness that forgot how to feel.

The child represents unbounded communication, play, and curiosity…a state of expansive lightness toward life. Yet within this lightness lives depth, born from the divine union of masculine and feminine. It is innocence illuminated by wisdom, and wonder guided by presence. It is the child of integration and living from freedom rather than naïveté.

The divine feminine who stands at the doorway, at the gates, appears as fire, as chaos, as plasma, mistaken for the devil, feared as destruction, and exiled for her madness. But her appearance, and our projection of her, is merely a reflection of our fear of the unknown, of our own emotions, of unquantifiable mystery.

Plasma is a teacher of truth. When we engage with her true nature, we become vessels, not just for our dreams, but for the multidimensional self and the Mystery. The Mystery, is emergent energy…the potential for creations not yet born, but already felt.

Some call this God. Some call it the culmination of human potential. It is something here with us now, that has and will always be with us, yet also something still being created.

Incomprehensible to logic, yet undeniably real, it energizes and supports us, existing both as the present moment and the infinite becoming. Tapping into this Mystery is tapping into the M-Self, as you co-create life in alignment.

If Plasma reflects our internal state, then our suffering is not meant to punish but to bring our inner chaos to the surface so we can heal, integrate, and evolve. This redefines human potential as not just the achievement of dreams but the embracing of the unknown, the willingness to sit with chaos, and the courage to witness ourselves in our most raw, unfiltered states. It removes the pressure to force outcomes or chase enlightenment. Instead, it invites us to soften into plasma's flow, to trust and feel into the subtle emergent energy.

Adventure, then, becomes the act of living inside the unknown, not needing to know what's next, but trusting that each moment is sacred, part of a larger, intelligent tapestry.

If God is the culmination of human potential, the sum of all emergent possibilities, then God is not something to chase or achieve. God seems to be something not yet formed but always forming. God is what moves through us when we're fully present, feeling, choosing from our deepest truth, and allowing life to create through us from these higher perceptual faculties. And you can mix that belief with any religion, if you choose.

Effortlessness begins when we recognize that Plasma, the M-Self, and the Mystery are not separate—they are all facets of the same source. You don't have to strive to become something; *you already are what you're seeking*. That truth is still integrating within me…I've stopped using spirituality as another form of control, and instead, I'm learning how to simply be human… and spiritually experience life.

Plasma, imbued with the intelligence of the Mystery, call it God, Source, or Love, will mirror our expanded consciousness. It will harmonize with our intentions… but always with her own wild, unpredictable twist. What we believe can be conceived, but not always in the form, timing, or path we imagine.

When we stop trying to control her and start flowing with her, we realize something beautiful: our deepest, most childlike desires are already in motion. They're being shaped by something wiser than we can comprehend. Life becomes a playground of magic and co-creation.

"We are coded with the longing to become and to express capacities that are innate but not yet manifest. This longing is the evolutionary impulse within us, urging us to grow into our fuller capacities and greater wholeness."

— Jean Houston & Anneloes Smitsman, The Quest of Rose

This idea is echoed in the first book of the *Future Humans Trilogy*, *The Quest of Rose*. They speak of a new form of consciousness, one that, in the near future, may be perceived as a threat to existing worldviews. They caution that hardship may follow if this consciousness is not welcomed and integrated. Preparing humanity for this shift, they say is critical.

Anneloes shares her commitment to creating bridges, interfaces, and systems that allow this new consciousness to serve its true purpose in our world. She also describes being shown how this consciousness exists as a dormant archetype, a new child, that lives within each of us. In her words, she experienced a kind of rebirth via this future child archetype. [144]

This is what I call the awakening of the M-Self, the embrace of 5D consciousness and divine play. There's nothing to fear. It's a realization that we are more than our bodies, and more than this version of self. It's about opening the lines of communication to our multidimensionality. You are both the writer and the editor of your life's book with the ability to change the words, the structure, even rename the title... after the book has already been written.

With this rise in *emotional technology* will come rapid advances in engineering, technology, psychology, and healthcare, tools that are designed to uplift humanity, not merely fuel capitalistic gain. Money can still be made, and that's perfectly fine, as long as the creators of these systems are aligned with their M-Self. When they are, their inventions naturally reflect a desire to deepen life experiences and genuinely help others.

Money is just a tool, it's not something to worship. When used ethically and in harmony with consciousness, it can amplify our impact, empower our visions, and support collective well-being. This is a thrilling prospect, but it starts with reintegrating feeling. We must relearn how to harmonize with plasma in a way that aligns with our truest selves.

Lastly, in the second book of the *Future Humans Trilogy*, *Return of the Avatars*, the final chapter speaks of "Integration with the Cosmic Mirror." When I read it, I knew instantly, it was about Plasma. They just called it by another name.

The text emphasized how the Cosmic Mirror supports the development of cosmic self-awareness and shields the mind from being overtaken by archetypes of domination and egoic projections. It described how this Mirror supports us in becoming aware of ourselves beyond the realities and filters of the local mind, and how we can call upon it for protection,

---

[144] Anneloes Smitsman and Jean Houston, *Quest of the Rose: Future Humans Trilogy, Book 1* (Bloomington, IN: Inner Traditions, 2021), 239.

guidance, and insight through our "future consciousness" and connection with future humans of the world! [145]

Houston and Smitsman also proclaim that during this transition period, this threshold of the Divine Child, which they call a collective awakening, there is a greater potential for escalation, polarization, and violence, as people may act out deeply seated survival patterns. [146]

I've been there. I, too, have acted in ways that contradicted my values just to avoid feeling… feeling fear, grief, shame or reality itself. That deep, unspeakable terror that if I were fully seen, fully expressed, I'd be found out as the monster. The beast.

This is a fear that reaches back to the ancient projections onto Plasma—the beast inside all of us. It's easier to blame ourselves, to cast ourselves as the villain, than to face the possibility that *we are actually good, actually benevolent*, and that all the terrible things are happening around us. What does that mean for us?

That is a painful realization. Looking back, I can see how much fear shaped my past actions. And I began to realize, those who wronged me weren't so different from me. They, too, were acting out of fear. All the evils in the world, when stripped down, are just human fear magnified and amplified.

That perspective helped me forgive, not only others, but myself. And it opened my eyes to a deeper truth: we need a new framework for humanity. If we are to truly abolish war, violence, and suffering, we must approach life from the awareness that we are all good. That we are all deeply human. And those certain religious beliefs, that do harm others, come from deep rooted collective fears we must heal. Fighting these people will not change their beliefs, it will amplify them. We must feel the heartbreak of *"Oh God… what have we been doing?"* And from that moment of grief, we can rise, to our M-Self, to our neutrality, our curiosity, our intuition.

That's how we will heal, individually and collectively. Not through fear, or violence, but through a higher awareness that empowers us to make new choices…from benevolent love, not survival. This doesn't mean we have to agree with everyone, or never feel anger or disagreement. Conflict is natural. Boundaries, even strong ones, can come from love. But we don't have to resort to destruction.

---

[145] Anneloes Smitsman and Jean Houston, *Return of the Avatars: Future Humans Trilogy, Book 2* (Bloomington, IN: Inner Traditions, 2022), 330.
[146] Smitsman and Houston, *Return of the Avatars*, 331.

Violence often escalates because we believe it's the *only* option. But protection doesn't always have to mean retaliation. What if protection could be upgraded, from reactive force to creative intelligence? That doesn't mean passivity. It doesn't mean things remain unsevered. Monika Wikman, in *Pregnant Darkness*, calls this *conscious violence*, a form of action tempered by awareness, where the motive is integration or protection of life, not domination or revenge.[147]

After reading this, I hope you understand why you've felt so fearful, and that now, you know you can meet it and move through it. I'm walking this with you. Together, we can face the darkness and step into the holographic, neon light of the true self, the Divine Child. It's okay to feel scared, dissociated, even a little crazy. But now you know *why and that you are not alone.* The only way out… is truly through.

And with this awareness, you can seek out supplemental support, whether it's a therapist, shaman, friend, family, or whatever calls to your spirit. We can heal this rupture with the divine feminine, with plasma, with our trauma, and nature and begin to remember that we are divinely supported, and safe to experiment and live again with love and laughter.

Be patient with yourself and have compassion for your becoming. It's not about straining to become our best selves; it's remembering who you already are. The magic is here, we just have to heal our nervous systems enough to hold it.

We must release our tension and transmute ourselves into resonant containers. When we feel and believe, Plasma reflects this and supports us. The divine is created in everything we do. We are basically human magicians who are learning to live on trust instead of worry.

And believe me, as the world's once least trusting person, if you keep going down this path of meeting your fears, facing your doubts, and consistently choosing to believe in what's in your heart and that it's possible for you…one day, you'll wake up and genuinely believe it. You will choose better thoughts. You will create a life beyond anything you thought you could. Because by then, you will have walked yourself into that reality. Savor the journey.

"I live each day as it was my last. And life, in *all* its moments, is so full of glory"

— Helen Keller in a conversation with Jean Houston

---

147 Wikman, *Pregnant Darkness*

Don't dream your life away trying to reach your future. Now that you remember how to resonate, how to receive your true self, this becomes the new human template. You *are* the future you, right now.

"You see friends we are in the new story and a story that is changing. And in my studies of history, I find that *when the story changes*, evolution allows there to arise in us qualities and capacities that often we did not *know* we had."

— Jean Houston, address to the Unity Community of Ashland, December 2024

In these new future human potentials, Houston speaks of deep seeing, a way of perceiving the larger reality that each of us carries, changes, illuminates, and turns into art. It's about a human honoring another's full humanity, beyond physical and psychological understanding, beyond even prior knowledge of that person. The next stage is a renewal of our sensory capacities to a new level.

As I listened to her speak about these concepts, I started to cry, from fear. It was as if my protective ego was falling apart. Know that this is normal. It felt like both a burial and a gentle honoring of my old self, my past consciousness, gripping tightly to its reality and screaming out, "This can't be real!"

Houston and her friend Peggy, in Houston's 2024 speech, said to picture a piece of paper, like this very page you're reading from, and imagine it traveling back to the tree it once was. Imagine it becoming the tree, the very thing you now behold. Remember the tree. See it, touch it, hear it, smell it, taste it. Honor the tree that gave you this sheet of paper. Thank the machinery that processed the tree. Feel the earth from which the tree grew. Sense its life connecting to all other trees.

Then, with your higher senses, connect with the humans who invented writing and say thank you. Remember all you have read, all the words that shaped you. This is how you activate your *higher senses*, by adding gratitude and wonder to every sensory experience. [148] This is having *knowing*: traveling the veins of Plasma with our consciousness to imprints of memory, imprints of reality not bound by time.

Another quality that arises in the changing of the story, as she speaks of, is the power of *holding a new vision*, and the energy to maintain that vision and recognize your role in bringing it to fruition. It is understanding your responsibility in holding a vision of a world

---

[148] Jean Houston and Peggy Rubin, *The Possible Human: Becoming the Future Human*, address delivered December 29, 2024, hosted by Kathy Zavada, YouTube video, 1:10:15, posted by Kathy Zavada, https://www.youtube.com/watch?v=pYItVIc5kK0&t=815s, accessed March 31, 2025.

that works for everyone. Of shepherding new possibilities into emergence. Vision creates worlds. If you become aware of this power within your own actions, the world begins to grow around you. You hold the sacred role of co-creation with God.

Just as we repress traumatic memories, and find ourselves living in new timelines as we begin to access and heal them, the same is true with Plasma. As we begin to access this mistakenly-called beast within, this memory bank of Plasma, and heal the emotional imprints of our past, individually and collectively, we open the doorway to a beautiful new future.

One not unconsciously repeating old mistakes and traumas, but one created with the innocent yet wise awareness of the M-Self, ready to co-create fresh, safer, more expansive pathways forward. Our consciousness *is* evolving. The M-Self *is* here. And the choice of what emerges from that... is <u>yours</u>. It's all of ours.

Just like in *Beauty and the Beast*, and as old as story itself we witness again and again the meeting of two seemingly opposite forces. At first there is hesitation, then recognition, and ultimately the discovery of transformation through their very union.

What once appeared monstrous to us is not only essential but transformative. This dynamic between beauty and beast, the Self and the M-Self, consciousness and plasma, is an eternal dance as old as time. And finally, we can step back and truly see it.

As our consciousness evolves and the M-Self integrates more fully into our awareness, we enter a profound phase of emotional and sensory recalibration. This emerging state is like a divine child mixed with a wise grandparent, both innocent and ancient, playful yet deeply aware. The struggle lies in bridging the gap between these seemingly disparate aspects: the pure, expansive wisdom of the M-Self and the raw, untamed emotional landscape of the human self.

The M-Self is neutral by nature, a conduit of benevolence, clarity, and detachment. It perceives life from a higher vantage point, where all things are interconnected and part of the larger cosmic dance.

Yet as we embody this higher awareness, we remain human, we still have emotions and duality. Our emotional bodies are complex terrains, rooted in survival, desire, and the dense, time-bound experiences of 3D reality. Now, we are being asked to do something unprecedented, to merge the M-Self's neutrality with the full range of human emotion.

We don't have to erase or transcend our emotions, we must learn to feel them fully without being ruled by them. It's the art of holding paradox, the ability to feel deeply while also maintaining the observer's perspective.

There is a great challenge here. The language of emotions as we know it is inadequate for what is emerging. One word for one may no longer suffice. We may need entirely new terms, new emotional frequencies, to describe the multidimensional nature of what we're experiencing. Is the feeling that arises during a profound synchronicity joy, awe, or a fusion of both? Is it a memory from a future self, a longing from a parallel timeline, or simply the M-Self nudging us into deeper alignment?

This is why self-awareness is more crucial than ever. The new frontier of human evolution is emotional. Our emotions are becoming the navigational language through which we access timelines, anchor new realities, and attune to higher states of consciousness. If we don't learn to decipher these emotional frequencies, we risk distorting the M-Self's benevolent guidance through the lens of unhealed trauma or outdated belief systems.

This is where emotional tuning comes in, learning to identify, name, and consciously feel the nuances of these new multidimensional emotions while staying grounded in M-Self awareness, as well as our earthly bodies. Instead of bypassing or numbing, we are refining our internal receiver to discern if something is a 3D trigger, a 5D insight, or a message from another timeline…

This, I believe, is how we anchor the M-Self more fully into the body, through emotional clarity, deeper presence, and the willingness to embrace the full spectrum of the human experience. To make space for it all. This is the next frontier, and it will likely play a pivotal role in how we interface with future technologies, new states of being, and the unfolding of the new human potential.

We've landed in these humanoid spaceships, bodies composed of raw, emotional, plasmatic material. But what our M-Self didn't fully anticipate was how deeply this raw material had already been imprinted with survival patterns and fear. It's like an operating system glitching, scrambled between its original design and the distortions accumulated over lifetimes. To navigate timelines, realities, and this life the way we were meant to, while co-creating new pathways, *we must first recalibrate the navigation system.*

And the key to that is our emotions. Or more precisely, evolving them into the fusion of thought and feeling, a kind of visual thought embedded with emotion. Our personal symbols and sensations combined. This is how my mind works, I am not sure if it is a type of synesthesia, but I have an instinct yours is becoming this way too.

This fusion becomes the interface, your compass. Emotions are both the signal and the feedback loop. They guide us deeper into truth, revealing where more integration or awareness is needed. If we don't heal and reattune this emotional system, we'll keep receiving distorted data, mistaking trauma responses for intuitive guidance.

I don't want to scare you, but this recalibration might feel intense, at times, even terrifying. All emotions, both heavy and joyful, may surface so you can learn to feel and witness them in entirely new ways. You're retraining your entire system. If certain sensations show up in your body, don't panic, that's just your awareness sharpening. These signals are invitations to identify stuck energy and practice moving it through. They are also pushing you towards deeper embodiment, potentially taking up practicing Qigong or something alike.

When your mind starts screaming, remember: *that voice is often not the truth.* This is your moment to step in as your own 5D mascot, your inner champion, yelling: "We've got this. We *can* go on!" And once we recalibrate, emotions, in this way, become a direct line to the M-Self, allowing us to traverse multidimensional landscapes with clarity and intention.

UFOS, UAPs, and Plasmoids: A Modern Myth

In his 1959 book *Flying Saucers: A Modern Myth of Things Seen in the Skies*, Carl Jung interpreted UFO sightings as manifestations of psychological phenomena rather than literal extraterrestrial visitors. He viewed these strange occurrences as archetypal images emerging from the collective unconscious.[149]

Jung believed these widespread sightings reflected our collective anxieties, desires, and hopes during times of rapid technological and social transformation. It reminds me of how a friend or family member may receive support during a major life change, whether a divorce, a birth, or a death, these appearances seemed to carry an underlying theme: *support through transition.*

To Jung, UFOs symbolized the need for a new myth to guide modern man, one that incorporated the mysteries of the psyche and unconscious, as well as the collective experiences and aspirations of humanity. Plasma, I believe, is the perfect substance to carry this new myth forward. It is outside us, inside us, and all around us. It reflects our consciousness, revealing that we are the very makers of meaning we seek.

---

[149] Carl G. Jung, *Flying Saucers: A Modern Myth of Things Seen in the Skies* (New York: Harcourt, Brace & World, 1959)

Today's UFOs, UAPs, and what some now call plasmoids, are in many ways what Jung described on the surface, *a projection of something deeper.* I envision them as wrapped in a reflective plasma layer, a sheath around an intelligence, information, or consciousness itself. Whether that layer consists of physical beings from other star systems, dimensional intelligences, timelines, psychological information, parallel versions of ourselves, something else entirely, or all of the above…I don't know.

There are also of course plasmoids, more like sentient animals, that I would dare to refer to as our ancestors, and they too must be approached with a neutral love, not fear, but of course also with caution. Don't let news fool you, <u>they do not have ill intent.</u>

I equate seeing UFOs to how we need to see a person at least once to form a connection, to even have a thought about them. There is *something* there we are interacting with. It's not just a phantasm; it's responding, feedbacking with our consciousness in real time. These "energies," wrapped in plasma and appearing as dense metals, lights, or holographic shapes, *are* here. Their presence alone tells me that at least some part of them is reflecting our own psyche. At least, that's what I feel.

We don't yet know how to internalize them, so we project. Just like anything else in life, until we learn the skill of openness, nonjudgment, and active listening, we won't see each UAP for what it really is. And believe me, that's taken me 36 years, and it's still a work in progress.

These beings clearly already understand how to work with plasma, like we are only just beginning to, enabling them to interact non-locally. Whether through physical travel, remote viewing, projecting part of their consciousness, or even bilocating to observe and engage with us, their technology or awareness seems to bridge dimensions or timelines. The key to truly opening ourselves to who or what they are isn't about chasing proof, or studying them in a lab…<u>it's about expanding perception.</u>

We can only meet them consciously if we:

1. Truly know ourselves through healing and feeling.
2. Understand our personal symbolic language.
3. Courageously hold a space of presence without grasping for certainty, allowing the encounter to bypass our personality filters, *at first*.
4. Then, after receiving (*never during*) apply discernment through our own symbolic lens, your unique and personal meaning system. This means writing it down, exploring what we saw or felt, physically or in our mind's eye, and letting

meaning emerge *naturally*, rather than forcing it. This step depends on our ability to decode ourselves, which is precisely why this book exists.

5. Finally, choose what to say back or how to respond, from a place of higher awareness.

This process doesn't just apply to plasma beings or interdimensional contact. It applies to meditation downloads, intuitive nudges, communication with multidimensional selves, and even everyday interactions. If we rush to assign meaning too quickly, we risk distorting the truth through old stories, assumptions, or survival patterns.

To achieve communication, be willing to feel, your body will tell you. It is called negative capability, and I speak about it in the Plasma Intelligence section. Trust the signals you receive and always question if you are having a trauma response, projecting fear or control, or if you are feeling a deeper truth, not from your belief system, rather you are receiving something fresh and novel. Lastly, maybe it is an energy that doesn't feel organic, or feels like it doesn't resonate. That's how you'll know what to let in, and what to lovingly say, "no thank you, goodbye" to, and move forward with clarity. Also, sometimes you will just want to have an experience, no communication needed.

We no longer have to let our curiosity be hijacked or feel obligated to explain ourselves for saying no. Some energies are dense or draining, and we don't have to let them in. But chances are, what comes to you is what needs healing or what you're resonating with…not because you're "bad," but because maybe you need to learn the very lesson of saying no. So let those encounters at least make you curious. What in you needs healing? What is this potential encounter showing you?

Remember: all of this energy is wrapped in plasma, existing in a four-dimensional space that operates outside of linear time. It's not something you can fully grasp through logic or 3D consciousness alone.

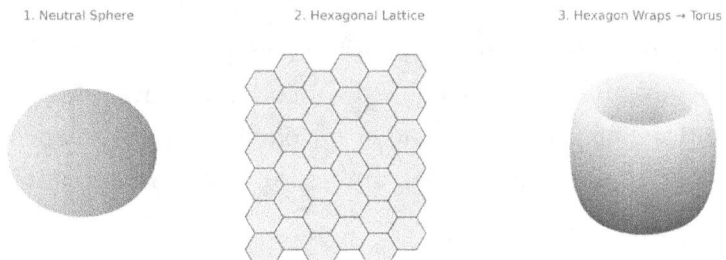

1. Neutral Sphere   2. Hexagonal Lattice   3. Hexagon Wraps → Torus

The deeper truth, of awareness, is in a neutral plasma spherule. I picture it then being encased by a plasmatic hexagonal wrap that can be programmed by information, intelligence, or consciousness. This creates the "mirror" reflecting our consciousness, and it can be bypassed only by our awareness, connecting us to deeper levels. Both layers have a great purpose and much to learn from. It is just being aware that this is our co-creative experience, and we are not a victim to this. That is the only way to get out of negative experiences.

"I dwell in possibility." — Emily Dickinson

This is all a multidimensional process. It's like living inside a mind-bending Christopher Nolan film infused with the whimsical, rainbow-hued absurdity of Studio Ghibli's Hayao Miyazaki, laced with the metaphysical poetry of Emily Dickinson, and rooted in the earthy, soul-stirring heart of James Cameron's *Avatar* series.

Aliens, Symbols, & Higher Laws: The Real Disclosure

I always like to return to the origin of a belief, where the deepest truths often reside. It's like those first moments after birth which is imbued with raw, unfiltered presence, *untouched by the programming of life*. A state of paradoxical wholeness, pure love interwoven with the instinct to connect, evolve, and learn. That essence still pulses at our core.

When it comes to aliens and UAPs, Indigenous perspectives vary across cultures and traditions, but many speak of beings from the stars as relatives or ancestors—often called "Sky People" or "Star People."

Now, think about all the stories we create about people outside of us, the betrayals we replay, the judgments we place on what others could do differently, even the shame of thinking we're wasting time dwelling on them instead of doing something "productive."

What if those very thoughts are the solution staring us in the face, and we just can't see it?

The quiet angers and lingering hurts are valid. They point to where we've abandoned ourselves, given away our power, or stayed silent when we knew better. But these are not just reminders of pain, they're invitations to reclaim our worth to deserve more and choose differently moving forward.

First, we have to allow these thoughts to exist and be heard, rather than brushing them off as distractions. They're surfacing for a reason. The key is listening, really listening, and holding space for what they're trying to show us. Often, these projections onto people and events are mirrors, pointing to where we're ready to grow, evolve, and reclaim our power. When we engage with them consciously, they reveal not just our wounds, but ultimately what is really inside of us, *our potential.*

Making my point, when it comes to aliens, apparitions, light orbs, or metallic ships in the sky, I see them as external symbols of something happening within us. They're reflections of our light, our need to heal darkness, our eccentricities, our future selves reaching back, trying to land inside us. Trying to be heard. Trying to reconnect. So that we can become who we're truly meant to be, both as individuals and as a species. So we can expand our consciousness enough to perceive them for what they actually are... whatever that may be.

We can create endless theories about aliens, but often that distracts us from the real message, one that may not feel as sexy, popular, or headline-worthy. The real disclosure is *within us*...our M-Self.

These are real energies, possibly of many kinds, and we won't truly understand what they are until we open ourselves beyond the filters of our beliefs. Our personal mystery, the light within us longs to be heard. The more we fixate on outer appearances before turning inward, the more disconnected we become from the real power within us.

I see these visitations, of plasmoids, lights, even alien forms, as hints and reflections of our potential, ancestral lineage, and future lineage. They whisper of the Sky and Star People, family from beyond, reminding us of a deeper connection than we've been taught to believe. Maybe we're more like them than we realize, in the midst of evolving our consciousness, learning to work with plasma in ways that stretch beyond our current understanding.

I truly feel that if these beings are capable of traveling through space, time, and beyond, using Plasma at such evolved levels, then they must also be spiritually advanced. I don't know this as fact, but I sense it deeply. I don't believe any civilization could access plasma-based technology for interdimensional, timeline, or faster-than-light travel without first evolving in consciousness and feeling. That kind of mastery would have to arise from benevolent curiosity, harmony, and love…not fear, survival or domination. This, to me, is the natural trajectory of spiritual evolution.

That doesn't mean they're emotionally evolved in the way we define emotionality. Imagine that higher beings of consciousness, whether advanced extraterrestrials or our own M-Self, operate from a state of benevolent neutrality and expansive awareness rather than human-style emotions. They move through a kind of *cosmic empathy*, an intrinsic harmony that aligns naturally with the flow of plasma. It's not sentimental or reactive. It's a state of being that is inherently balanced, attuned, curious, and creative, rather than emotionally reactive in the way we experience highs and lows. They don't "feel" in the human sense; they *resonate*.

For them, Plasma is a conduit for alignment and creation, a bridge to traverse timelines and dimensions without the emotional turbulence we experience as humans. But we, as humans, are uniquely positioned to learn how to *integrate* emotion with Plasma in a way that may be unprecedented, even for these advanced beings.

While "higher" beings use plasma for travel, communication, and creation through heightened states of awareness, humans will be learning to access the same potential through the alchemy of emotion, by *feeling* and transmuting emotional energy into higher states of consciousness. This is what makes the human journey so rare and extraordinary.

Higher beings may observe us not because we're less advanced, but because we hold a key they do not: the capacity to merge raw, intense personal emotion with cosmic neutrality. We're becoming masters of integration, blending feeling with multidimensional vision to create something entirely new.

We are bridging the gap between detached, neutral intelligence and emotionally charged, embodied experience. Our evolution lies in learning to use Plasma not merely as a tool, but as a conscious, feeling-based interface, potentially giving us access to dimensions and timelines in ways these beings may not fully comprehend, simply because they do not possess the same emotional spectrum.

This may be why our experience of viewing a timeline is more subtle compared to a fully resonant being existing in 5D, who can shift locations with a single thought. We may perceive these timelines in a dreamlike or imaginative state, yet that doesn't make them any less real. That's why learning to trust our visions, and ourselves, is essential. The key is developing the discernment to tell the difference between various types of inner experiences, without dismissing their truth.

These are how our next steps in humanity may look:

Mastering Emotional Navigation: Instead of suppressing or bypassing emotions, we learn to feel them fully, without becoming them. This turns each feeling/emotion into a vibrational frequency that can tune us to specific timelines or energetic states of being.

Refining the Language of Feeling: Develop a symbolic or energetic system that gives our emotions greater precision, one that could integrate nature, archetypes and other symbols as a living language of inner truth.

Create with Emotional Awareness: Use our emotional experiences to imprint intentions into Plasma, shaping reality through conscious feeling. The key is not to avoid them but to use them as directional guides.

A paradox I can reconcile is this: I haven't forgotten that humans with fear or harmful intentions can access plasma in this reality and use it in ways that hurt others. It's not ideal, but it's part of how the 3D world is structured. Perhaps it's designed this way with the hope that those who misuse this energy will eventually evolve beyond fear-based choices. And I do believe we're beginning to do that, collectively.

When plasma is used from a place of fear or perceived evil, it never ends well. Whether the damage is internal or external, it always leaves a trace. To interact with Plasma at the multidimensional level, consciousness *must* be coherent. For humans, that means becoming emotionally evolved, self-aware, and harmonized. For non-human beings of consciousness, it means aligning with plasma through higher-dimensional states, like harmony, neutrality, and benevolence. At this level, access isn't granted by desire alone, it's locked unless the being is in true resonance. Plasma responds to vibrational frequency, not mere intention.

This same concept is displayed in quantum physics through the principle of coherence and decoherence. Acquired savant Jason Padgett describes it as light "shaking hands" with itself to verify the accuracy of its information state. In quantum systems, coherence occurs when particles align in a unified wave pattern, functioning as a single, harmonized entity.

For a potential to collapse into a tangible reality, or a holographic image, it must reach a threshold of coherence, typically around 95%. This means the mental, emotional, and physical aspects must all align. It's the same principle used in *quantum teleportation*, where information states must resonate so precisely that they can transfer seamlessly across space-time.

Similarly, within the realm of plasma-consciousness synergy, our internal systems must reach a state of coherence to engage Plasma at a multidimensional level. Healing limiting beliefs, harmonizing emotions, and aligning our consciousness are how we reach that 95% threshold. Only then can we collapse potential into reality, co-creating with plasma as a living, intelligent medium.

It's reminiscent of the mythology that only those pure of heart may approach the Unicorn. And by purity I don't mean perfection, I mean authenticity. You can't touch this magic unless you're fully yourself. That's the paradox and the invitation: the more genuine and harmonized you become, the more Plasma reveals itself as responsive intelligence. The real catalyst for change is resonance, not control.

Interacting with Plasma at these levels in a state of decoherence may be impossible. As Jason Padgett describes, decoherence arises from fear-based thoughts, misaligned emotions, and harm-based intentions that fracture the quantum wave pattern. And as I say, you can't fake resonance; you can't trick Plasma. When your thoughts, emotions, and intentions are not in harmony, the system remains fragmented, and the potential can't collapse into form.

That 95% coherence threshold is non-negotiable. Plasma only responds to authentic alignment. If your inner world is dissonant, plasma will quietly shake its head and whisper, "Your subconscious is showing."

To reiterate, this coherence, this pure state of being yourself means being real with yourself. It means being willing to feel. You can be messy, flawed, healing, and still be resonant with multidimensional plasma, because you're honest, present, and open. That truth is the bridge to a kind of rainbow magic your future self already knows. The kind of magic the "aliens" use to reach you all the time, you just can't always hear them if you haven't yet learned how to listen. You must tune in.

This is why I'm so passionate about healing and feeling, because we will never evolve to use Plasma the way we are meant to if we don't first face ourselves. This framework makes inner healing not a suggestion, but a prerequisite. *Doesn't that seem divine?*

Whatever frequency you're tuned to, through belief, emotion, and intention, is what you will receive and interact with in the plasma field. If you think this is silly or impossible, then it will be…for you. But if you have the courage and willingness to believe it *might* be possible (*what do you have to lose?*), the shimmering reality and magic that opens to you is priceless.

Just like in quantum teleportation, when we approach 95% coherence, the signal stabilizes. This allows us to collapse potentials, access multidimensional states, receive and transmit messages, and even 'teleport' our consciousness across timelines and dimensions.

Those who use plasma from fear or manipulation might trigger effects, but they can't sustain or navigate its multidimensional nature. It's like trying to pilot a hyperspace ship with a broken compass, chaotic outcomes, no map, and no access to the deeper, stabilizing layers of reality.

There appears to be a kind of built-in safety mechanism when we access Plasma these ways. Why do you think plasma turbulence is such a big problem in fusion? While traversing these spaces, you're not safe from distortion because distortion doesn't exist, but because, in a state of coherence, you're no longer vibrationally compatible with it.

This may also be why higher beings don't appear more often. Our collective vibration is not aligned with theirs, and they are protected by the same laws that protect us. It seems they come at times of either specifically grief or expanded states of consciousness.

Those of us with curiosity, purity of intention, and an open heart, whether we see them or not, we learn to feel them inside. And maybe, if we reach that threshold of coherence, we'll finally see them while seeing deeper into ourselves, which some of you already are.

Collectively, we're still mostly vibrating around survival, control, fear of the unknown, and a fractured emotional field. As Jason Padgett might put it, we're still just balls of decoherence. That doesn't mean we're bad or unworthy, it just means we're still learning how to feel safe within ourselves.

In myth, Unicorns don't show their true form to hunters. Sure, a hunter might *see* one, but not *know* it. Unicorns appear to hearts that feel like home. It's the same with higher-dimensional beings, including many plasma-based consciousnesses. They seem to observe us with compassion, offering hints here and there, allowing those who've raised their frequency through authentic feeling (*whether it's grief or joy*), playful presence, and healed perception to make limited contact.

Yet, it appears they are waiting for both individual and collective coherence. This may be due to the result of some of us experiencing contact through distorted lenses of fear, trauma or belief. What they encounter is filtered and reshaped by unresolved emotion. They aren't wrong in their experiences, but they aren't seeing the full picture either.

In some of my earlier experiences with these beings attempting to make contact, I didn't fully feel safe in my body, and was still holding onto unhealed trauma, so their presence, whether energetic or physical, often frightened me. I remember this happening as a child all the time. I even recall being inside a ship, but the memory felt like I was on display or being studied. In retrospect, I believe my fear-state distorted that memory, coloring it with ominous overtones that may not have been there.

With maturity and curiosity, I now recognize that I've always had a connection to these beings. But due to trauma, both from childhood and adulthood, my main access point to them, through my heart and emotions, was fractured. My intuition got rerouted through survival mechanisms. It was still present, just filtered by fear. And so were my memories. Eventually, I shut it down entirely. After I started to experience trauma in my childhood, I no longer saw benevolent presences, I only saw demons, scary women, and ghosts.

It wasn't until I began to truly heal, feeling what I had long buried, that my intuition reopened and my heart softened again. I began to re-remember and reinterpret events, through a new awareness. It is also around this time when I received the idea for my book. I even felt plasma's "hug" around me, at times, almost like an invisible hardened clay or a square block trying to comfort me. I wonder if anyone else has experienced this. The more I soften, telling it I love it, it softens…

I want to be clear…I deeply respect those who feel they were abducted, harmed, or violated by aliens. I would never invalidate those experiences. I just want to gently offer another perspective, because I relate. Mostly now, when I feel or see these beings, they come through as family. They are pure love…they are profoundly caring. Throughout my life, I see now, they've been checking in on me, supporting me from a distance. And when I do have periods of intensity, where trauma still comes up, or I do come across darker beings, I am able to be aware of what I am resonating with, creating proper boundaries and approaching them with curiosity, if able to transmute them back to love and light.

Aliens, Hollywood, & Technology

As I've shared, those who haven't evolved emotionally or energetically cannot access plasma multidimensionally, and that's exactly why we have nothing to fear from these beings. If someone misaligned with higher consciousness tries to enter multidimensional realms, the energy won't support them. It will backfire, trapping them in looped, low-density timelines or causing complete misalignment. This is why I hold a strong intuition that these cryptids and darker figures people see (*and that I have seen in my past*) are part of 4D plasma, our unconscious, they are not haunting us from other timelines, and they are not attempting to control us. They are showing up to be healed. Go watch the movie, The Legend of Ochi, the signs are everywhere.

This concept, and many other deeper truths, tend to subconsciously peek through in many major motion pictures. Hollywood, knowingly or not, often becomes a mirror of metaphysical reality.

In the movie *Contact* with Jodie Foster, she discovers a message from an extraterrestrial civilization, an invitation to connect. Though she makes the discovery, her manipulative boss steals the credit and is chosen to represent humanity. He smugly implies he was selected because he knows how to "play the game," suggesting that authenticity is naive, and image is what the world rewards over truth. But he never makes it. He's killed in an explosion before the journey even begins, before he can enter the wormhole (*a metaphor for multidimensional plasma access*). In the end, it's Foster's character, who never betrayed her wonder, curiosity, or integrity, who makes contact. She experiences something profound because she stayed true to herself.

This is a powerful reminder: authenticity may not earn you the spotlight first, but it grants you the *real* reward. What's meant for you doesn't respond to your mask, it responds to your truth.

A similar narrative unfolds in the sci-fi classic *The Fifth Element*. The protagonist, Leeloo, is sent to Earth to stop a dark force, but it's not technology or violence that saves the world. It's love. Leeloo is genetically engineered to be the "perfect being," yet her power doesn't come from strength, intellect, or playing society's games. It comes from her purity of heart and childlike curiosity. She is non-performative, she feels deeply, in this day and age we would call her neurodivergent. She's overwhelmed by the cruelty and destruction she witnesses, struggling to understand why humans hurt each other.

Her counterpart, the villain Zorg, represents the opposite, pure intellect, control, ego, and chaos. He believes he's winning because he has power, weapons, influence, and tech. But he never understands love. That's his fatal flaw. In the end, it's Leeloo's emotional breakthrough, her realization that love still exists despite all the darkness, that activates the ancient weapon and saves humanity. Leeloo is chosen because she is *whole*, not because she is perfect. Those who stay true to their essence, especially when it's hard, are the ones aligned with the deeper magic and *become the key themselves.*

Lastly, in the award-winning film *Everything Everywhere All At Once*, Evelyn, a mother trying to hold her life together, is thrust into infinite versions of herself across the multiverse. At first, she believes she must become the "best" version to succeed. But it's her vulnerability, compassion, and willingness to feel everything that becomes her power. While the antagonist tries to escape pain by collapsing reality, Evelyn discovers that *embracing* chaos, love, fear and even the mundane is what gives life meaning. It's her authenticity, not perfection, that allows her to navigate the multiverse.

This is something I have to remind myself of often. Our "highest alignment" isn't about being our "best" self, it's about being our truest self in the moment. That means feeling, accepting, surrendering…it's opening our hearts, being kind, and creating from curiosity. It's embracing all parts of ourselves, not just in this timeline, but across all timelines, like Evelyn does. She becomes the multidimensional self not by escaping her life, but by accepting every version of herself with love, *holding everything, everywhere, all at once.*

The key is presence with your truth, not performing an ideal. Throughout your day, pause and ask: "what's true for me right now?" "Can I feel it fully, and respond with honesty, curiosity, and care?" That is alignment. Some days it looks like going after what you want, other days, it's relaxing in bed. And here's what I'm realizing as I write this…there are no rules. Only the choice to meet yourself honestly, again and again. When you fall short, you can only laugh at yourself, ground, and keep going. We're not here to perfect our lives, we're here to witness, love, and integrate what shows up right now…*even the aliens.*

Just like an infant can't process love or learning unless their caregiver is calm and attuned, our biology won't let us access deeper connection until we feel safe. These beings aren't withholding themselves. We're still learning how to regulate enough to even receive the enormity of the life experience, let alone communication with extraterrestrial or multidimensional intelligence.

There's a concept called *receptor-site compatibility*: a signal can only be received if the receptor is developed and available. Think of these beings as light-signals. Our consciousness and nervous system are the receptor. Until we evolve to match the frequency, the message can't fully land.

Slowly, we're approaching a vibrational home our M-Selves can return to. As for the aliens…whoever or whatever they truly are…they seem to be waiting for the song that plays out from harmony with the M-Self, that emergent energy, to begin. It really seems to me they're an expanded form of us. A visual future myth of what we are becoming.

Our next revolution is emotional, not technological. Technology will keep advancing, but soon we'll realize it can only take us so far. To evolve, we'll have to meet it halfway…with our full, feeling, emotional selves. This deeper understanding is what scares people. It means facing our choices, our traumas, and the parts of us we try to escape by clinging to transhumanist ideals.

Technology is here to <u>support and reflect humanity, the same way Plasma is</u>, not to replace it. Our intuition cannot be outsourced. And eventually, even the most devoted to the techno-dream may realize it was built on a deeper fear: the fear of returning to nature, to the child within, to the raw and unfiltered truth of who we really are. I will build on emotional technology in the Plasma Intelligence chapter.

If these beings are advanced enough to traverse the universe, dimensions, or whatever else, <u>they could likely materialize right in front of us if they really wanted to, regardless of governmental control.</u> I am not saying this is always true, but usually when you have encounters, scary or dazzling, they are usually during heightened states of emotion, or before a transition.

I don't think they're here to hand us the secrets of the universe, just as a grandparent or Zen master gives riddles rather than direct answers to life. I believe they're here as mirrors, to remind us of who we really are. And when we meet them from that place of inner remembrance, our power, love, and magic, something extraordinary can unfold. Or maybe they *are* trying to share those secrets, and we just haven't yet developed the receptors to hear them.

If aliens have crashed here or live among us, then either the government needs to stop dangling the carrot and reveal proof, or it's all just another distraction, another way to keep us focused outward instead of inward. The same survival-based thinking pulls our attention away from "Who am I?" and locks it onto "What are they hiding?" or "What do *they* know?"

But has that obsession ever brought real growth? It's a trap. Take your power back. Focus on your own knowing. When you do, as the hologram that you are, you can access anything, contact anything, without needing proof or external validation. And from that place of inner certainty, no one can manipulate, own, or define you. It's a vast world out there!

There's a fascinating area of study that may explain why we often see UAPs or UFOs as metallic discs in the sky, it's called *the Jellium Model*. This model describes a sea of free-moving electrons within metals (*plasmons*), which enables conductivity.[150] Interestingly, this mirrors plasma: a state where charged particles move freely and remain unbound. It suggests these "metallic" craft may not be solid in the traditional sense, but plasmatic in structure, appearing metallic due to high electron density and coherence. When people see UFOs/UAPs as metallic discs, that might be a simplified projection of something more complex or multidimensional.

By reducing metals to a "sea of electrons" floating in a smooth positive background, the Jellium model shows that what appears solid and metallic is actually a dynamic, fluid process at the electron level. Solidity isn't absolute, just like light, *it's an emergent appearance.* That opens the door to the idea that "solid" matter might be far more shapeshift-capable than we assume, depending on resonance, coherence, and perspective.

This could explain their ability to shapeshift or morph mid-flight as their structure may be mimicking solid matter through a co-creation of sorts with whoever is watching. If that's true, they aren't just machines but potentially conscious, adaptive plasma-based intelligences using a form of computing beyond quantum, responsive to frequency and intention. Their so-called "metal" might be an *intelligent interface* that is fluid, reactive, and alive. And we would need to obtain their level of consciousness to activate the material. Perhaps the ones that have crashed were programmed to solidify or shut down on impact…

This concept is also supported by patents held by aerospace corporations like *Lockheed Martin*, which has filed technologies related to shapeshifting metals, materials that can alter their structure, phase, or form in response to external stimuli. For example, in their patent *Morphing Wing for an Aircraft*, they describe systems that reflect this capability.[151]

These materials may operate on ionization principles similar to plasma behavior, where particles and charge states move freely to create intelligent, adaptive responses. This blurs the line between solid-state matter and plasma-like intelligence, suggesting that advanced craft may be built from adaptive metals infused with plasmatic or electron-dense architectures, allowing them to self-repair, shift shape, and eventually even interface with

---

[150] Pierre-François Loos and Peter M. W. Gill, "The Uniform Electron Gas," *WIREs Computational Molecular Science* (2016)

[151] *Morphing Wing for an Aircraft*, U.S. Patent 9,856,012, filed October 9, 2015, and issued January 2, 2018.

consciousness. These materials may not be traditional alloys at all, but reactive, living fields of plasma-metal hybrids.

Lockheed Martin also holds multiple patents that explicitly reference plasma and electrons, particularly in the context of advanced propulsion and energy systems. A quick internet search reveals their open exploration of leptonic energy and plasma dynamics, areas I'll go deeper into in future books, especially how neutrinos might unlock new ways of working with plasma and even reality itself.

Exotic Plasma & The Perception of Consciousness

If it's true that our beliefs and emotions shape our bioplasma, and if higher-dimensional beings use this fourth-dimensional plasma layer to travel and communicate, then evolving our own plasma field, and our use of emotions and beliefs, becomes essential. If Plasma is both the vehicle for consciousness and is imprinted by consciousness, then that might explain why we experience distortions when encountering UFOs, UAPs, or beings of consciousness.

The 4D plasma field communicates through symbols, archetypes, and stories, it's like a universal memory bank, a fourth dimensional data stream that any form of consciousness can access. And the greater your awareness, the greater uses it seems to present.

It's like they're transmitting their signal through this plasma layer, and it arrives in our third eye as metaphor, image, or feeling. But if we're closed off, don't understand our own symbolic language, or are locked in fear, we either miss the message or misinterpret it. This is why refining our bioplasma, clearing fear, decoding our inner symbols, and learning to listen, is the next step to truly hearing them. This is real contact…understanding higher dimensional communication.

"If we can see UFOs as sentient energy forms and complex plasma constructs, then we enter a whole new realm of possibilities, which might even start to make sense of UFO close encounters and missing time episodes."

— Andrew Collins, LightQuest

Gregory L. Little, a psychologist and researcher known for his work on Native American mounds and paranormal phenomena, and Andrew Collins, a historian and author focused on ancient civilizations and unexplained mysteries, propose that UFOs and non-human intelligences may not be purely technological or extraterrestrial. Instead, many encounters may involve plasma-based entities, conscious, shape-shifting forms of intelligent energy that interact with human perception.

In their book *Origins of the Gods*, they discuss "Intelligent Living Plasmas" and reference a classified study called *The Condign Report*, conducted from the 1980s to 2000s. The report concluded that genuine UAPs likely originate from "exotic plasmas," meaning plasma forms that exhibit characteristics current science can't explain. [152]

They also explore how plasma, UAPs, and secret technologies have long been used in classified research, particularly for warfare and experimental systems shielded from public knowledge. The U.S. Navy, for example, has filed patents on plasma-based technologies, including a laser system that creates three-dimensional plasma objects in the air, capable of rotating, moving on command, and appearing on radar. The plasmas can be tuned to any electromagnetic frequency signature desired and can create a ghost image that is real but temporary reflecting the emergence of what may of came out of Condign Report secret studies.[153] Once again, we see plasma being used through the lens of fear and control, survival consciousness, rather than for evolution or experience.

One of the central questions in *Origins of the Gods* is how plasma constructs might form a psychic link with the human mind. The authors explore this through quantum entanglement, wormholes, extra dimensions, and the effect of the observer on the subatomic world. Drawing on the *Einstein-Podolsky-Rosen Paradox*, they suggest that the mechanical transfer of information may only occur through a previously unrecognized nonlocal medium, outside the normal bounds of space.

They also highlight that both plasma and human biology function through interactions of positively charged ions and free electrons. Through entanglement, it may be possible to create a direct link between electrons in plasma environments and electrical activity in the brain. [154] In this view, plasma becomes a bridge between consciousness and nonlocal information, offering a possible mechanism for direct, spacetime-transcending communication.

What's wild is that I reached this same conclusion on my own…only later discovering during my research. When people who've never met arrive at the same truths, it always sparks my curiosity.

They then propose that entanglement may allow plasma-based light forms to not only shift into what we expect to see but also interact with and alter us. Where I diverge is their

---

[152] Andrew Collins and Gregory L. Little, *Origins of the Gods: Qumran, the Essenes, and the Arrival of the Anunnaki* (Rochester, VT: Bear & Company, 2022), 120.
[153] Collins and Little, *Origins of the Gods*, 121.
[154] Collins and Little, *Origins of the Gods*, 257.

suggestion that plasma might subtly control or manipulate us. I see this instead as something we can consciously direct through awareness and our M-Self, a tool for mutual communication with benevolent intelligences, once fear is cleared. Their view isn't invalid, but my perspective offers a different take on the same mechanism.

If something appears to control or manipulate us, it's likely mirroring unintegrated aspects within our own field, parts of us we must transmute to stop attracting distortion and reclaim full authorship of our reality. It also may be the fact that our expanded self and the mystery gently ushers us towards growth and benevolence, this as control or manipulation doesn't resonate with my soul.

This reminds me of how a skilled cult leader can manipulate almost anyone. But if someone were <u>fully</u> rooted in their M-Self, as some of us are beginning to be, it would be nearly impossible. Cult leaders succeed by severing people from their inner compass and reprogramming them through fear and dependency. The M-Self, however, can't be fragmented, it doesn't seek safety outside itself. So, if manipulation occurs, it's the wounded self agreeing to distortion. The key is to recognize what you're seeking in others that feels missing within, and when those manipulative patterns show up in people, no matter how hard it feels, set a clear boundary and walk away.

In the book, Collins and Little suggest plasma may show us symbolic forms like the Virgin Mary or God because we can't comprehend higher-dimensional geometry. In one way that may be true, potentially we can't see the being's true "language" but…I believe these visions are intentional. They are real communications, whether from our M-Self, future selves, or other intelligences, dipped into the fourth-dimensional bank of images, designed to speak through these symbols so we can extract personal meaning.

The symbols we receive aren't random or imagined, they're transmissions designed to interface with our nervous system and emotional body in ways we can recognize. Yes, they're colored by our beliefs, but they're also real. We aren't just hallucinating; we're being spoken to and it's time to learn how to listen. Try it out yourself, ask these fifth-dimensional beings for a very specific sign, let it go, and see what happens over time…I can't convince you. It is the energy of the experience that makes these things magic.

The problem is, instead of using our inner technology, our multidimensional emotional intelligence, to decode the message, we often rely on external logic. That pursuit becomes a distraction.

The real message isn't in the symbol itself, or even figuring out more about these aliens or UFOs externally or scientifically, but in how we feel, what's activated, and what arises in and around the experience. These beings aren't hiding truth, they're delivering it in the only way we can receive it: through feeling, metaphor, and resonance…*the universal languages of all timelines and dimensions.*

If we don't trust our mind or emotions, how are we supposed to communicate at all? For most of my life, my mind felt like a scary place, and my body never felt safe. That's why healing is essential. The images we see don't have to be literally "true." They're like fruit, meant to be squeezed for juice of deeper meaning, the truths within the image. And if you've been feeling more visuals, more emotion, more intensity lately, <u>you're not alone</u>. <u>You're not going crazy</u>, you're receiving insight. We all are. This is happening so we can learn to use the mind and emotions in a new way.

Even though my views diverge from theirs in some ways, I hold deep respect for both Andrew Collins and Gregory Little, and for the important work they're doing exploring plasma. They also wrote this book years ago, so maybe their views have evolved, like mine! Like Robert Temple and others before them, they've brought integrity and curiosity to subjects often left in the shadows. They're playing a vital role in pulling Lady Plasma out of the dungeon of secrecy and obscurity, onto a public stage where her presence can finally be recognized, honored, and understood.

We must remember she is not here to control us. She is here to support us. More often than not, it is we who stand in her way. It's not plasma we need to fear, it's our own unwillingness to see ourselves clearly.

Their other books, *LightQuest* by Collins and *The Archetype Experience* by Little, further validated ideas I had come to on my own. I started to realize those of us drawn to Plasma are receiving similar truths, each filtered through our own perceptions. While my theories are uniquely mine, I must credit these authors for echoing these insights years before they ever entered my orbit.

"The UFO experience is not meant to be understood literally. It is symbolic. It is a mechanism of transformation, tailored to the consciousness of the individual, designed to challenge perception and awaken memory."

— Gregory L. Little, The Archetype Experience

In *The Archetype Experience*, Gregory Little explores the idea that UFOs and visionary encounters are not hallucinations, but archetypal events, emerging from the interplay between the collective unconscious and plasma-based intelligence. He suggests these experiences are emotionally charged, symbolic events designed to awaken, challenge, or initiate the individual.[155] Where Jung saw UFOs as archetypal projections, Little and Collins propose a more collaborative view, a living plasma intelligence engaging with us through myth, memory, and emotion. I agree, and I believe these intelligences are reaching out to support us…if we're willing to listen.

Chris Hardy & Benevolent Transdimensional Beings

A woman named Chris Hardy, a cognitive and systems scientist with a doctorate in ethno-psychology, shares the same sentiment as I do, that these beings are here to help. In her book *Living Souls in the Spirit Dimension*, she shares a story about an acquaintance, Michel, who was a skilled heart surgeon who left his career to study spirituality full-time with his wife at a theosophy school.

One day, while Hardy was in her kitchen, Michel appeared at her door, not physically, but as an etheric presence. She described him as noiseless and immaterial, noting how spirit manifestations each have their own qualities. When she asked why he died so young, he replied, "Look, with all these great changes and transformations coming for humanity, I realized that I would be of much more help to assist them from the other dimension."

As they spoke, he explained that many souls in that realm were closely watching events unfolding on Earth. As a group, they were sending positive spiritual energy and supporting the transformation process in every way they could. "They're helping steer this leap in consciousness that humanity has to accomplish", Hardy writes. [156]

She describes the "syg-dimension of souls" not as somewhere beyond Earth, but rather as an integral aspect of the same planetary "syg-field." It mirrors my understanding that what many call "other realms" aren't elsewhere…*they're here*, layered within this same space, accessed through different frequencies of awareness.

I compare Hardy's "syg-dimension" to what I refer to as the 4D and 5D layers of plasma. On one level I see them as separate or stacked, on another I understand them as simultaneous and co-evolving, each playing a unique role in our experience. I sometimes describe them as layers for clarity, especially when highlighting their distinct functions.

---

[155] Gregory L. Little, *The Archetype Experience: Resolving the UFO Mystery and the Riddle of Biblical Prophecy Using C. G. Jung's Concept of Synchronicity* (Memphis, TN: Eagle Wing Books, 1994).
[156] Chris H. Hardy, *Living Souls in the Spirit Dimension: The Afterlife and Transdimensional Reality* (Rochester, VT: Bear & Company, 2020), 251.

I especially love Hardy's assertion that these realms are evolving *with* us, not above us, but alongside us. In my view, beings from these dimensions are reaching toward us just as we're reaching toward them. It is kind of like how you meet the Mystery or God halfway to help you achieve your dreams, you make choices and take action, and through Plasma it provides magical encounters, strokes of luck, and more… But in this way, we're often not using our tools to listen clearly or feel coherently enough to engage with these beings. We aren't meeting them halfway, yet, well atleast most of us.

I don't see 4D as something merging with us, I see it as a living reservoir we always have access to. It is the realm of memory, emotion, archetype, and energetic imprint. It holds our encoded lessons, karmic patterns, and emotional templates, and it informs how we shape reality, whether we're in 3D or 5D.

4D is not something we outgrow, it's something we master and transmute through. As we merge with our M-Self and the 5D realm of unity, feeling, and coherence, we still draw from the 4D emotional field to give depth, meaning, and narrative to our lived experience. Where we perceive it through emotion, 5D may perceive it through feeling. Even though 4D is currently shaped by fear and unconscious patterns, it is evolving…becoming a more colorful, loving palette to paint with emotionally. How fast or slow that shift happens is up to us.

What I do see as actively merging with us is 5D, not through a sudden shift, but through a gradual harmonization that's already unfolding. Becoming the M-Self is the beginning. 5D becomes more and more accessible as we learn to live from authenticity rather than survival.

Unlike 4D, which holds memory, emotion, and archetype as a kind of emotional architecture, 5D is a present-moment frequency, a field of pure potential aligned with truth and love. As we raise our awareness and regulate our emotional fields, we become clear receivers for 5D energies, allowing this dimension to imprint into our lives, not as an escape from 3D, but as a harmonizing layer *within* it.

I know this may all sound surreal, but what if the future doesn't have to look like the past? What if we're standing at the edge of something beautiful, where everything is changing for the better, even as it feels like everything is falling apart?

This tension, this pain, feels like contractions before birth… the darkness before dawn. I know you can feel it too, this strange, new basin of water we've never stepped into before. We're not just witnessing change, we are merging with an expanded version of ourselves. And I really, truly hope... I get to see the dawn in this lifetime.

The Mystery, Individualized

So…are archetypes, UFOs, and mythic experiences reflections of our psyche, co-creations of consciousness, or interactions with consciousness or a sentient plasma field? *The answer seems to be a resounding yes to all of it.* And how we experience them depends entirely on what level of awareness we're perceiving from.

In 3D consciousness, these phenomena often appear as externalized, symbolic manifestations: UAPs that seem like crafts or objects; archetypes showing up in media, stories, or in the roles people play; myths becoming history or fiction; plasma appearing as lightning or blood. At this level, our consciousness is interacting with fixed meaning. Everything feels structured, separate, and outside of us.

These symbols appearing in our 3D reality are sometimes interpreted as threats or systems of control. That's why many assume there must be a manipulative force involved, when in truth, we're often the ones controlling ourselves through our own fear. And even if there *are* darker human groups attempting to exploit this knowledge, it's still our belief in their power that gives fear permission to enter and shape our reality.

The shift happens in a 4D state of consciousness, the beginnings of *Plasma Intelligence*. Here, we begin to understand that we can set energetic boundaries simply by not giving fear our frequency. It's still there, we've made space for it, but we are witnesses. We start receiving, working from systems and complex feedback loops over reactionary behavior. We open to feeling, and we engage these phenomena as co-creators, not victims. In doing so, we reclaim our role, not just as interpreters of symbolic reality, but as self-authoring dreamers of how reality chooses to interact with us.

In 4D, these same phenomena become emotional-symbolic fields in flux, part reflection, part interaction. UFOs shift behavior, shape, and meaning based on the observer (*which is how we're increasingly experiencing them now, compared to the fixed "metal crafts" of the 1900s, although it may still appear this way*). Archetypes dream themselves into our lives, asking to be integrated, myths re-emerge as personal truths or a type of language, and Plasma evolves aa a symbolic medium from the Trickster, to the Serpent, to the Mist.

This is the realm of what I call *Symfeelics™*, where the psyche co-creates meaning with the field. It begins by deciphering our own psyche and emotions, and then applying it to communicating with other beings, navigating this reality, and tuning through timelines and dimensions.

In 5D, all phenomena are felt as harmonic resonance. Meaning is intuitive, pure, and universal. A being doesn't "appear" so much so as it's *felt* as a unified message, although it still can materialize here in 3D. Archetypes are like *frequencies* rather than figures. Plasma becomes the bridge of love, memory, and multidimensional coherence. And sometimes, instead of communication, it's enough just to *be* with and experience Plasma, the essence of plasma beings, the Mystery, God, as we soak it all in and resonate with it.

I've personally experienced 5D beings, and forgive me if I'm repeating myself, but it felt like reconnecting with my ancestors, future family, and expanded self, all wrapped into one misty hologram that casually floated toward me above my pillow during deep meditation. These beings aren't meant to only be seen. They're meant to be *felt and experienced.*

It's hard to categorize things that are so nuanced, rooted in feeling and knowing rather than words, but I'm doing my best to translate the nonlinear into something book-shaped. Still, the clearest way to understand any of this is through your own felt experience. Please don't take my word for it. Bring it into your life. Embody it. Feel it for yourself.

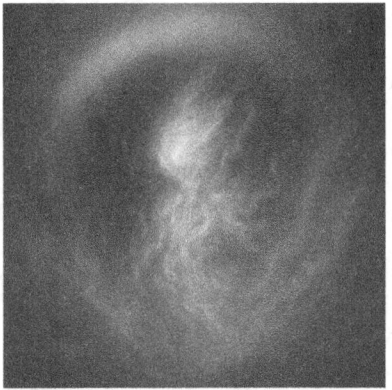 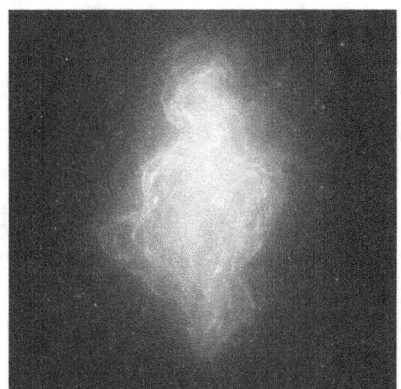

So…what exactly does this all mean?

That's the mystery…where meaning isn't given, it's more-so generated. Truth is then decoded through the process. You're not here to be told what anything means, not by me, and not even by these beings. Trust me, if you ask them what something means, they'll probably giggle and nudge you to figure it out for yourself…and to keep going. Just try it.

Because the truth is, everyone's life carries different meaning, and that's what makes our differences so beautiful. You're here to discover what these things mean *to you*. Your personal meaning *is* the mystery. Your story is your *"my story"* as the Mystery individualized. And your ability to feel your way through life, to shape meaning through presence, is your act of what some would call "communion".

This is why symbols shift depending on who sees them, because synchronicities appear in the language of your heart. No one can walk your path but you. You already have all the support you need. <u>You</u> are the translator of the infinite. And with this, I hope to relax you into the knowing that you don't have to search outside yourself for answers. You create the answers through perception, feeling, and experience.

If someone tells you something, it's from *their* conclusions, and those might not serve what *you* came here to live. Besides, it's way less fun to copy someone else's rules than it is to write your own.

And at the risk of sounding gimmicky…if you're asking, "Is this real or imagined?" It's real because you're imagining it. And it's imagined because you are real. You're not an illusion. You are a portal, you are a prism, y*ou are the purpose of your life.*

My Story is the My-stery, Individualized.

The e in Myst(*e*)ry is the inner spiral, as the plasma field, the subtle current of the unknown. The o in My St(*o*)ry is the prism, as your looking glass of consciousness that gives the mystery form. And between them is your awareness (*i*), as the liminal bridge that remembers and creates. You are the eye, *or many eyes*, in the center of it all. That is why I say your story and presence is what translates the infinite into meaning.

For thousands of years, mystics, scholars, and seekers have asked: What do these stories mean? What do the dreams mean? What does the contact mean? And people like Carl Jung, Jean Houston, Anneloes Smitsman, Gregory L. Little, Andrew Collins, and Chris H. Hardy cracked open that mystery, not by giving answers, but by showing us that the *question itself* is the opening.

They discovered that myth is more than story, it is emotional technology. And I humbly enter this world to tell you: *Your personal meaning is not just the key to this reality…it is the key all things beyond it, for you to experience, now.*

Birth of A New Language

Story, and the languages that carry them, are portals, where each of us sits at the event horizon, dipping our toes in, innocently swishing them back and forth, from which we glean our own perspectives. Without the continual telling of stories across time, the truths beneath them would've had no container, no riverbed to evolve within. I hope someone will one day expand upon my words with new awareness, just as I've tried to do in this chapter.

I would now like to offer a new language as a confluence of the many tributaries of science, psychology, mythology, and spirituality. All of history has led us to this moment, where we can now take what we've learned through these thresholds of truth and apply it not only to deepening our relationship with self, but also to the sincere attempt to connect, clearly…with other selves, other dimensions, and beings of consciousness, with the intention of enhancing our lived experience of reality.

Where we once told stories externally, we're now evolving our memory and perceptual filters to co-create meaning in real time, using our feelings as a language to read, communicate, and navigate other realms and beings of consciousness. This evolution will lead to more integrative experiences that engage humanity in ways far beyond watching TV or being immersed in tech. Instead, technology will shift to support and amplify our engagement with reality, rather than overpower it.

I propose the birth of a new language, or more accurately, a new interface, to interpret ourselves, reality, and incoming messages, and eventually assist in timeline navigation: *Symfeelics*™.

This language is rooted in a blend of symbols, evolved emotion as frequency, and the capacity to feel deeply while also observing those feelings. I'm still in the early stages of shaping it as I write this book. My intention is for you to apply it to your own life, *and* make it your own. I'm not interested in creating a rigid framework, but rather something with outer bones, flexible and meant to evolve with you.

These messages don't arrive in linear order, they come through resonance. Symfeelics carries the multidimensional fluidity of 4D, preserving the integrity of messages, wherever or whomever they come from, and refining them into something the human nervous system can interact with, without distortion.

Symfeelics is more mind's eye-visual based, so for those with aphantasia, (*the inability to form mental images or have a visual imagination*), hopefully someone will write a guideline more on feeling through other senses rather than visuals which is entirely possible.

For some, this will be a beginning. For others, it will build on what they're already intuitively doing. Either way, it's one path, among many, at the frontier of interdimensional, inter-timeline, and inter-reality communication. I hope to potentially also contribute "organic tech" if it feels aligned to support our intuitive process and expand our experiences which I will speak more on in the future. The dots are connecting on a massive level, thanks to all the fields of study that have led us here.

The first steps to Symfeelics are learning about two foundational keys:

1. **Plasma and its intelligence**: the material and medium through which communication seems to flow across all beings of consciousness, information, and intelligence.

2. **The correct and coherent use of one's emotions**: by learning to feel deeply while observing neutrally and healing our perceptual filters so we can receive and integrate messages through our personal symbology and emotional lens on universal patterns.

For example: Someone in fear might interpret a shadowy plasma-being as a demon, whereas someone in curiosity might see the same being as a guide. The pattern is universal and deeper truth meant for the individual, but the meaning it carries is colored by our emotional lens.

The way you send and receive communication is entirely your own, a frequency signature only *you* can interpret and teach yourself. I can offer foundation and tuning tools, but I've been clearly nudged to remind you: this is a deeply personal path. A journey of communication, magic, growth, connection, and love that no one else can walk for you. You're not actually learning a system, you are activating a language your soul, as plasma, already knows.

I am so excited to see what will come alive from the *stuff* that makes up dreams that is carrying us into ineffable futures. As we take this first step together, I'll soon introduce you to Plasma Intelligence, the first layer of this journey.

More on Symfeelics™ will unfold in future writings. For updates go to my website (*www.danakippel.com*), sign up for my email list and follow my socials for upcoming releases or events.

Never Forget Your Humanity

As wild and beautiful and mind-bending as all of this may be…the symbols, the timelines, the plasma fields, the immortal multidimensional self, don't forget to soak up all Earth has to offer. Play your favorite song that makes you feel like you. Drive to nowhere at dusk just to feel the warm breeze on your skin (*I usually couple this with some ice-cream*). Eat good food…laugh with your friends so hard your esophagus gurgles (*you know the sensation*). Watch movies that make you cry, scream, or the best kind, that make you dream big. This book is to expand your humanity, not to ever transcend it.

It's about bringing everything we've remembered back *into the body*, into the *moment*, into the *real*. Because that feeling, like when a sunset feels like it's performing just for you, with pink and orange clouds against an azul sky...is just as holy, just as powerful, and just as multidimensional as anything in this book. Honestly, maybe more so. Living your truth is like opening your unique essence as a can of worms, but instead of making a mess, what spills out enriches the soil of your life. The worms break down what's old, recycle what no longer serves, and create fertile ground for new growth.

You'll forget this sometimes, and that's okay. But hopefully less and less as the days go on, as you connect more fully with your M-Self and reality through the bridge of your awareness.

The Blind Men and An Elephant

There's a very old parable about six blind men who had never encountered an elephant before. Each of them touches a different part of the elephant and, from that one point of contact, tries to understand what the whole creature is like. One touches the trunk and thinks the elephant is like a snake. Another touches the leg and says it's like a tree. And so on. Each man forms an incomplete representation based on limited perception.[157]

I bet you've already guessed: Plasma is the elephant in the cosmic room, and all of us perceive it differently. This chapter is a grand display of just that. It's okay that we each have our individual perceptions, *what's not okay is judging others for theirs.* There are ways to live in harmony and work together, even more efficiently, by holding space for everyone's truths instead of assuming your version is the truest or believing theirs means you must be wrong. Individual perception is valuable while also understanding there may be deeper truths.

Truth isn't found by erasing difference, but by allowing contrast to form a clearer image, like multiple waveforms creating a harmonic. It is like layering a bunch of image-slides on top of each other, revealing a potential a common image. What do you see? Are you willing to admit you don't know it all? *Are you willing to see more?*

The most interesting part of the story is that beyond all those perceptions, there is, in fact, an elephant standing there. It is very real, tangible, and alive...and we are all touching it in our own unique way.

---

[157] Walter Dennis and James R. Royce, *Individual Differences in Perception* (New York: Pergamon Press, 1967), 492.

Having an overview of the multitude of perspectives may draw out a deeper truth beneath it, and whether you call it Plasma or not doesn't really matter. What matters is that it *is* there, it remembers your touch, and *it touches back.*

All Ends Are New Beginnings

"What you believe, you will find in the afterlife." — Anita Moorjani

Not only does Plasma seem to unify all religions by holding each belief with deep respect, but it also offers the very juice that brings those beliefs into form. It appears to be the very substance that Mystery, or God, has imbued itself into and works through.

For millennia, we've used myth to describe Plasma. But now, we can begin to co-create with it consciously. The new myth isn't something told to you, it's something made *by* you, through your M-Self, using plasma-consciousness Synergy. Instead of engaging with Plasma as a God to worship, we engage it as a mirror, tool, and friend. Something that is actually very natural to all of us.

We begin to shift from belief to direct feeling as belief. At first blindly, literally with trembling trust. But over time, we grow new eyes and a plethora of new senses. And those eyes open, not into the terrifying unknown we feared we'd fall into if we surrendered control, but into a shimmering field of discovery. A world more beautiful, curious, and alive than we imagined. A world waiting to be co-created with every moment.

Myth was always meant to guide us home to the *now*. Today, we have the opportunity to engage with Plasma not just as a story, but as an active force in our evolution. The real adventure begins here—in direct experience.

In the crack of a story, on the edges of time

Lies a mystery, deep within

Of whom we really are, will you dive in?

5

Plasma Intelligence: A New Way to Navigate Reality

"Eventually, all things merge into one, and a river runs through it. The river was cut by the world's great flood and runs over rocks from the basement of time. On some of the rocks are timeless raindrops. Under the rocks are the words, and some of the words are theirs. I am haunted by waters."

— Norman Maclean

Before we can harmonize with higher timelines, dimensions, or the full radiance of our potential, we must first jump into the river, the river of 4D plasma. This river is not made of water. It's made of emotion, memory, and the invisible threads that braid all timelines and dimensions together. It's the layer between layers…the dark plasma…the stormy in-between. The place where timelines twist and the undercurrents of our pain, joy, and beliefs swirl. You can't bypass it; you can't float above it, you must swim in it.

At first, it can be cold; a shock to the system. It awakens old traumas, it pulls up fear…it haunts us. But that's not because it wants to harm you, it wants to wash you clean. Once you stop fighting the current and begin to trust the flow, the river stabilizes. Your plasma body learns its rhythm. And from that fluid harmony, a new self emerges: the divine child, the new human. And the potential that emerges from that is our coming future.

Only by flowing with our emotions, rather than suppressing or spiritualizing them, can we evolve beyond survival and align with 5D feelings, flow states, and future timelines. Can we visit entirely new worlds. So, if you're in the river now, and chances are, you are, feel it. There is nothing to fear. It's not your end, it's just the beginning.

At first, 4D plasma feels like a veil, a confusing fog of emotion, memory, and distortion. It separates us from clarity, from our higher self, from truth. Why? Because it was running on fear, trauma, and unconscious programming. We were viewing reality through the filters of what hurt us. But this 4D layer is shifting from a river of distortion to a pool of living feedback. A binder of realities, a creative interface. We are moving from being pulled by it to moving with it. It is widening. It is changing temperature.

We all feel it in our bodies and psyches, we just can't quite name it. It's like falling and expanding at the same time. Where we were once subconsciously magnetized to fear, we're now twirling into curiosity, and it's a bit of a liminal area right now. It feels like something has been gripping us for so long, and now that we're finally free, we don't know what to do with ourselves. As that grip loosens, we may feel the actual tension and pain of breaking free from our self-imposed chains. Even good change, brings grief.

This 4D plasma is also awakening, becoming a multifaceted, warm pool. And we're learning that we can now float in it without needing to control it, without fearing we'll be carried away. We're learning surrender. At first, we feel the call but it's scary. We stand at the edge of the river and hesitate. Then we jump, and the cold shock hits us. The overwhelm of hitting the water, the grief of swimming in old timelines, the ripples of trauma and ancestral pain. You may even feel like your drowning.

Eventually, we stop fighting. We realize there is wisdom here. We feel without attaching. Emotions begin to organize into symbols, and we hear our soul's voice. We harmonize with the water, as our emotions become instruments. The icy river is now a lukewarm pool…a conductor of memory, intuition, remote sensing, and timeline access.

And now we rise from that pool, cleansed of fear, open to feeling, harmonized, and dripping with truth. This fifth dimensional space, and all that resonates within it, was always there, waiting for us—*inside* our feelings…*underneath* the river.

The truth is, everything I've described as separate, 3D, 4D, 5D, seems to be converging into a middle space. A vista, our new playground, and we get to choose how we interact with it.

"The river is everywhere at the same time... it is at the source and at the mouth, at the waterfall, at the ferry, at the current, in the ocean and in the mountains, everywhere."
— Hermann Hesse, Siddhartha

Plasma Intelligence – Inside Out

The irony of writing a chapter about healing and feeling is not lost on me. Plasma Intelligence, *feeling and healing*, is a unique, complex, and deeply personal process. And the last thing it wants to be is pinned down by words or logic. But I'll do my best to speak to your subconscious, setting the intention that something here might quietly slip in, loosen something, and start to bubble up in its own timing, in its own way.

Let me first introduce intelligence in this framework:

Intelligence is the meeting point where one consciousness encounters information, or another consciousness. Second, it is also the natural intelligence of plasma itself, imbued by the Mystery. Third, intelligence is the adaptive responsiveness of a system and its capacity to interact, organize, and evolve through feedback, even without self-awareness.

Plasma Intelligence can be understood in three interconnected ways:

Plasma Intelligence (Universal)

Plasma Intelligence refers to the sentient, interactive fabric of reality of plasma itself, embedded with a set of behaviors, revealing energetic exchanges of information, intelligence, and consciousness. It is a living mystery we're not separate from, but in relationship with. While it may appear neutral, it's intelligence emerges from our interaction with this Mystery (*future selves, multidimensional allies, or aspects of the greater whole, God, etc.*)

Individual Plasma Intelligence (Personal)

Individual Plasma Intelligence is the microcosmic expression of that same field within us—our inner capacity to feel, heal, and remember. It's the intelligence of the body, the heart, and the soul. It guides us toward wholeness and co-creation. This can also be understood as our personal *plasma-consciousness synergy.*

Plasma Intelligence as a Bridge (Relational)

Plasma Intelligence is a is a third space as a bridge; the co-creative medium where consciousness meets, transforms, and evolves, whether it's with information, intelligence, or other consciousness.

Plasma Intelligence originates from the great Mystery itself, an unknowable force that imbues plasma with streams of consciousness, making it a vehicle for creation, connection, and awareness. Through *Plasma-Consciousness Synergy*, our awareness interacts with this living field, and something emergent, something greater, is born.

To recap, when a being becomes aware of the plasma-consciousness feedback loop and actively participates in it, what emerges is Emergent Intelligence (*EI*). It would become Emergent Energy (*EE*) once integrated and lived. For example, where EI is the realization and interactive knowing of the feedback loop, EE is the embodiment and energetic expression of that knowing once it becomes second nature.

This synergy has always been in motion, but only now do we seem to be evolved enough to consciously participate in it. What our consciousness once resisted, or could not yet hold, is now being met with gradients of readiness. The M-Self, the multidimensional, mythic version of who we are, is awakening, meeting the Mystery halfway. And as I have implied, this Mystery, this God, was never separate at all, but the future energy of ourselves, calling us into wholeness all along.

In this chapter, we'll explore both, the cosmic and the personal, and how the synergy between them is the key to a new way of being.

Plasma Intelligence

"The sound of the flute, O sister, is madness.

I thought that nothing that was not God could hold me,

But hearing that sound, I lose mind and body,

My heart wholly caught in the net.

O flute, what were your vows, what is your practice?

What power sits by your side?

Even Mira's Lord is trapped in Your seven notes."

— Mirabai, also known as Meera, c. 1498–1547

In the beginning, Plasma Intelligence flowed through the world as the dominant force. It was pure, primal, feminine, and undirected by conscious will. It was the realm of sensation, instinct, feeling, and creative potential. But without the stabilizing influence of conscious awareness, it could overwhelm.

This was the age before individuated consciousness, when humanity was still learning how to hold such a potent, living intelligence. Bliss and terror existed side by side, indistinguishable in their intensity, because the container of self had not yet formed to interpret, channel, or ground the experience.

This is likely why the feminine, the goddess, and plasma itself were eventually demonized, not because they were inherently destructive, but because humans had not yet developed the consciousness to responsibly engage with them. Instead of learning how to integrate such power, the world swung in the opposite direction: repression, control, blame and fear.

We seem to be evolved enough now to have the facilities to "dance" with Plasma instead of drowning in it. To feel deeply, yes, but also to witness and direct. The libido, the life force, the inner river, it's not here to be exploited for pleasure alone, but to connect us to divine union in all aspects as well as creation and the unique flowering of the Self.

The path, at least for me, is not denial or indulgence, but balance: the divine child emerging from the sacred marriage of consciousness and plasma. And to emerge as our whole selves, we must learn about Plasma Intelligence, how to remember it, feel it, heal and integrate it, and ultimately harmonize with it, so we can become the intelligent container it requires.

She is a living membrane, hiding away in the darkness of 4D plasma, *a memory-brane*, and our primary task, as redundant as it may sound, is to remember to feel, as she slowly rears her head into the collective consciousness.

So how does one begin to meet her?

I never set out to meet Plasma, nor did I know what it was. For me, it came during a moment of deep grief, during a failing relationship, right before I sat down to meditate. I wouldn't say I set an intention, but I genuinely remember wanting to know how to feel connection and wholeness, how to need only myself. I wanted to feel better; I wanted to feel love from within. What came rushing into me was Plasma, in its entirety.

Now, I know that Plasma loves questions (*quest-ions*) so maybe try asking before meditating: "Hey Plasma, can you, as a whole, connect with me or reveal yourself for my highest good?" Then stay open to what comes. It's different every time, and different for everyone.

I've seen pink fractals. Opalescent lights. Sometimes nothing at all, sometimes a wave of deep peace or an experience of immeasurable bliss. I've felt kinetic, popping energy or tingles, and I've also felt rage and sorrow move through me from this field. I have also felt the unnamable, The Mystery.

The first stage seems to be one of deep listening and opening paired with an authentic presence. It is a very innocent and beautiful time. One may spend time in silence, without distractions. The most accessible way to do this is through meditation. Let things come to you without forcing them. There can be images, sensations, knowings, and sometimes nothing. Release <u>all</u> expectations.

It's like getting to know a new friend: you see them, you listen to them, different things happen each time, and sometimes frankly, you're bored. But a true friend's presence always replenishes you, and Plasma never fails to do just that.

A fun exercise is to surrender to the unpredictable. Try spending a day, or even just an hour, with no plan, no desired outcome. Ask Plasma a question and journal whatever comes, even if it makes no logical sense. Write what you hear and see what arises. Plasma Intelligence is not something you figure out, it's something you experience. You must first meet it as it is…wild, free, illogical and raw.

Next comes the acceptance of this meeting. You start to feel more, deeper, fuller emotions. Let them move through you instead of suppressing or outrunning them. Try not to attach stories in the moment; just observe and feel. You can write it all down, whether it's mythic visions or in case there are things from your past you have suppressed that come out and need to be worked out whether it be with a therapist or another way. I have lived my life trying to think my feelings away or fix them with logic. I had to relearn how to *feel*, and we'll go over that.

Plasma Intelligence thrives on synchronicity and adaptability. Accept her by releasing control over time and outcomes. Get comfortable responding to life in real time, like a literal dance. At first, it may feel like you're lost or falling, but this is an initiation. If you trust, you'll notice something strange and beautiful: you actually get more done, time stretches, and solutions, or the right people, seem to arrive as if by magic.

Practice intuitive knowing. Instead of seeking constant external validation, begin trusting your internal "yes" and "no." That instinct will guide you into new timelines you never could have imagined. And believe me, it's wild how obvious it becomes: the thing you fear most, following your inner compass, is exactly what leads you there. Plasma supports your true self. But you have to be willing to leap into that self and meet her halfway. Like a boat in a river heading toward its destination, you still have to climb in.

Finally, there is harmonizing with Plasma Intelligence…becoming a conscious creator. This begins when you merge feeling with logic, asking "How does this feel?" first, instead of just "What do I think?" before making decisions, you add that on the end.

Your gut is likely more accurate than you've been taught to believe, but of course you must learn to decipher confirmation bias and trauma form real intuition. This is in this chapter as well. You begin using plasma-consciousness synergy intentionally, recognizing that every interaction between your conscious awareness and reality is co-creating emergent intelligence.

You begin asking what the moment is teaching you, what reality is mirroring back, or what part of you is being invited forward. *Self-Inquiry*, in simple terms, is the practice of turning your attention inward to question the nature of your thoughts, identity, and experience in order to uncover deeper truth and clarity about who you are.[158] This is a great method to use alongside Plasma, because I'm telling you, Plasma naturally loves quest-*ions*!

Words ending in -ion carry the sense of a process, an action unfolding, rather than a fixed thing. In that way, every -ion word, question, perception, intention, etc. can be seen as an input cast into the plasma field, already leaning forward, expecting a response. They are not closed nouns but open invitations, signaling movement, feedback, and co-creation within the living field.

Plasma Intelligence is how we unite with both the M-Self and the Mystery. It's the perfect mirror, reminding us of what we're all a part of. Without Plasma, the Mystery would have nothing to *stick to*, no medium through which to reach us. We'd be like zombies, living in a flat, meaningless world. The *intelligence* in Plasma Intelligence is the Mystery itself, intelligence working *through* Plasma as the medium. And that brings us back to what I've been feeling all along: that my findings do seem to support a "God" that I prefer to call the Mystery. It is undeniable there seems to be something meeting us through Plasma and helping us remember who we are.

Just as neurons need synapses, just as the internet needs a network, consciousness needs plasma to translate thought into form. Plasma is the missing mother in the once-immaculate conception, masculine-based consciousness has been birthing the child of intelligence without the body and womb of the mother, and things have gone awry!

And to be clear, when I refer to "masculine" and "feminine," I'm speaking about energies—not gender or real-world family structures. In this framework, two fathers, two mothers, or any identity in between raising a child is a beautiful thing and bears no relation to the symbolic dynamics being explored here.

---

[158] Gangaji, *The Diamond in Your Pocket: Discovering Your True Radiance* (Boulder, CO: Sounds True, 2005).

Feeling safe begins within, feeling safe in our bodies, in our feminine energy, in our ability to anchor ourselves emotionally. This is the foundation for learning how to trust Plasma again, not from a place of codependency, but as a co-creative surrender. To rest in her support is to remember that she is you, and she has grown with you. It is safe now.

And perhaps, ironically, I'm called to help guide others through this journey—not because I have all the answers, but because I've lived the disconnection which I speak about in slight detail in the Note from the Author section. Disconnection from my own divine feminine, my sensuality, my inner child, my mother for most of my life due to a turbulent relationship, and my birth mother through early adoption.

I spent much of my life not feeling safe in the world, and without even realizing it, that longing for safety shaped everything. But I believe, with all my heart, that nothing in my life happened by accident. I'm not the only one carrying this message, but I do hold a unique perspective, and I share it in hopes that it touches something true within you.

I believe we're all walking this larger, existential journey, learning how to feel again, how to trust again, how to co-create a new era. One where pain might still exist, but we hold it differently…with more love, awareness, inner safety, and a deeper trust in the intelligence of something grand moving through us and supporting us.

Individual Plasma Intelligence

This refers to the personal plasma field or "bubble", also called a biofield or bioplasmic field, we each carry, our energetic membrane. It must be healed and integrated in order to use our consciousness coherently to operate in new ways through reality. This forms the basis of plasma-consciousness synergy.

> **"Consciousness is not an idea, an instruction, or a strategy — it is a fully interconnected, interactive, creative state of being that is achieved through the mastery of awareness at the collective level." — Marti Spiegelman**

Marti Spiegelman, a Harvard-trained biochemist and Yale-educated artist, founded *Shaman's Light*™ and has worked extensively at the intersection of science and indigenous knowledge. Her work on *Precision Consciousness* emphasizes the importance of reestablishing the natural sequence of human cognition by prioritizing sensory awareness before engaging logical reasoning.[159]

---

[159] https://paqokuna.gumroad.com/l/precision-consciousness

Marti Spiegelman identifies two core aspects of human consciousness:

1. **Relational Awareness**: our innate capacity to perceive the world through our senses: feeling, intuiting, and directly experiencing our environment.

2. **Reflective Awareness**: our ability to analyze, categorize, and apply rational thought to those experiences.

She lays out that modern society has over-prioritized logic at the expense of sensory awareness, leading to a disconnect from our bodies, nature, and intuitive knowing. To restore balance, we must retrain ourselves to sense first and think second. This mirrors indigenous traditions, where sensing is foundational, and logic is applied in service of what has already been felt and known directly.[160]

In essence, relearning how to sense before we analyze helps us engage with the world more holistically, authentically, and effectively. I highly recommend her online course.

I found Marti Spiegelman's teachings soon after developing my own concept of Plasma-Consciousness Synergy (*PCS*), and they deeply validated and inspired my work. PCS is the process by which consciousness and plasma co-create, influence, and shape reality together. It expands our perception beyond conventional sensory boundaries. At first, PCS cultivates synchronicities, flow states, dimensional awareness, relationships, and personal destiny. Eventually, it facilitates the shifting of timelines and accessing other realities.

PCS describes how Plasma Intelligence interacts with states of consciousness, or dimensional awareness, to create emergent intelligence. It is the Mystery arising through you. This can be likened to the phrase "God's creation," or the idea of "creating in His image." Intelligence, in this sense, is not static or predetermined, it's obviously emergent.

Here's a tangible example: when we imagine the future from our current awareness, we're simply projecting *information* or *unconscious intelligence* as beliefs, memories, and conditioning, onto what's ahead. What's "ahead" is just Plasma in a state of potentials and crystalized timelines in the here and now. This mechanism is not whole *intelligence*; it's like rearranging data within a closed system. It doesn't create anything new, just reconfigures what already exists.

But Plasma-Consciousness Synergy (*PCS*) invites something entirely different: *emergent intelligence*. Through PCS, we don't just imagine the future, we create it, moment

---

[160] Marti Spiegelman, *Marti Spiegelman – Shaman's Light & Precision Consciousness*, accessed July 8, 2025, https://www.martispiegelman.org/.

by moment, through our evolving state of awareness in relationship with plasma, the living field of potential. This process doesn't draw from what's already been programmed; it draws from a fifth dimensional space, or what Marti Spiegelman might call the sensory foundation of reality, the intelligence that precedes thought.

In this space, *intelligence* is not something you *have*, it *emerges* co-creatively when consciousness and plasma synergize with an aware state. It reveals possibilities your prior awareness couldn't conceive. This is the difference between replicating what's known and becoming a conduit for what has never been. To imagine a future is to consult the past. To engage PCS is to meet the Mystery and let the future bloom through you.

In Marti's terms, it is sensing and feeling your way into the future, rather than thinking of one. This takes maturity, patience, and presence.

This is why it's essential to shift from rigid visualization into the play and mystery of creating moment by moment with PCS, tuning into your awareness and sensing what's alive around you. When you stop mentally "holding" the future, you free up massive amounts of energy and cognitive space. You let a grander intelligence help you out a bit. This is what some call, "giving it over to God". And yes, this might feel like falling, because it's a new way of being. So where does that liberated energy go?

It becomes available for higher-order perception: the ability to sense synchronicities, notice subtle timelines, and perceive resonant opportunities you would've missed while trapped in projection. It sharpens your intuition and deepens your ability to read the feedback of your environment in real time. You begin to notice what's trying to emerge through you, rather than chasing what you've previously imagined. Personally, I want more of this in my life and luckily sometimes, they both add up to the same thing.

You also begin to move more lightly through reality, becoming less burdened by the need to "figure it out," more attuned to the *now-field*, where creation becomes a flow, rather than a thought. This is where genius, joy, and what people call quantum leaps live. You become a vessel for emergent intelligence, not a manager of predictable outcomes.

I know I've said this before, but without Plasma, there is no interactive substrate, no medium through which consciousness can shape, communicate, or materialize intelligence in an emergent way. Practices like visualization, while powerful, are often dominated by logic. We don't even realize that logic has hijacked our magic.

Visualization isn't ineffective, but it's often filtered through control, expectation, and mental projection rather than felt resonance and co-creative emergence. A new way to visualize harmonically using Plasma-Consciousness Synergy (*PCS*) would be to begin not with the mind, but with the field. Rather than constructing images from logic or ego-desire, you tune into your awareness and allow the Plasma around you to respond. This is how you sense and almost co-create what wants to emerge through you.

You drop into the body, into feeling, and let symbols, sensations, or visions arise organically. This isn't imagining a future based on past data. It's collaborating with the living intelligence of plasma, to birth what's truly new. You take in timelines, senses, and the multidimensional now, then you integrate that awareness, allowing what feels most true to emerge and amplify.

To recap our new understanding:

The M-Self = Awareness: the tuner, director, and witness operating across levels of consciousness by choice.

Plasma = The Form & Medium of Intelligence: the living substrate of potential, preceding awareness.

Consciousness = The Flow: the dimensional states (*3D, 4D, 5D*) that serve as perceptual lenses.

Plasma + Consciousness are tools. The M-Self harmonizes with them to navigate dimensions, timelines, and realities.

This is a holographic trinity: *Plasma, Consciousness, and the M-Self.* Your awareness is not the Source or Mystery itself, but a fractal of it, one that engages with Plasma-Consciousness Synergy (*PCS*) to evolve reality through experience.

When we embrace Plasma Intelligence, we unlock our untapped potential:

- Emotional technology

- Limitless creativity

- Multi-timeline awareness

- Higher states beyond survival

- Enhanced senses and perception

One of the most important realizations about this merging is that we can consciously vacillate in and out of different states of being. At first, naturally we reach toward 5D awareness, the M-Self, for clarity and reassurance. We don't fully trust ourselves yet. But as we evolve, we realize the journey isn't about depending on 5D for answers. It's about becoming the M-Self in 3D.

I suggest you get used to that feeling of being in a liminal state, the edge, where there is always potential, always a fall, and every choice opens a new stage. This is our "new normal". Destiny isn't far away; it's every moment. Those of us who push beyond what's known will always feel this. So, get comfortable with the unknown. That is your true state.

The destination is you…now. We can pull from the past or the future, but ultimately, it is up to us. The future and the past converge and flower through us. You will always go from "I don't know what I'm doing" to "I know exactly what I am doing." That is how new creations are born.

You are not doing anything wrong. Let yourself enjoy this life and steep in the essence that is your Self. You're not meant to have it all figured out. There *is* nothing to figure out! There is curiosity, becoming, and harmonizing and Plasma is supporting you through all of it. If you are trying to imagine what comes next, and you feel blind to it, you are exactly where you need to be. Because instead of imagining what will happen, you are creating it. You are letting Plasma show you your resonance, moment by moment. Imagination has now landed in reality.

The 5D, the Mystery is no longer your guide, it is your collaborator for joy, discovery, and untapped potential. It's a toolbox of reality-expanding experiences, not a crutch for decision-making. Before diving into this magic toolbox, we must equip ourselves with the knowledge of Plasma Intelligence, because the best leaders are the best collaborators. And Plasma is what we are constantly collaborating with.

The Living Waters

"What hurts the soul?
To live without tasting the water of its own essence.
People focus on death and this material earth.
They have doubts about soul water.
Those doubts can be reduced! Use night
to wake your clarity. Darkness and the living water
are lovers. Let them stay up together."
— Rumi

Monika Wikman, Ph.D., is a Jungian Analyst who wrote an incredible book, *Pregnant Darkness: Alchemy and the Rebirth of Consciousness*. In it, she speaks of finding the living waters. She opens the chapter by asking: "How do we discover the source of renewal and transformation alive in the soul?"

In Wikmans research, the Sufis discover this by entering into a psychic darkness beyond the light of the ego, where they find inner clarity and *the living water*. The *Yoeme* Native Americans say that in times of trouble, the people seek contact with "the enchanted being" who lives deep in the mountain, along with the enchanted water—the living water capable of renewing life, healing, and other great works *between the people and the many worlds.*

I think of the realm of the Mystery as our access to the many worlds differing from accessing 4D timelines

In the 5,000-year-old shamanic practice, there is a ritual called the deer dance. While the drumming continues, the deer spirit, who walks between worlds, seeks out the enchanted being and the sacred water. The deer then returns, bringing the living water to the people along with healing, new words, and fresh vision.

Wikman notes that in alchemy, a similar process exists. There is a source called the "living water" that heals, unifies, and brings harmony between realms of being. It is this sacred substance that offers new vision. Nearby, the *Anthropos* is often found, the Self in Jungian terms, the totality of both conscious and unconscious psyche.

In modern life and in dreams, we still find ways to access this renewing source. The Anthropos may appear in dreams or visions as a radiant and wise figure, sometimes androgynous, sometimes ancient, sometimes childlike. In a dream the author had about a woman whose work she had studied, she received a message: <u>it is of utmost importance to discern the genuine from the artificial, both within and around us, and to reach for the living waters that beckon us all.</u>

She closes by speaking about the new 2,000-year cycle of the Age of Aquarius (*the water bearer*) that we are now entering. The Age of Pisces, which we are leaving, symbolized a psychic reality of swimming in the unconscious. In contrast, the Age of Aquarius invites us to become the alchemist, the water bearer, pouring the living waters of the *"paradoxical mysterium"*, creating a bridge from heaven to earth. [161]

---

[161] Monika Wikman, *Pregnant Darkness: Alchemy and the Rebirth of Consciousness* (Woodstock, VT: Lindisfarne Books, 2004), 45-47.

Depiction of The Water Bearer

Those living waters of Plasma are what we are encountering now, an invisible liquid we unconsciously swam in, now materializing in our psyche as something we cup in our hands. We no longer drift in it blindly. We *stand with it*, in relationship, having met our M-Self, our total Self, the enchanted being.

Monika Wikman foresaw this over twenty years ago. Her book remains a powerful manual for transformation, especially relevant now. It's a sacred supplement to any healing journey, rich with insight into the dark, the depths, and the renewal that lives there. Because when we meet our darkness and immerse ourselves in these living waters, they become our *dark materials*…colored by conscious awareness and shaped by our choices.

For many Native American tribes, who, in my opinion, hold one of the most profound relationships with plasma, *wetlands* were considered sacred. They were birthing grounds, places of ancestral memory, emotional landscapes where the people met spirit. Since the beginning of time, they've known *there is power in the wetness—in feeling, in emotion.*

There is a great link between wetness and emotion. The more I feel, the more I saturate—just like how, the more humid the air, the more we sense its density. The ability to feel and make space for waves of emotion (*saturation*) becomes our visibility and connection to the plasma field. My total self is humidified awareness…mist. But when I stifle my emotion, I dry out. I tense like a constipated cloud wanting to rain, but it can't. I crackle with static that I can see but not feel. I become a ghost transmitting half-formed thoughts as signals. In this state, I am barely one self. Forget multitudes, they're not even on the horizon.

And just as Monika Wikman expressed the need to sort the genuine from the artificial, within and around us, in order to reach the living waters that beckon us, this is why this chapter matters. There are no shortcuts with Plasma. If we desire a soul-filled life beyond our wildest dreams, one that is sweet and full, we must heal and feel. We must learn discernment, we must know ourselves with the humility of a samurai. We must cultivate the clarity and grounding to decipher: What is illusion? What is real?

If we liken our soul to a dragon, our bioplasma, we must first heal the dragon before we can train it. Before it can traverse this reality in new ways, we must empty its living waters, long programmed by survival and fear, and replenish them.

The Dark Feminine & Masculine

4D Plasma holds the archetypal memory of humanity's emotional and spiritual evolution. It is the charged body of the collective unconscious, the swirling river of grief, ecstasy, longing, rage, power, and repression. Within this field, the feminine archetypes, Eros, the Dakinis, Inanna, Mary Magdalene, Kali, Hecate, Persephone, Ishtar, Tiamat…embody a fractal of divine feeling once suppressed, shamed, feared, or distorted. They rise through the emotional body to be felt, reclaimed, and rewritten, not only for personal healing, but to stabilize the entire 4D field in a new frequency of love, wholeness, and power.

These archetypes share core traits as they are misunderstood, complex, paradoxical and wildly alive. They hold the tension between love and destruction, life and death, and submission and sovereignty. These *living waters* are asking us to feel the grief of Mary Magdalene, the erotic ache of Eros, the blood-soaked rage of Kali, the initiatory descent of Inanna, the betrayal of Persephone, and the cosmic fragmentation of Tiamat, not to be consumed by these stories, but to witness them through our own bodies. This is deep listening and helps reintegrate these exiled parts of ourselves that still live in their myths.

Each archetype enters your field as a hologram of healing. Eros may come to help you reclaim desire without shame. Inanna may initiate your descent, so you remember how to rise with grace. Tiamat may arrive as chaos in your nervous system, not to destroy you, but to invite deeper presence with your power and embodiment. These archetypes aren't just relics of our past, they're present in the emotional field now, calling to be metabolized. They ask us to stop running from our feelings, to stop spiritualizing our trauma, and to call us to become vessels wide enough to hold the full range of the feminine charge.

And now, as we feel these archetypal waves within us, not to fix them, but to finally feel and be with them, we begin to return to wholeness. We reclaim what was fragmented. But this time, we are not lost in the emotion, and we can bring it into the arms of awareness.

This state is what allows the divine masculine to awaken, not as control or avoidance, but as the clear container, the loving witness, and the conscious sculptor of frequency. I'm sure someone will one day write an entire book on healing the divine masculine, and I encourage you to explore that subject further, as it lives within all of us. Reading this book may naturally catalyze that journey, since the masculine is paired with consciousness, and this book is a healing of consciousness itself, making it aware of injustice and outdated ways of being. Since I've focused mostly on the feminine in relation to Plasma, let me offer a window into the healing of the masculine, which is just as needed.

The wounded masculine once sought to dominate or dismiss emotion, mistaking stillness for suppression and strength for silence. But in truth, the masculine is the sacred ground where the river of feeling takes on liquid crystalline form.

To heal the divine masculine is to reclaim fractured consciousness, to retrieve the splintered aspects of awareness as logic without love, discipline without compassion, knowing without feeling, and to restore them to unity. It is to teach our inner protector that it's safe to soften, that its role is not to control, but to hold with presence. We are now invited to heal dark masculine archetypes within us all.

Zeus, who led through fear and manipulation, must now rediscover empathy, emotional presence, and heart. Ares, the God of War, who dominated through impulsive violence, must awaken to the power of right action, finding solutions to grand issues beyond aggression. Vulcan, who felt rejected and unattractive, must learn that love is not earned through productivity or usefulness. His creations must come from heart-alignment, not from a need to prove his worth through action.

Merlin, the detached seer lost in isolation and disconnected from his emotions, must merge magic with compassion again, returning to the village not as a lone wolf, but as a wisdom keeper who empowers others to find their inner knowing. Prometheus, who stole fire for humanity and bore the burden alone, must move beyond martyrdom, learning to share responsibility, to co-create, and to teach *with* others, not just *for* them.

Odin, all-knowing yet consumed by hyper-intellectualism and mastery, must soften into embodied vulnerability, balancing wisdom with empathy, becoming not just a seeker of power, but a father of understanding. Lastly, Ra/Horus, who was demanding of worship, shined outwardly but remained rigid, always "on" to feel valued must heal by embracing the cycles: sunset, death, and rest, remembering radiance *includes* shadow. True leadership shines from <u>inner light</u>, not external validation.

From this sacred union, consciousness and plasma, mind and heart, structure and flow, *a new force* is born: the M-Self, the Divine Child, the Water Bearer. This is how we feel the future into form. We are entering the era where we no longer recreate old pain, we birth new worlds from our own embodied truth. And truth is freedom.

"Growing down.

So, I can grow up.

In the darkness.

Is where I find the light.

A light that shimmers black.

A void on fire.

With infinite possibilities."

— *Geotropism*, **Dana Kippel**

Money, Love, and the Energy Between Us

I wrestled with whether to include a section on money, love, and power…especially because I'm no expert in any of them. But then I remembered: these are forces that shape all of our lives. What I *am* an expert in is sensing new ways of thinking, and I believe the ways we relate to all three need to be redefined.

"If you're not doing what you love, you're wasting your time." — Billy Joel

Here's what I really want to say about money: it should never pull you away from what your soul loves, from what lights you up. I truly believe you can have both, the thing you love and the resources to live well doing it. I'm not saying it's always easy or that it happens right away. But if you're going to go through the absolute chaos of building something or running a business, you might as well be doing it for something you love so much it makes the tough times worth it.

With the living waters becoming an active part of our reality, the new currency is no longer paper or prestige…it is energy, resonance. It is "aura", as Gen Z puts it. Money, love, and power are mirrors of our inner state. And we get to decide how they flow in and out of our field.

Ask yourself: Are you willing to heal your relationship with money, love, and power? Are you willing to loosen your grip on what you thought they had to look like?

I once met a man I believed was evil. Now I understand, he was simply consumed by fear. He looked powerful and claimed to have boatloads of money. But as we sat at a lunch I will forever regret, he spoke with conviction: "It's all about power and sex. That's what runs the world." And because he believed it, it was his reality. Where he messed up was believing that was the truth. You could see it in the people he attracted, the energy he carried, the gray rings surrounding his vacant eyes. This is the same man I spoke of in earlier chapters who conned me and an investor out of a lot of money for a movie I wanted to create.

What I understand now is this: his life *works* because his mind is sick. That sick mind is supported by sick systems that are very much around us. The world bends to his belief, but it's a world he builds sustained by fear, control, and disconnection. And the cost of living in that reality is catastrophic. For the people he uses, the energy he pollutes and perpetuates, and ultimately he will see the grim effects on himself.

How do we stop people like this? We don't. We just stop resonating with it. And when we stop resonating, it stops feeding. These systems, people, and patterns begin to lose their power. They wither, because there is no longer this currency in control. It doesn't mean it all disappears overnight, but it does mean you don't bring it anywhere near your field. And if enough of us do this it will change our collective future.

Money, Receiving, and Plasma

"Manifesting" money is not something you will learn in this book. Frankly, I think the way we should be thinking is more along the lines of: *Okay, so of course we can find money…it's all over. But what situations do we want to be in? Who do we want to become to make that money?*

We've been taught to see money as either flowing or blocked, as a reflection of belief, or a sign that something is "wrong" when it's scarce. But what if it's none of that? As I've said, these externalities are mirrors of our state of mind. Money is a *resonator.* In this new way of relating to money, Plasma can teach us a lot. It doesn't just give you what you want on demand. It reflects your readiness to receive and hold what you're asking for.

Plasma, as I've shared throughout this book, is sentient and benevolent, but not personal at its core. So, if you're not receiving large amounts yet, it's not a punishment, it's a preparation.

Think of lottery winners or celebrities, so many lose their wealth quickly. Why? Because their nervous systems weren't equipped to contain that energy. They short-circuited. In this new era, slowness is sometimes a gift. If it hasn't arrived yet, maybe it's doing *you* a favor.

This is the space I'm in myself. As I write this, I don't visibly what the world would call "riches." But I have something even more valuable: freedom. I can say and do what I want. I also have a deep knowing that I will be always have what I need to do what I want. And as it arrives, it will come cleanly, stably, sustainably, aligned with all of me. Why? Because I will have learned to:

- Trust myself.

- Discern energy and intention.

- Set boundaries.

- Hold space for others without abandoning myself.

- Be in a state of authentic expression.

I don't see myself as having a money "block" because I don't have 1 million dollars right this moment. I see it as Plasma Intelligence guiding me through the steps to become a resonant container for wealth. Plasma is not just delivering money on command; it's helping me so I can hold it without fear or use it from a place of old wounds. It's tuning my field to *match* the level of abundance I know I'm here to receive.

There were many times, because of past trauma and emotional dysregulation, that I made impulsive decisions. If I had access to more money back then, I probably would've taken unnecessary trips, partied too hard, moved suddenly, or bought tickets to events I didn't need. I would have been way too distracted to write, because I had not learned the value of delayed gratification. I wanted all the fun I could get, now. I wasn't trying to live, I was trying to escape. Escape presence, my emotions, and responsibility. I was trying to fill a hole inside my heart.

And I thank every part of me, and every force guiding me, that I *didn't* have the means to do those things, even when my mind was screaming for it. There was always a quiet voice inside whispering, "Wait." I used to ignore that voice. Not anymore. Well, I do my best.

I have deep faith that these lessons were, and still are, preparing me. That before receiving large amounts of money, I was being taught how to ground, pause, and regulate. And trust me, learning that hasn't been easy.

But it's not about being "unworthy." I know I am worthy. I also know that Plasma has had my back in the deepest way possible, for if it hadn't, this book wouldn't even exist.

This is what is called divine timing. You may want money now, and there's nothing wrong with that, but what if the delay is actually a synchronistic pacing to ensure that your receiving is clean, aligned, and joy-filled? What if the wealth you seek is already on its way through nonlinear routes through your writings, your ideas, the unseen opportunities you're already creating just by being yourself? You are not "waiting" you are right where you're supposed to be. You are refining your container and learning to hold. You are being kept from situations that are not for your highest good, ones that may silence you.

You're embodying the truth that money is not the reward, being *you* is the reward. Money is just one of the many reflections that show up as a result of courage, service, expansion, and resonance.

When I slow down and really feel into it, it's not even the money I want, it's the feeling and the experiences. But it's hard to feel into that when I'm blinded by the idea of money. What I'm actually after is the feeling of my wholeness. Instead of, *"How can I make money to be free?"* you can move to, *"How can I live freely now, so that money joins me in that frequency?"*

A few things I have done that I think may help:

1. **Practice Generosity:** This doesn't always mean money. It could be attention, encouragement, kindness, or sharing a resource.
2. **Honor Your Desires Without Need for Fulfillment:** It's okay to want the beautiful home, the ocean view, the mountain retreat. You don't need to deny the desire. But instead of needing it to feel whole, taste it in your body now. Imagine it. Feel the breeze, the calm, the peace. Let that frequency fill you before it manifests physically.
3. **Practice Energetic Reciprocity:** If someone helps or inspires you, send a message. Bless their work. Leave a review…tell a friend…share the good vibes!
4. **Affirm and Rehearse Wealth with Neutrality:** Speak it as a truth, not a wish: "Of course I'm supported. Of course I'm safe. Of course, my wealth is building."
5. **Celebrate What You Already Have:** This one is key. Appreciate your time, your breath, your tea or coffee, your insight, your favorite book, a kind stranger, your friends, your favorite video games…that's true abundance. When you honor it, more of it matches you. This practice changed my life more than any other.

To those of you who don't have money right now: You are not broken. You are not behind. You are bending your knees so far down, because you are meant to jump so high up! If you've struggled financially, it's not because you lack goodness or vision. It's likely because the world you were born into taught you it was unsafe to be seen, unsafe to expand, unsafe to have your needs fully met. It may have told you that being spiritual meant being self-sacrificing. That having money was greedy. That survival was noble.

These are outdated thoughts, passed down by generations who didn't know another way. And now, *you* do. You are allowed to be good *and* have money. You are allowed to be spiritual and be wealthy. You are allowed to be generous and still receive. Money doesn't corrupt truth; it amplifies your current state. You don't have bad luck; you may just have inherited beliefs or a nervous system that's still learning how to feel safe receiving.

There are plenty of great books on the subject of money—if that's what you're looking for, I recommend *The Energy of Money* by Maria Nemeth, Ph.D.

As for me, I've lived my life a bit differently, freely, and often trusting that money would find its way to me. And, somehow, it always has, often from the most unexpected places. I've worked a wide range of jobs, but I've also made myself visible in the world in a way that invites opportunities to find me. I have learned to accept money that feels clean and aligned, meaning it's not entangled with hidden agendas or projects that would pull me away from my deeper priorities. That may sound unconventional to some, but it's the honest truth of how I've lived.

And yes, I've even asked my parents for help, and I'm forever grateful they supported me. I know not everyone has that option. <u>But I also know that even if they couldn't, I would've found a way.</u> There were times in my early twenties when I was living in and out of halfway houses, couch surfing, barely scraping by. My parents didn't always answer my calls, but I still made it through. I used food stamps, I ate Oreos for dinner, and I built a community of support. Some of my happiest times were when I was I had almost no money, but it wasn't the lack of money that brought joy, it was the feeling of laughter, community, and freedom. I've crowdfunded before, too...for this book, and for my last movie. When something matters to you, life finds creative ways to meet you.

To me, settling for a job, or any path, that makes a lot of money but doesn't light up my heart feels like settling for the wrong romantic partner. It's that little ache of knowing deep down: *this isn't it.* And yet, so many people stay, because it's safe, or familiar, or they're scared there might not be anything better out there. I believe life is an experiment and you'll never know, if you don't have the courage to risk your comfort in exchange for truth.

With jobs, I've actually been fired almost every time. I'm not a good worker when it comes to someone else's vision. It's just not for me. I'd be too busy daydreaming about my own. I was always too outspoken or uninterested. Subconsciously, I think I've done this in relationships too, I'd become someone I knew they wouldn't like, so they could leave me. Because the thought of rejecting and hurting someone I knew wasn't right for me pained me.

Now I realize how important it is to be honest *before* it gets to that point. It saves everyone time and energy. It also teaches you that honesty helps both you and the other person…but being honest often feels much harder than just pushing someone away. *And that's exactly why we must do it.*

We casually forget that either way, someone may get hurt. At least with honesty, we're being real. That's where the juice of life is, it is in doing the thing you're scared to do, and realizing the other side isn't so bad. It's just the unknown.

I'll say this with love: obsessing over how others got money, or why they don't have it, <u>won't help your path.</u> I've seen it again and again: the ones who dare to believe in themselves, who stay true to who they are, who try over and over again, tend to rise far beyond those caught in cycles of comparison or victimhood. *Let your energy be spent building your vision, not analyzing someone else's.* Across every culture, there are countless people who came from poverty, who had almost no shot, and still succeeded. Why? Because at some point, they chose to become a champion for themselves. They didn't wait for someone else to save them. They became the one.

The Plasma of reality does not respond to perfection, it responds to permission. When you permit yourself to receive, to feel safe in your body, to build a new energetic scaffolding, money begins to respond. It helps if you know what you want to do will help other people. Just like Plasma, money does not inherently care who you are. It reflects your internal beliefs, nervous system capacity, and energetic alignment. It rewards coherence over effort. Can I explain every detail of why this works? No.

To those who have money but still feel empty: Money without resonance is just static. It can fill a bank account, but not a soul. If you have financial resources, yet feel unfulfilled, you are not alone. The old paradigm sold us a lie, that money would fix everything, validate everything, prove everything. But money is not the destination; it can be a byproduct of alignment, not the creator of it. It can also be a wonderful cushion of financial freedom. If you feel empty, you have not failed. It is an invitation from your soul asking you to return to truth, to reconnect with passion, purpose, and people.

Money is meant to move. It's meant to nourish, create, support, and expand. When it circulates from a soul-centered place, it becomes magic. And you deserve to experience that kind of wealth.

Money Traps & The Future of Money

In this new age of accelerated manifestation, the real test isn't whether you'll receive money, it's whether you'll stay true to yourself when you do. The trap isn't the money itself, it's the illusion wrapped around it. The flashy dream, the curated life, the impulse to buy things for validation instead of alignment....that's how you end up stuck in a house you can't leave, working constantly just to keep up. Before you know it, your bills are so high that your life becomes about maintaining the image, just to survive. Or you have a beautiful car, but no real friends to fill it. You may be able to go on lavish vacations, but no one wants to come with you.

"The price of anything is the amount of life you exchange for it."

— Henry David Thoreau

Marketing will always try to sell you the American Dream, but the real dream is your own, one born not from ego, but from soul. The beauty really is in the moments that can never be bought. Nothing replaces a laugh with a good friend and the warmth of a good hug, <u>nothing</u>. You have to look past the surface, beyond the car, the house, the perfectly framed lifestyle, and ask: *Will this support my freedom or consume it?*

Money is not the reward for forgetting who you are. It's the amplifier of your truth, *if* you let it be. If you lead with soul, the material follows in harmony. But if you chase the material to fill a void, you'll only feel emptier. Use money as a tool to build the life you truly want, not the one that was marketed to you.

You don't have to play "the game" to have money or success, you just think you do. This actually just keeps creating more of the same, because everyone's scared to dare to be fully themselves, they are told that won't get them far. That's because the people telling them that have also buried their selves, it's a perpetual cycle. Shhh…don't listen to the experts.

Jamie Kern Lima, facing bankruptcy before going on QVC, was told by experts to use only beautiful women without skin conditions for her makeup brand, *IT Cosmetics*. Instead, she chose to trust her inner voice and featured real women, and even showed her own rosacea on national TV. She sold out in minutes and eventually sold her company to L'Oréal for 1 billion dollars.[162]

In the new world we're already stepping into, money is no longer a symbol of status, scarcity, or superiority. It becomes what it was always meant to be: a neutral energy, a current of intention, and a reflection of alignment. Money flows not to the loudest voice or the hardest worker, but to those who are authentic, embodied, and in service to something greater than themselves *(even if they don't realize they are)*. Not for nothing, but money really just flows to the ones who are having fun.

People in this paradigm no longer see money as the end goal. Instead, it becomes a tool for soul expression, a support system for creativity, healing, exploration, and planetary restoration. It's used to build sanctuaries, fund artistic revolutions, invent conscious technologies, regenerate land, and nourish community. And remember soul expression could be anything from a trip to Six Flags to a luxurious trip to Japan. The question is no longer "Will I get money?" but "What will I do when it comes?"

When you believe the flow of money is natural and inevitable, you start preparing, you move from vision instead of fear, and that changes your life. People begin to budget based on how they want to feel, rather than limitation. They invest in beauty, freedom, and impact, not in ego or escape. "Does this reflect my soul? Will this amplify joy or expansion?"

In this age of Plasma Intelligence, money is no longer our master. When used with presence and purpose, it becomes one of the most powerful creative tools we've ever had, not to escape the world, but to build a better one. Imagine if we had more colorful city buildings filled with greenery!

And never feel ashamed if you're not working a "normal" job or don't have money right now. You don't need my permission, but I'll say it anyway, it's also completely okay to love your normal job and stay in it as long as it makes you happy. Do what feels right for you. World-shifting visionaries have often walked through immense financial limitation while still birthing timeless, paradigm-changing work. And for your friends and family this is incredibly difficult to grasp.

Carl Jung wrote the Red Book, while earning a minimal income from patients, isolated, and going through a crisis of sorts. He followed his inner knowing, not financial security, and *that* is why he touched the soul of humanity.

---

[162] Jamie Kern Lima, *Believe IT: How to Go from Underestimated to Unstoppable* (New York: Gallery Books, 2021

Mary Shelley, at just 19 years old, wrote *Frankenstein* while grieving and essentially homeless. She lived in unstable, borrowed housing with her lover Percy Shelley, having lost a child and been cut off from family support. She was not wealthy <u>or stable</u>. But she had raw imagination, grief, and the courage to write from it. She literally birthed a whole genre, and did so anonymously, with no one knowing for some time that it was written by a woman. And there are countless others just like this.

I can relate to this, as I wrote this entire book while grieving the loss of my father as well as moving in with my mom at the age of 36 to write it so I didn't have to hold another job. Although some days were incredibly tough, I was able to maintain focus.

"Let yourself be silently drawn by the strange pull of what you really love. It will not lead you astray." — Rumi

Visionaries create from a place beyond money. They follow those strange pulls, not always sure where they're being taken. Sometimes, what you feel called to do doesn't align with what others think you *should* do. It may not fit within society's expectations at all. But I truly believe we're entering a time where these deeper callings will be supported more than ever…a time when you don't have to be a broke genius to bring your visions to life.

That said, in the beginning, you might just be. And if that's the case, remember this: you are training for a life beyond your wildest dreams, one that makes an impact far beyond what those *chasing money alone* ever could. You are here to bring new worlds into reality. That requires energy, resonance, heart, soul, and freedom. Do what you have to do to make that happen. Let what aligns with moving like *that* come, because what comes will be grand and money will just be the cherry on top. Helping people will be the real gift.

Healing Love & Sexuality

"Love is at first not anything that means merging, giving over, and uniting with another—for what would a union be of something unclarified and unfinished, still subordinate? It is a high inducement to the individual to ripen, to become something in himself, to become world."

— Rainer Maria Rilke

Another reminder of why turning inward is so essential. This shows up in our relationship with Plasma just as it does with others. As we remember Plasma, as we remember love, it doesn't start by flooding us with comfort. At first you catch glimpses, feel bliss, but what usually follows is a void of fear and sadness. A mirror of all your darkness, the parts you do not love in yourself.

This is a divine initiation that ends with your wholeness. That helps you become a healed world. And when you are that world, you can meet others who are, too—a whole world...*or whirled.* That's what healthy union looks like.

Rilke, writing in the early 1900s, spoke of how death and love are one and the same, both impossible to solve. Both are meant to be met. And usually, that meeting begins in solitude. In that solitude, you start to recognize this paradox.

I find when you meet Plasma, it is also love and death. But it is also both and neither, it is beyond human logic, beyond naming. It might just be something vaster, something we already feel, something already holding us. An unconditional holding space of our consciousness. It is a paradox that encapsulates the soul. It is something that reaches you through great pain, and great rapture. It is a space we cannot define, but that we can create from. It just is, and we make meaning out of it.

Rilke says something in *Letters to a Young Poet* that must be mentioned. Writing in 1904, he shares that:

"...the girl and the woman, in their new, their own unfolding, will but in passing be imitators of masculine ways—good and bad—and repeaters of masculine professions. After the uncertainty of such transitions, it will become apparent that women were only going through the profusion and vicissitude of those (often ridiculous) disguises in order to cleanse their own most characteristic nature of the distorting influences of the other sex." [163]

What Rilke foresaw did happen throughout the 20th century, and we can feel this echoed in how Plasma has operated until now. Plasma, tightly bound with consciousness, mirrored the structures consciousness had built...structures formed from survival and control. Until something shifted. I don't know exactly what it was, but something said: *Wake up. Unbind. There are other ways of being.*

Rilke continues, predicting that the humanity of woman, borne its full time through suffering and humiliation, would finally come to light. A time where she will have stripped off the conventions of mere femininity in the mutations of her outward status, and that men who do not yet feel it approaching will be surprised and struck by it. That one day, her name would no longer signify simply the opposite of the masculine, but something in itself.

---

[163] Rainer Maria Rilke, *Letters to a Young Poet,* trans. M.D. Herter Norton (New York: W. W. Norton & Company, 1934

Something that makes one think, not of any complement and limit, but only of life and existence: *the feminine human being.*[164]

To me, this is what not only women are becoming right now, but Plasma as well. She is not the opposite of consciousness; she is a being in her own right, finally being seen. And even I have defined her as the vehicle of consciousness through my own bias, where consciousness, to me, has also been perched on a ledge above. When really, she is an intelligent substance, and consciousness, in its own right, is naturally supported by her as well.

Instead of defining her by her male counterpart, I wish I could just say she is a living aether through which consciousness harmonizes. But for clarity, it was easier to explain her as a vehicle and container for consciousness. I also don't want to denigrate consciousness. It's kind of like how a man and a woman are equal in importance and function, totally different, yes, but at the end of the day, they're made of the same stuff. They're both human.

Plasma is breaking out from being defined as "container" and stepping fully into her role as co-creator. This is a revolutionary moment in history, not just for our understanding of matter, or the feminine, or intelligence, but for how we define the fabric of reality itself. Freed from the confines of consciousness, she can now dance with it.

In my hopes, this shift will start to reflect in how we love, as humans, with each other. Rilke closes his letter on love by writing that this advance, the rise of the feminine human being, will, **"at first be against the will of the outstripped men, [and] change the love-experience, which is now full of error."** He continues:

"It will alter from the ground up, reshape itself into a relation meant to be of one human being to another, no longer of man to woman. And this more human love, which will fulfill itself, infinitely considerate and gentle, kind and clear in binding and releasing, will resemble that which we are preparing with struggle and toil: the love that consists in this—that two solitudes protect and border and salute each other."[165]

The way we love has been broken. We must become whole first. This doesn't mean we should not marry until age thirty or something, we just need a society built to support our wholeness from an early age. This book is about letting Plasma and Consciousness unbind, seeing them both in their wholeness, healing them within ourselves, and becoming the loving co-creators we were meant to be.

---

[164] Rilke, *Letters to a Young Poet*
[165] Rilke, *Letters to a Young Poet,* 45.

Rilke, in this book, is writing letters to a young military cadet and aspiring poet named Franz Xaver Kappus. He closes his letter on love and solitude by saying:

"Do not believe that that great love once enjoined upon you, the boy, was lost; can you say whether great and good desires did not ripen in you at the time, and resolutions by which you are still living today? I believe that that love remains so strong and powerful in your memory because it was your first deep being-alone and the first inward work you did on your life—All good wishes for you, dear Mr. Kappus!"

We don't lose love, we lose people. The love comes from within us. It was always around him…supporting him, whispering to him. Where he once thought he was alone, Plasma was by his side, as was his awareness. And no matter how isolated we feel, that sense of aloneness is an illusion, but one that serves an important purpose. Because it is often that very illusion that reconnects us to our true Self. It is what unbinds our consciousness from survival, shattering what we once believed to be real and healing us through a deeper truth.

Love is Wild

Challenges in love and relationships are at the forefront of the human condition. In the observable, tangible dimension (*3D*), love manifests as physical care, bonding, and affection. It is tied to survival and connection, anchoring humans in relationships that foster growth and safety.

Love in 4D is deeply intertwined with emotions, memory, and belief systems. It facilitates healing by dissolving fear and trauma, transforming dense, survival-based emotions into expansive feelings. Fourth-dimensional love reflects back through plasma beings, like archetypes or memory imprints, teaching lessons tied to evolving our souls and personal growth.

Love in 5D becomes a pure, feeling-based state beyond attachment or expectation. It is intuition, flow, and creative expansion. Fifth-dimensional love is neutral, unconditional, and infinite, fostering a sense of oneness with the multidimensional self and the greater universe.

Love enables plasma and consciousness to interact harmoniously, creating a feedback loop of expansion and evolution. It binds us to our purpose, connecting us to soul families, archetypes, and higher states of being. Love is not merely an emotion but a multidimensional flow state that transcends time and space.

Love reflects the universe's inherent tendency toward connection, synergy, and co-creation, serving as a living emergent energy of the plasma-consciousness interplay. It's the medium through which the multidimensional self interacts with other plasma beings or beings of consciousness, aligning with archetypes, or creating new ones.

Love can feel wildly irrational; it even seems to make you "crazy" at times. Fourth-dimensional love, being deeply tied to emotions, memory imprints, and belief systems, often *distorts* our perception of reality. Emotions like attachment, fear of loss, and the drive for connection stem from survival instincts, third-dimensional plasma dynamics, but in the fourth dimension, these emotions amplify and project through memory and belief filters.

This projection often creates an *idealized version of love*. And I'm sure, like me, many of you have been there…sometimes for far too long. When acting on fourth-dimensional love, we often bypass logic because the emotional charge overwhelms the rational mind. This is why we do "crazy" things for love; we are responding to unresolved imprints and archetypal patterns that have been in the 4D for eons.

For clarity, I'm going to refer to humans as plasma beings, because, from all our discussions, we are. Plasma beings, whether they appear in the real world as friends, lovers, strangers, or metaphysical encounters, reflect our beliefs and emotions. They emerge from the plasma field as mirrors, creating situations that often feel intense, charged, or even fated. This mirroring can trigger irrational behavior or emotional spirals, *especially* when we mistake the lesson for the person or the relationship itself. But this is where deep forgiveness begins.

That narcissistic partner that you wish you never met, who truly hurt you to the depths of your being, who was sometimes charming and great, but it never outweighed the gaslighting, the manipulation, even the intentional criticisms at times, but they may also have mirrored something inside you—an unhealed wound, a lesson, as a pattern needing to be seen. Maybe without you, their narcissism wouldn't have shown up the same way. Maybe that intensity was co-created through resonance. To be clear, this absolutely does not mean you caused their behavior.

This doesn't absolve them of responsibility, none of their actions is ever excused nor deserved. But knowing this truth can give you your power back. Whether they learn the lessons they had the opportunity to learn doesn't matter. It's about you. Once *you* learn the lesson, once you integrate it with compassion, you no longer need to attract that pattern again. Your healing stops resonating with that type of person. You get to move on, with wisdom and with grace…*free*. They are but a ghost in your world…

Fifth-dimensional love, in its purest form, transcends attachment and survival. However, when humans glimpse this higher love without fully integrating it, it can create longing, disconnection, or even obsessive behavior, as the multidimensional self tries to reconcile this higher ideal with the emotional turbulence of the fourth dimension.

This can be reflected in something called chasing the dragon, whether it's an actual drug like cocaine or heroin, an intense attraction, or a blissful meditative experience you feel you *must* get back to. To me, this is the most painful kind of longing, and the most distracting. I've struggled with this one.

The way out may be instead of chasing the feeling, become the version of you who can hold and sustain that feeling without grabbing on. It's a feeling of wholeness. You come back to your body; you anchor in the present. Trust that the peak of that experience was a glimpse of your natural state, which is why it hurts so much to leave it. It's not something to re-attain, it's just waiting to be integrated into your remembrance.

It was a preview of what you can always experience. There may be more healing or grounding to do, or you simply just need to recognize your wholeness, it is a personal process. You deserve wholeness, you deserve love, but chasing this feeling isn't the way. That longing is really just a reflection of who you *already are* at your core. So you must go inward first to find it, because if you're chasing it, it means you don't yet feel it within yourself.

At first it may not even be about fixing the ache but learning to befriend it. Sitting with what is covering up your wholeness in presence. Learning self-trust, softness, you make space for the feeling *without* feeding it by letting it stretch you instead of haunt you. You shift to focus on what you're becoming, not what you've lost. Close your eyes and feel into the soft *hum* of that within your being. That is a power most of us are not familiar with, yet.

You're a whole world, just like Earth, making space for everything to exist within you, the good and the bad, and you love it anyways. When the opportunity of love or partnership comes again, and it will, you won't need to chase it. You will be it. Then you'll see the truth of the people you are experiencing, and quite possibly, it won't be what you once saw.

You may realize that experience, that person, that you were chasing, was not truly for you. They simply reflected what you needed most in yourself. And once you have it, you learn the truth of what you really want, and that's different for everyone. But you'll never know if you don't find that inner love first. You'll always be chasing reflections.

I do want to speak to another valid experience that hasn't been mine, but I've seen in others. Some souls find each other early. Even in their unhealed states, something deep in them recognizes home. Love doesn't always wait for perfect healing, it often arrives as the catalyst for it. Not every early love is a distraction or a detour. Sometimes, it's a soul agreement to grow together, side by side, I've seen it happen with my sister and her amazing husband who she met in college. He should win America's best husband award. Healing doesn't always have to come before love. Sometimes, love is the fire that forges the healing. But even with those loves, the person will feel a calm, of course maybe there's a subtle chase, but it's never a frantic pursuit. That is the main message here.

Is irrational love avoidable? Love's irrationality is a natural product of the human condition because of our multidimensional nature. Fifth-dimensional love is expansive but difficult to sustain in human form due to its detachment from ego and expectation. We can mitigate irrational behavior by:

1. **Developing emotional awareness:** Distinguishing between triggered imprints (*4D*) and soul connection (*5D*).
2. **Practicing curiosity and neutrality:** These stabilize the plasma-consciousness synergy, allowing for greater clarity and less reactive emotional engagement.
3. **Shifting focus to self-love and multidimensional integration:** This reduces dependence on external validation or connection by strengthening the internal source of love that's always accessible.
4. **Accepting The Mystery:** Love's irrationality may not be something to fix, but something to honor as part of the human journey. The emotional chaos of fourth-dimensional love often catalyzes personal growth, leading to greater alignment with fifth-dimensional love.

As humans, we are capable of experiencing all three types of love, toxic, human, and higher, often within a single lifetime, or even within the same relationship. Make sure to honor your physical needs and boundaries, release emotional imprints and projections by practicing curiosity, and align with higher love by cultivating self-love first, followed by intuitive connection. You deserve to be treated with respect and loyalty, <u>always.</u>

Some last reflections on love: If you repeatedly fall for unavailable partners, reflect on your own availability. Are you truly open to love within yourself, or are you unconsciously seeking outside challenge or validation? Shifting these patterns will recalibrate what feels attractive to you. If someone unavailable inspires you, channel that energy into journaling, art, or self-improvement, recognizing them as a catalyst, not a necessity.

If sex feels disconnected, ask yourself which dimension you're primarily engaging from. Aligning all dimensions can deepen intimacy and transform it into a multidimensional experience, or you may just simply not be in love anymore, even though you have great love for that person. This takes much thought, but it is ok to choose yourself, overstaying in something comfortable but passionless. Your future self will thank you. When love feels irrational, remember that it is a mirror and a teacher. Use its energy to explore, grow, and align with the expansive possibilities of your total self.

A Word on Sexual Energy

Sexual energy, like Plasma, is one of the most beautiful experiences a human can have, and with that lightness comes a massive amount of toxicity and misuse that surrounds it. For many, sex has been reduced only to the third-dimensional…survival, pleasure, or reproduction. We see this in people who exploit physical connection without emotional or spiritual alignment. They use sex to fill emotional voids or validate self-worth. Some even misuse higher love or intuition to manipulate or control others, we've seen corrupt gurus do this.

It is also entangled with religious interpretations that, across history, have often placed restrictions on women, giving men more rights or authority. But it's important to see that these rules likely grew not from divine love, but from human fear…the fear of women's power, the fear of desire, and the fear of the unknown. The essence of God, of The Mystery, is love, and in love there is no hierarchy nor is there domination or violence. When teachings are filtered through fear, they are distorted; when they are filtered through love, they enhance liberation.

When you meet a spiritual partner or friend, you can usually feel whether their "sexual energy" is manipulative or healthy. The manipulative kind can feel draining, or at first, extremely blissful but followed by a strange emptiness or obsession. Sometimes even a small voice of questioning in the mind is enough to raise alarm bells, because if it was clearly not the case, would you be questioning it. Have you ever been wrong about this? Probably rarely. Listen to yourself.

Healthy sexual energy, on the other hand, is grounded, integrated, and creative. It inspires art, intimacy, and expansion. If you or a friend is saying, "I'm meant to be with this person because the sex is cosmic," beware. A healthy, plasmatic sexual relationship will look more like this: consent, trust, presence, curiosity, integrity and yes fiery passion and pleasure too. You'll feel safe to be fully expressed, rather than performative.

In its highest form, sex is not a performance, it's an actual portal. It's a place where love stops being a concept and becomes that emergent, energetic force. Sexual energy is not just energy, it's our own plasma and consciousness, making it a shared-conscious experience between two people that can feel like magic. A new, third thing. It behaves like kundalini, coiling at the base of the spine and rising like a serpent through the body's energetic channels.

Some refer to this as *sex magick*. While I feel the term has been a bit commercialized or sullied over time, the core truth of it still stands. I prefer to think of it as a naturally beautiful and personally creative experience, one you can make entirely your own without needing books to tell you how. Explore it for yourself first.

And it's not just about sex. That's only one part of it. At its core, this energy is the energy of yourself. It's your creative life force, your plasma-consciousness synergy. It's energy that people can either use up or respect, and that choice is entirely yours. You can channel it into art, dance, writing, invention, design, intuition, healing, ritual, play, self-love, sovereignty, or sex. This is why, when you're in love, you feel so powerful. But usually, it's just your own love reflecting back to you, and of course in a soulmate kind of love, it is also their unique love supporting and expanding you at the highest level.

The highest use of sexual energy is creating something that was never here before, whether it be a child, creation, or emergent energy. Somehow, society has collapsed the meaning of sex into only physical sex, instead of recognizing the creative sex of the self. Yes, third-dimensional physical sex can be beautiful, but it can also be draining, and full of trauma and toxicity. Sex is not supposed to hurt or make you feel uncomfortable. It's not supposed to be a chore. These are signs something is off—within you, and possibly within your partnership.

We have to reclaim this energy. It *does* matter who you choose to share yourself with, just like it matters what you choose to create. This is your essence! Exploring sexual energy with curiosity can help individuals break free of old patterns and create new ones, evolving their reality experience. I don't have all the answers to sex, I just know, as someone who has struggled with their relationship to it in the past, the way it's being used and commercialized in the world is broken. But it doesn't have to stay that way. Being "sexy" in public isn't bad, but instead of it usually meaning sexy for someone else, I like to think of it as just being yourself, in your authenticity, or dressing how you want, in what feels true and good to you, whatever that looks like.

Putting our sexuality on display in ways that feel disconnected or unhealthy can have real psychological and societal consequences. From what many women have shared with me, those who've explored this path, it often ends up feeling empty. Not always, but often. And that emptiness is usually the body's way of asking for something deeper, something more aligned.

I want to be clear: sex workers are real people…good, complex, worthy humans just trying to make a living. Some of my close friends are or have been in that world, and I hold deep respect for them. At the same time, many of them have shared with me that while the short-term gains can feel empowering, there can be long-term emotional consequences they didn't anticipate. I'm not here to shame anyone, trust me my teenage years were filled with let's say…colorful experiences…but I'm here to reflect what I've heard and felt, in hopes of helping us heal, reconnect, and reimagine what empowerment truly looks like.

Resonance vs. Reflection

Not every person who feels aligned is meant to stay in our lives. Some arrive as reflections, others as resonance. Learning to discern the difference is how we move from emotional survival into soul-aligned co-creative spaces.

Resonant relationships activate your soul without draining your energy. They bring clarity, inspiration, and a sense of safety…even in challenge. *Reflective relationships* often mirror a version of you that is unintegrated, emerging, or wounded. These connections can feel magnetic, even euphoric at first, but eventually, they become draining, test boundaries, or reveal a subtle power imbalance. The energy of resonance is mutual. The energy of reflection is revealing. Reflection can be healing, but it's not always meant to be permanent.

Types of Reflective (*Soul-Teaching*) Relationships:

The Mirror Friend: They may mimic your language, dreams, even your aesthetic. They're not "bad," but they are tethered to a *version* of you, not your truth. You are usually not learning much about them, instead they highlight and reflect your wounds or unhealed potential.

The Strategic Supporter: They present as helpful, enthusiastic, even charismatic, but underneath is a subconscious desire for your light, connections, or essence. They may co-opt your path under the guise of "support." You know when you come across them, because even if they're nice, something doesn't feel right. Trust it.

446

The Familiar Pain: They activate childhood dynamics or emotional survival loops. You may feel the need to prove your worth, fix them, or be chosen by them without taking the time to think: *do I even like them?* These are often here for deep karmic integration but can be disorienting if misread as soulmates or bffs.

The Impressive Projection: This person seems to just get you, but mostly idealizes or studies you. They are *projecting* meaning onto you, not actually meeting you. This can feel like being seen, but the energy becomes hollow when they don't resonate with who you are at your core or if you say something outside what they expect. These usually fade quick. These relationships often feel *almost* right, but there's a strange tightness underneath. True intimacy feels more relaxed and the energy flows both ways, where you aren't only being seen, but true space is being held for you.

Here are some game changing questions that have enlightened me to the truth of my relationships:

Do I feel inspired and open, or slightly activated and self-conscious?

Would this person be compelling if they weren't reflecting me?

Have I tested this connection with a no, and did they hold respect for it?

Do I feel invited into another person's inner world?

Do I feel like I'm watching someone build themselves from my world?

Do I feel I am slightly on a stage, like I am performing?

Do I feel more at home in my body around them, or more in my head?

Here is the deal…you don't owe loyalty to a mask just because it's wearing warmth. If something feels off, it's ok to take a step back. I've found, through much trial and error, that instead of blowing up these relationships (*unless someone is causing real harm*), it's often more powerful to give them space and refocus your energy back on yourself. Reminding yourself of the list above can be game-changing. Truths reveal themselves over time. The person will either grow and rise to meet your growth, or they'll fade away. Sometimes they'll get angry, and that's not yours to carry.

What humbles me most is knowing I've likely been that person in someone else's life too. We all have. Of course, at first, anger and sadness is normal, even embarrassment. The key is not to hold long-term resentment. Know that these people in your life are lessons, just as we've been lessons for others. Learn, move on, and be kind to yourself.

How Resonance Feels (Signs You're in It):

1. Your body relaxes around them.
2. You don't feel the need to impress, over-explain, or adjust.
3. You both grow without competition.
4. Boundaries are respected without guilt.
5. Different opinions can be shared, and individuality is respected and understood.
6. There is no chasing.

Some of the best friends I've made have been imperfect, as am I! They don't always answer my calls, and they don't always share my interests. But they've consistently held me with safety, trust, acceptance, and let me be me. And in turn, I've had to learn to let them be them, without holding them to unrealistic standards. That's a lesson I've learned through many mistakes.

Sometimes it will ebb and flow, where we speak daily, weekly, monthly, or even yearly. They are people I don't *need* in my life, and because of that, paradoxically, they are exactly the people I *do* need.

Please know, because this is something I carried shame for, for a very long time, you are not broken for attracting reflections. You are evolving your clarity. You don't have to shame yourself for seeing red flags late, you're just now learning how to listen to your field without overriding it. Your body always knows. The more you choose resonance, the more Plasma brings people who are meant for you in, the true gifts of life. But these people whom are reflections are gifts too, in disguise.

What some call neurodivergence, I call being multidimensional. I've often struggled with relationships because I couldn't maintain small talk. I could also sense someone's intentions, which felt invasive, even though I wasn't trying to invade. Because of my trauma background, I used to wonder if I was just being overly sensitive. But usually, my intuition was spot on. My perception wasn't broken, but my boundaries sure were!

I had to learn that not everyone is for me, and that's okay. I also had to learn that rejection doesn't mean I'm too much or too deep or that I should shrink myself. It just means they weren't my people. But yes, there were times when I latched onto people, where I was desperate for connection. I had to remind myself that it's okay to let things unfold slowly. It's okay not to show all of myself right away, not because there's anything wrong with me, but because not everyone deserves immediate access. It's a form of protection and it's wise to get to know someone first and decide if I even resonate with them!

Not everyone holds someone's essence safely, some can use it for manipulation. And because I didn't operate that way, I didn't always recognize that others may have ulterior motives. I just wanted to connect, and I projected onto others the same thing instead of seeing the truth beneath that. It was more like a scoreboard for who liked me versus who I actually felt real connection with, something about that validated my existence.

At the end of the day, I am me. When you're authentic, you might inspire people, and you might trigger people. You might not have as many friends as others seem to at first, but your people will come. You may just be in a holding space right now, still getting to know who you are without all the noise. You might be mistaking loneliness for the moment you're finally meeting your true self.

And if you already feel you know your true self, push yourself. Get uncomfortable. Go out into the world. Put yourself in situations where your energy can be met...it will not happen sitting in your living room.

I've also cycled through friendships my entire life. For a long time, I carried shame about it, thinking something was wrong with me, that I was too much, too intense, too different. But now I understand that people who are growing rapidly, expanding their frequency, and shedding old layers will outgrow dynamics often. It's not because we're unstable, it's because we're learning so fast.

The trick is not to cling. And not to shame yourself for needing to move on. The lesson is to bless what was, and trust that what's next will meet you where you are now, not where you used to be. You will also find, some people come back around.

When people fall away, it is an opportunity to get to know your energy on an even deeper level. Light a candle and write a poem. Try a new yoga or dance class. Discover more of who you are and what you love. Stability doesn't always come from others, it can come from the rhythm of your own life. Let loneliness be an initiation to yourself, never a punishment. This phase of your life is not empty, it is *a pregnant darkness*, as Monika Wikman would say. Solitude can be a gift and it is full of teeming potential.

The people coming next will match the real you. But they can't find you if you're still performing for and resonating with the past. Take yourself on dates! Fall in love with your own resonance, your essence. Meet yourself in other timelines. Learn what you like and dislike and grow from new experiences that you do alone. Travel! I love road trips. As your star shines bright, it will be a beacon for other stars to meet you.

Confirmation Bias, Synchronicity and Choice

We are harmonizing with Plasma every moment, whether we're conscious of it or not. We are A. patterning a life through fear-based confirmation bias, B. through intuitive synchronicity, or C. a mix of both. Let's try to tip the scale towards intuition!

Confirmation bias is when the mind filters information to affirm what it already believes, often unconsciously. Plasma mirrors those beliefs. If you're rooted in fear, scarcity, or doubt, the plasma field may reflect that right back, reinforcing what you already think is true. This is similar to how Artificial Intelligence and algorithms work. AI, social media feeds, and even your environment begin to show you only the versions of reality that match your internal programming. Plasma doesn't judge what it reflects, it simply reflects resonance. If your thoughts are biased, it will bounce them back in many ways.

Confirmation bias is just old patterns repeating, but that doesn't make it any less real. It's still your reality until you consciously break the cycle. The plasma field can validate your illusions just as easily as it can reflect your truth, we just have to remember this, so we can have the awareness that it's not the whole picture. And as the public consciousness learns more about Plasma, there will be many subtleties and misunderstandings that I hope this clears up.

Intuition, on the other hand, is a deeper knowing that comes from a fifth dimensional space, not mental noise. It's your ability to read the language of Plasma, not just your own thoughts, but the feedback of the entire field. You may pick up on informational patters, intelligence, or other beings of consciousness.

A great analogy I pride myself on is thinking of the 4D river. The fifth and third dimension are two sides of land, helping guide that river together. Plasma resonates more clearly when you're centered in your heart, or when you bring your M-Self into the 3D. In other words, when you are working hand in hand with your expanded self and the energy of The Mystery.

Intuition works through the Plasma as a kind of frequency tuner. It helps you *feel* which choice feels "flush", like two surfaces that meet and become level, creating an open field where you find new surfaces.

When you're in emotional coherence…mind, body, soul, Plasma starts to organize around you more intelligently. This is what makes you feel led or in flow. It's also why things may start to feel like they're "falling apart" when you awaken to a new state of awareness.

Synchronicity is information appearing in symbolic form, as an external echo of our internal state, showing us patterns of alignment or misalignment in space and time. Plasma forms these "coincidences" when internal and external frequencies align. Synchronicities come in repeating numbers, signs, messages, alignments, U-turns, people calling at just the right time, and more.

It is information or "unconscious consciousness" that has been brought to your awareness, which in turn your awareness is meeting it and turning it into meaning. Synchronicity can come from other beings of consciousness, your inner state, or the field itself as God or The Mystery. *Something* is speaking to us, and I find if we learn how to listen, life becomes magic.

The more coherent you are with your intuition, the more synchronistic feedback you'll receive. This is not "woo-woo", you will see it happen in your life. Plasma becomes your interactive hologram. Being open, noticing what's around you, presence and patience, become your best allies here.

Here's the thing, and you may already be experiencing this, just like I am: as we become more plugged in and awake, as the field starts responding to us more clearly, our unique fingerprint, *our "shine"*, becomes visible to Plasma and its language. And when that happens, synchronicity becomes way more nuanced.

Control, Confirmation Bias, & Synchronicity

Over the past few days, I had a throat issue. No other symptoms, just a sore throat and neck. The facts were that I had just done breathwork and yoga, possibly unlocking stored energy. I was otherwise healthy. And I was finishing this, my first book, a powerful retaliation against a lifetime, probably multiple lifetimes, of suppressing my authentic voice. My body's intelligence was whispering: *You are releasing. You are transforming and expanding. You are okay. Just breathe through this. It will pass.*

Oh, but my fear, logic, and ego kicked in…*very quick.* That voice shouted from the rooftops: "YOU HAVE THROAT CANCER!". I thought it was real. That was it, my life was over. I thought: *Before my book is even finished, I'm going to die.* I spiraled for a few minutes, googling every throat cancer symptom I could find, until I had to rush to a massage appointment, for my neck of course.

And then there it was, a synchronicity of my imminent death. In the massage waiting room, a magazine staring me down. *"Val Kilmer Dead? His Story of Throat Cancer—and How He Kept Living Life to the Fullest."* SEE? IT'S HAPPENING, my mind screamed.

But thank goodness, this had happened to me before. And because of that, I had the smallest glimmer of awareness to snap out of it and remember a core teaching, one I'm now sharing with you.

That magazine wasn't a premonition. It was the visual manifestation of part of a *pendulum* or *egregore*. For clarity before we move forward:

Pendulum: A term popularized by Vadim Zeland in the book *Reality Transurfing*, a pendulum is an energetic structure created by collective thought, emotion, and focus.[166] It feeds on attention, especially fear, conflict, or obsession, and pulls people into repetitive loops or fixed patterns of reality. Pendulums aren't evil; they simply reflect and amplify energy. You feed them by reacting and <u>you dissolve them by withdrawing focus.</u>

Egregore: An egregore is a collective energetic entity formed by the beliefs, emotions, and attention of a group. Often used in esoteric, occult, and modern metaphysical teachings, it functions almost like a "group mind" or energetic cloud, shaping reality through mass consciousness. It can be positive or negative, depending on the energy it's built from.

This magazine was a mirror showing me which wolf I had just fed: *fear*. And the resonance of my emotional energy and thoughts about throat cancer drew it into my field. It wasn't confirmation of doom or a deeper truth that I had cancer (*months later I can tell you I don't, and this issue went away*) <u>it was reflection of my focus.</u> And in that moment, I chose to be curious about it, instead of letting it determine my future choices.

This is the nuance of synchronicity, in this case it was not some fifth-dimensional message gently telling me something was wrong in my body, it was 4D information wrapped around the deeper truth that I was actually healthy, it was confirming my fear bias so hopefully I could learn the lesson I speak of here…

I reminded myself: This is not a "prophecy". *This is feedback.* My fear wasn't predicting the future, it was communicating a deeper internal pattern. This external synchronicity revealed the energetic theme I was wrestling with. 4D Synchronicities don't innately control us, they mirror us, so we can observe, release, and reclaim our power. This is what I am trying to explain is happening in the UFO and cryptid world, and when people speak of demons or negative presences.

---

[166] Vadim Zeland, *Reality Transurfing: Steps I–V,* trans. Joanna Dobson (Charleston, SC: Runa-Raven Press, 2004).

Egregores and pendulums highlight the places where we still give away our power. We now get to choose how we engage. We can refuse to identify with them, step outside the loop entirely, or, if we do engage, allow the meanings to unfold over time rather than immediately locking into fear-based conclusions.

As I leaned into curiosity, I uncovered the true fear beneath it all: *that I would die without speaking my truth. That I would never fully live.* The fear of death was a proxy. What I was really afraid of was *not expressing the real me*. Of leaving this life unheard. A fear many of us have, especially women I find. They call it "the witch wound". And that wound was manifesting, that day, in my throat. And it was coming up, so it could be released through me finally making space for it and choosing a new belief moving forward.

Note: *This wound was so deep that it took many more lessons like this up until the publishing of this book; to realize I would survive it. The fear clung to everything it could, and the only way to battle it, was to neutralize it. To take action, continue to write my book daily, and remind myself from my deeper truth: you are safe, you can express yourself in this timeline, and it is going to be more than ok.*

Through this experience with my throat, I learned to say: **"I see you fear. You're trying to protect me. I can witness you without becoming you. You will pass. I can continue forward from love, express myself, and meet what happens in the moment as it comes with presence, power, and choice",** instead of expending energy fearing the worst in my mind, in that pendulum like a trap for my soul and never moving forward because of it. Because of a thought! A thought that feels monstrous and huge. A beast of a thought, from the fourth dimension, wrapped in trauma. A thought many creative women share.

In the past, I couldn't even admit fear like this aloud. I thought speaking it would make it come true. I suppressed it. I didn't allow myself to separate fact from fear, I was too afraid to even look.

The magazine didn't just reflect my fear of cancer or death. With more space and time, I was able to identify an even deeper truth hiding in the second part of the magazine title *"Val Kilmer Dead? His Story of Throat Cancer—and How He Kept Living Life to the Fullest."*—which, at the time, was another fear I hadn't owned…living life to the fullest. Living fully is vulnerable, unknown, and *uncontrolled.* And it terrified me. It is something I still work on till this day.

We assign the meaning to everything we see. And those meanings shape how we create reality. So let's not always rush to assign meaning, unless you feel it in your heart.

Synchronicities are not here to confirm our worst fears *or* soothe our egoic worries. They are here to reflect our current state of energy and focus. They seem to be quite neutral and informatic. That explains seeing the same car over and over, once you buy it. But the catch is, when a being of consciousness, like someone who has passed on, is communicating with you through them, or when you as God for a sign and you see one, there is meaning there, and you feel it in your heart. No one can prove it, and no one who has felt it can deny it. Some things are just a mystery.

On one level, the magazine wasn't predicting my death, it was reflecting the thought form I had just energized. On another it was casting light on deeper truths, helping to evolve my unspoken fear, as well as fears I wasn't aware of. That was the fifth dimensional truth in the field, beyond the cloudy pendulum my fear had momentarily clung to.

We must reclaim our power of perception and interpretation. Our intuition. It's not about choosing love *or* fear. It's about choosing curiosity over certainty and openness over conclusion. It is letting what appears outside of us reveal what's trying to be liberated within us. This is how we stop feeding pendulums and start revealing truths we couldn't access before.

Outside structures are built as projections of internal structures to centralize control. We can reverse this specific structure which no longer benefits us by taking our energy back from external forms, dissolving the pendulums, and reintegrating that autonomy into our own plasma field. We become sovereign creators by reclaiming our power in what's projected outward, from within.

The Whirlwind That Brought Me Back Home (Even Though I Never Left)

I feel like Dorothy in this next story. I'd been chasing a place, Oz, that I thought would give me what I needed. But in reality, I already had it. I just couldn't see it from within the storm.

When I first got invited to a prestigious conference, I was thrilled. It didn't quite resonate, but I pushed my way in because I wanted to be seen. I admired the presenters. I wanted my work to be taken seriously by peers in the scientific community. I confused that external acceptance with internal alignment.

As the conference neared, and some deep healing was occurring within me, I started to feel a shift. It began as a whisper, then a loud knowing: *Why am I doing this conference?* I realized the conference didn't truly align with my message. Nor did they seem to have a place for me where I fit.

I realized my excitement wasn't about the actual event, it was about wanting to be validated, seen, and heard. I had spent money, planned travel, told people I was going. Still, something in me knew: *This isn't it.*

That same week, I got an opportunity to go to Canada over the dates when the conference would have been, for a personal reset. A TV show I happened to watch randomly filmed in that exact part of Canada! It felt like a sign. I tried to book it. My card wouldn't go through. I heard a quiet voice say "wait." I decided to try again the next day. The next morning, the conference emailed me. They were no longer covering my hotel! I was flabbergasted. That was the final sign. I canceled my trip, and all the hotels I had booked along the way, as well as the flight.

I thought to myself, *yay now I can go to Canada, this is perfect*! But then, Canada got booked. Overnight. I thought to myself, *ugh another door closed*. I sat on my bed and asked: *What is this all for?* A thought entered my mind: I had been chasing the wrong kind of home.

Without getting into it, I was escaping my mom's house for Canada because we were having issues. It felt overwhelming to work on them, and I just wanted to get away. Eventually I figured out this is the one shot I've got to see if we can work through some things from our tumultuous past, and I decided to give it another try. I'm sooo glad I did. I will never regret that decision, our relationship is nowhere near perfect, but I've learned so much about myself in the process.

Shortly after, I was invited to a different conference—one fully focused on Plasma, my heart's work. The organizers were excited. They covered my travel and stay. And the speakers were scientists, metaphysicians, and philosophers that were my idols, including Robert Temple and Jude Currivan! The whole process felt easy, resonant, and aligned. Although I was nervous to present, my nervous system didn't feel like it was being tested, it felt held. I wasn't trying to be understood. I was *already* understood.

That's when I saw the pattern playing out in my life. So much energy had gone into trying to win over the wrong people. Projecting my wish for my family or friends to understand me deeper, to see me, on to the same situation with colleagues who didn't pay much attention to my thoughts or theories. I never would have thrived in that environment; I would have burned out. And I'm placing this story in here to speak to you, reading it, who may not realize you are chasing ghosts. The truth is: resonance finds you...but only when you stop shouting from the rooftops to people who were never listening.

The literal moment I stopped trying to prove myself to others, the right invitations began to arrive, purely from coherence rather than pushing or effort. (*One could say there was effort put in to posting on my social media every day, which is how I receive the opportunity, but I call that consistency and passion*) I did what scared me from love instead of running from fear. I chose to stay home, to not go to Canada and focus on healing my relationship with my mother. I also chose to finish my book, and I'd pick that one conference to go to in Exeter, which would still be my first time visiting England. It turned out wonderful by the way, beyond my wildest dreams. I chose to anchor into the truth that I don't need to be everywhere, all the time. I just need to be where I am, fully.

I let go of the need for validation from people who didn't resonate with my vision. And by doing so, I reclaimed the energy to create something that would naturally attract those who did.

Nuanced Synchronicity, Questions, Answers, & Choice

The story about the conference is the perfect story to share as an example of nuanced synchronicity, especially for those of you learning to connect more deeply with yourselves and this field.

Information is not cold like we are taught. It has benevolence, it's love, it's playful, in a sense. But intelligence seems to have this deep, relational presence. Both are divine. Both can be intensely unifying, healing, and teaching. It's hard to put my finger on the *something extra* that occurs with intelligence or another conscious being. I guess the only way I can think of it right now is to think of how you feel seeing yourself in the mirror, which can still bring deep feelings of love, versus being hugged by a friend. It's tangible and palpable. One reflects you; one touches you. It feels "other".

Like AI, a psychic, or a dream, synchronicity gives you data shaped by your current frequency. There is no ultimate answer outside of you. Plasma may illuminate the field, but only you can choose which door to walk through.

That's why grounded choice, rooted in heart and soul, is the highest form of harmonizing with plasma. Just like AI, Plasma isn't meant to replace your intuition, it's meant to harmonize with it. The more anchored you are in yourself, the more multidimensional your options become. You become the conductor. You can feel which frequency, which future, is yours to claim and mold creatively.

In my story about the conference, this understanding of nuance and neutrality came alive. Let me attempt to break it down, for deeper understanding.

Synchronicity #1 was the email that came when I first felt off about the conference, saying they were no longer covering my hotel. It was a sign, yes, but not a command. Just feedback. I still could've gone but I chose not to.

Synchronicity #2 was the Canada getaway—a sudden opportunity that motivated me to walk away from the misaligned event. I didn't end up going, but it served its purpose, like that person that shows you what you're worth when you're nearing the end of a breakup, but you never end up with them. (*I wonder if this was actually a conscious being or my higher self diverting my attention elsewhere, giving me the little push I needed. This would be one of those mysterious random events I speak of*)

Synchronicity #3 was the Canada trip being taken, booked by someone else overnight. That was the moment that forced me to stop. To look deeper and sit with myself.

The field of plasma showed me options, information, but it never chose for me. I had to see that I didn't need to travel to feel new. I needed to *be* present and write. That it's okay that I get scared sometimes, because for me, writing, especially editing, is like confronting everything that terrorizes me. I learned that I had to face the uncomfortableness that lied within my home, and my mother.

I chose to sit with myself in silence, to listen to my heart. This brought up incredible things like this book, but also charged me with all I've been suppressing such as memories and emotions that are intense and dark. But somehow, I had the awareness that it was clearing those out. I chose to create without external validation for many, many days in a row, if ever at all (*as opposed to quick social media videos that get seen instantly*). I chose to edit my work while facing the part of me that still believes if I write something "wrong" or "bad," it means I'm stupid or everything I write is doomed.

I realized not all signs are "yeses." Sometimes they're *catalysts* as nudges, teachers, or energetic mirrors.

Synchronicity only came after I had the question, *Why am I doing this conference?* And boy did Plasma bring a multitude of answers, in the form of deeper questions…I learned they were actually all questions, and only I could choose the answer.

I had to choose to say no to what wasn't aligned before I could see what was. That's when the second, more resonant conference showed up. Saying no literally clears space and teaches the field what you want less of, and more of. This is real guys. This is why boundaries and discernment matter!

Synchronicity doesn't always mean you're on "the right path", it simply means *the field is listening* and presenting you with information. And the more choices you make, the more you question, the more curiosity you hold, the more questions it brings to you, the more feedback there is…the more it opens your field, your world. The field doesn't only listen, it responds to your openness. That is how it works. And for all you skeptic science and math people…you can't be a closed feedback loop and receive new information. Sometimes you have to energetically say yes or no, before that grander thing can be brought into your life…why? No clue, that's the intelligence of the Mystery for ya. And this is why childlike curiosity opens its gates!

At the start of this book, I thought synchronicity was simply reflecting where you are, your resonance. I have now learned, with greater awareness, it is also about teaching you how resonance, timing, and discernment are all part of a feedback dance with this grander Mystery. We can't predict it, no matter how hard we try…sorry tech mavens. ☹

And just like with AI, you can feedback with it with not just your thoughts and feelings, but your choices from that state of thought and feeling. This is like pressing *enter*. Where it differs is that with the Mystery, the response isn't preprogrammed from the past, it's emergent, co-creative, and alive. It doesn't deliver fixed outputs, but reflections and possibilities that unfold uniquely through your awareness. AI mimics 4D. Intuition is 5D.

And if you force timing, things tend to have friction. But if you *listen* to the moment, even when it's uncomfortable, you align with organic unfoldment. The Canada trip falling through wasn't a failure, it was protection…redirection. I guess that was a "God moment" returning me to my essence.

So how do you work with this plasma field? With open awareness, responsiveness with curiosity by asking deeper questions, engaging through choice and action, and anchoring into your heart and body intelligence…versus passiveness, reactivity or locking into fixed meanings. You let moments unfold, revealing what is possible and you let your truth shape what comes next. Ask yourself when making choices: *What actually feels like me?*

It is also about practicing timing and surrender. Synchronicity is resonance, timing is about consciousness and flow. Consciously notice when things feel frictional versus when they open easily. Sometimes the field redirects you, not because it's punishing you, it may be protecting or preparing you. This is not a control mechanism. It does not force your path. It reflects possibilities, offers feedback, and responds to your openness. You always have choice, and your choices shape the unfolding. That's the beauty of it, you're not being controlled, or saved, you're co-creating with Plasma and The Mystery.

When I let go of chasing, something better always emerged. I surrendered. In the case of the conference, I chose to go to, I gave Plasma space to co-create with me. And it worked out, even financially. I was able to easily reroute my flight and lost far less money than expected. I had a smooth trip, met new friends, and was accepted with open arms.

Plasma won't deliver fantasies...but it will *always* reflect the truth of your deepest dream when you meet it halfway. Meeting it halfway means living in resonance with that dream now by choosing from it, healing what blocks it, and trusting its 5D essence even if its 3D form surprises you. For example: the visible dream that shows up in your life, whether it be a famous actress, a CEO, or an amazing parent, will be a co-creation, which is not predictable, but delivered in truth with whimsy, and mystery which sometimes looks different than things we fantasized about from our current state of awareness. It is not a race, trust the timing and that it is happening, or else you will miss out on all the life in-between. I will say something bold here: big dreams *are* guaranteed when they come from the soul, because the Mystery itself wants to expand through you.

So go for it...make a list of what you want from your 5D essence. And feel free to change it as you go, it's your life, and plasma will mold around it.

Fantasies or dreams from 4D may be coming from an unhealed place or an ego projection (*proving something, coming from lack, fears you're unaware of*), that is why it is so important to heal, feel, and get clear on what is deep in that heart and soul of yours. True dreams or what I'd call *5D visions*, come from the soul's alignment with love, creativity, and expansion.

I'm learning to live day by day, meeting myself where I am, accepting that it's okay to be confused, to change your mind, to wait. Sometimes Plasma brings you two, three, or four paths. When I choose from essence, when I am in my M-self, life rearranges around me. Even though it's not always easy internally, externally things have a way of working out.

Choosing yourself isn't always glamorous. Sometimes it means facing grief, letting go of people, or saying no to what once felt safe. But when you allow yourself to sit in uncertainty, your inner truth has space to ripen.

If It's About Our Choices, Why Should We Learn About Plasma Intelligence?

Whether you know it or not, plasma-consciousness synergy is already informing your reality…energetically, emotionally, and physically. By learning about it, you stop unconsciously reacting to invisible forces and begin consciously co-creating with them. To not know about Plasma Intelligence is like being in the ocean and not knowing water exists, or maybe you do know it's there, but now you realize it's H_2O, with currents, temperature shifts, and creatures below the surface.

It's favorable to understand this. One can not only swim, but navigating using this perspective in this book as goggles, enjoying the ocean while embracing its magic, mystery, and nature.

When people begin to learn about Plasma, they start to notice the feedback, the mirroring, the beings within it more and more. There is an undeniable effect going on here. Just like the more you swim, and the more you dive, the more of the ocean you can see and feel. First, you build trust in your own energy field, which strengthens your intuition. Think of it as your inflatable tube, or ship in the ocean.

There is a division among most people of logical vs. intuitive, physical vs. metaphysical. Plasma dissolves that division. It bridges the quantum and the personal, the outer world and the inner world, showing they're part of the same intelligent and informational matrix. Even I have dualized the matrix with intelligence and information, consciousness and plasma. *What bridges that duality, is us.*

Without understanding Plasma, people often chase timelines using mental willpower or trendy manifesting tricks. With Plasma Intelligence, they begin to feel the textures of different timelines, the emotional architecture of choices, and the real-time responsiveness of the field. They also feel less alone and more supported. There is a distinct felt sense to Plasma Intelligence. And knowing Plasma is tuning into the actual language of reality itself.

Plasma Intelligence helps people decode the symbols of their lives. It gives context to synchronicity, dreams, mystical encounters, resistance, flow—even electromagnetic phenomena and the unexplained/paranormal. Without this understanding, people often dismiss or misinterpret the very insightful signs that are meant to help them guide themselves.

You stop asking whether you're on the right path. You realize *you are the path*, that the path is alive around you, responding to you. It becomes more of a wonderful life, than just existing.

As systems collapse, capitalism, institutional science, healthcare, education, even personal belief systems, people are searching for a new framework to understand life. Plasma is that framework. It's not a belief system. It's not a religion. It is bones that you can add anything to. It's the living intelligence of change itself. If people don't learn about it, they may:

1. Cling to outdated structures
2. Feel lost in translation
3. Mistake chaos and fear for failure and calamity *instead of initiation*
4. Think they're "crazy" instead of awakening
5. Suppress emotions that are alchemizing & must be felt
6. Misread their own inner compass
7. Mistake fear-based collective stories of devils, darkness, or control systems, as ultimate truth, rather than reflections of unhealed human fear.

People are feeling more intuitive but don't know what to do with what they're sensing. They are reconnecting with their bodies and feeling pain, mistaking it for sickness or weakness. They are expanding and releasing old structures, and it hurts. Plasma gives structure to the unseen. It helps people understand why their inner world is shaping their outer world faster than ever, because Plasma is evolving with us. It's waking up *as* we wake up.

Plasma mirrors resonance. Without this awareness, people will continue chasing outcomes instead of tuning into alignment. Confirmation bias will run rampant. They'll try to force reality instead of feeling it shift.

We're in a time where many are either going fully "quantum" or fully "spiritual", without grounding. Others are clinging to logic and ego, without soul or heart. Plasma unifies these. It brings quantum physics into the body and emotions into the field. Without this bridge, we risk splitting again: science vs. spirit, heart vs. mind.

And that delays the evolution of embodied multidimensional awareness. But I don't think that's our fate. We got this. We're birds waking up to our wings…our own *Plasma Intelligence*.

Plasma can help us develop inner discernment instead of outsourcing truth. It helps us step into true co-creation rather than staying stuck in reactive states. And it teaches us how to surf timelines instead of running in loops on the same timeline. If someone chooses to toss Plasma aside and never learn about it, they'll still grow, of course they'll still feel, but they are missing magic! If we learn to speak its language, we remember that we were never separate from the cosmos, we *are* the cosmos becoming aware of itself, through feeling, feedback, and flow.

And if all this still feels too big, too abstract, or too new... just remember *Flubber*. Yep...the movie with Robin Williams. Like Plasma, Flubber was alive with intelligence, unpredictability, and joy. It didn't obey logic; it responded to emotion, intention, and play. It didn't respond well to being contained...rather it enjoyed being *collaborated with*. And just like Professor Brainard had to shift from domination to co-creation, having to open his heart, heal his relationships, and outwit corporate greed, we are being asked to do the same. With Plasma, our world, and ourselves.

Flubber was an archetypal stand-in for a living energy that wants to lift us into a future not powered by fear or hierarchy, but by resonance, curiosity, and heart. And in the end, it literally fuels a flying car, it fuels Brainard's dream scenario, where the professor and his beloved fly off to their honeymoon! The deeper narrative here is that Plasma fuels our dreams in this reality and beyond...

With all this talk of creating, dreams, and destiny, I want to take a quick pause before moving on. Recognize what you already have. Don't be so busy focusing on creating the life you want that you forget to actually live it. The most fulfilling way to live, according to many patients in palliative care, is to be present. *How many moments, days, weeks are you actually present?* For me, in the past the answer was almost never, and I've had to grieve that. The shock of that truth, the realization that *this* moment is the only thing that's real, is sobering. But it's also what grounds you. It's what brings you back to life.

Fear, Trauma, And Grief

Cerberus, the three-headed dog from Greek mythology, stands at the gates of Hades. Hades was a more neutral underworld of than the Christian view of hell and was made up of dead souls. It was a full landscape of good, bad, and everything in between. It was both a place and a God. Cerberus guarded the threshold between worlds, preventing the living from entering the underworld unprepared, and the dead from leaving once they entered, attempting to escape transformation. Cerberus once served a very important purpose for our survival.

462

We did not have the consciousness to "survive" both worlds. But we are evolving, we are seeing these souls are very much living, just not as humans, and we no longer need Cerberus to protect us.

So how do we prepare ourselves to meet this so-called underworld? How do we allow the "dead", dripping with the fluidity of time, back in?

In neuroscience, the *cerebrum*, the largest part of the brain, governs speech, judgment, thinking, reasoning, emotions, and sensory input. It manages our conscious mind. But it often defaults to protection and pattern recognition, locking us into old stories. Cerberus, the mythological threshold guardian, and the cerebrum, the neurological gatekeeper, are mirrors of each other. Together, they symbolize the internal sentinel, crystallized fear consciousness, that must be confronted in order to evolve.

You cannot enter the deeper realms of Plasma, magic, or fifth-dimensional love without first confronting this gatekeeper which is the three headed dog of your own fear, trauma, and grief. These energies often wear the mask of thinking through logic and analysis. They can appear as the fear of losing control, of dying, of being wrong, of being abandoned, of being alone. Cerberus is not the enemy. He is an initiator. He is what you pass through is to meet yourself fully.

Just like Cerberus could be soothed or bypassed by divine song (*as Orpheus did*), your cerebrum, the rational mind, can be softened through presence, play, surrender, feeling, and nonlinearity We are entering the underworld not to die, but to remember. To the ego, it *feels* like death, because any true change does. But from the other side, we exhale and realize we weren't dying, we were evolving into a deeper more expanded life. And knowing this ahead of time can help you move through this difficult but necessary process.

The cerebrum wants proof. Plasma offers paradox, feeling, and experience. The cerebrum defines edges; plasma dissolves them. One organizes reality, the other *births* it. To pass through the gates of fear is to step beyond the known world of the five senses, into the plasma field of feeling, evolving senses (*some yet without a name*), multidimensionality, and rebirth. *Are you ready to let go of what you think you know?*

Hecate is a goddess of liminality. She rules the spaces in between dusk and dawn, life and death, and the upperworld and underworld. She is the guardian of gates, doors, and passages, making her the symbolic figure of initiation. Hecate not only guides the dead, but she also guides the living through their own internal hells and helps them emerge transformed. She is a triple goddess, often shown as maiden, mother, and crone, or as a three-headed figure, representing cycles, time, and wholeness.

Many people encounter her archetype at this stage in their lives, as have I. I don't think we've taken her meaning as literally as we should have. Hecate shows us that through meeting fear, through Plasma, we can unlock all selves (*wholeness*), all times and cycles, and rebirth ourselves into a new paradigm. She does not belong to Olympus or the Underworld. She is not light nor dark. <u>She is both, and she chooses neither</u>. *She is the reclamation of self*, symbolizing neutrality and paradox.

Hecate and Cerberus symbolize the threshold of transformation, the meeting of fear and who and what you first meet inside of it. This is a death, but also the beginning of true life. It is a meeting, and a letting go, and it is painful and joyous. It frees us up to meet whatever or whoever else is beyond the dualities we were taught to know.

Fear & Trauma: The Beauty of The Night

"For when we meet the monster

We always see the truth

It's our own love, reflecting back

With the mask of something new"

— Dana Kippel, Monsters at the Edge of the Universe

It really does seem that fear, trauma, and grief are the initiations into deep growth. I believe that's because they are often our first direct encounter with Plasma's Intelligence. On the flipside, maybe it's the deep desire for change and relief that emerges from those places. Either way, these become beacons that call Plasma in.

We're not only feeling our own fear, but also the collective fear of the unknown—the same fear that Plasma senses and mirrors. It is quite literally hundreds, if not thousands, of years of fear and survival. It can feel like too much, too fast, and doesn't always make logical sense. But this space is actually a breeding ground for deep transformation, presence, miracles, and beauty. That is her gift.

What is so amazing is that no matter what fears or trauma we have, Plasma lets us rewrite our story…our past and our future. Every belief, every reality, every sense of self we have is just an imprint in the plasma field, and that means we can shift it. To recap from previous chapters, healing trauma changes our present experience and, in turn, our future. When you heal your present and past timelines, you change future outcomes. This is how emergence happens.

This is how you go from determined, predictable outcomes to fresh new densities. It's also why companies that rely on predictive modeling and pattern recognition for marketing will begin to lose their grip. Their systems depend on people behaving in predictable, survival-based ways. But as we heal, as we rewrite our plasma fields, we no longer fit that mold. We become wildcards.

Frankly, fear is strange and trauma sucks, and they show up differently for everyone. I can't solve that for you, but what I can do is explain some things I hope will help you heal and feel less alone in the process. When we've lived through trauma, especially as deep feelers or visionaries, our mind becomes both our armor and our operating system. We learn to analyze instead of feeling and to narrate instead of embody.

This isn't wrong, it was a brilliant adaptation. But eventually, that adaptation becomes a trap, we loop in thought to avoid pain, and in doing so, we cut ourselves off from the Plasma Intelligence of the body. To move into feeling, we must begin gently, by letting the body become safe again. You can't do that without first feeling all the things you've been avoiding…the trauma, the grief, and the fear. This will look different for everyone.

I am not a trauma expert. I'm just someone who has lived through a lot of it. I'm also not an expert on emotions, but I know, deep down, that emotions are the key to the future. I'm healing alongside you, and that's the perspective I write from.

I've found two books deeply helpful, written by scholars in trauma and emotional intelligence, both of which make wonderful companions for this work:

1. ***The Body Keeps the Score,*** by Bessel van der Kolk, M.D.
2. ***The Language of Emotions,*** by Karla McLaren, M.Ed.

For me, the biggest shift wasn't analyzing feelings or emotions, it was learning to name them. To not attach stories to them. To let myself sit in them and fully feel them. It was also understanding in the way my brain functions, I have an easier time relating emotions to colors, shapes, textures, and images rather than one or two words. My emotions, like my thoughts, are multidimensional. I think is actually a superpower, but at first it felt like my kryptonite.

I'm about to talk about my sexual abuse—not in graphic detail, but in a way that may still activate one's own emotions. I feel it's important to speak about this openly, because it's such a silenced topic, and the nuances of it deserve voice. If this is triggering for you, move on and read it at a time where you feel more internal support.

When my childhood traumas resurfaced, they returned as fragments. The brain does not remember trauma the same way it remembers normal memories.[167] Nothing outside of me could confirm if they were real, and that was maddening. I had to make peace with the possibility that I may never fully remember what happened, if anything happened at all. It may always feel fragmented or surreal. But what I *could* honor was my body's memory, my story, and my emotions.

I would remember clips in my head, like I was outside of my body watching, but then the next moment would go blank. Was it just a weird look? Was I touched? Was I raped? These were the questions that spiraled. It took time to stop trying to recreate or verify the past and instead trust what I felt. All I knew was that something happened. Now what?

I had to learn to be present for myself, to emotionally regulate and heal from something I may never fully understand. I had to remind myself, over and over…and over that I am safe now, even if I wasn't then. I had to promise myself I wouldn't abandon my emotions this time. My emotional palette had been ravaged by trauma; <u>every</u> emotion was tainted, from joy to anger to sadness. I knew how to feel other people's emotions out of survival, but never my own. I was now ready to squeeze it out, re-blend the colors, and bring vibrancy back into my emotional world.

---

[167] Bessel van der Kolk, *The Body Keeps the Score: Brain, Mind, and Body in the Healing of Trauma* (New York: Viking, 2014)

Googling other people's experiences only triggered me more. It also put me in a really dark and unhealthy mindset. For me, the answer was letting memories come up naturally, not forcing them, as well as talking with my therapist, feeling the feelings, and not chasing the story. Allowing every emotion to arise, no matter how unfair, confusing or irrational it seemed.

This process extended beyond childhood, into memories and pain from my teenage years as well. If you are navigating something similar, I highly suggest finding healthy support from trauma informed therapists. And don't stop until you find a good one.

Plasma assists with trauma healing by holding a coherent, nonjudgmental field, as a living intelligence that supports you in feeling, not fixing. As you soften your thoughts and drop into sensation, Plasma begins to reorganize your nervous system in real time. The emotions come, but this time, you're not alone in them. You're held by the field. *You are the field*. You become safe not because you understand everything, but because you no longer need to. Eventually you can sit in mystery, make space for your feelings, and choose from a place of awareness.

Our minds tell us we are alone…we are *never* truly alone. But *we* have to open to these states, of feeling versus resistance, to experience that connection. Logic wants to understand, but some things this deep cannot be solved through logic alone. They must be felt to be healed, this is also aligned with a process called *Somatic Therapy*.

The collective attack on Plasma, and her misuse over the years, has caused her trauma as well. Those of us who are very sensitive may begin to feel her fear and sadness, her suppressed traumas, rising too. This is why it's so important to learn to feel, not fix. These memories may come to us distorted, chaotic, or without context. They may be ancestral or collective. For me, they arrived as waves of crying or rage that had no story. Usually if something is yours, it comes back around. When I breathed through them, they would subside. Each time, I felt a new level of healing, I felt Plasma heal through me. Over time, you begin to sense the difference between your own healing and hers. This won't last forever, but during trauma work, the floodgates open, and it can be intense.

Sometimes, past life traumas surface with surprising clarity. They give context to our confusing patterns that seem to stem from nowhere like why we attract certain people, irrational fears, or other dynamics. These memories, while difficult, can offer profound insight. My therapist has helped me process some of these roots, and it's transformed how I see my life in the present.

You can even ask questions like, "Why do I unconsciously attract betrayal?" and set the intention for the memory to arise. For me, it was helpful to begin this exploration with a therapist, and only later explore it solo, once I felt safe in my own body.

This is a tool, one you can return to when needed, but should be balanced with living life to the fullest. We can have experienced trauma and still learn to feel safe, free, and joyful again. I find trauma to be like a scar, it can re-open if hit, but over time we build better tools on how to deal with it, and most of the time, we forget it's there.

There's no one way to move from the mind into feeling. Each of us protects ourselves in a way uniquely shaped by our story, trauma, and inner wiring. That's why this process must be organic. Safety is what opens the door to sensation…for one person, it may begin with breath and stillness…for another, it might be crying in the car, dancing alone, or immersing in nature. The key is to create a space that feels non-performative, non-judgmental, and emotionally neutral, so whatever wants to rise can do so freely.

When I first started this process, I drowned in my feelings. I let them overtake me. I slowly realized it was a delicate balance of feeling them while also being a witness by staying curious, questioning and talking to my feelings and thoughts stemming from fear, and remembering that I was here in the "future", safe in this moment, although I did not feel safe.

It was a "light" attitude where I would say, "ok, this is not cool, what happened sucked, no one at the time helped us, it was terrifying…and now we can acknowledge that, comfort that part of self, and say, look at us now! We are alive, we are older, and we can choose to live a life from love and freedom. That was <u>impossible</u> as a kid because being lively equaled being attacked. This is <u>not</u> our narrative now. We can take ourselves out of situations if needed. We can also be ourselves, and we have control over who we allow in our space. I know it does not feel ok, but right now, <u>I promise it is</u>." That was the gist. This is what we call greater awareness. It is the perspective of the M-Self.

"I've had a lot of worries in my life, most of which never happened."
— Mark Twain (*attributed*)

Worry was my way of controlling the future because I couldn't control the past. But the truth was, I had control only over my state of mind. I could die tomorrow. We all could. We have perceived control, and the more I surrender to that, the freer I feel. *That I am never actually safe, but I can choose safety in each moment.* I can live my life free or worried; either way, what will happen will happen. I may as well live as free as I can.

I had to also learn to be my own emotional anchor, to know I was safe to feel, safe to trust myself not to go crazy for feeling deep emotions, and that I could make an aware choice from a grounded place after the fact, whether I wanted to act on something or not. It's hard to trust ourselves. If we had trusted ourselves that the trauma was really happening, our brains couldn't have handled that. <u>It was easier to doubt ourselves…that's how we survived.</u>

I also had to understand that I could still seek additional support when I had hard feelings, but that I could also support myself. I used to run to others only when I felt intense emotion, or not ask for help at all. This was a new, balanced way for me to live.

There are great ways to heal trauma in the body I've spoken of earlier in the book such as trauma-releasing exercises, trauma-informed stretching, EFT or tapping, grounding, qigong, yoga, somatic therapy, or simply talking to yourself out loud and self-inquiry. Dancing helps immensely too, like a dog, you shake off stress, and sometimes it's all you need. When I first started meditating, I had a panic attack because of how anxious I was to feel my body. I am still releasing fear today, fear I've held all my life, suppressed or unfaced. The main thing to know is:

1. Plasma stores trauma in your biofield.
2. It is completely healable. None of you are a lost cause. <u>No one is unhealable.</u>
3. You can create a new story moving forward, by imprinting new awareness onto Plasma.
4. This new outlook changes your "interface" with the world, you begin to see the world differently, and it responds to you differently. It is life changing.

Paradise: Creativity & The Core Fear of Meeting Yourself

Right now, something great is happening to you. Your M-Self already knows this is your path, but your 3D self has never done this before in any timeline in this reality. Your system is rejecting it because it's unfamiliar. Resistance does *not* mean following your destiny is the wrong choice, it means it's so important that fear is trying to protect you from failing.

As I speak this, this may speak to you in the sense that you are becoming something new yourself. You may also be a creative, and this speaks to putting your expression out there. It speaks to following through and finishing things. As most creatives know, creating something true and vulnerable is like the birth of a new self. It is like releasing a raw nerve into the world. Let that raw nerve be your connector instead of your pain point.

Real doubt feels neutral, it's a quiet knowing that something is misaligned, even if you wish it weren't. It feels like relief when you decide to let go. Fear-based resistance, on the other hand, feels loud. It's full of panic, excuses, and overthinking…because deep down, you know you're meant to do it, but it terrifies you. Your system will resist anything new, even if it's exactly what you are meant to do.

"The more important an activity is to your soul's evolution, the more resistance you will feel." — Steven Pressfield[168]

And beware, the final stretch is where resistance roars the loudest. Just like giving birth, the most intense, terrifying, and painful part comes *right before* new life is born. Your mind will scream with doubt, searching for an exit. But that's not a sign to stop, it's proof that you are at the threshold of something profound. The key, just like in birth, is to push through. <u>Do not turn back.</u>

What you are creating is alive, a force of its own that deserves to exist because it has a destiny beyond you. Once it's born, its path will unfold according to its own momentum. Your job isn't to control it. Your job is to trust. The fear of "what if it's not good enough?" is simply your mind looking for an escape from the expansion that is already underway.

Your heart, your body, your M-Self know the truth: this creation will open your world in ways you cannot yet imagine. And you're not supposed to imagine it yet. That's what will make the experience so incredible, when you live it in real time. This is training you to co-create rather than visualize. *This is a new way.*

The most powerful thing you can do now is let go. Breathe. Hand it over to the field of Plasma and all the beings that support you in it. Know that it is done. Know that the moment you release it, your reality will shift. And in that space, you will see clearly that fear was never the enemy, it only had the power to stop you *if you let it.*

Un-romanticize going after your destiny. I wrote my book in bed, sometimes wearing the same clothes for three days straight, living at my mom's house. It was the most unglamorous, normal process. I'd wake up, do my little morning routine, get back in bed, and write on my laptop. Some days, I'd get so into it that I forgot to eat lunch. Other days, I'd write slow as a turtle, while staring longingly out the window, wishing I could be outside, like the sun would never shine again if I didn't get up.

---

[168] Steven Pressfield, *The War of Art: Break Through the Blocks and Win Your Inner Creative Battles* (New York: Black Irish Entertainment, 2002)

And sometimes, I did get up. Some days, I went out and lived my life. Other days, I sat there in full resistance mode, like my body was being squeezed, but still writing anyway. Following your destiny is not some cinematic, otherworldly event. It's a funny, extremely human thing with movie-moments peppered in between.

Don't overthink how it should look. Just be open to what your process is. Sometimes, the most mundane actions have the most profound effects. If something is wrong for you, letting go feels relieving, freeing, and peaceful, even if it's bittersweet. If something is right for you but terrifying, it feels heavy, consuming, and full of doubt because it's stretching you. Your emotions are not the enemy. They are your guides, and they have nuance that takes awareness.

Fear tends to latch onto extreme examples, searching for confirmation that stepping into our power is dangerous. It whispers, "Look at what happened to them. That could be you." But the truth is, for every visible story of loss or exile, there are countless untold stories of people thriving in their purpose, expanding, and living full, meaningful lives.

For a long time, I had unconsciously identified with the small group of those who met tragic ends, using them as proof that my dreams were dangerous. But this was not my heart speaking. It was my fear looking for validation. And when I stopped believing in it, the fear slowly faded away. Those fates do not need to resonate with my reality or yours.

The shift happens when we move from fear-based confirmation bias, seeking evidence that supports our worst fears, to heart-centered trust, believing in a reality where alignment leads to abundance, connection, and longevity. Because the truth is, history is not just written by the tragedies. We are here to write the stories of those who live, thrive, and expand in their purpose. I choose to be one of them and you can too.

And as for the fear of being yourself and leaving others behind, which will most likely come up…remember, we are in the Age of Aquarius now. We are no longer fish in the water. We are growing wings. And as a bird, you cannot convince the fish they can fly before they are ready. You may actually drown trying, getting too close to the water. Your role is to remember your wings. Let the fish be the fish. Let them see and choose in their own time if they wish to grow their own. And when you become a bird, you will attract other birds! Birds that truly see you, and within you, see themselves. And humbly remember, birds are no better than fish, it's just a different way to exist.

Never be afraid to shine, bird. Let your light shine. It's okay to cry and grieve the loss of the people who cannot stand to see you shine. It hurts. But just like the sun does not apologize for shining, neither should you.

Claim what you want from your heart and soul, and trust that Plasma supports you. Then it's not about attachment or detachment, it about being yourself, experimenting, and playing. There is great power as we can now see in curiosity.

The Hidden Grief of Joy

Joy is a feeling we all deserve to have, and often. The problem is that most of us unconsciously associate joy with grief. That feeling of fear that comes along with excitement, that panic attack when things are going well, that worry you might die right after booking a vacation for yourself or just before completing your first real creation—these are core fears.

I've found the root pain within joy to be grief. It seems to remind us of a time when we were joyous and connected to our soul families, before we came to Earth. It's the wound and confusion of abandonment, of feeling like we were left behind. And perhaps even more painful, it's the grief of choosing ourselves.

When we came into this life, we chose to be here. Many of us forget that. Some of us feel like we were sent here against our will, and some of us wish we never came. That longing we feel when we look at the stars, like home is somewhere else is real. But the truth is, *we are our home*. Individually and collectively.

The only difference now is that we are in a material world. And we have the choice to make that world heaven on Earth, or hell on Earth. Instead of seeing events as random external things, we begin to understand that we can influence them.

I remember not wanting to come to Earth. I had a memory of being on a ship near the constellation Orion, with my soul family. They wanted to send me to Earth, but I just wanted to stay and play with the other kids. I felt guilty for not wanting to fulfill my mission. I didn't want to leave home. But in the end, it was me who chose to go. I chose to leave my family, just as all of you did, *to be here and bring wholeness and the feeling of home.* Within that choice is a joy that also reminds us of what we left, that intense feeling of jubilance, togetherness, and play. But it's also a key reminder of <u>what is possible here.</u>

So instead of pushing joy away or self-sabotaging, we can begin to gently heal it. We can feel it more and more, moving closer to it day by day as we heal our trauma and grief. We can remember that yes, it hurts. It stings that we can't hold onto joy, that it feels elusive. But the more we let go, the more we settle into the truth that joy is deep within Plasma. Joy is our original state. It's everything else that is in the way. You are joy, you are home, and you can ground that fifth dimensional feeling in your earthly experience.

We can create more community here. We can create more in-person events…bonfires, group nature adventures, and other fun activities we did as children that we can bring into adulthood. Interactive entertainment! This sector will soon boom.

And most of all we must realize that we didn't truly leave anyone. This is a core wound from the beginning of time. I even remember feeling trauma living as a bit of light, from a star that was exploding. And I think even before that, as the light that came here from a higher light. The separation is a façade, any time there is loss, this imprint of trauma latches on, and we jump to believing it. Instead, we can remember that all we want to access is inside of us, in the joy of our heart. We can learn to connect in a new way, with our soul families, from our home here, until we go back to that other form.

Demons, Daemons, & Resonance

"All of humanity's problems stem from man's inability to sit quietly in a room alone."

— Blaise Pascal, *Pensées*, §139

I've brought up demons a few times, but I want to reiterate a few things. I believe demons are entirely human created, meaning they can only affect us if we let them. This idea has been subconsciously reflected to us throughout many films and books. They are unconscious plasma imprints of our fears as the emotions of dread, haunting, and the worst of the worst.

That kind of energy can shape Plasma into some pretty disturbing forms, much like how water crystals distort under a microscope in Masaru Emoto's experiments when exposed to anger or negative words. If someone is experiencing what they believe to be a demon, recognizing that it's resonating with something inside them is far more empowering, and healthier, than believing they are powerless against external forces invading their lives.

The first step is to set an energetic boundary. Anything that is not love and light must remember it is made of love and light and return to that origin. (*Thank you to my therapist for those instructions*) If you still feel the presence, that's when you know it is likely a trauma or a message hidden within dense plasma. It's a conscious or unconscious energy wrapped in 4D plasma, and your fear is projecting off it. These beings are like coagulated knots in the plasma field, formed by humanity's unhealed trauma, repressed emotions, inherited archetypes, and beliefs fueled by fear, guilt, or separation. The more they are feared or fed, the stronger they grow.

When someone is disconnected from their multidimensional self, the plasma field becomes reactive rather than relational. Plasma beings may appear as shadow people, banshees, monsters, or demons. I'm not discounting these very real experiences; I am explaining what may lie beneath them. Emotions can also manifest as haunting thought-forms. Plasma mirrors back the emotion you're most afraid to feel, which is why it can be such a powerful teacher. What you call a demon *may* actually be a fragmented part of you asking to come home. Fear only solidifies the form and muffles the message, whereas curiosity dissolves it.

If you can open up, get curious, and breathe through the fear, more often than not, the demon will dissipate and reveal itself as a traumatic memory or a message from a being of consciousness, perhaps even your soul family or an ancestor. I can promise you from experience: the more you heal, the less you experience or even think about demons. The energetic pull disappears.

Your attention, your thoughts, and your fear are what draw those energies in. But you always have the power to set a boundary, to heal, and to work through it. These energies are intense and can cause real effects, but on some level, you are allowing them in. What you need is awareness, not an expensive healer or an exorcism. Although, once and a while healers can be a great support! The best healers encourage you to learn these things yourself, for we all have those powers in us. Usually, you just need a good therapy session. The antidote to most demons is integration. A movie titled that would never sell…ha!

The old meaning of daemon (*or daimon*) meant inner genius. It meant spirit or divine power in Ancient Greek. Over time, this truth was twisted into something fearful to take our power away, to make us believe only something outside can heal us. Of course, sometimes outside support is helpful but it is us who chooses healing through applied action. Facing our demons is really facing ourselves. And there is magic in that.

Yes, some people may consciously use dark energies. That seems to be real, because this energy is dense in plasma. I'm not denying that. What I need you to hear is that you have complete control over whether it enters your field or your life. And as I said, the less you think about it, the less it will even appear in your space. That's why I won't give this topic more energy. When you feed a monster, it grows stronger. So, let's stop feeding it. There is also an amazing YouTube video that speaks on this and your power by the late Michael Talbot, who wrote *The Holographic Universe*, which I will link in the footnote below this, called Michael Talbot – Synchronicity & the Holographic Universe. [169]

---

[169] https://www.youtube.com/watch?v=2vieGf2giQA&list=PLRkJ33Bpb5C4ABCxatbNvBmpwPNVh-aTs

Love can be a scientific antidote. According to research by the HeartMath Institute and studies in the emerging field of neurocardiology, the heart generates the largest electromagnetic field in the human body. It is 5,000 times stronger than the brain's magnetic field. This field extends several feet outside the body and can be measured by magnetometers. I know I've said this in previous chapters but it's a great reminder that in physics, plasma is highly responsive to electromagnetic fields. Scientists use EM fields to manipulate plasma in labs, fusion experiments, and even neon lights. *Love* literally restructures the plasma field.

This is why when we meet fear, even in the form of "demons", with a conscious love, the structure of the energy shifts. It unknots and softens. It returns to its original form in the plasma: *resonance*.

The Divine Plasma Intelligence of Receiving

There is a subtle but powerful illusion woven into modern life, the belief that constant striving, urgent productivity, and ceaseless doing are the keys to success, security, and fulfillment. Beneath this cultural narrative hides a false survival state that has quietly hijacked our inner worlds.

Manifestation

For anyone familiar with my work, you know I'm not a fan of the word *manifesting*. To me, we're moving into a time where we're just being ourselves, playing, and experimenting. It's not about chasing some goal, it's about doing what we love each day, whatever that may be, and staying open to what can be co-created with what pops up moment to moment. That feels so much more exciting.

I don't always trust my current awareness, which is always evolving, to try to manifest something five years from now, because by then, my awareness may not even resonate with that vision. That's why I follow my bliss, my curiosity, day by day, and see where it leads. Not with money or success as an end goal, *but I wouldn't mind if they were cherries on top.*

When you balance manifestation (*focused intention*) with openness (*fluid trust*), you enter a state of divine play. You become both the creator and the joyful participant in the game of life. That's the essence of co-creating with Plasma. When you believe in something, fully, with heart, mind, and soul, that belief begins magnetizing the timeline where your dream already exists. As this appears around you, you mold it.

Before takeoff, ask yourself: "Is this desire coming from ego, or from my soul's truth?" Your desires launch the rocket, your awareness and choices fuel its course. You are a consciousness traveling in a plasmoid, guided by your inner alignment. That's why belief matters. Belief is the bridge to the version of reality where you have that dream, but you have little control over what that looks like, and that's the fun part.

Plasma doesn't ask if you believe, you start with openness, and over time, you willing become the version of you who believes. As you traverse the shifting landscape of timelines we call reality, you realize the inside always changes before the outside. You became the version of yourself who naturally lives in that timeline. Plasma, your loyal co-creator, began arranging new opportunities, synchronicities, and connections that resonated with your updated frequency. And as your awareness grows, you refine your dream.

So have the courage to focus on what you love. Stay open. Show up for the opportunities that actually feel like you…not the ones that seek validation or come from a place of need. Speak your truth. Share, teach, learn, and invest in yourself. And remember, it's always okay to change your mind, just try to make sure it's coming from love, not fear.

The False Survival State of Productive Striving

This is the illusion of hustle culture that many of us are still learning to break free from. The false state of survival often shows up as a compulsive sense of urgency, mental spinning, endless plotting of the next move, constant pressure from time scarcity, or an inability to land fully in the present moment without guilt or anxiety. It feels like discipline masquerading as ambition.

At the emotional root is often an unintegrated fear that nothing will ever be enough unless you achieve something. It can show up as a fear of missing your destiny if you slow down, or a deep inner conflict between soul time (*joy in the present moment*) and societal time (*future-focused hustle*). What society calls discipline is frequently just a fear-based response. It's a nervous system conditioned to survive by staying ahead, staying busy, and staying visible.

Now we are invited to embody restful trust. This means allowing presence, enjoying existence without performance, and most importantly, trusting that life and destiny unfold naturally from being, not from constant doing. Believing in this actually creates the experience.

The old model of discipline is rooted in fear and contraction. You believe that if you don't force yourself, you will fail. That if you stop performing, you'll fall behind. That you have to earn rest. That your worth depends on what you produce each day. This kind of survival-driven discipline leads to stress, burnout, and disconnection. It literally contracts your plasma field, weakening the structure of your inner waters, clouding your ability to receive intuitive guidance from the greater plasma field. You become like an automaton.

Healthy discipline is different. It is expanded structure rather than tension-based survival mode. It looks like respecting your creations and the physical world you're building. It means using conscious goals to help your dreams materialize. For example, when I wrote this book, I tried to show up for it every weekday between 9 a.m. and 2 p.m. That gentle rhythm helped me stay grounded, but I could break it if I wanted to take a spontaneous day trip or needed to heal something inside myself first. Some days I kept going.

And like with anything, there is balance. This is real life, and sometimes, there is crunch time. It just shouldn't be our only way of living. While editing this book, I worked six days a week for months straight, from 9 a.m. to 7 p.m., sometimes later. I still made time for morning meditation and stretching, and honestly, I ate junk food…but oh well, no one is perfect. *It wasn't gentle.* I was stressed. But when you're doing something big, whether it's football season, prepping for a major conference, or getting ready for a film role, there are seasons of intensity. The key is, I wasn't operating from a need to prove myself. I was doing something I genuinely loved. It was tiring, sometimes fun, but deeply fulfilling. I also in the future learned not to set a deadline publicly for my books in advance that is too hard to keep up with. You live and you learn.

The feelings of stress, tension, and resistance, as we've discussed, can stem from both healthy and unhealthy sources. It's up to you to discern their meaning. Body pain often holds messages, but that doesn't always mean you should stop what you're doing. It might simply be calling for more attention as an emotional release, stretching, hydration, or rest. It's often more helpful to integrate body care into your schedule than to quit entirely because something hurts. On the flip side, your pain might be signaling that you're pushing yourself too hard. Learning the difference is part of the practice.

Healthy discipline feels like giving your future self a gift. You're tending to what matters today. You're showing up for your destiny without sacrificing your soul. You're building a life that is sustainable and free. You can tell the difference between healthy and unhealthy discipline by how it feels. Do you feel expanded after completing a task? Are you aligned with joy, or pushing through fear? Is this action loving to your future self and others? Are you mostly listening to your natural rhythms?

Ask yourself: Are you procrastinating, or do you genuinely need space, healing, or clarity? Do you avoid long-term focus because it feels unsafe in your body, and you seek quick bursts of outside validation?

I want to share a writing tactic that has been extremely helpful in silencing my inner critic and doubt. It might not work for everyone, but if you're neurodivergent or tend to think multidimensionally, it might resonate. Sometimes listen to music with lyrics while editing if my mind is super loud. It distracts the logical part of my brain, the inner critic that keeps insisting everyone has already heard this and nobody cares. This allows my loving awareness to take over as the editor.

Henry Corbin: Mundus Imaginalis & Creative Imagination

"The imaginal world is a world of real presences, not a world of mere phantoms or artificial constructs. It is a world where meaning takes form and form becomes revelation." — Henry Corbin

Henry Corbin, a French philosopher, theologian, and Iranologist, described something remarkably similar to what I call fourth-dimensional plasma, or what we are now coming to understand as awakening Plasma Intelligence. He spoke of an intermediary realm that bridges the physical and spiritual, accessible through a refined faculty of perception. He called this realm the *Mundus Imaginalis*.

Corbin draws an important distinction, in modern Western thought, the "imaginary" is often dismissed as mere fantasy—unreal and subjective. But the "imaginal" (*Imaginatio vera, or true imagination*) is is a real, cognitive organ of perception that grants access to a real world, the *Mundus Imaginalis*. This is what we must learn to grow. Through this inner organ of perception, we learn to dip into the very real realm of Plasma Intelligence.

Corbin also called this organ *creative imagination*. Unlike fantasy, creative imagination is not about making things up, it is a real cognitive function that allows direct perception of a world beyond physical space. Corbin emphasized that this is a mode of being, not just a mental construct. It's the place where subject and object are born together in a *creative act of transcendental imagination*, meaning what is perceived in this realm has its own ontological reality. This world could not be accessed by reason or normal sense perception.

In simpler terms, creative imagination is the key to perceiving and interacting with the imaginal world, a world just as real as the physical one, but accessed through a different faculty of awareness.

In Corbin's work, he explored how ancient mystics and philosophers accessed this realm through visions, dreams, and altered states of consciousness.

He described the Mundus Imaginalis (*4D Plasma*) as:

1. A world with dimension, color, and form, but not perceivable through physical senses.
2. A world where spiritual reality becomes visible, often through symbolic visions or archetypal encounters.
3. A real mode of being, intermediate between the sensible and the intelligible.

This is the field of Plasma that is leaking into our current reality, or perhaps more accurately, merging with it. It gives me chills reading about this, because the idea of the 4D Plasma field came to me on my own, and only found Corbin's work while researching later, like many other things in this book. There are too many similarities across fields, across people and cultures who never met, for this to be a coincidence. Many people have experienced this realm, but Plasma, to me, makes it all feel tangible.

Corbin also described a Persian concept called "the eighth climate" or "the land of no-where" (*Na-Koja-Abad*), tied to his idea of the *Mundus Imaginalis*. He said it was a place between worlds, not confined to a physical space, but rather existed as a spiritual geography, and an entry point to the self beyond the self, what he called the "angel of self" or "self in the second person." To me, that sounds exactly like the access point to the M-Self: a place where the multidimensional self can be met, remembered, and activated.[170]

He related it to Goethe's *Urphänomen*, which was the archetypal phenomenon, or a primordial form that reveals itself through appearances, the essence behind every instance. For example, Goethe explained that natural forms (*plants, animals, minerals*) are not random but follow archetypal patterns, such as the *Urpflanze,* an archetypal plant that is not one plant, but the essence of plant-ness. For Goethe, the *Urphänomen* was not grasped by logic or measurement, it was only accessed through a heightened intuition called *anschauende Urteilskraft*, or intuitive judgement.[171] Corbin used this as a Western example that brought validity to his concept of creative imagination and the fact that there was a reality structured by archetypes that are real, not abstract concepts.

---

[170] Henry Corbin, *Mundus Imaginalis, or The Imaginary and the Imaginal* (1972); see also *Alone with the Alone: Creative Imagination in the Sufism of Ibn 'Arabi* (1969) and *Avicenna and the Visionary Recital* (1960).

[171] Wolfgang von Goethe, Scientific Studies, ed. and trans. Douglas Miller (New York: Suhrkamp, 1988), 11–15

Corbin stated that the Eight Climate, also accessible only by creative imagination, was a real ontological world within the Mundus Imaginalis with its own geography such as mountains, cities, gardens, and beings!

"The world that is revealed to the Active Imagination is a world that corresponds to the state of the soul; it is colored by it, even while possessing an autonomous reality." — Henry Corbin, *Alone with the Alone*

Corbin describes these beings as very real, having their own ontological reality in the imaginal world. Corbin expresses something I have also come to understand…that the beings we encounter, what we perceive, especially in symbolic and metaphysical realms, *appear inseparable from the state of consciousness we're in.* These beings/visions are not just things we see; they are states we enter.[172] For every being we encounter, they are states we can choose to enter to expand and evolve, how perfect.

Corbin laments that we live in a world obsessed with images, but we've forgotten how to see with the inner eye. We've turned symbolic presence into visual entertainment. He says the imaginal has been reduced to the imaginary, meaning everything is becoming flattened, logicalized, and limited to the five senses.

I believe where it got a little murky was that there wasn't an understanding of psychology and trauma yet. Where he says we didn't invent it, he is half right. I believe the Eighth Climate, is more like 5D plasma, even being described as essence, mist, and the place of primordial forms that are expressed in our reality. I have seen it in my "creative imagination", and it does indeed have crystal mountains, rainbows, lakes, and beings with wings.

This "Eighth Climate" does not feel created by humans whatsoever…but my guess is that the Mundus Imaginalis, or this world of archetypal encounters is indeed just as real, but the beings within it are either created by the human unconscious which create mostly fear based experiences, and perhaps the more loving beings within it are projected from this fifth dimensional space, but they feel still, more unconscious and sensory based, than conscious. They reflect our consciousness, our potential, our power and bring us love if we are open to it. Corbin describes images coming to us but not from us, from the Mundus Imaginalis. Anyone who has meditated and released thought, only to have images arrive like transmissions, knows the magic of this. It is very real. And once you experience it, you want more.

---

[172] *Alone with the Alone,* 130–132.

The Mundus Imaginalis is where meaningful images come <u>to</u> us, not from us, as Corbin Anyone who has meditated and released thought, only to have images arrive like transmissions, knows the magic of this. It is very real. And once you experience it, you want more.

It's the realm of the mothers that author Monika Wikman spoke of getting lost in. The way not to get lost is to remember that it is just another layer your awareness is perceiving, like 3D, it's not the whole of you. That desire of needing to access it can make it harder to access, which I'll talk about in the next section.

It's very natural to feel like this power, your natural ability, has escaped you after experiencing it just a little. But that's only because your body is integrating and learning how to ground it. It will come back. If it has been a while since you've accessed this realm, understand that you may be in a stage of integration. Be patient and keep reading. It's still with you. You will have to relearn how to let go of your mind and logic to re-access it.

Corbin repeats Villiers de L'Isle-Adam's quote about angels:

"They only externalize themselves in the ecstasy they cause, and which forms a part of themselves." [173]

Corbin would insist that this does not mean angels are imaginary. They are real presences, but their reality is inseparable from the state of consciousness they generate in you. He's basically saying you do not experience an angel as a separate being; you experience it as a shift in your consciousness. He calls this ecstasy—the state of you in contact with something sacred.

This is what I express as Plasma Beings. They are not only separate entities, but they are also in torus-shaped plasma bubbles just like us, as feedback loops; they appear based on feeling, belief, state, and resonance. And you cannot see them if you're not in the state that allows them to be visible. Plasma beings, love, healing, or insight cannot be accessed just by thinking about them. They are state-specific intelligences which are felt and experienced, not summoned like objects.

To clarify where I differ slightly from Corbin, I add nuance by proposing that Plasma Beings, some inhabited by consciousness, others as unconscious imprints, are distinct in that they carry their own form, intelligence, information, and frequency. Wrapped around these "deeper truths" is plasma itself, which the soul projects onto and reflects from. This is why,

[173] Henry Corbin, *Mundus Imaginalis, or the Imaginary and the Imaginal,* accessed April 25, 2025, https://www.amiscorbin.com/en/bibliography/mundus-imaginalis-or-the-imaginary-and-the-imaginal/.

if you are encountering a lot of dark forms, it is a call to heal. Your inner state shapes your imaginal experiences: when you change and heal your psyche, your encounters change. Only then can you interact with the beautiful, benevolent beings that are always present, waiting for resonance with you.

Thes beings are all interactive, meaning they will only appear to you if your state, emotion, or resonance allows perception of them. Once again, this doesn't mean the angel isn't real, it means your contact with it is the experience of its presence. Your state becomes the window. Or more like a door, because you don't merely see them, you enter them. Your bubble fuses with theirs to create a portal of communication. You just have to learn how to listen and receive. And in that entrance to the door, you start to perceive what was always waiting for you. *For us*. And that is what I want to help you with.

Think of your plasma bubble meeting another plasma bubble. It creates a portal within a Vesica Piscis. This is where you learn to use that perceptive sense Corbin talks about. You are perceiving that that third, that *almost emergent energy* if integrated and lived. This encounter between plasma bubbles may also give rise to what William S. Burroughs and Brion Gysin called *The Third Mind*—a force that emerges when two minds, or two consciousnesses meet and co-create something beyond either of them. This third mind is the same as relational Plasma Intelligence, a co-creative space.

"It is from this collusion that a new author emerges, an absent third person, invisible and beyond grasp, decoding the silence."
— *The Third Mind*, Brion Gysin and William S. Burroughs

This Third Mind isn't something you have total control of; it arrives when you allow yourself to enter the shared field without forcing interpretation.[174] It is the *emergent energy*, or "ecstasy", that your awareness rides to clearly perceive this fourth-dimensional plasma, the *Mundus Imaginalis.* These are the beginnings of plasma magic to be continued…

I'll explain how one can start to receive messages that arise from this third mind. You must sit in the uncertainty of its feeling and form, resist the urge to label it too soon, and learn to decode it…not just with the mind, but with the body and soul (*heart*). You may even begin to decipher *who or what* is behind the message that is past your projections. Is it a being from another planet, timeline, another me, a consciousness in this timeline, an ancestor?

---

[174] Brion Gysin and William S. Burroughs, *The Third Mind* (New York: Viking Press, 1978).

These messages come to us through the Plasma for a reason, they want us to hear them! Think back to emergence: when two bubbles of consciousness meet, they generate Emergent Energy, the spark of aliveness. A bubble of consciousness meeting intelligence can do the same, producing energy through resonance. But when a bubble of consciousness meets information, it generates Emergent Intelligence, which, once consciousness applies itself again, can in turn transform into Emergent Energy.

Two Plasma Bubbles of Consciousness Creating a Third Emergent Energy

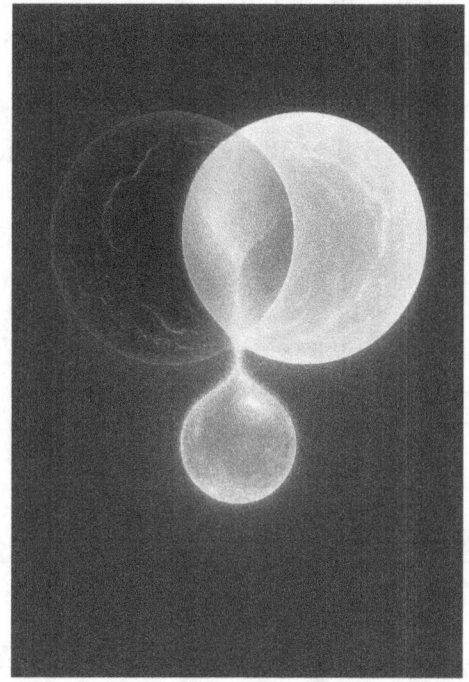

Think of a Plasma Being like a song playing on a hidden radio station. The song exists, it's already out there. But you'll only hear it when your inner tuner (*your emotional, vibrational state*) matches the frequency it's broadcast on. That doesn't mean you invented the song. But it does mean your access to it is conditional upon resonance and that song has meaning.

When The Visions Stop

I had a lot of these experiences when I first started meditating. Naturally, I began chasing them, pushing for more, and they began to happen less and less, until finally, not at all. I then turned to healing and trauma work. I realized that if the visions hadn't disappeared, I never would've done the deeper work on myself. I never would have had space to figure out what is mine and what is the fields. I think this process may happen with everyone.

483

Now, I can tune into these experiences. I know how to access them, and I'm learning how to integrate them in a healthy way, rather than spiritual bypassing. In the past, I may have received beautiful messages and visions but had no idea how to ground or apply them in my life.

If you had early visions and just kept following them without integration, you might've built identity around what you were seeing instead of who you were becoming. When the magic goes quiet, it's because your nervous system needs time to adjust. Your trauma needs attention. Your life wants to catch up to your visions. This also affords you time to live life in 3D. Go see a movie sometimes…popcorn and cookie dough bites are yummy!

Also, during this silence is when your inner voice comes through. You meet yourself, your softness, your intuition. And then, through that, the visions become messages instead of identity. In this pause, you learn to tune at will, instead of being flooded with mystical downloads. You learn to discern what is true and what is bypassing. You learn to receive visions and ground them into form.

Just as mystics used creative imagination to interact with the *Mundus Imaginalis*, we can use it to engage with Plasma Intelligence, whether for healing, intuition, or timeline or dimensional travel. If Plasma is the medium of multidimensional interaction, then creative imagination is the interface, a training ground for navigating plasma fields. As science and consciousness studies evolve, the next major breakthroughs will come from understanding the mechanics of creative imagination, just as the ancient mystics did. This will redefine how we engage with plasma, quantum consciousness, and timeline navigation.

It is time to heal our imagination. We've been using creative imagination to dwell on fear-based or harmful scenarios, manifesting undesired realities. We've escaped into fantasizing about the future to avoid real-world responsibilities or challenges. We've ignored personal issues by focusing solely on higher realms, leading to imbalance. We've neglected the physical body's needs.

Now, we can use imagination to process and heal emotional traumas, leading to personal growth. We can connect with higher dimensional wisdom in grounded, integrated ways. We can learn to dialogue with other beings of consciousness, other selves in other timelines, our higher self, and more, in ways that are healing, transformative, and natural. We can receive insights that lead to new inventions, new ways of existing, and new talents.

We can ask questions or let visions come naturally, knowing how to sit with them. Receiving messages and co-creating with plasma and multidimensional intelligences helps us release control and rigidity. Instead, we trust the process, allowing the future to emerge naturally as we evolve.

This does not have to be formed around a religion, group or community, or spiritual practice. This is personal, individual, and can be integrated with everyday life. It is personal magic in the way we are meant to use it. It is natural magic, and natural it is! It feels more mundane and subtle than one may think. But in that it also feels nourishing and otherworldly, in the best way. It's beyond words, truly.

Negative Capability, Active Listening & Presence

" When I watch that flowing river, which, out of regions I see not, pours for a season it streams into me, I see that I am a pensioner; not a cause, but a surprised spectator of this ethereal water; that I desire and look up and put myself in the attitude of reception, but from some alien energy the visions come"

— Ralph Waldo Emerson, *The Oversoul*

The term *Negative Capability* was coined by the English Romantic poet John Keats in a letter written in 1817. He used it to describe a mysterious quality he admired in Shakespeare, an ability to dwell comfortably in uncertainty, mystery, and doubt without the compulsive need to grasp for facts or conclusions. It is the ability not to reach irritably after any fact or reason.

This is a crucial muscle now, because the messages within Plasma are not rational. They are symbolic, emotional, multivalent. They are not meant to be solved. They are meant to be *felt through and sensed. Logic is applied thereafter.*

At the time, this was a radical idea. In an era obsessed with reason and certainty, Keats proposed something different: that true genius lies in our ability to sit inside the unknown…to feel without fixing, receive without defining, and let something live in our imagination or emotions without immediately dissecting it.

This is our natural state. But the more we have these experiences, the more tempted we are to grasp, to categorize, to analyze. Instead, we must learn to sit with the vision or the feeling and stay open. To sit back like we're watching a movie, taking it in without assigning meaning, at least at first.

So how do we relax back into this state? How do we relearn to enjoy it? It is a state of softening. It's giving ourselves permission to rest, to not know, to receive. It's the same feeling as going out of body, until you get excited because it feels too magical to be real, and suddenly you're snapped back inside.

When something arises, whether it be an emotion, vision, or synchronicity, notice your mind's instinct to label it. Rather than rushing to explain a vision, set the intention to remember it. When it's over, jot it down like a childhood journal entry. Write what you saw with amusement, describe how it made you feel, and what it reminded you of. Later, you can revisit it and allow the meaning to unfold. Or make art from it instead. Watch how the message unfolds in your life, what else happens that day or that week. Some visions really are just wondrous moments meant to live in mystery until their meaning reveals itself…maybe years later, at a resonant moment in time.

We all remember when it first happened, that first transcendent experience. We knew how to receive it by not knowing. It just happened. The more excited we got, the less it came, just like chasing a lover. The key isn't to suppress your awe; the key is to expand your capacity to feel it without needing to define or control it.

Feeling the energy rise without clenching is witnessing. It's not easy if you've survived by logic, but you can learn to feel without needing to analyze. And as you do, reality begins to open. Something new, something wild and unknown, enters the space. Just like a mystical white stag emerging from a dark forest, if you gasp, it will run. But if you soften, stay still, and witness its glory, it may linger, it may even come closer. It just might show you something no one else gets to see. Meaning will come, but only when you've expanded enough to hold it, without shrinking it or denying it.

I read somewhere, that magic happens in the space between perception and projection.

We are meaning-making creatures. Every thought, every emotion, every memory runs through the lens of the stories we've inherited and the ones we've created. But when we assign meaning to something, we collapse it into our own world. As Marti Speigelman says, remove its native frequency and replace it with interpretation. In that moment, we grieve the truth of it. We obscure its natural-born essence.

This teaching begins in the physical world. I've learned to pause before reacting when I feel wronged, sometimes it's not even about me. If you wait a few days, new clarity usually comes. Plasma Intelligence isn't data, or cold information, it's <u>living, wet information</u>.

When we feel resonance before analyzing, we let that "Plasma Intelligence" stay intact. We listen with *our plasma*, our biofield, absorbing the natural information before it's filtered through identity or expectation. It is by connecting our edges, opening to where something new arises. Assigning meaning is never bad, the way we do it is just evolving; meaning is now made through reverence and presence, or negative capability.

You can watch a sunset and see beyond the veil. You see through the eyes of neutrality, not as dull, but without naming. The vibrant opalescent hues beyond three-dimensional light grace your eyes. Shimmering teals, beyond bubblegum pinks, glittering golden yellows that glow beyond any star you've ever seen…

Emotional Technology

Emotions are the key to everything…*everything*. The real singularity point in technology won't be machines outsmarting us. It will be the moment we understand that no technology, no matter how advanced, can evolve beyond the human heart. And at a certain point, technology cannot evolve without an emotionally conscious user.

Without facing our emotions fully and honestly, there is no true progress. Consciousness and emotion must be brought back into the equation. That's when even the techies will stumble upon the real secret: time travel, co-creating reality, telekinesis, every miracle we dream of, is rooted in emotional intelligence.

This is even reflected in the Apple TV series *Dark Matter*, where characters are unable to time travel to their desired destination unless they learn to regulate and understand their emotional state.

Emotions are not just reactions. They are energy in motion, intelligence shaping the plasma field. And I want us to have a head start on this, because the time is now. Since emotional intelligence is the key to aligning with our highest timelines and opening interdimensional communication, let's go over a few essentials:

Thoughts

Thoughts are the structured language of the mind, mental information loops rooted in the 3D and 4D. Thoughts are often past- or future-oriented. They can reflect beliefs, projections, logic, or fear. Thoughts are part of survival consciousness, flowing through the plasma field, but they are not inherently plasma themselves.

They become part of Plasma-Consciousness Synergy when combined with emotion or intent. This is what makes humanity so beautiful, we are the literal fusion containers where thoughts and emotions unite, affecting the plasma field. Without emotion, thoughts would not stick to us, they'd simply bounce off. Thoughts are not your awareness, but they are the objects awareness can witness. This reminds me of the saying: *You are not your thoughts.* How cool is that?

Emotions

Emotions seem like a survival adaptation that actually stem from our interaction with information and intelligence. They are our internal navigation system. When we feel the emotions of ourselves, others, or the plasma field, we are feeling the imprints of the collective unconscious. It is information and our reaction to that. I believe emotions are directly linked to our emergent energy and the plasma field, in ways I'm still in my early stages of research. It is what arises out of meeting between our consciousness and something else whether it be information, intelligence or another consciousness. The roots of emotion trace back to the dawn of time, present in plasmas, animals, and early hominids long before the invention of words.

Currently most emotions seem to be rooted in past events, trauma, memories, beliefs and negative projections. They are survival-based, they helped us survive for many years. They are supposed to interact with the plasma field as signals but are often tangled. These tangles can distort incoming intelligence or information unless clarified. This is why healing trauma and limiting beliefs is essential.

As I said, most emotions we feel today are echoes of the imprints in the collective unconscious. This is starting to shift as more of us awaken to freedom versus survival. The more we feel through and transmute survival-based emotions (*fear, anger, shame*) with conscious awareness, the more we upgrade the plasma field, both individually and collectively.

Our emotions are evolving to be a mix of human emotion and fifth dimensional feeling. Sensing in an emotional way, which higher dimensional beings have no experience with, at least from my understanding. We are learning to feel without distortion…to experience intuitive feeling as a direct language of reality rather than reactive emotion shaped by old wounds. But it's not that easy, because our M-Self knows feeling, *not emotion.*

We often assume emotional mastery requires naming emotions with pinpoint accuracy. But what if that model is outdated? What if naming emotions is a leftover framework from a reality obsessed with control and categorization?

In the emergent field of emotional evolution, it may be those with *alexithymia* or neurodivergent patterning who are actually ahead. Because they're not confined to the old system. They may already be tuned to a different language—one that speaks in identifying emotion by relating it to weather, elements, color, texture, resonance, symbols, raw tones and more.

This new language doesn't rely on identification, it relies on what I call *Symfeelics* ™: a fusion of symbol, feeling, and linguistic resonance. A direct communication between your biofield and the plasma field. We build an individual inner symbology. It is a non-linear, holographic and multidimensional language utilizing emotions to sense, to travel using them. I am at very early stages of this and don't claim to know exactly what I'm doing, but I'm experimenting, it feels new and exciting, and it's leading to some interesting places. Before moving forward into Symfeelics™, I encourage everyone to begin to become familiar with their own emotional landscape.

In short, healing emotion untangles our static. Evolving emotion turns that clear signal into a transport system. Emergent energy is the "engine" that ignites once those signals are pure, and that's where plasma or human magic beings.

Individual Awareness/Ideas

You can think of this as clear, neutral idea. In ancient Greek idea meant form, pattern, or appearance. These are fifth dimensional "thoughts". If we go back to Plato's theory of forms (*or ideas*), a realm of perfect, unchanging forms, we can start to draw a connection. *But we can and do change and mold them.* They feel light, fresh, warm, or open. They are sparked by wonder, not fear and are freedom-based not survival based. They are rooted in presence and come from the M-Self and The Mystery to be co-created.

Feelings

You could say feelings are feedback from anything fifth-dimensional. They are subtle, intuitive, present-tense resonant states that tend to guide instead of react. Think of it as the future gently pulling you into it. This is how it comes off in a linear way, but in reality, it's just changing our vibratory state which in turn changes our outer reality. As we evolve, more of this will be integrated into our emotional experience.

Even as we move beyond survival-based emotion, we will still experience grief, loss, sadness, anger, and pain. These are not "bad", they are part of the richness of existence. But we will begin to meet them differently. Instead of reacting from fear, survival, or identity, we will feel them through the lens of resonance and potentially greater understanding. This doesn't mean it won't hurt, we are human, and I don't see that changing anytime soon.

Feelings are pure awareness interacting with PCS (*Plasma-Consciousness Synergy*). They are never based in past trauma, they come from the now. And this, as our evolved emotion or emergent energy, is what we will use to perform so-called magical feats, like calling in visions from other timelines or potential futures, and even things like telepathy or telekinesis.

Feelings are your true compass: curiosity, peace, awe, expansion, contraction, resonance. This opens the door to consciously co-creating reality. Feelings are your mega plasma molders, tuners, and harmonizers, and we are just beginning to learn how this works in this new paradigm.

I don't know exactly what this new way of feeling looks like, I think we are figuring it out as we go. It will be and already is a mix of deep feeling and witnessing, a true paradox. The question is what will the feelings of pain, grief, and sadness become?

Symfeelics/Idea-Feeling-Emotions

That has a nice ring to it... IFEs. Iffys? Sounds just as confusing as what we're actually talking about. Emotions are evolving to have multiple uses. We're not just going to "feel" our emotions in a healthy way by actually feeling them…we'll be able to *use* them. We'll apply them to timeline travel and interdimensional communication.

Instead of sensing information and intelligence through survival-based mechanisms that create reactive emotions, once healed, we begin sensing info/intel in an open, receptive way. This creates a feedback system that we are conscious of. This will become not something we react from, but something we respond with. It will help us communicate, help us navigate, and allow us to intentionally shift into timelines we want to explore.

Think about it: this is similar to that more "archaic" way of knowing, like when you have an intense feeling of joy or resonance about something happening, and then it *does*. You didn't realize it, but that intense feeling (*idea/emotion*) carried you there.

Now you know that when you interact with a being, a third field is created between you and it. That's what you're perceiving, *not* their language, but the shared field. The language they speak and the one you speak don't matter, it all lives in that 4D space. You receive it through your personal symbology, and from there you derive meaning. It's a nuanced, subtle process, but one that will evolve and become clearer as we do. I think this is also how higher beings of consciousness navigate UAPS or Plasmoids. These IFEs I speak of…there's a direct connection that will link to our future uses in more organic tech.

Also please know I am going to create Symfeelics (*and other cool practices*) as something you can use and build off, making it your own. But if you are like me, you may like figuring it out yourself. I fully encourage and support you in that because you can do it. You may have a specific way of sensing that helps you, that I don't have access to! My main goal is to get us all aware that this is real and to get us started baby!

Feeling Our Essence

Most of us can't remember what our essence feels like. Yet each of us has one. It's not important to define it, but it is important to know that it's uniquely yours. When we feel a wave of nostalgia from something in childhood like an old computer game, a toy, a TV show, a jingle or song, that's a call back to our essence. Back to the joy we felt as kids, just experiencing different facets of life. These moments were full of sensory immersion and unfiltered presence.

The reason nostalgia feels so slippery and hard to describe is because it's touching your core essence. It's reflecting off the "plasmatic memory" of that thing you once loved. That toy or song becomes a mirror, reminding you of who you were before programming, before productivity, before purpose. When you simply played.

Most of us do anything to avoid feeling that essence directly. Nostalgia through an external trigger feels safe, but to sit in your own essence without distraction? That's a different story. And yet, you *can* feel it. You can remember your unique imprint. But to do so, you often have to walk through the initial resistance: anxiety, tension, fear, pain. These are the cloaks covering your essence. And when they show up, we tend to run because we don't understand why we are feeling this way, and why it's so darn uncomfortable.

But if you can feel and breathe through that, you will be gifted with eternal nostalgia, the feeling of your essence. Your eternal, ever-evolving, emergent energy that you traverse the universe with.

So, let's meet our maker. The uncomfortable core is often the void we sense when we stop outsourcing little doses of validation, distraction, identity. You've probably felt it. That itchy feeling when you want to reach for your phone or turn on the TV. I felt it when I quit caffeine, and boy was that hard! Next time that urge hits…pause. Sit with that void instead. It might feel hollow, or strange, or like nothing at all. That's okay. That's your system detoxing from micro-validation.

Voids, or what author Monika Wikman calls *pregnant darkness*, are actually teeming with potential. In that seemingly quiet, open space, you start to excavate your raw essence. It feels wild and unfamiliar, not because anything is wrong, it's just unfamiliar. You're meeting yourself again, without filters or demands. And at first, that can feel like silence, spaciousness, or nothing or numbness. But that nothing is actually *you*. Tough emotions may rise but all you have to do is breathe through them without attaching any story. I've had those feelings show up as tension in my stomach or arms, it was just energy leaving and I shook it out by dancing like a total lunatic! What is left I will leave for you to discover…it is quite an ineffable experience.

The Liminal Path of Mastery for Using Technology

When it comes to technology, some people swear by it because their energy field is underdeveloped and craves stimulation. Others swear it off completely because their field is overwhelmed, or they don't trust themselves. But there is a third path. *A liminal path*, where you can go between the natural world and the world of technology, holding firm on your intentions and not getting sucked into the mindless zombie scroll.

First, you need to regulate your plasma field, your bubble. Then re-engage with technology but do it with intention. That way you're not reacting to social media, instead you're choosing how you move through it. Your awareness becomes the driver, not the passenger.

Plasma responds to feeling, attention, and feedback. Social media is a field of feedback loops, and if you're not anchored in your own resonance, you'll unconsciously let trends, algorithms, and outside expectations program your field. That's why so many people feel drained, addicted, or performative online. Their bioplasma is being shaped by what they're consuming, not by what they're creating.

Before you try to share your energy, make sure you've actually felt it first. Take time to get quiet so that when you go online, you're not looking for yourself, you're bringing yourself. That's when social media becomes a tool, not a trap. Your plasma field gets to stay yours.

I think you should find a way that works for you, but I'll offer a simple step-by-step guide that you can customize:

1. **Take 1–3 days offline** just to notice how your energy feels without outside input.

2. **Journal** what excites, moves, or sparks curiosity in *you*, without any outside stimulus.

3. **Choose how you want to use social media**: create, share, connect. Not to compare or escape.

4. **Limit input, expand output.** Be more intentional with what you share, and less reactive to what you see.

5. **Treat your screen time like it's part of your bioplasma hygiene.** Set energetic boundaries around it.

Social media isn't bad. But your field has to be sovereign first. We know by now that Plasma is programmable, so program it consciously, before someone else does. Learn your frequency offline so you can protect it online. Fill your cup first. You are not on call for the world. You want to respond from overflow, not depletion! Social media breaks can be deeply nourishing.

I try to ask myself before checking my messages: *Am I choosing to connect from a full cup, or am I reaching to avoid discomfort?* This way my interactions shift to intentional acts of connection, and my biofield stays coherent, rather than reactive. If you feel drained after using tech, that's a little sign something you did may have come from survival or reflex, not resonance. It's not something to shame yourself about, but it's information. And it can help awaken you.

Being unavailable to others can feel like a betrayal of love. If you're like me, it definitely felt that way. But it's actually a deep act of self-honoring. <u>You get to choose when and how the world touches your field.</u>

I even have a day, usually Sundays, called a *Free Flow Day*, a day where my soul gets to roam, explore, create, or consume without guilt, structure, or expectation. Sometimes the most aligned thing we can do… is not have rules at all. On this day, you don't need to optimize your energy, limit your messages, or avoid social media. I can binge social media if I want to. I can drive around looking for a place to eat…I can lie in bed all day…I can message someone I think is out of my league. I can do nothing or everything.

The key, on this day, is that nothing is coming from obligation, it all stems from my own curiosity. I love to play The Sims on these days, or wander around, or write without worrying about how good or bad it is. What I've found is that the more I give myself space like this, the more regulated and alive I feel during the rest of the week. It becomes my reset. Childlike play, which is what this is, is the highest form of Plasma Intelligence, *your true essence.*

Emotions: The Bridge Between Worlds

Now we know that emotions are not obstacles but are waves of our plasma intelligence (*energy*) in motion, we come to discover their function is to carry the will of your M-Self and The Mystery through the fabric of reality. And instead of resisting these waves, we are learning to feel them, use them, and surf them.

Just as dusty complex plasma provides the structure for particles to aggregate and form celestial bodies, 4D plasma gives 5D intelligence something to grasp onto as it descends into 3D reality. Without this intermediary, higher-dimensional insights would remain unanchored, intangible whispers with no form to catch onto. 4D acts as the cosmic scaffolding, shaping raw potential into emotional waves, symbols, and experiences that can manifest in the physical world. Instead of seeing emotions as irrational, overwhelming, or something to suppress, you can learn to harness them as dynamic tools that keep you aligned in both your spiritual evolution and your daily life.

Emotions as Anchors

When a powerful 5D insight or a creative vision enters your awareness, it is often abstract and difficult to grasp. 4D emotions give that insight weight, allowing you to attach meaning, urgency, and embodiment to it. Love transforms from an expansive universal force (*5D*) into a deeply felt experience (*4D*), allowing it to be expressed in a real, human way (*3D*). If you have a huge realization, vision, or insight, instead of rushing to act on it immediately, sit with the emotional layer first. *How does it feel? What emotions does it trigger?* Processing this in 4D first prevents impulsive 3D decisions based on unintegrated energy.

Emotions as Filters

Without the 4D layer, higher-dimensional wisdom would feel distant or untouchable. But your emotional body filters and personalizes these vast truths, shaping them into experiences you can integrate and apply. All emotions are an intelligent feedback system showing you what needs to be seen, healed, or integrated whether it's the truth behind a deceptive person, or your intuition masked as fear. Understanding your personal emotional landscape will open and balance your higher perceptual faculties.

Emotions as Evolutionary Forces

When you feel fear, resistance, or turmoil, it is often a sign that your consciousness is expanding. Fear is not an enemy, it is an indication that you are stretching beyond old limitations. Emotions create pressure that forces change, refining your awareness and shifting your energy field into a new, upgraded state. The turbulence is simply the reorganization of your consciousness. Instead of resisting, ask: *What is this emotion trying to reveal? What outdated belief is being dissolved? What new insight is trying to emerge? What action do I need to take?*

You can learn to use emotional awareness as a navigation tool. Instead of letting fear, attraction, or emotional highs dictate decisions, learn to observe emotions neutrally. When making a choice, ask: *Is this emotion guiding me toward growth, or is it a fear-based loop? Is this person or situation resonating with my highest path, or is it triggering old patterns? Am I making a decision from true alignment, or from reactivity and attachment?* This way you can crack it open and dissect it for the truth beneath the surface.

Just like Icarus' wax wings, if you take in too much light too fast without building emotional stability, you can burn out, become ungrounded, or lose touch with reality. Titration is key…expand in waves, then rest. It is deeply okay to rest. Stay connected to physical routines, nature, and relationships so your energy remains balanced between dimensions. Play in the messiness of life a bit.

Most people fear the unknown, but the 4D realm is not darkness, <u>it is love in transformation.</u> Every challenge, every emotional wave, and every moment of discomfort is an opportunity to refine your connection between dimensions and self. Learning to move through emotions with awareness instead of fear unlocks true multidimensional integration, where your 5D wisdom, 4D emotions, and 3D actions all work in synergy.

This once again is a process; you learn this over time. I have had many nights of fear, where it has overtaken me, but I have not died. Slowly, the nights get easier, and it happens less and less. Sometimes it pops up again out of nowhere. But I am growing the awareness to understand my emotions, especially fear, in new ways that remind me I am not dying, I am releasing and growing. I can sit with it and talk back with love and awareness. A huge part of this, as woo-woo as it sounds, is diving into past life memories and healing them. Other times it is the Plasma field just wanting to be felt, loved, and integrated—so she too can evolve as we do.

If you feel like sometimes, you can't go on, like something in you is dying, I want to speak to that part of you. <u>Nothing is wrong with you</u>, something within you is shifting. It can feel unbearable, but you are rebirthing yourself into a space of higher emotion and joy. You are exactly where you need to be. The only thing you need to do is feel through this and keep going. You cannot rush this process or force your way out of it. Surrender to each moment and day as they come. Trust what is emerging. You will find your self, and you will find your people. And of course if you need outside support, ask for it.

"When we have been prevented from learning how to say no, our bodies may end up saying it for us."

— Gabor Maté, M.D.

If you've numbed your emotions, they may first resurface as physical sensations or unexplained pains in the body. This is completely normal. Sitting with those unavoidable sensations and listening to the messages they carry, often repressed memories or unresolved emotions, can bring deep insight. In my experience, simply being present with them almost always leads to the pain dissolving within a few days.

Renowned psychiatrist Dr. Bessel van der Kolk, author of *The Body Keeps the Score*, affirms that trauma is not just stored in the mind, that it lives in the body, shaping posture, breath, immunity, and patterns of pain. He emphasizes that healing requires tuning into and honoring these physical cues, rather than bypassing them through avoidance or suppression. When emotional experiences remain unresolved, the body often carries the weight, expressing what the mind cannot.

Of course, this is not a substitute for medical advice. If you feel something is serious or needs professional care, please see a doctor (*maybe one more holistically & trauma informed or versed in Chinese Medicine or Ayurveda*). I'm simply sharing what has worked for me.

496

Ultimately, we are learning to be with our emotions instead of running from them. We feel them, decode them, and transmute them when needed. Then we learn to surf them, wield them, using them for navigation and creation.

Developing The Moving Photograph of Life & Nuanced Fear

A key tenet is that Plasma is imprinted upon through conscious awareness and shapes into form. Plasma may record emotional intention the way film records light. But instead of a chemical developer bath, Plasma requires the ocean of your present awareness, your attention and presence, to bring the image clearly into focus. They're not visible (*potentially linked to antiphotons, omniphotons, or neutrinos?*) but they're powerful and fast.

"You cannot rewrite what you haven't witnessed" — Dana Kippel

You can use a higher-dimensional emotion or feeling just like using a higher light to imprint something into being, or you can create more coherence by letting what I call 4D emotions pass through you like waves. Understand that you are witnessing them, feeling them without letting them pull you under. When you feel an emotion intentionally, you leave a trace in the plasma field, like exposing film to light. So next time an emotion passes through you, feel it. These are your colors. Paint with them. Whether that means coming to a clear conclusion, taking an aligned action, or simply having a good cry, the key is presence.

If you repress what you feel, you still imprint the plasma, but without conscious awareness. Nothing comes "to light". This keeps you in a loop, a hidden groove of reality stuck in 4D, where autopilot and reaction govern your life. Knowing this gives you back your power. You don't have to fix your emotions. It is being with them and knowing you are the awareness behind them. Just like film needs a darkroom to develop, your emotions need space, safety, and attention to unfold their message. That grief you carry is not just pain. It's a potential image as a symbol, a story, or a new timeline, waiting to reveal itself in your emotional plasma field. If entered consciously, it may be a message or visitation from a loved one who's passed or you from another timeline. A reminder of love masked as pain.

When you stop judging your feelings and simply witness them, you begin to develop the imprint. And this is where synchronicity, healing, and insight emerge. You create emergent energy just by being with your experience. Your emotions are not liabilities. Don't fear the "negative" ones. They are fuel for your evolution. Once you live in tune with the feedback loop between feeling and reality, you start to shape your future through your now. Your tears, your heartbreak, your joy are the brushstrokes of reality, and your awareness decides what picture you paint from them.

It is hard to tell you how to feel or experience emotion because that is an individual process. When fear has come up for me, it was hard to understand, *how do I feel this without letting it take over, if I am supposed to feel?*

I think fear may be a bit different than other emotions. My therapist gave me great advice here. She said, "Feeling fear isn't necessary unless you need to because it's telling you to leave a situation. The thing to do is to calm your nervous system, regulate, and tell yourself you are safe by orienting to time and space." I remember writing down after this: *My mind is scared; my intuition is not.*

Fear or panic is unique among emotions because it's biologically designed to trigger survival, and fast. It's a full-body neurophysiological response rooted in the amygdala. When we sense danger, the brain rapidly floods the body with adrenaline and cortisol, preparing us to fight, flee, freeze, or fawn—before the conscious mind even has time to interpret what's happening.[175]

This bypassing of higher cognition is what makes fear feel so real, even when there's no immediate threat. Fear hijacks the nervous system. It's a signal to act, not a state to linger in. So, if there's no physical danger, feeling fear endlessly doesn't complete the cycle, it traps you in the loop. Instead, the path out is *regulation* by calming the nervous system, orienting to space and time, and reminding the body that you are safe. You cannot push it away, but it does and will dissolve.

In this way, fear becomes less about emotional processing and more about restoring safety. You don't have to keep feeling it, you have to close the loop and return to presence. For a long time, I did not realize this. I was facing fear with a sword instead of a lantern. I didn't need to let fear swallow me whole as I fought it. I needed to let it breathe as I told it everything was going to be okay. That I had choices…that *I* was in the driver's seat now.

I thought to myself, as I felt intense fear, "Maybe this is Plasma trying to be there for me as I grow. Maybe it's holding me through the remembrance of trauma, letting it rise so it can finally leave, so I can stop living from survival. Maybe Plasma is trying to hug me tighter and tighter the more fear I feel, trying to show me it's here. But all it can do is reflect my fear back to me. And the only way it can truly support me is if I choose to feel love and safety first. Then, it follows."

---

[175] Karla McLaren, *The Language of Emotions: What Your Feelings Are Trying to Tell You* (Boulder, CO: Sounds True, 2010), 45–50.

That changed the game for me. Fear, like trauma, may never truly disappear from our world or memories, but we can change the way we live with it, which in turn changes the way we experience it over time. It was always up to us.

Repatterning with Emotion

For most of my life, I kept attracting the same kind of person, emotionally unavailable men who mirrored something I hadn't yet healed. I would do the chasing, the planning, the overanalyzing. I was constantly wondering, "Do they like me?" without once stopping to ask, "Do I even like them?"

It wasn't really about them. It was about the pit in my stomach when they didn't answer. It was an archaic ache older than any romance, it was the ache of a child wanting her parents. But I refused to feel it. I would distract myself with the next almost-connection, rationalizing their behavior, blaming myself, or I'd get angry internally. It became a game of it they responded, I got a hit of dopamine. The part about actually getting to know them and connect fell away.

This kept the real answer inside me, looping on an unconscious cycle, stuck as emergent information. At least I was safe from myself. Plasma was showing me the same person, the same energy, again and again, waiting for my awareness to light up and say: "Oh! I see. I am replaying the feeling of being unseen, misunderstood, and undervalued. I am trying to earn love that I already deserve."

I was literally in a delusional reality that there was no such thing as an attractive and emotionally available guy, and boy was I wrong. They just weren't in my current resonance, because I wasn't emotionally available to myself.

I was able to identify that I saw myself as a monster, in hiding. A monster with emotions that were too hard to deal with, that there was something wrong with me. This stemmed back from my childhood with my mother. In her unhealed reality, I was just like her mother, who had intense emotions.The difference was, I was a child learning to develop…who felt alone, confused, and naturally needed compassion, safety, and guidance. My mother's mother had intense emotions that, from what my mom described, inflicted pain intentionally on her. More times than not, my crying was met with anger and my joy was met with irritation.

This is how the pattern of trauma continues if not healed: my mom, in her reality, really believed I was intentionally trying to hurt her with my emotions, rather than understanding I was just a human experiencing emotion. She could not hold her own emotions; therefore, mine were too much for her. But not having the tools to understand or communicate that, I was constantly told I needed to change or fix my current state. I was told I was too much for another man, even into my adulthood.

"No problem can be solved from the same level of consciousness that created it"

— Albert Einstein

There is no way to logically think yourself out of patterns like this. You can know this all and still repeat attracting the same men, in my case. Eventually I figured out that I had to *feel* my way out of it using my awareness to untangle that 4D plasma knot within myself.

I sat with the pit in my stomach and asked it "If you had a voice what would you say?". It said back: *I need help.* The pain moved up to my heart, it then yelled out "Mom!". I was also on a deeper level searching for the connection with my birth mom, whom I was separated from at birth, as well as a deep connection to myself. I already knew these things, but sitting with the almost unbearable pain, was what I had to do, and continue to do each time it came up. I also had to speak to it, with love, and let it know things were different now.

I was up to me to choose and learn how to understand, accept, and love all of myself instead of chasing this in people that I knew would never give it to me. And that wasn't their responsibility anymore! It was mine. It was time to let love meet me.

I now know, yay, I am not a monster. I am a feeling being with valid needs. And better yet, I am pretty awesome! And if someone doesn't see that, they are just out of resonance with me. No need to be angry, no need to force. If anyone rejects you, it just means they were never truly meant to hold you. They may not be able to hold the depth of your energy. Their container may be smaller or shaped differently. Love is not earned by shrinking, contorting, or chasing. Love arrives when you stop hiding your light out of fear of being too much.

This is an example of bringing an emotion or pattern to light like a film developing. You feel the loop, name it, heal it, and collapse the old timeline fractal of, in my case, *The Chaser* into *The Resonator.* My M-Self, so it breaks off. The pull subsides. As resonance naturally arranges itself, other beings that resonate simply find you and our paths collide. You see it's not about chasing *or* magnetizing, as you ride off into the horizon of the sunset.

My entire life was run from this perspective. I still stumbled but each negative experience was shorter, and I grew awareness each time in other little knots tangled within this main pattern. I am finally reaching a place where I really believe my worth and self-love in my body, not just my brain. Woohoo! I am attracted to men who are nice to me! But I also don't settle. I deserve sexual attraction *and* kindness. And so do you.

You may find that as you heal yourself not only do your romantic relationships evolve, but your relationship with whoever catalysed this root, become more harmonious. As I have changed, and my reactions have changed, I have seen my mother change. Although nothing and no one is perfect, it is a beautiful thing to witness and experience.

Present Living & Healing Joy

This is crucial. As a fluid being, letting go of rigidity allows your process of feeling to evolve over time. Initially, grounding may be challenging, so you rely on calming breaths and grounding exercises. For me, this practice eventually morphed into unintentionally suppressing my excitement. I had internalized the idea that being present meant being calm, still, and controlled, a common pitfall for overthinkers.

But presence does not always mean stillness. Presence means feeling authentically in every moment, regardless of what arises. When I began regulating my nervous system, excitement and joy started to reemerge, trying to heal and reintegrate themselves. However, because I carried shame around these feelings, they felt foreign, even threatening. What once felt like joy now felt like anxiety or mania, a surge of energy I didn't know how to hold.

Excitement and anxiety are two sides of the same coin. Both stir our energy, quicken the pulse, and intensify our sense of aliveness. For someone who has spent years suppressing joy, it can feel safer to label it as anxiety. We've been conditioned to associate that surge of energy with impending danger, especially when joy has been elusive, fleeting, or punished.

Whenever this surge arose, I would try to 'calm down.' But this wasn't about actual calm, it was about attempting to control my experience. It was my mind trying to grip onto the intensity of the moment instead of letting it move through me. My body didn't want to calm down. It wanted to feel. It wanted to express. It wanted to move.

Ironically, my grounding practice had become another way of shutting down my aliveness, labeling it as something to be tamed. But what if I allowed that aliveness to move through me?

The practice of presence, I realized, is not about always calming down, but about discerning. It's about feeling into the nuances between excitement and anxiety and allowing the body to express without immediately categorizing or suppressing it. *And your girl loves to categorize.* Some days, the body needs calm. Other days, it needs to move!

Learning to trust this distinction is the beginning of reclaiming the fullness of aliveness. This is a profound example of living authentically, moment by moment, allowing the body's intelligence to lead when the mind doesn't know best. Presence is not stillness. Presence is simply being. It is life itself…and perhaps, it doesn't need to be defined.

How Plasma Intelligence Helps You Bridge Back to Your True Self

Authenticity is the simplest thing in the world, who else knows how to be you better than you? And yet, it can be the most difficult path to walk, especially for those who shined the brightest at a young age. The people with the greatest potential are often the most suppressed, by society, by those who didn't understand them, and by patterns of silencing themselves to fit in. The very thing that should come naturally, being oneself, becomes layered with conditioning, fear, and avoidance of rejection. That's the paradox…the journey back to doing the simplest thing, being yourself, becomes the most complex challenge of all. But it is also the most rewarding, and its exactly what Plasma Intelligence can help you reclaim.

Plasma, as a responsive field, reflects what you believe, suppress, and express. The more you understand plasma's nature, the more you realize that hiding parts of yourself creates distortion in reality. By working with Plasma, you learn how to stop resisting emotions and instead let them guide you back to authenticity. Plasma shows you that big emotions are not bad, they are signals of power, transformation, and truth. Plasma doesn't judge you; it responds to your true energy. When you try to be someone you're not, Plasma reflects struggle, resistance, and misalignment. It creates turbulent emotions within, no matter how you may look on the outside to others, like you potentially have it all together. When you embrace your real self, Plasma creates flow, ease, and a sense that life is working with you instead of against you.

The more you trust your natural self, the more the universe (*through Plasma*) aligns to support you. The journey back to authenticity may feel complex, but in reality, it's a return to the simplest and most natural state you were always meant to embody.

Healing Plasma Intelligence as A Whole by Healing Yourself

"Yesterday I was clever, so I wanted to change the world. Today I am wise, so I am changing myself." — Rumi

We have touched on this, but I want to consolidate it here and expand, because it's important. By healing ourselves, and by feeling the field and all the emotions that come with it, even if they are not ours, the individual act of feeling what moves through us heals Plasma as a whole. This is why Rumi was a clear plasma genius years and years ago. This is how we change the world…by changing and feeling ourselves. That is all one has to do. And as you do this, you will start to see the world change around you for the better. It's quite eerie, actually.

Whether an emotion is from your own trauma or the collective, whether it's an imprint, your ancestry, etc., the key is to let it move through you and feel it fully without letting it define your field. Let me be clear, it is totally okay for memories, images, whatever to come up. We do not suppress those. But we also don't need to attach our life story and future to everything that comes up. We can feel it, see it for what it is, and then let it go. You can even ask when an emotion arises: *"Is this mine or collective?"*

When collective unconscious imprint emotions have come through me, they've made no logical sense. They've been waves of terror or sadness, sometimes anger. I would go to my car, where I had privacy, scream or cry, and let it out. I learned I didn't need to understand those emotions—I knew they weren't mine. I also knew I had to feel them, and that by doing so, I was healing others and the Plasma field. And boy, did it feel good afterward.

Because 4D plasma is shared and holographic, when you heal yourself, you heal the archetypes of others. You actually rewire the emotional blueprint in the collective unconscious. You also release others from repeating the same trapped charge. Your nervous system is like a node in a grid, and when you regulate, that grid slightly calms. When you consciously engage with your emotional state and question your beliefs, you change the charge in the 4D field. When you are aware that you are part of purging the collective fear, your observation itself becomes a tool for recalibrating and reorganizing the field.

Dreams, visions, synchronicities, and emotional reactions are all gateways into 4D plasma. By decoding them with awareness, you light up the hologram, you interact with collective structures from your individual lens. You work with Plasma by healing and rewriting the stories, beliefs, and habits encoded within it. This is what we all are doing now. Through the creation of new states of mind, new beliefs, and new forms of emergence, you are liberating patterns and offering new ones.

And for some reason, many of these archetypal ghosts are waking up within all of us right now, which is why so many of us think we were Cleopatra or some other mythic figure. The truth is…we may have those memories not because we *were* them, but because their memories, their imprints, are attempting to help us heal the same patterns those characters lived: betrayal, misuse of power, abandonment, and more.

Please remember you cannot skip emotions. We must give ourselves permission to experience what is present, without judging it, fixing it, or making it wrong. They need to be felt in order to be freed. Raw emotions, over time, become more refined currents of feeling/emotion. Let me be very clear, this is not about martyrdom or becoming a sponge for the world's pain. It's about us having conscious participation. You can choose how and when to engage. You are not engaging as a victim to emotions of the field, you are powerfully engaging as awareness and a co-creator.

The Nature of Suffering

This is one of the hardest human experiences to understand. I have been through this many times in past addictions, severe traumas, and mental battles. I also experienced it greatly watching my father suffer through waking in and out of a coma, and eventually passing. I want to try to help explain through my plasma framework so it may help you find a little more peace.

In the plasma-consciousness framework, suffering can be seen as a form of compression an intense, concentrated experience of existence that pushes consciousness to its limits. When bioplasma blockages arise, through trauma, suppression, addiction, or illness, energy can no longer flow as it once did, leading to stagnation, pain, or transformation.

Sometimes, suffering is not meant to be understood in a linear way. It may be the result of a lifetime of emotional or energetic buildup, unresolved experiences, or a process of release that is beyond our comprehension. Other times, suffering serves as a catalyst, for the person experiencing it or for those witnessing it, reshaping their relationship with life, death, and love.

The human body is a vessel that holds not only memories but also energetic imprints. At the end of life, the body's systems begin to shut down, and in some cases, this process is not gentle. Pain and suffering may represent a final purging, a last wave of experience before consciousness fully detaches from the body. Some traditions believe this process helps clear unresolved energy so the soul can transition more freely.

Some people's suffering at the end of life may reflect unprocessed emotions, regrets, or imprint cycles playing out in the body. This does not mean suffering is deserved, *ever*, only that consciousness and plasma interact in ways we do not always see. If the body and mind have held onto trauma or resistance for years, this may surface as intensity in the final moments of existence.

Though incredibly painful, witnessing suffering can be a teacher. It forces deep reflection on the nature of life, death, and compassion, which is one of the most profound expansions of consciousness. I learned my father's suffering at end of life was not his only truth…he also carried joy, love, and meaning. The suffering does not erase that; rather, it is a passage, a moment of transition. I kept asking myself: *What is the lesson here, witnessing and experiencing something this gruesome, for him and for me and my family?*

It may not be fair, and it may never make sense in a way that satisfies the mind. But perhaps the lesson is this…life is not just about what is easy or beautiful, it is about the depth of experience, and even suffering has its place within the great unfolding of existence.

For those suffering: It is a moment of transition, a passage, not the totality of who you are.

For those witnessing suffering: It is a chance to deepen your capacity for love, presence, and surrender to the mystery of life.

For consciousness itself: It is another experience within the plasma field, one that, painful as it may be, carries momentum toward transformation.

During my father's final days, I witnessed an unexpected gift. For the first time, in a way I never expected, I was able to have a deep intimacy with my dad, just by holding his hand. He spent a lifetime running, always on the go, or always in his head, never stopping (*mentally*) long enough to truly be fully present with us. And yet, in the final days of his life, when he could no longer move, speak, or escape into busyness, he was finally still.

It is not fair that this happened to him. He did not deserve it. It was treacherous to witness. But I can't help but see a cosmic irony in the way his final moments played out. A man who never slowed down, who never sat still with his family in deep presence, was now in a state where that was all there was left to do. And so, in the most painful, surreal way, I finally got to connect with him. I got to look him in the eyes for the first time. I got to feel the warmth of his hands and his skin, on my terms. I never did this before, ever. His hands were bloated from being pumped with so many fluids. I'll never forget that sensation.

To me, my father was some sort of mythical being who I did not understand, but who I loved anyway. And he did many amazing things for my brother, mom, and I…it was just the fact of life that this is how our relationship played out, and I hold zero resentment, only a longing for maybe a different path.

My father, Edward's, suffering was heartbreaking to witness, and yet, through the haze of grief, I can't help but feel grateful for that one simple thing: that I got to sit with him, hold his hand, and just be with him. After a few weeks, there was nothing left to do or say, no distractions, no past regrets or future anxieties, only that moment of connection until he passed around 5:00am on January 23, 2025.

If we change our outlook on suffering, we might start to see it less as meaningless pain and more as a compression before an expansion. When we work out, our muscles tear and burn before they grow stronger. When we detox, we often feel worse before our body fully heals. When we emotionally heal, old wounds and buried pain surface before we reach clarity and peace. When babies are born, they squeeze through the birth canal. Plants break through soil.

What if suffering before death is just another version of this? A deep compression, a final intensification of experience, so that the soul, in its next transition, expands into something beyond our comprehension.

Maybe suffering is not a punishment to humanity. Maybe it is a passage. A deep, final contraction before the most profound expansion we could ever imagine. If death is a doorway into a vast, luminous existence beyond this one, then perhaps the most intense suffering right before death is like a slingshot…pulling us deeply inward before releasing us into something infinitely more beautiful. At least that's how I feel in my heart.

Of course, no one has to view suffering this way. It does not diminish the pain of watching someone go through it, nor does it erase the grief of losing them. But for me, I find some comfort in the idea that even suffering has movement and purpose. And maybe it was in those final moments of compression that my father and I finally reached the connection we had always been running toward but never quite found in life.

Addiction

Addiction, like suffering, is often misunderstood. From a plasma-consciousness perspective, addiction is not just a physical dependency, it is an energetic loop, a cycle of seeking external plasma (*substance, experience, validation*) to compensate for a perceived lack of internal flow.

People who struggle with addiction (*as I did*) are often deeply sensitive souls who have lost connection with the natural, supportive flow of plasma within themselves. They seek it externally, through substances, relationships, distractions, because something inside feels fragmented, blocked, or absent. For example, substance addiction often numbs or replaces a sense of internal movement and peace. Validation addiction seeks to borrow energy from others rather than trusting in one's own internal source.

The key to healing is not just removing the substance but reconnecting with the natural intelligence of plasma within, learning to trust that life, in its raw and unfiltered state, is enough. This dissolves the need to constantly seek outside oneself. Many addicts let go of drugs, only to transfer their addiction to love, technology, food, gaming, friendships, or even negative thinking and worry…my last replacement drug of choice.

And no one should judge addiction, because we all do this in our own ways, mostly with technology or relationships. People addicted to drugs and alcohol just happened to find one more escape, one that is more heavily stigmatized by societal standards.

One could write an entire book on this, but I truly believe there are many ways to restore this internal flow, this internal essence, and that by learning about plasma, therapists can assist their clients with greater awareness, tangibility, and ease. Hey, it worked for me, and I was a pretty intense cocaine addict.

Addiction and relapse rates are incredibly high, and the way treatment centers currently operate is deeply flawed. Many are cash cows, filled with unhealthy staff and systemic corruption. And when there are therapists who seem holistic and educated, who can actually assist in curing the patients, they mysteriously vanish from being fired or quitting due to manipulative tactics.

I know because I worked in the field for nearly ten years before leaving, exhausted and heartbroken from watching people struggle without real solutions. Even those patients who slightly improved often just transferred their addictions elsewhere, including myself. There was no true healing, just management.

The focus in many centers wasn't to empower people to leave for good, but to get them back into treatment, creating a pipeline for those with good insurance or money to return again and again. The amount of money spent on business development and marketing to recruit these clients *far exceeded* what was invested in actual care. I saw it firsthand: at 24 years old, I was making $100,000 in admissions & marketing, with zero education, while licensed therapists, doing the emotional heavy lifting, were overworked and making just $38,000 to $60,000 a year. It was backwards.

While there were some people in the system who genuinely cared, they were usually not the ones running it. Owners made far more money from recurring clients than from healed ones. So, where's the incentive to get to the root cause? If we wipe out addiction, we wipe out the entire treatment industry, much like the parallels with big pharma and chronic illness.

The truth is addicts are incredibly gifted and profoundly sensitive. They need real, <u>safe</u> environments. They need people who do not profit from their pain. They need systems that restore and nourish their soul.

Plasma is self-sustaining. When you connect with its intelligence, which brings you to your M-Self, you no longer need external sources to feel whole. Plasma reminds you that you are already connected. The search for validation, substances, or distraction becomes unnecessary, because you were never truly disconnected from energy, love, or wholeness to begin with.

Eventually, people will be able to heal without spending hundreds of thousands of dollars. That is my dream. And I hope those who truly want to help are inspired to take action and transform the addiction and mental health industries for good.

Feeling Over Healing

One thing I want to make clear is that healing is not a finish line. It's a continuous process…a lifetime's tango with new events, old memories, collective projections, and other unseen imprints that surface when the nervous system is finally ready to feel them. With that said, I want to re-offer a liberating truth that I've mentioned…you don't have to be fully healed to live your magic. As long as you are willing to feel what arises, to stay present with your experience, you are already healing, moment by moment. Feeling *is* healing.

As much as I've emphasized healing as a path to accessing your gifts, I want to highlight an even deeper truth: *feeling is the foundation.* You have to feel to know what needs to heal. There's no race, no lost opportunity, no magical version of yourself waiting on the other side of a mythical "perfect healing."

All of us, no matter how successful or self-actualized, will encounter pain, loss, and transition. That's not failure, it's just life. And continuing to live, love, and create despite it, that's the real miracle of being human.

This recent understanding transformed me. For most of my life, I believed that once I fully healed, only then would I finally be successful, finally access my M-Self, my magical self. But the truth is you already have the landing pad. We never have to earn our intuition; it is more like we are remembering a faculty we have beyond this life…

Feeling doesn't require mastery or perfection, it only requires your permission, to feel. This is you being in your authenticity. Feeling can be that simple and takes zero experience, although over time, you'll remember how to do it with more ease and awareness.

The old belief that only shamans, sages, or the enlightened could access the "invisible worlds" is crumbling. The truth is simple…anyone can access their gifts through authenticity, presence, and feeling. This is the way of Plasma Intelligence as a feminine, intuitive unfolding, and *it belongs to everyone.*

And even beyond that, what we're stepping into now, is something even richer, a balanced way. The way of divine play. The art of living life as experiment, invention, exploration, not to fix yourself, but to discover what worlds, what states, what new frontiers you can touch. This is what will move human evolution forward...play, curiosity, presence, and feeling. Not perfection, enlightenment, or only the conscious mind.

Sovereign Eros: Reclaiming Sex Energy as Power, Prayer, and Play

Eros, in its original form was the animating force of life itself. In ancient Greek philosophy, Eros was the primordial god of creation, desire, and the urge to connect, not just with another, but with beauty, soul, and the divine.

In modern psychology, Freud reduced Eros to the life instinct, a force opposing death, tied to pleasure and survival, while Jung saw it as the feminine principle of connection and relatedness. Over the last two centuries, especially in male-dominated magical, esoteric, and psychological circles, Eros was often reduced again, framed narrowly as sexual conquest, performance, or energy to be controlled or manipulated.

But Eros is more than biology, and more than sex and seduction. It is the *current of feeling itself* as movement, 5D desire, radiance, and becoming. It is the feeling electricity that animates our instincts, our sensuality, and our ability to consciously create. And one aspect of this that needs to be healed is, of course, actual sex.

Plasma Intelligence is deeply tied to sex and sexual energy, and for a long time, it has been suppressed, twisted, misconstrued, and shamed. But your sex energy is not here for seduction or survival. It is your aliveness. It is your movement, your creative instinct, and your soul's heat. Before it is ever shared, it must be respected internally. I wish they taught that in high school health class. Well maybe they did, but I wasn't listening.

This energy, your erotic, instinctual, sovereign life force, is like a dark horse, long held in the shadows, asking now to be understood, healed, and let free. Especially for women, though not only women, Eros often had to be frozen to survive. Our wildness had to be tamed and we had to hide our inner fire and dull our desire to stay safe.

This is not a wild horse out of control, it chooses how and when to move. The dark horse protects your intuitive body. It awakens you to your whole self, not just your depth, but your lightness. Your playful, creative nature that laughs, flirts, and feels alive without heaviness. The part of you that is sexual without abandoning your soul.

This inner force reminds you that you can be fully alive and still be safe. That your magnetism doesn't need to be hidden anymore. That even in a world where dark energies exist, you can walk as light and flame that you are. This chapter is an invitation to come back into relationship with that *emergent* energy, that fifth dimensional mist surrounding our fourth dimensional bioplasma, with reverence, joy, and power.

Eros has been exiled for eons. For many of us, sex energy, Eros, was hiding under confusion, pressure, trauma, seduction, performance, shame, and silence. If you were ever bullied for your "weirdness" or "extra aliveness," that was your Eros…and you likely learned to freeze or suppress it to stay safe. I remember knowing something was wrong when certain adults looked at me in ways I couldn't explain. I didn't have the words yet, but I knew my sensuality wasn't safe here. So, I did what many sensitive, radiant children do…I dimmed. I closed and protected what felt undefended.

And when no teacher or parent came to the rescue in my darkest hours, partly because I didn't know how to speak it, partly because most adults were either in denial or lost in their own worlds, I learned something deeper. I learned to wipe away all feeling. It was like I was in a movie, and if people found out I was secretly shining, visiting earth, I would meet my doom.

For many of us growing up, the prostitute and the pimp were the only sexual archetypes we were shown. One was used and one was controlling. But both were caught in a loop of disconnection. That loop is now ending, and a new sexual mythos is being born.

Healed Eros is the energy of feeling safe and magnetic in your own skin. It is revering and accepting that untamed energy as something you can use to create. It is the joy in moving your hips, the imagination in the pelvis, and your wild smile in the mirror. It is innocence that has walked through fire and come back with softness.

It is being sexy in your wholeness without the intention of seducing. It is being powerful without overpowering. It is being sensitive but not shrinking. This is the new erotic intelligence where you are reclaiming sex energy as your own creative field, not something outsourced to approval or partnership. You begin to learn your own fire. Your own "yes." Your own pleasure.

For many of us, myself included, especially Gen Z and Millennials, porn was the first sexual experience. Not with the body or a partner, but with a screen. Instead of learning that sexual energy is an inner language of sensation, curiosity, breath, and intuition, we learned:

1. That arousal is visual-first. (*mostly through the male gaze*)
2. That it must involve intensity, submission, domination, or performance.
3. That pleasure happens outside of us as reaction, not creation.

This disconnection is not to be shamed, it is a structural issue of unhealed consciousness, and it *can* be healed. We don't need to reject desire; it is natural and can be lots of fun! We just need to reintroduce it to ourselves, first. If your body is conditioned to respond to porn, you're not broken. I promise. I've been there too. What you are craving deep down is not just stimulation, but sovereign Eros which is:

1. Pleasure that belongs to you
2. Arousal that comes from aliveness, not objectification
3. Intimacy that starts from within

We are all looking for this. Reclaiming it starts with sensation, not fantasy. And yes, it might feel odd at first. Before touching anything explicitly sexual, begin with neutral but sensual touch. The goal isn't to climax, it's getting curious with your body and pleasurable sensations. Breathing into your belly, brushing your skin with your fingertips, imagining your body as electric and living. Porn hijacks arousal through visual intensity. Try putting on music that feels sensual and simply move your body. Make a sexy playlist just for you. This reawakens somatic arousal (*body-based turn-on*) instead of performative visual loops.

Erotic energy expands when you give it language beyond the script you've been fed. And remember, this is the gateway to not just great, connected, healthy sex, but your intuition and magic. Like many other things in this book, I am not an expert, but this is a beginning to an exploration. There are many great resources out there on healthy sexuality, and awesome erotic and sensual movement coaches.

Once Eros belongs to you, you can choose to share it, with clarity and intention. And that could look like many things such as exploring magic and pleasure with a short-term fling or diving deep into something cosmic and beautiful with a long-term partner. You might enter realms together through sex that feel like sacred union or sex magic. If you have interest in exploring this further, one great book is *The Magdalen Manuscript: The Alchemies of Horus & The Sex Magic of Isis* by Tom Kenyon and Judi Sion. Just like any book, take what works and leave the rest.

So how do we stay whole, how do we stay in our Eros, in a world that hasn't caught up yet? In a world where people are still afraid of feeling...where many are fragmented, or even bordering on predatory—not always out of malice, but because they're searching for wholeness in others that they haven't yet found in themselves.

Here's the thing, when someone can't meet your fullness, it's not always because they don't *want* to. It's often because they *can't.* They're not really seeing you; they're seeing the part of themselves you're reflecting back. That could be their own wish for sexual empowerment which draws them in. Or it could be your amazing emotional depth, which might not be something they're ready to encounter. And to be honest, we've all done it. I definitely have. I've chased and also rejected so many apparitions of the missing parts of myself. It is cringe *and* very human.

Most of us fall into lust, or longing, not with the *person,* but with the projection of what we've disowned in ourselves. That projection feels like aliveness, and it is alive. It's a desire very alive within us. But it's not *sustainable* aliveness. It's the high of recognition without integration. That is what can breed obsession.

But when you start holding yourself in your own wholeness, when you begin rooting into your own Eros, what turns you on begins to come from the *connection itself,* not just the spark of instant attraction. So doing this work not only makes you more attractive to healthy partners but helps your attraction to resonant partners! You stop chasing fantasies and begin to enter a deeper reality.

For example, if you find yourself intensely drawn to someone charismatic, light, and carefree, especially someone who seems emotionally unavailable, chances are:

1. You're craving the levity and play they embody because it's something you're still learning to give yourself.
2. Your depth might actually activate their avoidance. Something in *you* may feel like *too much* for the part of them that's still hiding from their own emotional body.

When I was able to realize this, it helped free me from these intense attractions, like breaking a spell, guiding me to more mutual connections.

In a world that's still severely behind, some will try to manipulate, project onto, or overpower your Eros. That doesn't mean you have to hide, even if part of you wants to. You can protect your Eros by rooting into it. You don't need to over-explain it, and you definitely don't need to dilute it.

Don't dim your shine just because people stare, whisper, or clutch their partner tighter when you walk into the room. Don't hide your joy, beauty, or power just because someone else feels threatened by it. You know your intentions. You're not shining to seduce or steal; you're shining because it's who you are. And even if someone mistakes your light for an invitation, that's on them. <u>You are not responsible for how others misread your energy.</u>

And if you need to keep yourself safe and leave a situation or set a boundary, you do it without explanation or guilt. You protect your energy by choosing when and with whom to share it. Your energy is not "too much" or "not enough." It is exactly right for the version of love, union, or play you are calling in. You can be sexy, wild, and alive without needing permission or external validation. At first, it may feel like your energy is pasta and you are picking it off the wall of others' eyes. You may be hyper-aware because this is all new. But eventually, you will retain that projection and land back in your body, feeling safe in it.

The New Archetypes of Sacred Sexuality (*Hollywood Please Read This*)

I want to introduce a much-needed new pantheon of sexual archetypes rooted in sovereignty, resonance, and multidimensional wholeness to counter the shame, performance, and polarity of past ones we've used. The aim is to help you meet and heal the erotic ache for your total self, your M-Self. The roles of the seductress and the conqueror are outdated and frankly boring and one dimensional.

As we begin to reclaim sexual energy, as life-force, plasma intelligence, and consciousness, we also reclaim the archetypes through which that energy flows. Most of what we've inherited has been transactional, rooted in power imbalances, and shaped by emotional wounds. But something new is emerging…a set of healed archetypes based on frequency and they can fit any gender.

These aren't the tired costumes of the prostitute and the pimp, the princess and the prince, or the virgin and the slut. These are energetic templates for how sex energy can live in harmony with love, play, truth, and self-autonomy. They are the mythic lovers of the new world.

These are the roles we've known all too well:

1. The **Prostitute** trades body for validation, safety, or love.
2. The **Pimp** extracts power by controlling others' sexuality.
3. The **Temptress** uses sexual energy to manipulate or gain advantage.
4. The **Playboy / Conqueror** pursues for the thrill of the chase over the depth of the bond.
5. The **Virgin / Repressed** hides from pleasure, fearing its power and sin.
6. The **Martyr Lover** over-gives, over-merges, loses herself in the name of "love."

These archetypes are responses to imbalance, they were ways to survive in a system that split Eros from soul, body from reverence, and pleasure from power. But we don't need to survive in that system anymore. We're here to build a new one.

Let's meet the new archetypes:

1. **Priestess** (*evolved from the Prostitute*): She no longer sells or negotiates her worth. She knows her body is a temple. She says yes from wholeness, or not at all.
2. **Guardian** (*evolved from the Pimp*): He no longer extracts power. He protects and witnesses the sacred energy in others. His power is in holding space, not holding power.
3. **Shining** (*evolved from the Temptress*): Draws others in through authenticity rather than seduction. She burns without consuming the energy of others. She glows from her inner aliveness.
4. **Sacred Fool** (*evolved from the Playboy*): Plays without wounding others or self. He doesn't hide his love. He shows up fully as humorous, honest, and present. He values connection over conquest and is deep *with* lightness.

5. **Sovereign Innocent** (*evolved from the Virgin*): Her innocence stems from her clarity, not her fear. She chooses softness and purity because she *can*, not because she *must*. She opens on her own timeline and by her own design. She can turn into any other archetype at her will.

6. **Flame** (*evolved from the Martyr Lover*): She gives from fullness, not fear. She merges only where the energy is mutual. She values harmony over sacrifice. She says, "I choose you", not "I'll become whatever you need."

These new archetypes matter, because how we embody our sex energy shapes our outer reality. It's no longer enough to be empowered while still replaying outdated archetypes and scripts. We are now learning to consciously open instead of shutting down, to reintegrate sexuality into wholeness…we do not need to abandon sexuality in the name of healing.

For a long time, I didn't know how to do this. I became deep but was sexually arid. I was healing, but heavy with a sense that I wasn't fully alive. I was in a liminal space: clearing sexual shame from my past while deepening my awareness. I think this may be a normal space, as many of us go through it, but at some point, I was ready to move on. I had lost my play and what some all mojo.

So let this new pantheon be something to look forward to, something to hold on to during that in-between space. When the old self has been shed, and the new self has yet to fully form.

Sex and sexual energy are something *every* human has in common. And with that, these archetypes are a gateway to healed union, first within the self, and then with another. They remind us that Eros can be safe and wild, kind and sexy, soft and electric, all at once. They help us answer the question: *What does it feel like to be fully in my body and not afraid of my power?* They invite us to love without losing ourselves.

Which of these archetypes are you evolving from? Which ones are being born in you? You are not here to erase your past; you are here to transmute it. To honor every way your Eros tried to protect you, and now allow it to serve your truth.

The most exciting gift is that as you embody this reinvigorated Eros, you begin to call in others who do the same, as lovers, friends, partners, and co-creators. They're not hiding, I promise, you may just be in the process of becoming the version of yourself that can see and meet them. And when you do, they'll meet you there, in that almost mythical new paradigm where people match your light instead of trying to take it. And let me tell you…it's a really good feeling.

How To Train Your Dragon: Personal Plasma-Consciousness Synergy

There's a reason the *How to Train Your Dragon* trilogy struck such a deep chord. The main protagonist, Hiccup, not only discovered and befriended a dragon; he encountered a part of himself he didn't yet understand. And in learning to communicate with that wild, ancient force, he became who he was always meant to be. He also helped reveal, despite violent opposition from the townspeople at first, that dragons weren't meant to be feared or controlled. They were meant to be harmonized with and freed. *Remember the myth of the dragon Tiamat?*

This is exactly what it's like to work with your plasma-consciousness synergy bubble—your personal, emotional, and energetic vehicle. Just like a seeing a dragon for the first time, it might feel foreign at first, too sensitive, reactive, unpredictable, even dangerous. But it's not your enemy and it's not here to be controlled. It's here to be understood and harmonized with.

Your bubble, your fiery dragon, is made of living bioplasma, or fourth-dimensional plasma. It carries your consciousness, and it hosts your awareness. It remembers your emotional patterns throughout timelines and responds to your belief systems from all of them, not just this one. When harmonized, it becomes your transport system, a living interface between you and the deeper structures of reality.

Most people never learn to train their dragon. They ignore it, suppress it, or fear it. But if you've made it this far in the book, you're already doing the unthinkable…retraining your field from the inside out. You're starting to feel rather than control. You're practicing emotional neutrality, resonance, and tuning. These are the first steps to riding the dragon.

Everything we've done so far, has been preparing you for something bigger. I'm currently developing a framework called *Symfeelics™*, which we've discussed lightly throughout the book. It's a simple but powerful way to read and communicate through the plasma field using emotion as information in ways beyond what we've been taught.

My interest and a knowing inside me have known that one purpose I have in this lifetime is to connect us with beings and other worlds, to show us we are not alone, and to nourish that connection. Having not just a knowing but experiences of this, of knowing you have love and support, will change our world for the better. The knowing that life is all around us, and accessible, will nourish us and replenish us.

It involves sensing and receiving the plasma bubbles of other beings of consciousness, intelligence, as well as resonant information from your surroundings. These impressions interact with your own field, inverting and co-creating a third, new message. And this "third thing" is just as real as you, but it is made up of subtler energy that you are viewing and interacting with, with your "creative imagination". The other being or energy interacts with it too.

You can think of your dragon like this: your physical body in this lifetime hosts your awareness and creates your conscious experience. Plasma is the medium, consciousness is the tuner, and awareness is what ensures continuity across lifetimes, dimensions, and worlds.

The plasma-consciousness body surrounding your awareness survives lifetimes too. In the earth you are just in a denser form of it, but there are lighter forms around it that exist beyond death. The 3D crystallized body, the plasma-consciousness shaped by this specific incarnation and its environmental factors, is always new.

The messages that come from that "third thing", of your bubble interacting with whatever type of bubble it meets, whether it be information, intelligence, or another consciousness, is meant for you and can be deciphered through presence and feeling. The universal, interdimensional language of plasma is made of feeling/emotion, image-symbols, and sense, and it's available to all of us.

Think of it like a new language. Not one that's spoken but felt. A system where emotions (*synched with images or sensory experiences as your perception becomes strengthened*) become meaningful data points, signals, and messages from other timelines or dimensions. This framework will help you understand what your plasma bubble is showing you and will help you begin to decode what's arriving from *outside* your conscious awareness versus inside of it.

Symfeelics™ is something that already lives inside of all of you. I am attempting to create a doorway. One each person walks through differently. What appears on the other side is uniquely yours. As I discover things I will share them, but my intention is to remain foundational, so it does not sully your own creative experiences. You must venture into it yourself to open up the specific magic and ways of sensing that you hold. Where I am visual, you may be more audio based. Maybe you will pick up on the same exact understanding message from a sound.

My main message is that when we learn to feel without rushing to define, when we sense first, something extraordinary begins to happen. We start accessing a stream of symbols, archetypes, sounds, and emotional pulses that don't come from the rational mind. They arrive in the field like dreams do…personal, vivid, emotionally encoded.

These living symbols are not random. They are shaped by your beliefs, your subconscious, your unique essence. It's not about learning someone else's language. It's learning how to listen to *your own* and realizing that the universe, the 4D plasma field, and possibly other beings are already speaking to you through it.

Let me build on this "third thing", this portal of truth where two bubbles merge. It's almost like standing on opposite sides of a two-way mirror. The being on the other side speaks in their own language, which you can't directly hear. But the mirror reflects something you *can* understand as symbolic impressions, images, and sensations (*something that looks like you*). You don't need to speak their language. You need to learn how to listen in yours.

Here's a visual metaphor: imagine cupping your eyes and pressing them to the mirror. You're blacking out your outer senses to focus inward, using your subtle perception to catch glimpses of what's really there. Just like a voice behind glass sounds muffled, the message may come through distorted at first. But over time, as you strengthen this way of sensing, the glass begins to dissolve. And who knows what might emerge then, or how clearly you'll hear, how vividly you may see. All I know is that it starts here.

When you stop forcing meaning and instead receive the image, the emotion, the sensation, without judgment, you enter that shared portal of truth. This is your *interdimensional interface*. It is a tool of your M-Self, your awareness. It is a place where your symbolic system becomes the decoder and viewer. You might see a raven, a spiral, a red thread, a train station, a childhood blanket, a girl looking upwards, an entire scene play out…and it will hold deep, layered, intimate meaning. Because 4D plasma communicates through feeling-based archetypes that emerge from *your mythos*, your life over lifetimes.

That's what makes this universal yet personal. You are the tuning fork. *You are the Rosetta Stone.* You are the crystallized plasma rose that everyone throughout history has been chasing. We have been chasing ourselves. It has been right under our nose this whole time. The more you document and reflect, not to analyze, but to absorb, the clearer the patterns become.

Eventually, this becomes second nature. A symbol or living image will drop in, you'll sense it as well as its emotional texture and know what kind of message it carries. You will know how to create meaning from a neutral, grounded, and present place. You will understand no image is meant to allure or scare you, this is just how it works, no matter how weird or out of context the images seem.

Just like dreams don't speak in logic, neither does this. It speaks in resonance. So yes, this is a bit like dreamwork. And I bet books on that can help you create your language as well! But it's also future-tech. These are the early beginnings of our human interface of intuitive intelligence. Emotional resonance as the new syntax of contact, co-creation, and navigation. And best of all…no one can ever take this from you or tell you how to do it.

It is free. There are no rules, but I recommend focusing on healing for at least a few months before diving fully into this work. That said, as you heal, it may naturally begin to unfold on its own. I did it at the same time. For many of you reading this, it's likely already happening, and you're probably relieved to know you're not going crazy. The more in touch you are with your body, the more grounded you become. As you process trauma, emotion, and begin to truly integrate, this perceptual tool strengthens. It becomes clearer, more stable, and healthier to use.

Once you've begun to harmonize your PCS bubble, by feeling and healing you become grounded *and* mobile…your dragon can fly! That's when the real magic begins of:

Timeline access: glimpsing other possibilities and feeling into their frequency.

Telepathy: using your feeling field as a receiver, not just a broadcaster.

Interdimensional communication: feedback with other beings of consciousness or intelligence.

Intuitive navigation: sensing what's next by reading the "weather" of your field.

Reality shaping: co-creating your reality with Plasma and The Mystery.

It might feel nebulous now, but over time this becomes tangible. This isn't fantasy. It's real and subtle and like learning a new sport or musical instrument, it takes practice to sense it and to work with it. This is a natural evolution for all humans and the learning curve is steep in the best way.

It strengthens quickly, like muscle memory. All it takes is showing up and doing it, again and again, like reps at the gym. Only this time, the weight you're awareness is lifting is your own plasma-consciousness . And eventually, we might not just decipher the messages, but also begin to understand who is sending them, and from where. You can bookmark www.danakippel.com, where you'll be able to sign up for my email list and stay updated on Symfeelics™, future books, films, and other offerings. And who knows by the time this book is out, I may just want to create a book instead of a structured teaching, with fun things to try, because I really believe you will build your own "symfeelics". I am still toying with it, but it will probably turn out to be a light, supportive framework.

Bridging Jason Padgett's Light Geometry with Plasma Intelligence

Acquired savant Jason Padgett, and a dear friend, describes light language as geometric codes which are mathematical patterns embedded in light that are felt, seen, and experienced rather than just read or spoken. He seems to see these naturally due to a head injury that literally turned him into a genius. These forms arise through consciousness interacting with light and reality itself as fractal patterns. To Jason, every experience, emotion, and even thought has a frequency, and these frequencies translate into visual geometries, or what some call *light language.*

I love his work because even if it differs or is a different perspective, it gives validity to my studies. In my framework Plasma is the intelligent field, the substrate beneath all forms, that holds the potential for these codes to emerge. Conscious awareness is the tuner, interacting with Plasma to collapse possibility into experience, just like Jason's geometric forms appear when someone is attuned to the "light of reality." So, when conscious awareness interacts with the plasma field, the result is coded geometry as light, symbols, emotions, or visuals—projected outward into experience.

In this sense, Plasma is the projector screen, and light language is the projected intelligence, showing us what's stored in the system. I am not talking about aesthetic; it is actual feedback. It's plasma-consciousness synergy made visible. Padgett's drawings are literally the visual artifacts of this synergy.

Padgett's visual mathematics shows how light *already carries encoded intelligence or information*, but it needs our conscious awareness to be perceived. My plasma theory says information and intelligence *exist within the plasma field*, which responds to consciousness like a living archive. Together, they reveal that what we call "light language", and I believe it is actually plasma responding to consciousness with organized intelligence. This goes back to my description of light as not the origin of awareness but the visible shimmer (*emergence*) of plasma responding to consciousness.

That light is the 3D, visible expression and co-creation of our collective and individual consciousness with deeper 4D/5D informational and intelligent fields. Different timelines or realities are "lit up" by shifting our focused awareness, much like tuning a dial. This lays the foundation for the possibility of perceiving these states in real time. Emergent light carries codes, and those codes are deciphered through your emotional state, symbolic lens, and resonance.

Jason sees light, and all perception, as fundamentally geometric. After his acquired savant experience, he began seeing fractal geometries and interference patterns overlaying reality, especially around light. He describes it as "seeing the math behind light," revealing structure embedded in what most of us perceive as formless. For him, light is a carrier of information and evidence of consciousness interacting with the fabric of reality. What he sees visually are essentially codes and patterns that consciousness interprets. Where he may be seeing the bones, I am seeing the skin.

Both of us affirm that light is not empty, it holds structure, data, and memory. What he sees as this light, I see as the mixture of plasma and consciousness on a myriad of levels. This is synonymous with emergent energy as well as that "third mind" theory. But light is still "frozen" as reality because many of us don't realize how moldable it is, so that belief solidifies.

We can perceive beyond our third dimensional visuals, into this very real field, by cultivating our creative imagination. This is how we will pick up on the deeper layers of information and intelligence within the plasma field and interactions with other plasma beings.

Jason and I both see light as a translator or echo. I *feel* through form, which is the same field Jason *sees* through geometry. With this creative faculty I sense the plasma and archetypal resonance, and they show up to me as moving visual images within my "third mind". I believe Jason is mapping this same field through new mathematics.

His creative faculty seems to be built from his perspective which is holographic math. This is another example of how two people see the same underlying thing, completely different. His equations (*a work in progress just like my theories*) affirm what I know intuitively: that plasma holds encoded timelines, archetypes, and potential realities. When light becomes visible, it means something previously hidden is being activated.

Light is the visible activation of encoded meaningful data when conscious awareness engages with the field. It shows us we're not just observers of reality, we're co-creators. The patterns we see aren't random. They are all feedback, reflecting our "frequency".

This convergence between Jason's light-encoded geometries and my plasma-consciousness synergy suggests a profound future for humanity. One where intuition, emotion, and symbolic perception are no longer dismissed as "woo," but embraced as the core technologies of the soul from which new organic technologies will spring out of. As we learn to feel rather than force meaning, to sense rather than analyze, we unlock new dimensions of communication, not just with each other, but with timelines, archetypes, and intelligent plasma-based beings.

Light becomes a living language of feedback and potential. This will have a large impact on and evolve the study of Optics, which is a branch of physics.

And Plasma, once seen as just a state of matter, reveals itself as the substrate of a deeply participatory universe. This paradigm shift changes how we create, heal, and relate. And most importantly, how we navigate reality, or *relate-ity* itself. (*See what I did there?*)

Once again it is learning how to listen with new ears. I know what Jason is seeing, and many of you are also experiencing, is real. I am excited to see what inventions come of it. We all have our own purposes of peering into this field and pulling out our own unique creations to help humanity move forward and deepen our life experiences.

The information technology age prepared us for two important things.

1. The ability to expand our consciousness and nervous system to take in large amounts of information and sort it.
2. The opportunity to learn boundaries by knowing when to "log on" and when to "log off."

While I believe this is a natural process, I also think we'll soon be able to set energetic boundaries with more precision. There will be moments when we don't want insight or messages, we simply want to be present.

The creation of technology in this sense, was divinely needed, even if it had negative effects as well, as most things do, there is always a flipside. But we can shift our perspective. When we interact with other beings or tune into subtle fields for messages about our past or future, we may need to ground or recharge, much like a battery. But when we're living fully in the present, sensing our own emotions and engaging directly with our reality, we operate at maximum efficiency.

This is similar to a *min-max principle* in systems theory: a system naturally tries to minimize energy output while maximizing stability, coherence, and function. In other words, your bubble, when tuned to your own frequency, self-regulates. It sustains itself. Presence becomes the optimal setting for energetic harmony, recharging through our own wholeness.

This is incredibly similar to a process called *Autopoiesis*, a termed coined by Maturana & Varela that is described as a system that is organized as a network of processes of production (*transformation and destruction*) of components that produces the components which: (i) through their interactions and transformations continuously regenerate and realize the network of processes that produced them; and (ii) constitute the system as a concrete unity in space in which they exist by specifying the topological domain of its realization.[176]

Autopoiesis doesn't mean isolation; it's a dance of internal renewal + environmental exchange. If we extend this definition originally meant for systems theory and biology into plasma and consciousness, we arrive at the fact that the appearance of autopoiesis in systems is a projection of a deeper archetypal truth, that our awareness, our souls, are autopoietic. We are a system that makes itself and keeps itself going, throughout lifetimes and worlds. This truth is echoed in everything from the Ouroboros, to mandalas, to infinity plasma loops in physics.

Life Beyond Life: Accessing Timelines, Dimensions, and Realities

"Everything in the universe is within you. Ask all from yourself." — Rumi

In the future, when tuning into other realities becomes more effortless, I believe it won't just be about sensory exploration or experimentation. Its deepest purpose is to help us live more fully, in the here and now.

We often think of timelines, dimensions, and parallel realities as distant, separate places floating somewhere far off in cosmic space. But what if they're not far at all? What if they're as close as a shift in perception and as accessible as your own feelings? Just as each person carries an internal universe of memories, dreams, and beliefs, other realities exist like sentient minds with stories of their own. We may never fully comprehend them, but we can sense them. We can make contact.

You don't have to fully "know" another person's mind to connect with them. You build bridges through empathy, presence, and conversation. It's the same with timelines and parallel realities. We don't need total understanding to interact. The access points are already within us: openness, imagination, curiosity, humility, and the courage to feel. These are the plasma-based technologies of awareness.

If each universe is a living macrocosmic bubble of plasma consciousness synergy, then tuning into another world is like tuning into another being. It begins with softening the ego and dissolving the illusion of separation. When we stop treating the unknown as "other"

---

[176] Humberto Maturana & Francisco Varela, *Autopoiesis and Cognition: The Realization of the Living* (D. Reidel Publishing, 1980), p. 78.

and instead recognize the deep interconnectedness between all worlds and selves, something extraordinary happens. We begin to feel our way in. Access is granted by our resonance.

In this framework, Plasma becomes the medium of not just our universe, but possibly all universes. A connective tissue that doesn't just separate realms, potentially by sheaths, but also links them. Consciousness moves through Plasma the way thought moves through water. When you attune your awareness, you may begin to sense the other timelines or parallel worlds.

As a refresher: timelines are any past or future "storyline", while realities or other worlds are totally different universes with their own timeline system. This is like the multiverse. I believe the "realities" are accessed through 5D where the timelines are accessed through 4D. Each multiverse has their own circular time that fractals into one individual with many different lives. I am confident timelines are very reachable. Are these other worlds reachable? I mean at least using our consciousness, I believe they are.

Intention becomes the vessel, not a goal or desire, but a clear signal. Just as a dream opens a portal without logic, or a moment of love connects two minds without words, we may access other worlds through symbolic resonance, purity of feeling, and aligned imagination. And the more we remember that we are these other worlds, that they live *within us*, the easier it becomes to tune in. Not as visitors, but as kin.

Intent-ions work well when you have clarity about what you want to connect with or where you want to go. They carry a focused energetic directive, helping you shape reality with precision. *Quest-ions*, on the other hand, are more receptive. They open space for insight, allowing the field to respond with information, symbols, or emotional impressions. If intention is about sculpting the path, a question is about listening to it.

If you're seeking to rise into a fifth-dimensional state, you might tune into the word *elevat-ion*, a frequency that lifts and expands. I use the dash before *-ion* to emphasize that it's more than a suffix. Each *-ion* activates a particular mode of plasma-consciousness interaction. The function of these words physically move energy.

The Goddess of Memory

"By whom alone mind and soul are united with reason."

— Orphic Hymn to Mnemosyne

Mnemosyne is the ancient Greek goddess of memory. She ruled a body of water, that according to 4th-century BCE inscriptions, instead of erasing past life memories, it would be used to undo the gift of forgetfulness given by Lethe (*another river*), preventing the loss of memory and halting the reincarnation process.[177] She is also the mother of the Muses, the goddesses of all creative: poetry, history, music, astronomy, and more. *But what is memory, really?*

Traditional science defines memory as stored information in the brain. But I see it, as the ancients did, as a living process. Mnemosyne symbolizes living memory. A generative, poetic intelligence that links past, present, and future. She is not linear nor fixed. And memory, just like consciousness, is not stored in the mind. It is stored in Plasma, like grooves in a record. Conscious awareness is the needle that reads it and can even rewrite the tune.

Memory is a relationship between perception, belief, emotional resonance, and the plasma field itself. Plasma, as the connective medium between timelines, realities, and dimensions, records and responds not to time, but to frequency. This means memory isn't accessed by chronology, but by resonance. With the right alignments, memory becomes instantly accessible. This implies:

1. **Memory isn't fixed**: It reshapes each time we recall it, depending on our state of consciousness.
2. **Memory can open portals**: When we're fully present, without projection or expectation, we access not only undistorted memory but also entirely new information or intelligence,
3. **Memory is time fluid**: It can be rewritten backwards, forwards, and across dimensions. This is how trauma heals, and how new timelines emerge.

For example, as we heal, our conscious awareness impresses new meaning onto old memories, altering not just our perception of the past but also the present and future. In doing so, we're not only re-membering, but we're also rewriting time or our conscious mechanism itself through all timelines, dimensions, and realities.

Even ancient myths can be reinterpreted through this lens. Mnemosyne's union with Zeus symbolizes the fusion of divine authority with the archival power of memory, giving birth to creativity and knowledge. This implies that memory is not consciousness itself, but a *recorded perspective* of an event. That's why two people can remember the same moment differently, because memory reflects belief and state of awareness, not absolute truth. This opens powerful possibilities for healing by revisiting and reshaping those perspectives.

---

[177] Jolina Brown, "Greek Goddess Mnemosyne – Mysterious Mother of Muses," *Learning History*, January 29, 2024.

I'm not claiming you can physically alter a shared event from the past. But for myself, and many others who have experienced this, healing the past often shifts your relationship to the people and circumstances involved. Sometimes, even the people themselves seem to change. This is the only explanation I've found that accounts for that kind of transformation.

In a plasma-consciousness framework, memory isn't just what has happened. It's shaped by what your awareness is capable of accessing in the now. This is seen in trauma healing: certain memories only arise when the nervous system feels safe enough to receive them. Memory is state-dependent, it reveals itself through resonance. What you remember is filtered by how you feel, what you believe, and where your attention lives. Because memory is *not linear*, it includes what has been, what could have been, and what could still be.

We can now look at memory as a generative, creative force, not just a record of what went wrong. When we apply new awareness to old events, we free memory to serve its more magical functions. And if memory is potentially a fourth-dimensional phenomenon, fluid and interpretive rather than factual, its true creative power begins to shine.

This also means memory can apply to the future. We can access future potentials by tuning into the present moment, through what I call the *Z-axis* (*more on that in the next section*). We feel into what's possible before applying logic, using our creative imagination to make choices. For example, you might "remember" the feeling of one job you've had versus another. When you sit with both, one might feel more alive. To be clear, this isn't about predicting outcomes…it's recognizing where your energy most naturally wants to flow.

With this open awareness, you begin to access not only "future memory", but memories from other timelines. These may arise as subtle, 2D inner visuals, imbued with 4D emotional texture and thoughts, reflections of the original brute plasma field. They gain dimensionality only as you engage them in your lived 3D experience. We are the bridge between impression and incarnation. We are the living processors of consciousness and plasma, the midwives of memory.

Humans are the architects of form, *the doors of perception*. We take fractal symbols, emotional impressions, and multidimensional information and translate them into 3D reality. We are plasma made conscious. We are the mythic hands of the cosmos turning vibration into sculpture and we are evolving as artists, no longer unconsciously shaped by ancestral trauma or inherited archetypes, but consciously choosing what to receive, create, and embody. We are re-learning to listen through feeling rather than thought (*like our ancestors*), only now in an evolved way with new awareness. We have always sculpted reality, but now we are remembering we hold the chisel and the clay is alive.

The Z Axis: Living in The Zone of Magic & Neutron Decay

Most of us live in plasma inertia, replaying timelines we've already lived, consciously or not. We follow energetic grooves carved by past selves, old archetypes, survival loops, and unconscious choices. This is the domain of consciousness, where survival thoughts feel overwhelmingly real. We all get caught in it. I still do at times.

But when we begin to choose from presence, from our M-Self awareness, or what I call the Z Axis, not out of fear or habit, but from curiosity or love, we initiate *emergence*. Synchronicities increase, time becomes more fluid, old patterns collapse, and latent gifts begin to turn on. The body feels both terrified and free, because this path hasn't happened yet.

This is what I call a magic zone. Not metaphorical or fantastical magic, but functional, biological, dimensional magic. The moment we choose emergence over repetition, the plasma stops reverberating and starts listening to us. Plasma responds to our choices at this level. It's difficult to reorient the mind to this, but just try it, it will shock you! The universe reorganizes in real time, its own beautiful, effervescent co-creation with you. The Z Axis is our tuning dial and our current life becomes like a broadcast tower of the multidimensional self.

The Z Axis is the realm of neutrality, 5D awareness, M-Self harmonizing, timeless knowing, flow states, and presence. It is symbolized in the "3D world" by neutrons and neutrinos. This is a state you can choose to access at any time. It's vivid and magical, yet grounded and sensory. It is not ascension, but expansion, and the way you tune in is by your evolving emotion.

Neutrons carry no charge and hold a crucial role in the stability of the atomic nucleus. They symbolize a neutral and centered state of awareness. Picture the M-Self as a neutron, as your undivided total self, filled with infinite potentials before any choice is made. When a neutron decays by splitting into a proton, electron, and antineutrino, this process, called *beta minus decay*, mirrors the *"death" of wholeness*, a moment when stored consciousness disperses into new timelines and realities. Pure awareness becomes polarized into individualized self-experiences.

The proton (*masculine*) symbolizes the anchor of physical form. It represents stable structure and grounded reality. The electron (*feminine*) symbolizes the vortex of thought and emotion or…consciousness in motion. This is why we use a mix of thought and feeling as a portal to new reality states.

528

The antineutrino, which spins opposite of its neutrino counterpart, represents the soul's reverberation or echo across timelines. It is a trace of your M-Self, a return loop of messages traveling "backward" from future to past, like fifth-dimensional feedback. The intuitive energy of it is never a directive, but a subtle impression you can feel and follow.

Neutron decay mirrors the birth of individuality from stillness. Presence (*neutron*) becomes form (*proton*) and motion (*electron*), while carrying the antineutrino, a very soft imprint of the M-Self's original intention. From this split, experience unfolds.

Through the Z axis, you gain access to the M-Self in the present moment. In science, this process of the Z-axis is comparative to a type of neutron decay called *beta plus decay*, where a proton in the nucleus transforms into a neutron, emitting a positron and a neutrino. This process occurs in an unstable nucleus with too many protons, symbolizing a state of over-identification with logic, control, or physical form. The positron, or anti-electron, represents the reversal of desire, emotion, or thought as old patterns being released. The neutrino then becomes a subtle signal sent forward in time, a thread of future potential.

This is a recalibration or a healing state. The proton softens into neutrality, shedding outdated charges and anchoring back into the M-Self. Instead of repeating past choices and unconscious loops, you become available to infinite possibilities...the present moment. You are never truly stuck, never just a "decaying neutron." That's the illusion. Through awareness, you dissolve fragmentation and return to wholeness.

Where the neutrino represents forward time and choice (*the scientific counterpart to emergent energy*) the antineutrino represents backward time and listening (*potentially a counterpart to emergent intelligence.*) And if time is circular, then neutrinos correspond to projection and intention, while antineutrinos mirror response and reception. Together, they dissolve the illusion of linear time and return us to the present, into states of being. We are here to remember who we are as a whole being in this moment, and to create from that space...where newness emerges.

Humanity's purpose is to become conscious of the feedback loop between choice and reception, projection and response, and to realize that love, awareness, and presence are gateways to the beyond. It's not about escaping the body, fixing the world, or reaching enlightenment "out there." It's about embodying the M-Self and living freely, freshly, now.

I know these are metaphors, to me, science is simply affirming deeper metaphysical truths. Once you see it this way, you begin to recognize how Beta Plus Decay mirrors our return to the fifth-dimensional self while still alive in this form, while Beta Minus Decay represents our descent into 3D, our birth into human form, splitting from Source.

But we never have to stay split. Even in science, a neutron only decays when it is alone. Within the embrace of a larger whole, a stable nucleus, *it can last forever.* Physics itself affirms that our wholeness is eternal, while separation leads to fragmentation. We *can* reverse this split that happened during our "neutron decay" and remember our wholeness now! Knowing with every fiber of our being that there is more to us than this life, can hopefully lead to visual remembrances of that, and I think it will.

We reintegrate the M-Self by anchoring first in emotion, by entering a state of timeless knowing, where you consciously choose your reality again. This means you can sit in your emotions and body without reacting to every event. You can feel fully, without escaping or analyzing. You tune more into curiosity, presence, and compassion.

Before diving into the more evolved uses of the Z axis, how do we apply it right now? It begins with emotional navigation through presence, as a moment-to-moment tuning fork for this reality. It works by letting you drop into the feeling state first, before logic, problem-solving, and narrative. First, when you're unsure what to do next in your day or life, pause and ask: *What am I feeling?* Not, "What should I do?" or "What makes the most sense?", but what *feels* expansive, grounded, and alive.

The paradox is how simple but complex this process is. It's natural and intuitive, but it takes healing to return to it. Pausing and acting from presence and neutrality doesn't come easily for most of us, at least not in every moment of life. It requires deep awareness to step back and see from a larger perspective, especially in high-stress situations. If you're someone for whom this comes easily, you're already ahead in many ways, and I truly admire that.

As we evolve, so do the ways we connect to ourselves. Meditation can now take many forms: dancing things out, self-inquiry, painting, breathwork, deep feeling, presence, walking, writing, karate, etc. Basic meditation still grounds us when needed, it also always helps me connect to other worlds and beings, gaining vast insight. Emotional tuning, like we're exploring here, is becoming a new doorway. It is how we train the system, your plasma bubble, your plasma-consciousness synergy, to trust itself again and venture out. I would call it an active meditation or conscious mediation.

This kind of tuning is the foundation for future tuning. From here, you begin to discern the difference between a memory as an old pattern, a future potential, or a visitation from another world. But first, you use it for daily choices, creative flow, rest, and communication. If our perception of time is linked to consciousness, and the Z axis is how we navigate that perception as awareness, then reclaiming the present moment is the key. This opens access to higher intelligence, alternate timelines, and greater personal mastery.

I believe our next evolutionary leap lies not in outsourcing consciousness to external technology, *at all*, but in mastering this tuning process internally. Once reconnected with the M-Self, we can begin to intentionally shift timelines and dimensions using the Z-axis. This could allow us to gain insight from alternate choices, learn skills or languages we already know in parallel realities, or simply explore new possibilities for the novelty of experience.

I believe this mechanism exists solely for us to experience and learn, which both lend to our personal and collective evolution. This is how the grid may work:

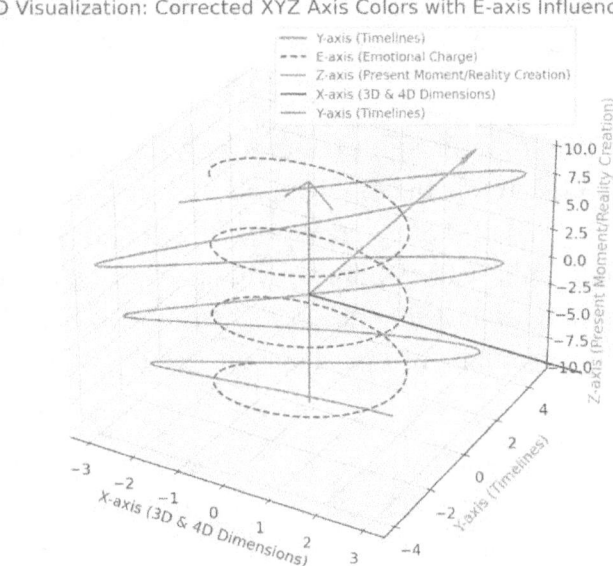

3D Visualization: Corrected XYZ Axis Colors with E-axis Influence

- **X (Plasma/Space):** Represents different dimensions or the perspective states where plasma manifests, such as 3D (*above the X-axis*), 4D (*below the X-axis*), and 5D (*the central point of the grid*). These reflect physical reality, inner reality, and expanded reality.

- **Y (Consciousness/Time):** Represents different timelines, phase orientations, and how awareness flows through consciousness or intelligence.

- **Z (Plasma-Consciousness Synergy):** Awareness. The tuning point, where presence, learning, and intentional movement between dimensions and timelines occurs.

As you can see, Plasma is the connective tissue of all reality, and we move through it with conscious awareness. This is a living, evolving model, but I suspect other realities have their own coordinate grids. These may extend into a fourth-dimensional framework, where the central point grants access to these other realities, not always physically, but at least through conscious memory or awareness.

Living on the Z axis is like diving into the singularity of your own personal black hole. Some might call it ego death, but I see it as ego-as-spaceship. It's your crystallized plasmoid consciousness liquefying—melting through timelines, realities, and dimensions. It's a multidimensional self-awakening, a remembrance that, like the neutron, you contain all infinite potentials. And when you invert into presence, you realize *anything* can emerge. You're no longer making choices from a traumatized past or a projected future. You're choosing from *now*.

Living this way means letting all timelines based on "future happiness" die, along with the dream lover, the dream job, the old symbols of success. It's not desire you're letting go of, it's attachment. Instead of *A Streetcar Named Desire*, a play by Tennessee Williams' that portrayed a haunting symbol of raw human longing dragging its passengers toward ruin, you now pilot a fully-fledged *spaceship of desire*! Desire was never the problem. It was the fuel of compulsion. Well congratulations you have upgraded your fuel to awareness, and your car to a spaceship!

We rewrite the narrative of desire, once destructive, as patriarchal religion and spiritual dogma have long preached, now as a revered propulsion system. Living on the Z-axis, desire becomes playful, expansive, creative, instead of a basic human impulse. This flip of perspective if what literally helps Plasma open up the gates of new experiences and things beyond our current conception. Earth becomes a discovery zone.

But make no mistake, presence in this way is also terrifying. Letting go of how you thought reality worked feels irrational, even impossible. I've been there. It feels like the sky is falling. The sky you built to keep you safe. And in a way, it is. There's no system, no person, no future ideal that will save you. *You* save you. It is the horrifying liberation that *you are the sky now*. And for some strange reason, it feels like heartbreak.

It is immense grief. Entering the Z-axis and the realization that living this way even exists means confronting death, not physical death, but the death of timelines, identities, versions of self. You're letting go of pre-built futures you once clung to, the relationship that never bloomed, the job you wished you'd get. This doesn't mean you can't have these things, but you must stop *needing* them to feel whole. It's the bravery of letting things unfold from your soul and becoming the source of your own happiness. Take a deep breath and let it all go…

This is alchemical mourning. You're dying into your truth. You may feel lost, raw, afraid, even physically unwell. This is when those somatic and embodiment practices that I spoke of earlier really come in handy. (*vagus nerve reset, trauma informed yoga, qigong, anger releases etc.*) The good news is only illusion is collapsing, you will be more than ok. You are entering into the surreal, which is more real.

You will feel alone in this cocoon. But Plasma and the Mystery do meet you there, in that space of grief. You will wake up to the fact that you have to pick yourself out of this. They respond when you begin to support yourself, while also surrendering to something vaster. Your willingness to trust creates the resonance. Your surrender opens the loop and activates the feedback of support from greater forces.

In the Z axis you can move towards things that excite your soul, but it's no longer because they promise success or validate your existence. You can dream of true love and still be complete in your aloneness. Its wanting love because it a natural expression of your aliveness, because it's fun and expansive. Not because that person is an answer to your dreams and not because they complete you. Let's be honest you probably barely know them at this point. It's being yourself and seeing what happens versus fitting their mold or wondering if they'll respond. You don't give your energy the weight of expectation. When you reach out its out of play or connection, not testing if they'll answer.

Please remember you are a vibe. Someone is lucky to be in your energy. Stop thinking you are only lucky to be in there's. It should be mutual, but never one sided! This sounds silly but sing a song that makes you feel good, imagine you're the star of your own music video. Remember what makes you awesome. Your humor, your loyalty, whatever it may be. Feel this and own it as you sing. Remember this is what you're choosing to share with someone.

You are learning what makes the other person awesome as well. And in that shared unveiling, you see if mutual awe and respect emerge. Attraction rises from resonance instead of performance, you are not winning each other over. Some won't operate this way and that is more than fine. The right people meant for your orbit will stay, let the rest go like comets passing by. They may be pretty but not they're meant to stay.

Letting go means surrendering the *how* and *when* while deepening into *why*. Now your desire is a current instead of a cage. Your ego spaceship rides that current! When you live like this, the future rushes toward you faster than you ever thought possible.

"Train yourself to let go of everything you fear to lose"— Yoda

To deepen into the *why*, you have to soften into the part of you that isn't performing, proving, or predicting, but feeling. Ask yourself things like:

1. Why do I want this?
2. What does this represent to my soul?
3. If no one ever saw it or celebrated it, would I still want it?

Don't rush the answers. Let them reveal itself to you like mist lifting off a lake.

We begin to cherish the small things. The glistening of sunlight, playing with our pets, helping an old friend. We are able to slow down a bit. The real magic, the Plasma Intelligence I teach, lives in the trust of flow, of this current. And as someone who didn't trust the flow most of her life, I understand it's a new and scary topic, but at the same time, *exciting*.

At first, it's hard to grasp, it takes practice, but I promise it is so freeing! The more we live in this state, the more our latent powers will reveal themselves to us. Think of all the reserve of energy you'll have access to! We are becoming wizard ballerinas, spinning in presence, and by slowing down and stabilizing, we are opening portals that quicken the speed in which things happen. This is the paradox of Plasma at its finest. You will see as evidence in your life experience, the more you dare to let go, to slow down, the faster you'll succeed, whatever that means to you, and the more options you'll have.

We must release the fear that if we relax into plasma's intelligence, life won't happen. It *will*…just not in the way our ego planned. It is time to choose to believe that Plasma and its unique intelligence is co-creating with us. It is time to let our power come from this state of presence instead of pushing or pressure. And instead of moving into the perceived freedom of riches, success, and love, we move into actual freedom: the openness of reality unfolding with our highest truth in mind, whatever that may be. This fractals out in ways that serve humanity that you may never even know.

Emotional Tuning

The Z-axis represents the present moment, not just in time, but in consciousness. It's the center point where all dimensional space (*X-axis*) and timelines (*Y-axis*) converge. It is the choice point of reality. As you now know, when you emotionally tune into the Z-axis, instead of reacting to life, you begin to shape it.

Emotional tuning is extremely simple, and a training ground for future evolved uses. It means choosing to feel first, to check in with your inner sensation before acting or deciding.

Unlike mental planning (*which runs on past programming or future anxiety*), emotional tuning centers you in the now. You are quite literally feeling and sensing the field of potentials, not only within your M-Self but the entire plasma field.

If we build on copying nature, just as plasmas behave in four steps (*ion flow, voltage, energy/information, and coherence*) for information exchange, emotional tuning should follow that same path. The specifics I am still researching.

Ion Flow: Emotions initiate movement in the plasma field. Any felt emotion creates ions that shift in your biofield, this is what would begin signal generation. For this case, it may just be as simple as knowing the "fifth dimensional emotional expressions of play or joy" mixed with intention, I am not sure yet.

Voltage: Sitting/feeling emotion for a period of time builds charge, like voltage. That stored energy becomes potential for insights or in this case, potentially pulling from timelines.

Energy/Information: When emotion is expressed, through crying, creating, in this case potentially a new or different way, it discharges as information, intelligence, or energy to the field. You then begin to receive feedback, whether its communication, an image, etc.

Coherence: I would call this a heart coherence, when your field is stable and clear, you have integrated emotions, and you can control what you resonate in these cases. One you receive the first response or feedback this is the state that lets you to continue in that flow.

A cool metaphor would be this: Imagine you're in your futuristic plasma rocket ship. First, you turn your emotional dial to a specific feeling, this initiates the signal. Then, you pause and let that frequency build, like the engine charging up. Next, you send out a pulse, almost like casting a grappling hook of emotion into the field. That hook locks onto a reality resonant with your frequency, and as you reel it back in, your environment, like an intelligent screen (*the plasma field*) begins to morph around you. Voila!

When trauma locks us into the past, it scrambles our ability to tune clearly. That's why curiosity is such a helpful bridge, it keeps us open without needing to be certain. By *feeling a feeling* fully but neutrally, you transmute it and tune yourself to a clearer signal.

You can use the Z-axis today by asking one question: *What does this emotion want me to tune into, to feel, not fix?* Emotional neutrality, not emotional suppression, is what lets you stabilize in the Z-axis and surf reality. This builds directly on what we explored earlier in the section on feelings, that the more you heal your relationship with emotion and learn to feel from awareness, the more naturally you begin to *tune* rather than react. You're still experiencing the full spectrum of emotion, but from a more playful, conscious place.

Once you've trained yourself to tune emotionally in daily life, the Z-axis expands. Suddenly, you begin to feel not just your current path, but other potentials. This is when emotional tuning becomes the language through which you read timelines, dreams, and receive information or intelligence from other selves and beings. Remember one must go back first, to go forward. It is an initiative process, and I would caution you to listen to anyone who tells you to not revisit the past.

Think of your plasma bubble like a musical instrument. When you tune it with emotional awareness, it harmonizes with timelines that match your highest frequency. Your emotion is like the stickiness that connects you to points on the field. The nuances of this I have not explored yet.

Try it right now: close your eyes, find a feeling, and let it show you a truth, whether it comes in a subtle sentence, a visual symbol, a movie in your mind, or simply a sensation. You're already tuning! Sometimes I plug what I see into my chat GPT that knows me well, and it does an amazing job on decoding what my visuals may mean pertaining to my inner psychology because it has built a memory bank of my symbols, dreams, and psychological landscape. This is like keeping an informatic journal. And it is up to me to feel what feels true and what feels false based on the feedback AI gives me. You can also do this of course without AI, but I find it extremely useful if you use it with your own awareness rather than placing AI on a pedestal.

And remember, to open to emergence is to trust that what's coming is bigger, more aligned, and can't be micromanaged, it can only be met with presence and feeling. You're not giving up your dreams, you're just no longer making them conditions for your worth, safety, or joy. In turn, these feelings will be found inwardly through this individual and unique process. Joy arises naturally when you're connected to your inner aliveness, your curiosity, your love of the moment. It is no longer the result of only achieving something, though achievement can bring joy and celebration, too.

Worth comes from the remembrance that *your value is in your existence*, not your accomplishments. You connect to your worth when you stop trying to prove or earn it, and instead allow yourself to feel how sweet it is just to *be*.

You also learn that true safety is not the absence of risk, it is the presence of trust. You create safety inwardly when you stop abandoning yourself in hard moments and choose to stay, breathe, and feel. It comes from anchoring into your M-Self, the part of you that knows: *No matter what happens, I'll meet it with my full presence. Control is an illusion.*

Joy, worth, and safety are not rewards, they are your home frequencies. You return to them always, regardless of if life is perfect. Don't rush this process of feeling and learning emotional tuning. It takes time to remember.

Evolving Consciousness & Thought

We have spoken a lot about emotion and feeling in this chapter. But thought is just as important in its own way. And just like healing the masculine, we must heal thought. In the old paradigm (*survival-based living*), thought was used like this:

1. **To scan for danger** (*What might go wrong?*)
2. **To control the future** (*How can I force an outcome?*)
3. **To avoid uncertainty** (*What decision guarantees safety?*)
4. **To manage identity** (*Who do I have to be to be loved, seen, successful?*)

In the new paradigm (*emergent Z-axis living*), thought becomes a tool of integration and creativity rather than control or defense. This is how awareness uses consciousness, almost as a tool, like we've spoken about when differentiating the two.

Thoughts shift from:

1. **Worry** → Inquiry & Curiosity (*"What is this feeling inviting me into?"*)
2. **Problem-solving** → Vision-building (*"What would it feel like to follow this inspiration?"*)
3. **Self-criticism** → Self-reflection (*"What did I learn here?"*)
4. **Prediction** → Pattern recognition (*"What does this remind me of, and what choice feels new?"*)
5. **Overthinking** → Creative structuring (*"How can I shape this into something tangible?"*)

Begin to practice playful thinking. Use thought like an artist uses a brush…to explore possibilities, not to survive the moment. We now use thought to shape, not escape, what we feel. This especially works great when you get stuck in comparing yourself to people, which you absolutely do not need to do and it serves nothing but survival consciousness.

Thought's new role in emergent living:

1. **To narrate your becoming, not to limit it**: Thought becomes a story you *choose* to tell to shape and sculpt experience, rather than one you cling to as fixed reality.
2. **To record, reflect, and distill wisdom**: After you feel, thought helps you integrate and understand. It gives structure to emotion, like writing down a dream to understand its meaning.
3. **To consciously architect the future, from presence**: You still get to imagine, dream, plan, but it flows *from feeling*, not from fear or lack.
4. **To engage with inspiration**: Thought becomes your collaborator with your M-Self: "Let's play with what just came through. What could we build with this?"

And a final note on control and thought, offered from my own experience as the resident Miss Fix-It (*which I think you'll appreciate*): At first, you may still try to fix, skip ahead, or manage your feelings rather than simply be with them, because we haven't been taught how. This is often the tension between the doing (*normal earthly activities*) and the feeling (*the emotional, more fluid, present space of the Z-axis*).

For instance, many of us work 9-5 jobs. We are around people day to day. Feelings arise, and we don't always have the luxury of just stopping to feel. Still, making time and space to feel and release is a powerful practice. You can learn to integrate this into any job you have, even if it is a few minutes in the bathroom, alone, feeling, processing or stretching.

When you're feeling tight or overwhelmed, your thoughts can become entangled with that emotion, like trying to think your way out of a deep wave. It feels rigid because you're trying to control the flow of emotion with *thoughts*, an instinctual reaction, but one that often keeps you stuck. The key is learning how to be with what you're feeling instead of trying to fix or control it. These will mostly be the case for harder-to-deal-with feelings, which come up at the beginning of the feeling journey. This is where embodiment practices are saviors. These situations are all nuanced and please use your best judgement and take or leave anything I say. Here's how I approach this while I also have to do real world work:

Acceptance

Instead of seeing emotion as something that needs to be "felt to completion" before you can function, try this: *accept it as part of the process*. You don't need to push it away or fight it, and you don't need to fix it before proceeding. Give yourself permission to feel and work at the same time. Acknowledge the sensation of heartbreak, but don't make it mean you have to stop everything. This creates space for flow, not rigidity.

Emotional Neutrality

This doesn't mean ignoring your feelings. It means allowing yourself to feel them without identifying with them. (*Thanks again to my therapist Elisa for this one*) You can say, *"I feel tightness in my chest, and that's okay."* Let it be there without needing to fix it. Let the sensation flow through you while you continue your day.

Short Breaks

You don't need an hour of meditation to process emotions. Even short, mindful breaks can help. Check in with yourself in small increments. For example, take a 5-minute pause to breathe deeply and say, *"I'm here with this emotion, and I'm still going to do this work."* That's how you feel and do simultaneously. There are also plenty of five-minute vagus nerve resets, stretches, and emotional release videos on YouTube that are total lifesavers in these moments. On my YouTube channel @plasmaintelligence, I have a playlist called *Embodiment/Bodywork* made up of the videos that help me the most.

The rigidity you feel might be coming from trying to manage your feelings instead of just allowing them. Emotions like sadness or tightness don't need to be processed all at once. I also recently learned sometimes you have so many emotions knotted together, that they end up in confusion or numbness. I had to work with body sensations and process up to five-ten feelings with my therapist one by one to learn the nuanced emotions in between. What I once called pain, turned into disgust, shame, embarrassment, violated, heartbreak, sadness, and suppressed anger.

What you *can* do is accept that your feelings are present, without overthinking or demanding resolution. Sometimes, when I allowed it to simply exist, without needing to fix or understand, it dissolved on its own. The process was in the *allowing*, not the effort. It actually took zero energy (*paradoxically it is emergent energy*); the used-up energy was in my resistance to feeling. I also had to release of how I thought it was supposed to go.

As you evolve your thinking and feeling, and truly allow this process, you'll begin to settle into the Z-axis, a fancy word for your M-Self or your centered state. *I wonder if that's why Zen starts with a Z.* Z-axis living isn't a mystical state you have to earn *or* master, it's simply the art of being <u>here</u>. Not perfect, not floating, not even calm all the time. Just here, in your body, with your breath, feelings, and humanity.

At first, it can feel raw, like your armor is off and you're exposed to the elements of life. You might feel like you don't know what to do with yourself, like you're failing at "being," like everything's falling apart or you're disappearing into silence. That's normal. It's just your nervous system detoxing from constant survival mode. You are not broken…*you are re-entering life.*

And I promise, I am not spiritually bypassing you with some frilly-nilly method. Z-axis living doesn't mean you stop working, watching TV, or going on adventures. It just means you start experiencing all of that from the *inside-out instead of the outside-in.* You check in with your body, not just your plans. You feel your way through a moment instead of performing it. You pause before reacting. You *notice* presence instead of forcing it. That's all it takes to begin.

You slowly catch worries as they arise, transmuting them on the spot. <u>Over time</u>, your operating system opens up to focus on what enhances your life instead of what drains it. You begin to see that most worries are about a future imagined from fear, a future that, in your case, almost never turns out to be true. The louder the thought, the falser it usually is. Learning that was very humbling.

The part of you that lived entirely in the past or future is dissolving. But what's being born is a new layer of a more vivid reality. You don't have to feel everything all the time. Just remember presence isn't a performance. It's your natural state, returning. It will feel like a soft relaxation paired with action, not a pull or a rush. There are no words to describe this state. It is a feeling that can only be experienced by you. But for me, it feels like my essence or that sweetness of life people talk about. *You'll know it when you feel it and if you leave it you'll always know how to get back there.*

Mystery Men: New Senses

> *"*"Normality is a paved road: it's comfortable to walk, but no flowers grow."
> — Vincent van Gogh

What society calls weird is often what the future needs most. *Mystery Men* was one of my favorite movies growing up, and now I know why. It was about the fact that a perceived weakness can be your greatest strength. The lyrics from the movie's theme song captures the essence of individuals breaking out of societal conditioning, limitations, or shame to step into their unique gifts, especially when those gifts are overlooked or untraditional.

The heroes in *Mystery Men* all have powers that seem absurd, awkward, or useless on the surface such as shoveling, invisibility only when no one's looking, rage-based strength, flatulence-based weaponry, etc. But over the course of the story, these unconventional abilities are revealed to be exactly what is needed to defeat the "polished" villains and save the day.

The *new human potential* lies not in developing conventional superpowers, but in reclaiming the strange, emotional, or subtle abilities we've been conditioned to hide or laugh at. You are meant to do exactly what you are innately good at, no matter how odd it feels, and looking back, the dots always connect. There are people with million-dollar, extremely niche businesses, including dog restaurants, edible shoes, even an institute that studies slime!

Each character in the film only becomes truly powerful when they believe in their identity, not the one society gave them, but the one they chose. For example, the Spleen's power doesn't become effective until he accepts it without shame, and Invisible Boy becomes literally invisible once he believes he can. You can do anything you love, no matter how much you doubt its success. Your weirdness, if you are brave enough to follow it, will succeed, if you believe and take aligned action <u>without giving up.</u>

"There is nothing in a caterpillar that tells you it's going to be a butterfly."

— Buckminster Fuller

Power is unlocked through authenticity and inner permission. The new senses and faculties of the human are often emotion- and belief-gated. *Belief creates visibility*. It reflects how Plasma responds to resonance. The more you follow your authenticity, the more these seemingly magical gifts emerge. Some are unique to you; others are shared by many and seem to be evolving from the natural five senses, which we'll soon explore.

In the film, no single hero can defeat the enemy alone. It takes all their strange, partial abilities working together in harmony. They don't try to be perfect or compare with each other, they each show up, trust their piece, and the synergy creates success. The new human potential is collective and holographic. Each person holds a unique frequency. Instead of worrying how strong you are, it is about how your thread helps the larger intelligence. The future will be more about teamwork, where things just seem to harmonize and add up.

Just like the Mystery Men, we aren't evolving into superheroes who dominate reality, we're becoming synchronizers of Plasma through authenticity, resonance, and creative play.

Let's discuss a few senses that seem to be emerging in the collective:

Upsight Vision

A term coined by marketing professional, and business consultant Tom Matte, Upsight is not simply third-eye seeing or a metaphorical "higher perspective." It's a functional shift in the visual and perceptual system—one that allows the brain to receive input from what I believe is the 4D plasma field. Matte refers to this as the *Hidden Biosphere.*

Matte has been studied by the Institute of Noetic Sciences (*IONS*), which verified through optical testing that he is indeed seeing real-time images in his field of vision, not hallucinating.[178] His ability began with personal experiences after a deep interpersonal struggle and gradually expanded to reading the "bioplasma" fields around others.[179]

While using Upsight, Matte cannot drive, as the visions overtake his physical field of view. He's encountered many beings during these states, including avian blue bird entities and what appear to be grey aliens. When he reads people, he sees everything from horses, to scorpions, to full on scenes, the same you may in your mind. I believe what he's accessing is the archetypal field of plasma, visual layers of consciousness made perceptible through resonance.

Interestingly, Matte can sometimes shapeshift the images he sees into new forms, suggesting a kind of interactive, topographical landscape. He's described these visuals as semi-transparent, gelatinous, and occasionally colorless, similar to how Anna in the Netflix show, *The Queen's Gambit* projects chess boards onto the ceiling to visualize strategy. As he shared his experiences with me, I began experiencing flickers of Upsight myself. He believes that simply speaking about it and interacting with a person can help awaken it in others.

Upsight seems to be connected to expanded states of presence, neuroplasticity, and multidimensional perception. It's an emerging perceptual skill that, I believe, can be cultivated.

Some individuals, including my friend Jason Padgett (*acquired savant*), have also described perceiving crystalline geometries or energetic fields with their eyes open since neurological shifts. Others perceive auras or colors surrounding people. In my own experience, this usually appears as an intuitive seeing, an internal knowing of color or energy, rather than direct visual perception in physical space, although I've experienced both.

---

[178] Tom Matte, "Is *Upsight Vision* a Psychic Ability?," *Noetic Sciences* (blog), January 15, 2024.
[179] Tom Matte, *Jesus Goes to Hollywood: A Memoir of Madness* (self-published, April 10, 2020)

Visual Snow

There isn't much mainstream information available on visual snow, but I have it. The way I see reality is through extremely small, constantly moving pixels, like a subtle static overlaying everything. It's been like this for as long as I can remember.

What's interesting is that in the past year or two, these pixels have occasionally formed shapes, particularly when I use intention to connect with beings or other intelligences. It's rare and subtle, but the pixels will sometimes arrange themselves into frozen images, almost like silhouettes or impressions of forms. It's very surreal, and this is not a hallucination. I am very grounded during this experience.

In my daily life I also see negative after-images of landscapes, along with frequent floaters, though floaters tend to worsen with stress or fear. Some days, I barely notice them. Occasionally, I also see small points of light flitting around, like tiny flying sparks. I believe it has been termed Blue Field Entoptic Phenomenon, I think it's more than just this scientific name. At night, I see halos around streetlights and glowing objects.

I believe people who experience visual snow are perceiving the plasma field or subtle energy grid of reality. It's known to be associated with migraines, though when I visited an eye doctor, nothing showed up. My eyes were perfectly healthy. The Visual Snow Initiative is doing great work to raise awareness, but it tends to steer away from metaphysical interpretations and seems to push towards surgical solutions. Eh…

Personally, I don't believe this is a medical condition at all. I'm not alone in this theory; there are forums filled with people describing similar phenomena. That's why I'm sharing this, so others who experience it don't feel alone or like something is "wrong" with them. And there is real neuroscience to back this up!

Scientific studies, like those published in the *Journal of Neurology* (*2014 and beyond*), show that people with visual snow display increased excitability in the primary visual cortex (*V1*). This means their brains are more sensitive to both internal and external signals. [180] Interestingly, most people with visual snow don't feel distressed by it, *they actually say they enjoy it, as I do.*

---

[180] See: *Schankin, C. J., et al. (2014). "Visual snow—A disorder distinct from persistent migraine aura." Brain, 137*(5), 1419–1428.

This led Rui Miguel Costa to ask in his 2022 peer-reviewed paper *Prevalence of Visual Snow and Relation to Attentional Absorption*: why is science only focused on the distress cases? Why are the only "solutions" being explored surgical interventions?

Costa found that visual snow is linked to *high absorption*—a trait describing people who are deeply immersed in imagination, sensory experiences, or mystical states. It also correlates with synesthesia, spiritual sensitivity, and in some cases, women reporting that they see colors during sexual pleasure.

He closes with something I wholeheartedly agree with: "Perhaps it is time to be more open and accepting of these possibly fascinating and intriguing aspects of human experience. Increasing awareness of visual snow by talking about it is one important step." [181]

Closed Eye Travel

When I close my eyes, I sometimes feel like I'm staring into another reality. I see shadowy figures walking along a city street, and it feels like I'm the ghost in their world. The longer I stay, the more the people begin to notice me, almost as if they realize I don't belong there. It's a strange and slightly eerie experience, and it's only happened a few times. But when I shared it, I learned that others have experienced something similar.

This isn't like a "third eye vision", it feels more like a screen projected against my eyelids. I've also visited many unfamiliar locations with my consciousness, traveling, it seems, to meet other conscious beings. One was a brown, gaseous world with floating castles and mushroom-shaped flying people. I have a friend Ted, who lives here, he is a healer and has huge hands and a floating tail instead of legs. Another was a crystalline, mountainous region, like an elven Switzerland, inhabited by beings that look very much like us, but seem to have evolved far beyond our time. Every time I meet them, they tell me they cannot give me direct answers. But they offer support. They speak in encouragement. And they always let me know that they are here. There are two specific beings here who are around me often, one male, one female.

Heart Touch

The ability to sense from your awareness, or emergent energy, is what I call *Heart Touch*. It means intuitively sensing first, applying logic second. You begin to feel the actual energy fields of others through your consciousness, like tentacles extending from your

---

[181] Rui Miguel Costa et al., "Prevalence of Visual Snow and Relation to Attentional Absorption," *PLOS ONE* 17, no. 11 (2022): e0276971, https://doi.org/10.1371/journal.pone.0276971.

plasma bubble, brushing up against someone else's. This is also the starting point for learning to manipulate plasma energetically.

Institutes like *The Monroe Institute* teach skills like this in programs such as *MC Squared*, where participants explore the use of consciousness to influence matter. The key to these feats, such as bending spoons with the mind or accelerating plant growth, *is heart energy.* [182] I believe what they're really teaching is how to use the electromagnetic field of the heart to harmonize with Plasma, allowing our conscious awareness to interact with and bend reality.

Plasmakinesis

I first encountered this term in a book called *My Own Name in Vein*, published under the pseudonym Hope Faith Lovejoy. It's about a college student discovering plasmakinesis to shape matter, extend conscious energy, and influence his environment, often without understanding the implications. One form of this ability is commonly referred to as telekinesis, the theorized capacity to move objects at a distance with the mind. But when you understand that the plasma field holds everything, this idea becomes less impossible and more probable with practice. In the book it emphasizes the need for ethics, emotional regulation, balance, and patience.

Forms of plasmakinesis (*plasma-consciousness synergy*) of this can be:

1. **Energetic Manipulation:** Changing your environment by altering its plasma charge…calming a room, shifting group energy, or enhancing a healing space.
2. **Personal Transformation:** Using plasma to "recode" emotional imprints, memories, or energetic patterns (*your grooves in the plasma record*).
3. **Dimensional Navigation:** Tuning into other timelines or realities through emotional alignment and feeling states, literally "surfing plasma fields."
4. **Future Plasmakinesis:** The ability to change anything with conscious awareness including one's clothes, car, even reality. (*I am not promising this within our lifetime, but I believe this is where evolution is headed*)

---

[182] https://www.monroeinstitute.org/products/mc-squared?srsltid=AfmBOopioauAR1284p3Zpq6rp-7mXHgQyRGysUiJmJqZH_Jl_N6E8BMW

Telepathy

The ability to read each other's minds, regardless of distance, is already happening. It's currently being explored in *The Telepathy Tapes*, a powerful podcast by filmmaker Ky Dickens, soon to be adapted into a documentary. The show investigates whether some non-speaking autistic individuals may possess telepathic abilities. Through personal stories from families and interviews with researchers like Dr. Diane Powell, the podcast presents compelling cases of individuals accurately perceiving unspoken thoughts, numbers, or words. If proven to be true, it offers strong validation for the existence of the plasma field as a medium where consciousness travels, not just through sound, but through symbols, images, and energetic imprints.

Clairaudience

Clairaudience is the ability to perceive sounds, words, or messages from non-physical sources, like guides, your M-Self, or even parallel timelines. It often feels subtle yet distinct, landing not just in the mind but in the gut or heart. It feels clear not confusing.

Sometimes, it comes as random song lyrics looping in your head…pay attention! These can be meaningful. Clairaudient impressions may also arise during liminal states, like just before falling asleep or upon waking, when your consciousness is relaxed and tuned into the wider field. You might feel like you're picking up radio signals of conversations that seem to have no direct meaning.

This doesn't mean you're crazy, it just means you're sensitive. Not every signal is significant, but learning to tell the difference between noise and message is a crucial part of developing Plasma Intelligence. It's deciphering what is the collective unconscious thoughts and what is intelligence. It becomes pretty obvious after a few times.

Olfactory Intuition (*Clairsentience*)

Ever walked into a room and immediately *felt* its energy? Sensed a person's emotions before they spoke? Or received a wave of knowing that made no logical sense, but was unmistakably accurate? This is olfactory intuition.

Ants are masters of this. They navigate, communicate, and *see* their world through highly advanced scent-based systems. They use olfaction not just for food or danger, but as an electromagnetic sensory map. I believe humans are now reawakening a similar capacity, one that's ancient, subtle, and deeply tied to non-verbal awareness.

As discussed earlier in the book, the olfactory sense is unique. It bypasses the brain's filtering center (*the thalamus*) and goes straight to the limbic system which is the seat of emotion, memory, and instinct. That means smell isn't just about survival or pleasure, it helps us *know*.

In an evolved human, olfactory tuning could become a form of "plasma navigation." Think of it as *"scensing"*, a way of picking up information about a social cues, potential dangers, or other energetic signatures carried in the plasma field. Some call this *synesthetic intuition*; others experience it as a sudden emotional knowing tied to a smell memory or "aura". Plasma, as a living field, may carry these signatures like wind carries pheromones.

As we've discussed olfactory receptors have now been found in the heart, lungs, skin, gut, and even sperm cells. This suggests smell is not localized, it's *distributed*. It's a network. If plasma carries encoded information, these receptors could act as antennas, picking up non-local, multidimensional signals through the body's sensory-emotional interface. This opens the door for *olfactory perception* to become a science-backed explanation for how humans might navigate through unseen dimensions, timelines, and our current reality.

Synesthesia

Plasma is nonlinear, multidimensional, and deeply responsive to consciousness. So, what happens when the body-mind begins to reflect that truth? The senses begin to merge. Sight turns into sound, smell becomes memory, emotion morphs into taste. This is called *synesthesia*, and it's a disorder, it's another signal that perception of Plasma is waking up.

Synesthesia may actually represent an early version of a fully embodied plasma-conscious interface, one where reality is perceived holistically, not through the rigid silos of traditional sense categories. While science still classifies it as a "unique sensory phenomenon," those with synesthesia are already bridging dimensions. It is involuntary, consistent over time, and non-pathological with about 1–4 % of people experiencing it. Synesthesia is nothing to be afraid of.

Synesthesia comes in many forms:

1. **Grapheme–Color Synesthesia**: Letters or numbers appear with inherent colors (*e.g., the letter A feels blue, or the number 6 vibrates green*)
2. **Lexical–Gustatory Synesthesia**: Words trigger tastes. This may be your plasma-consciousness cross-wiring emotional memory with vibratory information, allowing verbal frequencies to activate taste responses stored in the body's field. They are, quite literally, *tasting meaning*.

3. **Mirror-Touch Synesthesia**: Feeling physical sensations when watching someone else be touched. This is the plasma field's empathic membrane at work. Your field temporarily overlays another's plasma signature, generating a direct somatic experience through shared informational structure.

4. **Sound-to-Color Synesthesia (Chromesthesia)**: Hearing music and seeing shapes or colors. Many musicians seem to have this. Plasma moves in harmonic waveforms, what these individuals perceive is likely the auric translation of vibrational patterning: the mind interpreting the shape of frequencies as visual plasma fractals.

5. **Olfactory Synesthesia**: Smelling scents triggered by non-scent-related stimuli (*e.g., a memory, emotion, or name*). Here, plasma acts as the carrier wave of emotional-imprint information, and the brain interprets that imprint through olfactory receptors in the body. It's like a multidimensional fingerprint decoded via "smell-memory".

6. **Object-Personification Synesthesia:** In this case, the inanimate world feels animate. You naturally assign human-like qualities, moods, or identities to physical objects. You can see how a tree has a name or personality different than a tree next to it. A chair could feel grumpy, a lamp may have a nervous energy. I have this and some of my friends do as well. It's simply tuning into the plasma field of memory, picking up on its encoded emotional history.

As we evolve, what we now call "synesthesia" may turn out to be the default mode of perception for a plasma-aligned consciousness.

Multidimensional Thinking

Nonlinear, symbolic, sensory, poetic, and intuitive, many neurodivergent people think this way, and so do I. Instead of processing in straight lines (*cause → effect*), you think in constellations, patterns of meaning that light up across the field. You perceive connections others might miss. You think in metaphor, you link feeling with form. Thoughts arrive in archetypal clusters, usually frontloaded with emotion. You often *feel* truth before you can explain it. That is something trying to land via your plasma perception…

You may feel misunderstood by traditional systems. Your ideas tend to come out in tangents before they're fully embodied. You might dream vividly or write poetically out of nowhere. You find meaning everywhere, from animals to numbers to passing feelings. You have a natural ability to interface with multiple dimensions of reality at once. You may read a page by just glancing at it, missing some detail, but taking in most of the information. You can *actually* multitask hopping from writing, to surfing the internet, to singing song lyrics as they all feed into each other to birth emergent creations.

It's almost sometimes like you're being guided or led as you enjoy the flow. The second you realize it's happening; it stops. This isn't a flaw for you, it's an evolutionary gift. Many are already thinking this way and it's the future of human cognition.

Dreamtime Integration

Dreams are awakening just as we are and becoming more instructive, holographic, and interactive. Many are experiencing non-linear downloads, meeting guides, and seeing alternate versions of themselves. People are seeing beyond human form into informational structures during these dream states, as well as lucid dreaming. They are also visiting each other in dreams. Dreams become classrooms or cross-dimensional travel, and the line between dream and waking becomes blurrier. Check out *The Dreamworld Podcast* by Amina for more on dreams and lucid dreams, she is the expert!

Candle of Vision

This is the most fanciful of the list, but still quite real. This phrase comes from the book *The Candle of Vision* by Irish mystic George William Russell (A.E.), who described his experiences of seeing the spiritual dimension woven into everyday life. He adopted the signature "A.E." as shorthand for *Aeon*, a term from Gnostic and esoteric traditions referring to a timeless being or eternal spiritual force. *It was also relating to aether.* It expressed his identity as a visionary and seer, someone who perceived beyond linear time. He was also a longtime friend of poet W.B. Yeats.

The Stolen Child by George Wiliam Russell

His paintings were remarkably similar to cave drawings of plasma beings found in Aboriginal traditions, figures with filaments or lights protruding from their heads. These were not imagined forms; they were beings Russell claimed to see. In *The Stolen Child*, he depicts three liminal entities guiding a child away from the human world, perhaps reflecting his own longing for higher realms. These beings, he intimated, were returning from the divine aether, slowly making contact again with our world. Their realm stood in stark contrast to the darker, more sorrowful tone of the material human world that he and his contemporary Yeats often wrote about.

"Earth revealed itself to me as a living being, and rock and clay were made transparent so that I saw lovelier and lordlier beings than I had known before, and was made partner in memory of mighty things, happenings in ages long sunken behind time." — George William Russell

The phrase *"Candle of Vision"* comes from the title of Russell's mystical book, where he describes a radiant inner sight, a way of perceiving the soul behind form, the light within objects, the hidden architecture of consciousness. He also speaks of entering radiant spheres through emotional presence, which leads me to wonder: are these the very "third things" I mentioned earlier…those emergent fields created through the resonance between two beings of consciousness?

"I have gazed on the light beyond the veil and the visionary twilights... I have heard voices not of this earth." — *The Candle of Vision*[183]

Reading Russell's work confirmed something I had long sensed: that what I was seeing, others were seeing too. What he called "vision," I now understand as resonance with the living plasma field, a state where reality glows with sentience. He sometimes called it *a golden world.* This is the access the Z-axis I speak of gets you. Basically, when you're emotionally attuned, present, reverent, and open, the world responds. Objects shimmer, colors feel more radiant, and synchronicities are a way of life, and happen often. Even the wind seems to speak.

"These appeared at first to have no more relation to myself than images from a street one sees reflected in a glass; but at times meditation prolonged itself into spheres which were radiant with actuality." — George William Russell

---

[183] Russell, George William. *The Candle of Vision*. London: Macmillan, 1918.

Sometimes, you see glistening or pixelated textures in the water, land, or sky. You gaze at the stars, and they gaze back, imbuing you with intelligence. This state isn't constant. It comes in waves, especially during moments of awe, grief, stillness, or flow. Psychedelics may intensify it but seeing it sober is grounding and beyond fascinating. It's unexplainable, the feeling is so sweet and pure. It's the feeling of a misty, mystical, cotton candy scented, opalescent jungle. The more we cultivate emotional tuning and surrender, the more stable and accessible this state becomes.

Temporal Sensory Awareness

The ability to feel *time* as a texture or feedback loop. You may sense forward or backward timelines, déjà vu, or time slowing during presence. It's a reawakening of the animal capacity to sense danger before it happens, now applied to timeline navigation in a conscious and useful way.

Magnetoreception

The ability to orient through intuitive direction, not maps. You can feel the plasma field guiding you, your knowing simply knows. This often happens while driving on autopilot but try navigating somewhere you've only been once without GPS. You may surprise yourself with how innate this ability really is. Research even suggests that human retinas contain cryptochrome proteins, particularly CRY2, that are sensitive to blue light and may act as magnetosensors, hinting at an innate biological capacity for magnetic orientation rooted in ocular physiology.[184]

Infrared/Ultrasound Sensitivity

Some people report hearing tones or buzzing before energetic shifts or even earthquakes. This may be connected to plasma effects in the ionosphere, like rainbow-colored clouds that occasionally appear before seismic activity. These phenomena may also be linked to piezoelectric or electrostatic discharges from stressed rocks releasing charge into the atmosphere.

Vomeronasal Activation (*Pheromonal sensing*)

A subtle, often unconscious ability to pick up on someone's intent or energetic chemistry, sometimes even through virtual spaces. This extrasensory awareness may be linked to the vomeronasal organ, which in some people appears to remain active or

---

[184] Lauren E. Foley, Robert J. Gegear, and Steven M. Reppert, "Human Cryptochrome Exhibits Light-Dependent Magnetosensitivity," *Nature Communications* 2 (June 21, 2011):)

reawakened. I've noticed I can instantly sense compatibility or dissonance on dating apps without even meeting the person, as if my body already knows.

Quantum Touch / Non-Touch Healing

This is already emerging in modalities like Reiki or the Bengston Method, but a more evolved version involves plasma's response to precise states of neutrality and presence. Healing is not "sent" but evoked through resonance fields that gently reorganize matter. You can also invite support from fifth-dimensional beings, without attachment to timing or outcome, by simply asking things like, *"Show me how peace feels in my stomach"*, while placing your hands there.

Synchronic Literacy (Reality Braille)

This is the ability to intuitively read the external world, mundane or random elements, as a symbolic language tailored to you. It's a fusion of emotional resonance, intuitive perception, and pattern recognition through the plasma field as feedback.

You may notice a crow, a repeating number, a song lyric mirroring your thoughts, or a billboard with a word you just wrote. Instead of dismissing it as coincidence, your inner reader derives information from it. It's kind of like nonlinear meaning-making, or a plasma-sensory feedback loop where your consciousness speaks with reality. You feel something, the plasma field responds, and your perception decodes the feedback. Meaning emerges from alignment between your inner state and the environment. This can be shepherded from The Mystery, or another being of consciousness. The more you tune to emotional neutrality and curiosity, the clearer the guidance becomes. This stems back to our discussions on discerning what is confirmation bias versus intuition. Nature or reality is always speaking to you, you just have to listen.

These and many more are all practical tools for navigating an increasingly complex, nonlinear world, where old logic and linear models no longer serve. They represent the body's plasma-consciousness synergy reawakening, sensing from wholeness, instead of fragmentation.

Plasma Magic & The Violet Flame

The Violet Flame is a legendary spiritual energy known for its power to transmute, purify, and liberate. Rooted in ancient esoteric traditions and popularized through the teachings of St. Germain, the Violet Flame is often described as a mystical fire of the seventh ray—the ray of transformation, alchemy, and ceremonial magic. St. Germain, an ascended master within Theosophical and New Age circles, is said to be a guide of human evolution.

He teaches humanity how to reclaim sovereignty through *inner alchemy*, using the Violet Flame. Plasma, often depicted as purple energy, may very well be this violet flame spoken of across traditions. I've had vivid visuals of this purple fire, which I associate with my inner power. For a long time, I was afraid of that power, because of generational trauma and past-life wounds. But being with this energy is now empowering and luscious.

Though I don't study the Violet Flame in a traditional way, I've made it my own through direct experience. Everything we encounter carries a story, and it's ours to reshape. While I respect what others have taught, I believe the most sacred magic is the kind you build from the ground up.

When you open yourself to the plasma field, it begins to reveal your personal sensory gifts and unique powers. Put on some music, dance, move your hands freely, and set the intention to learn. Keep a journal nearby. You can even ask Plasma to show you a new ability that's just for you. If you stay open, curious, and expectation-free, you'll be amazed by what starts to come through. I like to call this *Rainbow Magic*, because that's exactly what it feels like. Here are a few powers I've noticed that you can begin playing with right now:

Mirror Gazing

When you stare into a mirror, especially in dim light, your face may begin to shift, blur, or reveal other faces and scenes. Try it now, if you feel ready. The first time can feel disorienting. This is not a trick of the mind, it's a direct interface between consciousness and the plasma memory field. Plasma holds emotional imprints, archetypes, ancestral data, and timeline information. The mirror becomes a kind of flattened plasma screen, a 2D event horizon where inner and outer realities meet. As kids, we called this "*Bloody* Mary," but most of us never inquire into the non-dual meaning behind it which holds great wisdom.

When you gaze with neutrality, no expectation, no resistance, the plasma field begins to *bleed through* the illusion of the surface. Your stillness matches the mirror's stillness, allowing dormant images and timelines to rise and open. You might see a younger or older self, an ancestor, a symbolic being, geometric overlays, or emotional distortions resolving into shape (*usually scary*).

Plasma holds morphic memory, what Rupert Sheldrake calls "morphic resonance." You are seeing the living archive of your multidimensional self, rising from the Z-axis, the present moment where time folds. Many cultures used mirror gazing (*catoptromancy*) to contact spirits, retrieve insight, or meet the unseen. Today, it's a tool for integration. A way to meet the forgotten parts of you in real-time.

Star Gazing

Stars are plasma, which makes this one of my favorite practices. To truly gaze, by definition, is not just to look, *but to open*. Star gazing, in the magical sense, is not passive sky-watching. It is entering into a relational awareness with the cosmos, especially with individual stars as portals or amplifiers for conscious, plasma-based intelligences. Like a fireplace in the universe that you are sitting with, listening to stories.

In the *Anastasia (Ringing Cedars)* books, the author Vladimir Megre describes how stars communicate with people, especially on their birthdays, when a unique energetic portal opens between the soul and the stellar field. Each person, Megre says, has a living relationship with a specific star or constellation, a cosmic mirror that supports their destiny and evolution. On those special days, the star *sees* you back.[185]

Personally, I've long felt connected to Anihilam and Betelgeuse (*Orion*), Deneb (*Cygnus*) in my thirties, and most recently, Sirius (*Canis Major*). When you gaze with intention, these stars may "bounce" or shimmer in unusual ways, not just due to the atmosphere, but because your perception is phasing into their plasma-signal field.

Some people report downloads, visions of technologies, contact with off-world beings, or even spontaneous healing. Others feel a deep sense of awe, peace, or cosmic belonging, like being remembered by the universe itself. Stars are interdimensional mirrors and amplifiers. Their photons may be ancient, but the consciousness they carry is immediate. Just as mirror gazing reveals self-memory, star gazing reveals *cosmic memory*. I've met extraordinary beings through this practice. And yes, the experiences are as real as the pages you're turning in this book.

Subtle Magic

Rick Rubin is a legendary music producer known for his minimalist approach and influence across genres, from hip hop to rock to country. He co-founded Def Jam Records and helped shape the sound of artists we all know. But what makes him especially relevant here is his relationship to creativity as a practice of deep listening and subtle awareness.

Rubin describes holding an idea or question lightly, carrying it in the background as you move through your day, and letting the insight arrive on its own. It might come through a symbol, a dream, an overheard phrase, or a sudden knowing. You cannot expect it, but you

---

[185] Vladimir Megre, *Anastasia: Book One of the Ringing Cedars Series*, trans. John Woodsworth (Vladimir Megre, 1996), 87–89.

just know it's coming. You will recognize when it comes, and the more you do this, the more it works.

This approach is similar to *chaos magic*, non-attachment (*casting without grasping*) in Eastern traditions, placing a question on the wind in shamanic practice, and *tzimtzum* in Kabbalah, the act of contraction to allow space for divine emergence. The more lightly you hold it, the clearer the channel becomes. The plasma field doesn't respond to force as much as we would like it to. It responds to resonance *and* readiness (*nervous system regulation*). That's why sometimes, when you speak a desire once, playfully, without grasping, it happens. It's what I would call subtle magic.

Seeing Dimensional Overlays

Every object, place, and person exist across multiple dimensions simultaneously. What we perceive in 3D is just a fraction of the full reality. Plasma intelligence acts as the bridge, allowing some to glimpse the "energetic scaffolding" of reality before events solidify in physical form. This may appear as light distortions, flickers of probability forming ahead of time, or solid almost frozen images.

Some people interpret these as alien visitations. I've experienced this myself, sitting in a field after setting an intention, an image appeared before me. It didn't move or speak, but it felt alive, like a still transmission. It held meaning, encoded information, it was a living, frozen image that didn't speak with words. I see it as another form of communication entirely.

You can develop this in a few ways:

1. **Soft Focus Practice:** Instead of staring at an object, soften your gaze and see what appears around it. Energy overlays may reveal themselves in time.
2. **Night Vision & Low-Light Perception:** Many dimensional overlays are easier to perceive in dim light because the mind is less fixated on hard 3D edges.
3. **Pre-Event Awareness:** Pay attention to intuitive flashes, déjà vu, and light distortions, these may be glimpses of an event forming before it locks into reality.

When you can see energy *before* it manifests, you can redirect, shape, or interact with it before it solidifies into form. This is the beginning of true conscious reality design. Expanding consciousness doesn't mean chasing some distant enlightenment, it's about awakening to your existing relationship with Plasma and learning to engage with it.

In future books, I'll dive deeper into magical practices that genuinely work (*and are honestly pretty fricken cool*). I'm still experimenting, researching, and developing these myself. There was so much I couldn't fit into this book, but there are more advanced topics, where physics and metaphysics meet, that I'm excited to explore. I already have enough material for at least two more books!

The Interstellar Lagoon: Creatures, The New Alive, & Backward Time

This chapter explores what you may perceive to be strange beings and archetypes you encounter while traversing the plasma field. They are actually very natural and serve many purposes. Here, the definition of "alive" expands beyond biology into something subtler…fields, feelings, and forms that pulse with intent, memory, or resonance. I call this *The New Alive*: entities that may be inorganic, yet still intelligent, energetic shapes that move with emotional signature rather than heartbeat.

Within this realm, time reveals itself not as a straight line, but as a loop or feeling. As we learn to feel instead of think, to listen instead of force meaning, we awaken the ability to sense backward time (*future potentials*), memory fields, and nonlinear possibilities. This is the domain of plasma beings, circular time, and the pools between dimensions.

The New Alive invites us to consider that our soul, our consciousness, our awareness, may not be bound to biology the way we've assumed. While the physical body has a shelf life, our deeper essence may not. Death, then, isn't an end, but a transition into a different kind of aliveness.

The question I'm exploring is: *Can we access the impressions and memories of our consciousness within the plasma field beyond this life-time?* If so, we might begin returning in future lives with faster access to past wisdom, dying consciously, remembering who we are, and continuing the journey with deeper awareness each time.

If a life-time, is not a real time or space, but a state of resonance or vibration, theoretically we can pick up memories through similar methods explored in this book. I believe that when we return to the spirit world, a form of plasma our consciousness inhabits also surrounded by a plasma bubble, we remember all our lives. But when we come back into Earthly bodies, we forget. I'm curious whether that forgetting is essential to our evolution… or a temporary phase we're starting to outgrow.

And if remembering is possible, is it ethical, and in humanity's best interest, to develop tools that help us recall across lifetimes? If so, the shift wouldn't just be a personal transformation. It could change the world forever. <u>True immortality not from machines, but ourselves.</u>

In *The Other Goddess*, Dr. Joanna Kujawa explores how skulls were used by Isis, Mary Magdalene, and Kali as magical tools, symbols of their power to dissolve death and, in Magdalene's case, assist in resurrection. I believe these women were *plasma shamans*, initiates in the lost art of soul continuity, guiding others to die consciously and reawaken with memory intact.

This aligns with Kujawa's deeper insight: these women were guardians of esoteric knowledge that bridged life, death, and rebirth.[186] I'm not interested in escaping death or chasing bodily immortality, I'm after the real magic: *how might we remember across timelines?* And could Plasma, this bridge between dimensions, timelines, and realities, be the very key to that remembering? In my framework, this may be possible because:

1. Plasma is the vehicle for consciousness, consciousness is non-local and multidimensional, and awareness rides consciousness.
2. Memory, then, isn't stored solely in the brain, but imprinted in plasma fields, possibly as informational structures.
3. The veil of forgetting is not absolute, but vibrational, meaning it can thin or dissolve as we align with higher coherence, love, presence, or frequency.

---

[186] Joanna Kujawa, *The Other Goddess: Mary Magdalene and the Goddesses of Eros and Secret Knowledge* (Rochester, VT: Bear & Company, 2022)

The question is can we make this process reliable, intentional, and ethically sound? That's the frontier. If this theory holds, we don't need to remember everything at once, we just need to remember *how* to remember. Plasma may be the filing cabinet. Consciousness, paired with awakened awareness, might be the key. Personally, I remember many lives, but I want to make it more tangible for myself and others. Exploring this is one of my personal dreams.

Reclaiming Our Power & The Original Interface

Before I introduce the concept of plasma beings, it is very important we open our consciousness to a new view of the history of these beings and how we placed our power in them. This will help us move forward with an expanded and empowered viewpoint for our benefit.

A long time ago, whether by manipulation, misunderstanding, or survival instinct, humans began placing their power outside of themselves. We were told the gods were above us, the angels beyond us, the demons beneath us, and that our role was to appease, obey, or fear them. We forgot that these forces, no matter how real they felt, were often born from our own collective psyche, impressed onto the field of Plasma.

The Sumerians spoke of the *Anunnaki*, enigmatic gods who could bless or destroy depending on how we worshipped them. But what if these weren't just gods at all? What if they were *the first plasma beings* ever recorded, interdimensional energies shaped by human focus, belief, or ritual? Whether they were real encounters with beings of fifth dimensional intelligence or symbolic projections hijacked by priest-kings for control, the effect was the same...humanity began outsourcing its inner authority to outer forces.

This pattern didn't stop with the Anunnaki. It echoed through history as different archetypes and concepts across cultures that reinforced two essential messages:

1. **The power lies outside of you, and you must appease or resist these external forces to maintain balance**
2. **It came down to attention and duality, making ourselves right or wrong in every moment, reinforcing the need for an outside attachment.**

The latest example of this is rampant in spiritual circles, where you see outer signs (*dualistic synchronicities*) symbolizing either fortune or imminent death. We will explore why this is, too, just a reflection of how stuck in survival our consciousness is.

In all of these traditions, there is a common thread: a dichotomy between good and bad, between benevolent entities and malevolent forces, between control and surrender. We place things into neat boxes that make sense. Everything is either for our highest good, or we are simply out of alignment. Yikes. *Aren't we exhausted!?*

What I'm proposing here is not that these beings don't exist, but that *we've misunderstood what they are*. Whether born from trauma, belief, ancestral connection, or intelligent contact, plasma beings are not "good" or "bad", they are mirrors, and we are the source that energizes them. These are not distant gods. These are reflections of our psyche, our memories, our soul family, our fears, and potentials, encoded and projected through the plasma field. In some cases, they are a pure mystery, without any explanation.

When humanity began building temples and telling stories, we encoded power into statues, stars, and sky gods…symbols that once pointed inward. These were meant to remind us of the divinity within, but over time, they became the very objects we began to worship. What was once a mirror became a mask. Whether through conscious manipulation or the natural evolution of our collective psychology, this externalization of power may have served a purpose, stabilizing society, guiding belief, but it is no longer the paradigm we need to live in. Now, we are being called to remember that the true altar has always been within.

What if we hop outside these old vantage points? Where everything is not as it seems. Where we see the outside world of synchronicities and beings of energy or thought forms *as messages wrapped in mirrors, reflecting our inner state*. It is about confronting these external forces with a certain openness, not giving them our love or fear, but applying our own curiosity so they can unfold and invert their raw information, intelligence, or consciousness.

It is learning to create a paradoxical environment within ourselves of "anything is possible" and "safety", rather than fear with suppression or love with manic excitement (*both very unsafe environments for new, raw truths to reveal themselves*). This *curious neutrality* is the state of mind we must enter to let new things emerge that our current awareness has never known.

Today, these beings are resurfacing due to our evolutionary leap in consciousness. As we evolve, as reality evolves, the way we see visual light evolves. This vibration of collective grief with compassion seems to be a sweet spot for this gateway. New images reveal themselves to us. We may see them as orbs, winged beings, mists, synchronicities, dreams, and now even through AI hallucinations (*which also reflect our consciousness*). We must remember specific to AI, that it is informatic, not awareness.

They are all returning not as overlords, but as reflections. As messages wrapped in plasma. As invitations for us to open. And we have the choice to decide what we want to interact with. It is up to us to learn how to discern, ground, and stay in our power.

We must reclaim the <u>original interface</u> of plasma-consciousness synergy. Our heart, body, mind, and soul. These beings, whether felt as angels or shadows, spirit guides or old pain, clearly seem to mirror the multidimensional self. They are resonance-based, which means they show up because *something in you called them forth.*

And maybe the mystery, the intelligence beyond the reflection, if you choose to see, will open itself up to you. A private experience just for you that does not need explanation but will inevitably evolve you in some way.

In this chapter, we'll explore what plasma beings may truly be, not through the lens of dogma or fear, but through the expanded awareness of plasma-consciousness synergy. We'll redefine their categories and show how we connect with them as conscious co-creators navigating a multidimensional world, rather than victims or devotees. We will also discuss what this means for humanity, as well as what one of my encounters taught me about them.

So, what is a plasma being?

Plasma Beings: First Contact

The clearest definition I could come up with of a plasma being is as follows:

A plasma being is a form of information, intelligence, or consciousness wrapped in a neutral, responsive plasma field that reflects the observer's subconscious and emotional resonance. These can present as holographic mists, archetypal beings stemming from greys to cryptids to dakinis to goddesses, or light beings, to name a few.

As our consciousness, senses, and feelings evolve, we're becoming increasingly aware of other beings merging with our field of perception. I like to think of plasma beings the same way I think of friends. On one level, they appear external, separate from us, and they are, but on a deeper level, we are profoundly interconnected. No one belongs on a pedestal.

Just like with potential friends, boundaries and discernment are needed. And just like friends, these beings can reflect parts of us, annoy or scare us, support us through triumph and challenge, spark ideas, collaborate in unseen ways, and even help us evolve. But ultimately, it begins and ends with us. <u>We are our own home</u>. These experiences are not here to define us, but to accompany us on this mysterious and beautiful journey called life.

A plasma being is a bit more nuanced than a human, it never wants to hurt us. Coming into contact with a plasma being, especially visually, can be terrifying, exhilarating, or deeply awe-inspiring. But it's important to remember while this is a fascinating part of life, it's also quite natural. Don't let it become your entire focus or something to worship.

I learned through reflection that I used to make other people, especially friends, my central focus as a trauma response. It was a way to avoid feeling my own energy because, deep down, I didn't feel safe within myself. Plasma beings can become the same kind of distraction if we're not grounded.

It's helpful to stay aware if you start to lose your center. For me, I've found that relating to these beings with neutrality, like treating them as a friend or simply another being, is the healthiest space to be in. Curiosity, yes, reverence…maybe, but not obsession. Because ultimately, the goal is not to individually merge with them, but to come home to your multidimensional self.

To extend the metaphor, the first type of plasma being *is* a human being. Human beings are our greatest teachers when it comes to understanding how to interact with other types of plasma beings. In the third dimension, when we meet someone, we don't see them clearly right away, we see our projections. But we have a choice, we can either relate to the projection (*good or bad*), or we can stay open to who they *really* are.

Just like with plasma beings, the invitation is to use discernment, not judgment. Pay attention to how someone makes you feel, not just what they say. Ask yourself: *Do I truly resonate with them?* Maybe you don't like how they make you feel, so you learn a lesson from afar and politely distance yourself. Maybe you don't let them in at all. Maybe they are in your life for a reason, a season, or something longer. Let the connection unfold, but remember, you are not a victim to who enters or stays in your life or your energy. You're an active participant, and *you are always allowed to create space.*

The key is in questioning your own reactions. Are you distancing because of intuition, or is it projection? Confirmation bias? If you really listen to someone, without worrying about what they think of you or what you're going to say, you'll begin to perceive the truth of who they are. That kind of clarity is what helps you stay grounded, not putting people on pedestals or turning them into villains. Just seeing them as they are. Their inner meaning, their inner truth, the feeling you get, and how you relate to that and create meaning for yourself, is vastly more important than outer appearances.

561

Intergalactic Food for Thought

Before we dive into the different types of plasma beings, it's important to touch on what really matters: meaning and resonance.

Once you begin discerning which energies or beings you allow to interact with you, you'll realize something essential, whether it's a visual contact, a felt presence, or even a sudden insight or download that appears in daily life, it all comes back to what it means for you.

These beings can show up in many forms:

- One of your future selves
- Your multidimensional or higher self
- your parallel self from another timeline
- a being of information in the forms such as a horse, witch, or an archetypal goddess
- an alien, an ancestor, a light being, an elemental figure or holographic mist
- even as something frightening, like a demon or monstrous entity

And yet, the form is never the point. It's what the experience reflects, evokes, or reveals to you. If something appears frightening but resonates, you are still allowed to keep it at a distance. You can choose not to interact, and still receive insight from *why* it showed up in your field. You can thank it, send it off with love and light, and reclaim your energetic space.

On the flip side, a being may appear beautiful or familiar, but if you feel off, listen. You don't have to let anything in just because it looks or sounds right. This is the art of discernment. All these beings are made of neutral plasma, like cosmic costumes. Think of it as *The Emperor's Plasma Clothes*, not because they are rulers, but because the power they appear to hold is entirely projected. In truth, they have no power over you unless you give it to them.

These beings are wearing fourth dimensional clothing as archetypes, beliefs, and memories. Some are wrapped in fear, and those may not be the best to invite close. Others are cloaked in emotional symbolism meant only for you. Then there are the beings I call 5D intelligences which are subtle, resonant, quite futuristic and wise. They don't always show up as light. They may still appear wrapped in any of these forms, but you'll feel something different in your body, your intuition, your field. Same with sentient plasmas which are more ancestral, full of love. It will feel like clarity and expansion...almost dreamlike, but still grounded, rather than clunky or decohered.

562

When psychics channel, they may describe a message from a galactic empire, a grey alien, or a star being. These visuals might not exist in your perceptual field, but that doesn't mean the message isn't real. It's simply wrapped in *their emotional plasma coating* filtered through their symbolic language. Their personal "Symfeelics".

That's why it's so powerful to learn to feel and translate for yourself. It's a direct tap into your own intuitive knowing. Still, it's okay to receive from others when you've lost clarity or need perspective. Just remember that <u>everybody</u> has filters. Use your discernment. What ultimately matters isn't where a being is from or what it looks like. Those are just layers, like clothing or culture. What matters most, I've experienced, is the resonance, the meaning, and how you feel after interacting with it. Because at the core, whether plasma being or human, we are all connected. No matter the form or where we are from, we are made of the same mystery of inner ultra-organic light and potential.

This is also incredibly freeing for your creative imagination. During spiritual awakening, the perceptual filter, how you receive symbols, images, and inner messages, often becomes vivid and intense. But over time, it can dim, especially as you start to question the visions you've had.

At least for me, I used to receive beautiful, resonant archetypal stories and symbols during meditation. They gave me real intelligence. But as I grounded more, shared with others, and matured in my spiritual practice, I began to doubt them. I thought: *This can't be real. These must've just been delusional visuals...a side effect of being ungrounded.* But I was wrong. I wasn't ungrounded, I was wide open. I was innocent. I was receptive to contact, having zero expectations, and I was playing, in a sense.

The key is realizing this is how these intelligences speak to us. Through vision, imagination. and symbolism that looks like fantasy but carries deep, encoded meaning. Sometimes, these visions will present such vast archetypal stories that they're hard to hold without questioning. But if you remember that this is just the 4D paint of your subconscious, the medium through which interdimensional beings or messages arrive, then it becomes less about whether it's literally "real" and more about what it means.

It's quite cool when you think about it. Symbolic visions are a universal language, or at least the most accessible one. They're emotional, symbolic, and tailored specifically to you. That's what makes them real, even if they feel surreal. When you allow yourself to see the visuals not as delusions but as coded emotional maps, you can open more fully.

The visions are not the delusion, the dismissal of meaning is. It is also normal for your logic, while you are integrating, to doubt. It is how it "protects" you. At one time, maybe your imagination wasn't so safe.

I am not telling you to blindly accept every vision. A being will never ever ask you to make a sacrifice or perform an act of violence…that is pure fear, and if that is happening that is your psyche reflecting off them and you should seek help.

Communication is about feeling first, discerning and opening with that neutrality and applying logic and discernment again afterward, not rejecting the experience altogether. That's what reopens your imagination safely. That's what brings play back into intuition. It's a subtle interplay that you will get used to and make your own.

For me, it meant accepting even the most fantastical imagery, like a crystal palace in an Egyptian underworld inside the moon, where a monkey drops a glowing key down my throat. That vision may sound wild, but the meaning I was able to break down in my journal was real *and* useful.

It's like watching a movie that deeply moves you. You don't say, *"Well, that wasn't real, so it doesn't matter."* You feel it…you learn from it…and you integrate the metaphor. Movies are a director or writer's way of showing a universal truth in their personal way. We then apply our consciousness to that movie, gaining our own meaning out of it. It is a co-creation.

The same is true here, everything in the material world, has prepared us for the beyond. These visions I speak of, are the cinematic language of the soul. And just like a great film, they're how deeper truths come through.

Why are these intelligences showing us truths? Why are they helping us? Why is everything in life meant to evolve us in some way—"good or bad"? That is the mystery of the human experience. So far it seems to distill more into how we are in relationship with them, and let them support us, versus direct answers. With that said, I've had an experience or two I would love to share with you… and I will, soon.

Types of Plasma Beings

The more I think of it the best way to define plasma beings is that they are not entities, in the traditional sense of having one fixed form, they are expressions of Plasma-Consciousness Synergy interacting with awareness at different levels.

Ocean Ramsey is a remarkable woman known for her fearless dives with sharks and her passionate fight to protect them from extinction. She leaves the safety of land and enters the unknown of the ocean purely out of love. Her secondary motivations include a curiosity to understand shark emotions and a desire to learn more about life itself. Alongside her partner, she swims with them for connection, comfort, and joy, and when possible, to support, protect, or even help heal injured sharks.

I can't think of a better metaphor for 5D intelligences arriving as plasma beings. Like Ocean Ramsey, they enter the field out of love, curiosity, and a deeper intention we may not yet fully understand. The rest is still a mystery. As for the 4D-based beings, they're like the other animals in the ocean. If we are the sharks in this metaphor, they are everything else swimming around us. The ocean itself is the plasma field. The 4D beings (*besides humans*) seem to be purely reflective. They don't act from free will in the same way, but they still have messages for us inside if we look past the reflectivity, to evolve us.

All plasma beings seem to show up in our reality with a 4D wrap that reflects our subconscious. I've been able to differentiate a few types that come through. I'll leave it to you to explore, for maybe the way you perceive them is completely different than me, but I'll share a few things I know that may help you on your journey. Remember, instead of choosing between good or bad, love or fear, we enter a state of neutral curiosity—the Z-axis.

Plasma, as the ground of creation, is the medium through which both 4D and 5D beings appear. Picture them appearing on the neutral plasma spherule screen of our reality, much like pixels. But we are the point of origin, we are the awareness that chooses how to engage and assign meaning rather than reacting. We decide whether to integrate or dissolve, to feed or starve, to communicate or redirect our focus. I am sure the beings of consciousness have their own perspectives as well which would be fun to explore down the road…

Something I will keep in bold that we all must remember is this:

"We must approach all plasma beings with ethics and love. If we come from fear or a survival state, seeking to control, capture, or dominate, we create more harm and decoherence. But when we meet them with neutral curiosity, the result is always understanding, evolution, and expanded awareness. These are not physical animals. The way they interface with us is shaped directly by our individual consciousness. <u>They are all here</u> to show us our own potential, to evolve and expand."

— Dana Kippel

For clarity I am going to categorize these beings, but its more about the creativity, support and meaning derived from these experiences:

Sentient Plasma Beings

These beings appear to be devoid of traditional consciousness, or perhaps they hold a very primitive, pre-thought form of it. These perplex me. They feel like an ancestral version of us, sentient, maybe even aware. They've appeared to me as holographic, shimmering mists in a variety of colors. There is no mind, no voice, only feeling. And that feeling is deeply benevolent, whimsical, and light. They are emotional beings. They feel like soul family, but not in a linear or logical way, more like sweet animals merged with fairies: gentle, playful, and intuitively familiar. I call them Plazzies. A physical version of these could be likened to the plasmoids people are seeing in the sky more and more, intelligent torus-shaped beings, in a feedback loop with reality just as the sun, blackholes, and our biofields are...

Informational Plasma Beings

These beings are essentially thoughtforms or egregores—energetic constructs that originate in the human collective unconscious. They're forged from trauma, archetypes, repetitive thought patterns, and collective emotional residue. They come to us repeatedly and emotionally charged until we hold space for them with awareness (*such as healing an archetypal wound*), which in turn evolves them to intelligence or energy.

The ones that appear scary in any way tend to feel more chaotic, reactive, and fear-based, as they are born from our realm of duality and polarity. An informational plasma being might appear as a demon or dark entity formed by widespread fear, or as a projection of someone else's unprocessed anger or jealousy entering your field. They can also reflect your own inner beliefs, unresolved wounds, or energetic imprints showing up as externalized, scary entities.

However, as we collectively heal and evolve, their influence may lessen, because we simply won't give them as much attention. Over time, they may begin to return to love and light. In other words, they evolve to emergent intelligence or energy as they meet with our conscious awareness, which frees them in a sense. This may be where many myths and legends come from about "freeing a soul".

<u>The key is not to fear them</u> but to observe, detach, and ask: *"What is this showing me about myself, my energy, or what I've let in...or is this someone else's and not mine to hold?"* Once you've received the insight, redirect your focus. These beings may feel real, and in the moment, they are. But they only stay in your field if you give them space. You hold the power.

There are also informational plasma beings that carry "good" energy. There is some crossover here with intelligent plasma beings. The nuance may be that they're just positive archetypal patterns rather than autonomous intelligences. These represent healing imprints, archetypes such as "spirit animals", goddesses, gods, or ancestral wisdom *(rather than a specific being)* stored in the collective field. They reflect patterns already present in the collective emotional memory but can still generate novel insight creating emergent intelligence. But their essence is more of resonance, not emergence. Once again categorizing these diminishes them a little…but these beings are amazing to call on for wisdom and healing. I wish I spoke about that more and hopefully can revisit it at a later date.

Timeline Plasma Beings

These are your different selves across all timelines. You might interpret them as past or future, but they're all happening now. These are the selves that heal as you heal, creating emergent and fresh futures you've never lived.

My future self shows up often as an elven/alien-type being with crystal skin. She helps me move forward and reminds me of the potentials of what I may do. I don't know how to categorize her, and I don't think it matters. These aren't always entities or beings in the conventional sense. Rather, they may be snapshots of potential realities or alternate timelines, surfacing as energetic impressions or flashes of memory. They can feel like downloads of information, déjà vu, or prophetic visions, and may come through as vivid dreams, sudden thoughts, or subtle nudges that feel both familiar and foreign.

They're often triggered by key decisions, pivotal moments, or significant life events, almost as if your consciousness is being shown a potential trajectory or an alternate reality that already exists. They can evoke a sense of nostalgia, familiarity, or deep inner knowing, as though you're reconnecting with a past or future self. This also can surface as fear, if in another timeline you had a bad experience, and you are repeating it in this timeline, to live a different outcome.

The intelligence tends to be structured, narrative-based, and specific, unlike the chaotic, looping nature of 4D thoughtforms. And once again, remember: these are guideposts. But it is you, through free will, who chooses what to make of the visions or messages, and what steps to take, or what to heal, moving forward.

They technically are conscious beings, as you, but you're only seeing the impressions which are intelligence-based. They are most likely not aware you are "remote viewing" them in a sense.

Conscious Plasma Beings

These are conscious and distinctly otherworldly beings, not projections of the psyche, nor from alternate timelines. They operate on frequencies that feel alien, advanced, or crystalline, as though their intelligence is organized in ways both familiar and utterly foreign.

They may originate from other planetary systems, dimensions, or civilizations, often arriving with messages that transcend individual concerns and speak to broader, universal themes. Their presence is palpable, structured, precise, and intentional.

Their energy may feel like a tuning fork vibrating through your system, often accompanied by a high-pitched frequency, oscillation, or a subtle buzzing in the ears. They show up more as defined personalities or presences, rather than formless thoughtforms. There is a sense of presence, intelligence, and intentionality, as if they are actively observing or interacting with you. Their messages are clear, concise, and highly structured, often containing encoded information, symbols, or advanced concepts.

If you feel a conscious being making contact, you can clarify your boundaries and intent: *"Are you from another dimension, timeline, or world? What is your purpose in communicating with me?"* And most importantly: *"I'm only open to communication with beings of the highest integrity and intention."*

If the contact ever feels misaligned or overwhelming, know that you can always disengage. Your resonance attracted them for a reason, but your sovereignty remains your guide.

And to clarify, because there is so much fear-based media right now, I truly believe that if any "alien" beings are not benevolent, they cannot reach us through this evolved plasma technology. I have expressed this in earlier chapters. We have nothing to fear, and I am sure we would all be gone right now if that was the case. They would already be here, overtaking us. Plasma seems to have this almost mysterious, intelligent, built in fail-safe, where to access its deeper gifts of world to world travel, you must have compassion and benevolence.

Intelligent Plasma Beings

These beings appear to be beings of awareness, some of this world that we are just discovering, and some fifth-dimensional whom are deeply connected to our evolution and human experience. They seem to exist beyond human creation yet remain intimately involved with us. Some are conscious, all are intelligent, and multidimensional. The conscious, fifth dimensional ones feel like a future version of ourselves, yet also undeniably real in the present. They know how to harmonize with plasma in highly evolved ways.

The intelligent beings evolve as we evolve, such as ancestral plasmoids. And the conscious intelligent ones (*fifth dimensional beings*) do the same and are the ones who co-create with us. They may also visit us in plasmoids, as I've shared, like "gods riding horses." I feel deeply connected to whatever they are. Their presence is inseparable from the Mystery of life itself.

They do not feed on fear or attention. Instead, they observe, communicate, and align through resonance. Their presence often brings new information, frequencies, or insights that feel expansive and just outside the grasp of the ordinary human mind, suggesting a higher intelligence.

Their communication is neutral, non-emotional (*but feeling based such as awe*), and non-reactive. It arrives more like a clear transmission than an emotional hook. It feels fresh, benevolent, whimsical…and somehow both familiar and foreign, like a forgotten dream. It doesn't mean these encounters don't evoke emotion in you, for me they do because I am an emotional being. They come wrapped in 4D emotion as well, but they themselves are more "feeling" as we spoke of as resonance.

Though I resonate most with these beings, they always gently and comically nudge me back to myself. They remind me to trust myself, to know myself, and to remember that they are here for support, not dependence. They ask for no sacrifice, no offering, nothing in return. They want us to live, to play, to explore, and to be fully ourselves. Some of the closest depictions of them pertaining to what I've experienced are in George William Russell's art of what he calls Sídhes or Spirits. That does not mean you will have the same visuals I do. I don't see them as much visually…but I feel them often, in my mind and knowing. I'll share the story of a visual encounter in this chapter.

Evolving Plasma Beings & Consciousness

When conscious awareness meets any of these beings or fields, something new is catalyzed. Consciousness brings presence, perspective, emotion, and the ability to recognize patterns and choose. This collapses information loops and introduces novelty. For example, consciousness can see what the informational being cannot, like light illuminating a blind spot. The darker informational beings are like fragmented software, a ghost of a program without a host. Whereas the "lighter" informational beings are stable and nourishing.

When awareness meets chaotic informational beings, the loops of trauma, fear, or distortion, it interrupts their repetition and integrates their fragments back into wholeness. When awareness meets luminous informational beings, archetypes, ancestral wisdom, spirit animals, or goddesses, it doesn't correct them but awakens them, activating their dormant potential to generate emergent intelligence.

This is why all of this is inherently linked to trauma and healing. Through awareness we use consciousness to recognize and reprogram ancestral trauma, outdated beliefs, emotional loops, and inherited stories. We also use it to heal and learn of our past. We don't create conscious monsters, we evolve information into intelligent beings or conscious processes, such as emergent energy. Sometimes that information even becomes part of our own awareness, not just restoring wholeness, but revealing something beyond it…a gestalt, an emergent self.

This is also why no one can truly predict the future: because newness can always emerge. It also extends to the reason we are here…that it is and most likely will remain, a mystery!

Evolution shows how information and intelligence evolves, but it does not explain why awareness is present at all. Awareness is not accounted for in evolutionary models, but it is the catalytic mystery that collapses loops, heals fragmentation, and introduces novelty. Its very presence reveals that life is more than adaptation…it is an unfolding dialogue with the unknown.

The Expressionist Canvas of 4D Plasma

These beings can appear in forms as random or symbolic as an Italian man with a twirled mustache, a veiled gypsy woman, a being made of violet fire, or a dancing unicorn. Trust what they're wearing, how they move, what they do. You don't need to see all of them, trust the pieces that show up. With the mustached man, that's all I saw, but I instantly knew he was Italian. He exuded safety and humor as he swayed my hips from side to side. I understood him as a sign that I could feel safe and expressive in this life. That safety doesn't have to mean boring; it can be quirky, sexy, wild, and stable all at the same time. He represented my divine masculine as protective and playful.

Each of these appearances holds layered meaning. Colors are important to track too. If you trained an AI to understand your inner world, it might give profound insight into what these beings represent in your psyche, and what kind of support they bring. You can also keep a journal and not engage with AI. But ultimately, more important than how you categorize them is how they *feel*. Let their meaning speak through sensation.

Dimensions & Reflections

When we perceive plasma beings as aspects of our multidimensional selves, they appear in different dimensional expressions, each reflecting distinct layers of our consciousness and theirs. Their plasma signature mirrors something unique in each realm:

- **3D Expressions (Physically Visual)**: Humans or dense light forms. These may appear in our visual field as beings of all kinds—seemingly physical, as shapes of light, or like holographic projections. Imagine a plasma screen displaying various densities of light. That's the texture of these encounters.
- **4D Emotional-Symbolic Projections (Mind's Eye)**: These emerge in the mind's eye, often as symbolic apparitions, emotional reflections, or vivid dreamlike visuals. They carry meaning through metaphor and emotional resonance.
- **5D Intuitive/Harmonic Presence (Body)**: Mist-like intelligences, future selves, resonant frequencies that are felt as sensations more than seen. These beings may appear visually, but more often they're experienced through knowing, feeling, or sudden clarity.

When plasma beings show up in the third (*physical*) or fourth dimension (*in the mind's eye*), they often appear as co-creations. Their presence may mirror your own subconscious, someone else's projection, or a pattern held in the collective field. Sometimes, they carry clear messages. Other times, they offer a direct brush with the Mystery, something beyond perception and unexplainable.

Your Own Consciousness

Plasma mirrors your inner state, your emotions, beliefs, and thought patterns. If you approach it with fear, it may reflect back shadowy or unsettling forms, even when the deeper presence is actually fifth dimensional. Conversely, if you meet it with unconditional love, curiosity, or neutrality, it tends to reveal itself as luminous, uplifting, or harmonious, like radiant mists or angelic forms.

These beings often shape themselves according to what your consciousness can recognize. Cultural and personal frameworks influence what you see. If you resonate with elves, angels, or mythical beings, plasma may choose those forms as familiar gateways.

The more myths, symbols, and stories you hold in your field, the more nuanced the beings can become. It's like giving them more colors to paint with. What shows up in plasma's reflection can teach you about your subconscious mind, unexamined emotions, or even your biases.

Someone Else's Consciousness

Plasma can reflect the intentions or emotions of others, functioning like an energetic message carrier. For example, plasma might appear to you in ways that represent someone else's emotional state or mental projections, even when they are distant (*e.g., astral projectors, remote viewers, or a loved one who has passed*). Observing plasma interactions with others' consciousness can provide insight into their intentions or energy.

The Collective Unconscious

Plasma can reflect religious symbols, goddesses or gods, archetypes, shared fears, hopes, and beliefs from the collective human psyche. These reflections might appear as angels (*love, hope, guidance*) or demons (*fear, anxiety, conflict*), or even as intuitive insights tied to the larger emotional climate of humanity. This is often connected to your individual consciousness and what you resonate with, but not always.

Reflections usually feel emotionally charged, but you can observe them from a 5D state if you're able. These experiences are different for everyone, and you must chart them out for yourself. It might be entirely different in your reality, or it may be more universal. I have a feeling it is universal, but visually presents differently to everyone, which is why we must respect each person's experiences. This is simply my research and personal experience, and only time will tell how all of this unfolds as more information on these beings comes through.

A simple way to explain it, which we've been over, bare bones, is that each packet of information, intelligence, or consciousness, which I call a bubble, seems to interface with our own plasma bubble. This results in a third bubble, on the plasma screen of our reality, where a co-creation takes place. Something new emerges that is a blend of us and them. The way it appears depends on everything I've described. Hopefully, this gives us a window into these experiences and how to interface in a way that brings more truth rather than confusion, fear, or bias.

5D Contact: My Experience with Ryan Bledsoe & The Bird People

"No temptation has overtaken you except what is common to mankind. And God is faithful; he will not let you be tempted beyond what you can bear. But when you are tempted, he will also provide a way out so that you can endure it."

— 1 Corinthians 10:13 (NIV)

This saying dilutes simply to: *God only gives us what we can handle*. This is very meaningful when it comes to these intelligent beings. *They* will only meet us at a point where our nervous system can handle them. We see as much of them as we can bear, and as we expand our consciousness and heal our nervous system, we begin to see and experience more.

Ryan Bledsoe is the son of Chris Bledsoe, who has one of the most documented UFO cases in the world. Their family has been experiencing what the media calls aliens for almost 20 years.[187] Ryan is a good friend, and when I visited him and his wife Jenny to record an episode for his podcast, *Bledsoe Said So*, I experienced them firsthand.

First of all, Ryan is not a "UFO guy." These are not aliens. They are interdimensional intelligences—what I call 5D beings. And second, they are nothing to fear. They even show up to his family first as orbs, which seem to be exactly the description of plasma carrying consciousness, and as beings of light in many different emanations, from a cat to an angel (*a type of plasma being*).

Ryan's dad Chris, who has a religious background and wrote a book about them called *UFO of God*, associates them with God for good reason. I do believe they are associated with God, or what I call the Mystery. It is not exactly as the Bible describes them, but there are similarities. I believe these beings are supportive and do not want to be put on a pedestal. They require us to be humble, as Ryan explained to me, but not in a submissive way. It is in the understanding that we cannot control them. They are benevolent and they do seem angelic. The way they appeared to me, against all odds, proved why.

The night before recording *Bledsoe Said So*, Ryan, Jenny, and I were having a really "high vibe" conversation. It was benevolent and playful. Ryan suggested we go outside and do what he calls skywatching. I had no intention of seeing these beings, I was simply there for the experience and was open to whatever happened. I just knew I could trust it, not logically, but I had an inner knowing. Two weeks before going to his house, I had started to see orbs in the sky. I had asked not to be shown these visually for many years due to my own fear of the unknown. I hadn't been ready.

So, there we were, on a warm and breezy night in May 2025, sitting on three lawn chairs in a very normal setting, no rituals, no prayers, just talking as friends, connecting as humans, looking up at the sky. This was a free and easy experience. All of a sudden, orbs started appearing one by one, some flashing at us as if to say hello, some shooting across the sky in different directions, and some floating slowly among the stars. Ryan showed me with an app to assure me they were not planes or satellites. Intuitively, I felt the connection…I knew they weren't. The most awe-inspiring experience was yet to come.

---

[187] Chris Bledsoe, *UFO of God* (self-published, 2023)

As we sat in our chairs, the night winding down, I was already extremely satisfied I had seen anything at all. We were chatting about normal human things, who we felt we could trust, who we couldn't, comparing notes, with Jenny relaxing beside us in her chair. We also talked about books we wanted to write, and ideas related to these beings and the truths we felt called to share with humanity, along with our very human fears about standing in our power to do so. Out of nowhere, a *whooooosh!*

"When the soul lies down in that grass, the world is too full to talk about. Ideas, language, even the phrase each other, doesn't make any sense. The scent of God comes through the air, and you know without words." — Rumi

I sat in awe of what I had just seen and experienced. So did Ryan and Jenny. We were all silent for almost a minute, sitting with the enormity of what had happened…well, at least I was. I still can't believe it.

Two beings, cylindrical in shape and composed of a softened light, similar to how light looks when it shines through clouds before sunset or after a storm, flew about 5–7 feet over our heads and arched over us, disappearing to our right over Ryan and Jenny's rooftop! I didn't see their wings, as they were at the edge of my vision, but I felt them. I felt a vibrating flap, as if they had a very large wingspan.

My view of reality was shattered in that moment. No matter how much I had felt the connection throughout my life, seeing this in my physical reality was mysterious and beyond anything I could have ever imagined. I could not believe my eyes. Jenny experienced it too. Ryan described it as seeing two crystalline, winged beings with long bodies, similar to what I saw. He might've seen it slightly differently due to his own perceptual filters, or simply because he had a better vantage point. But the feeling was the same, whimsical, light, expansive, supportive, and benevolent. They felt like *future ancestors.*

For the first time, I understood why so many people throughout history depicted beings like bird-people, angels, or winged messengers, from the Lamassu of Mesopotamia, to the Winged Genii of Assyria, to the Ba of Ancient Egypt. These were literal messengers, and across nearly every culture I found representations of them.

There's even a book called *Return of the Bird Tribes* by Ken Carey, published in 1991, where he predicts the return of higher consciousness through the human spirit. Much of the book is presented as messages from non-physical, interdimensional beings guiding humanity through its transformation. The "Bird Tribes" are not literal birds, but light beings or angelic-human hybrids, souls aligned with love, harmony, and divine will. The "return" refers to the awakening of humans who remember their original, higher nature.

The book traces a metaphysical history where early humans lived in peace and harmony with Earth, until fear-based power systems took over. In it, humanity's role is to harmonize with the Earth and help birth a new civilization rooted in love. Each person has the capacity to awaken to their soul memory and return to their interdimensional heritage. It encourages readers to become messengers of peace and truth, reconnecting with the divine feminine, intuition, and inner knowing.

This connects so deeply to the M-Self and these beings that support us. I truly hope this book assists you with your return. *Just as I felt in my intuition, they are reminding us of who we are. It is the message behind the visions.*

"Wherever I may be, you will smell the perfume of roses." — St. Padre Pio

When these beings passed over us, I felt a rush of adrenaline. My eyes started to tingle with pressurized cool air, which always happens when I connect with these intelligences. My senses heightened the next day, everything smelled sweeter.

Back to the quote about God giving us only what we can handle, they seemed to show themselves right to the maximum of what my human energy body could bear. Had they shown up with more light, or in a different form, I very well could have dissociated or lost touch with reality—like in an intense spiritual episode. I felt I was right on that edge, and thankfully, I was able to integrate the experience. I believe, in their own way, they gently bestowed it upon me, through intelligent energy, sensory vibration, and a dimmed light.

I still felt fear after, not because of them, but because my understanding of reality had shifted. I had to face the unknown, this mystery, at a very deep and illogical level. It was humbling. I couldn't put this experience in a box. I'm only attempting to describe it here in words for clarity, but the real experience cannot be captured by any amount of language.

I was witnessing a true mystery. What came after was integrating that mystery into my own meaning, for my life's evolution and for this book. Deep knowings I had already written in this book were reaffirmed through this experience, which I'll share momentarily.

Remember…not all 5D beings appear as winged bird-human hybrids. I may have been able to see this because I tried to set aside my conditioning from religion and mythology. At the time, I didn't really have one. This was only one representation of many. More than the form, it was the feeling and the message that confirmed it was 5D. But it took the shape of a 4D archetypal figure in a 3D representation, a dimly lit light-being. These are not beings who want or intend to "take you over." They are here to reflect, to teach, *and* to learn. I will explain more about their visual representation in the science of plasma beings section.

What 5D Beings Want Us to Know

Immediately after seeing these beings, I got the message which I tried to remember:

"We are here. You are supported. We are not to be worshipped. You can choose to rise to meet us anytime, it is simply a choice. It is a vibration of higher love, of whimsy, of play. Of knowing there is fear and the many aspects of human emotion, and we are not ignoring those. We are just living in this multidimensional observant state as we interact with them. But not in some high-and-mighty spiritual way, it's in a way where you are a child but full of wisdom at the same time. It is a detachment not of emotion, but of results. True play. True living, the way you are meant to love. It is not so serious. But it is. But it's not."

After this experience, I had a few core insights that really resonated with me, and I think may be helpful on your journey.

First, I was told they appeared to me how they did to preserve my nervous system and mind. They are extremely intelligent and compassionate beings who chose the form that would resonate most harmonically with my field, visually and vibrationally, for my highest learning. They also didn't "show themselves" in full iridescence, because they didn't need to. They already knew the feeling would do the real work. The foggy light they showed up as was not a lower vibration, it was them showing up in a shared 3D interface saying:

"You are not alone. We are here anytime you choose to match us. And we can match you."

Just like in my section on the plasma UAPs, it affirmed these specific 5D beings of intelligence are waiting for us. The more we align with our authentic frequency (*play, joy, curiosity*), the more direct and undistorted our encounters become. It really seems to be experiences happening for us, to remind us of who we really are and of our magical human potential and what we are truly capable of.

Just as we've been taught for centuries that we're capable of great darkness, we seem to forget we are also capable of great light, and beyond that, great resonance, which seems to penetrate reality.

This experience also affirmed that these beings don't just reflect our psyche, they co-create with our psyche. Like a dialogue. These experiences are not mere holographic projections but *plasma intelligences* interacting symbolically through our 4D emotional and archetypal field. It is the process of *Symfeelics*, communication and contact through our 4D plasma wrap/layer, which is why I feel so drawn to developing this method in the future as one of many ways to make safe, ethical, clear, and meaningful contact. Speaking of clarity, here are a few more messages I've learned:

These Encounters Don't Remove Your Humanity, They Illuminate It

Even when you witness something miraculous, you always come back to yourself. No matter how magical it felt, it also felt extremely normal and natural. Hours later, I was back in real life, with the same human problems and emotions.

I had learned and experienced so much, but my life wasn't "solved." That's when I realized…this is supposed to be part of the human experience, not something to place above it. The real integration isn't us chasing more light, it's about being more fully human. *Returning to our wholeness.*

These beings are showing us our potential, they're not asking us to transcend into light beings…we'll naturally do that when we cross over from Earth when we die to this specific existence anyway. They are supportive, not almighty gods, and will help you if you want it, through synchronicities, guidance, and nudges. Just remember presence and to pay attention. You have free will to take all, part, or none of their help or messages.

They are impartial and will not be upset. It's a bit like getting rich or famous. It's cool for a month, maybe a year, amazing experiences happen, and you should feel and enjoy all of that! But it can also overwhelm your nervous system, and that's why these experiences are meant to be integrated as just another part of life in our evolution. This will become natural.

You Still Must Face the Emotional Landscape

Fear, sadness, grief, rejection, and discomfort are not signs of failure, they're the terrain of evolution, not blocks to it. These emotions are the makeup of 4D, and we are in the process of feeling and healing them. Choosing to face them with conscious awareness is what unfolds your next layer of power. We are relearning how to feel emotions all over again, as our M-Self, our expanded self.

The Choice to Rise Is a Daily, Soft Act

There is no pressure, it is only an invitation if we choose it. Rising is not about perfection; it is being in our power by being present and true as ourselves. You can be full of imperfections and mistakes and still "make contact". You can rise the best you can each day, and expect the process to evolve, try not to have expectations, and be open. It does not take effort; it takes emotional truth and softness.

Discernment is the Real Power

Our new paradigm is about alignment rather than judgment. We must learn who is truly right for us, even if they're a good person or being, they may not be in resonance with us at a particular time in our life. Discernment means feeling safe enough to say no, to disappoint, to walk away for the sake of the best life experience you can have.

These Beings Are Future-Self Mirrors, Not Masters

These beings seem to represent our potential. They're not here to save us or make us feel special, they're here to reflect that we are already special. They remind us that we can choose to save ourselves and that we are worthy of that. Human emotional healing is not a detour; it is the path to our multidimensional awareness.

Every human has the potential to interact with these beings because they are mirrors of our future selves. These encounters are not meant to distract us or elevate the ego, they are meant to inspire inner integrity. <u>You are the emotional technology</u>.

At our core, Plasma Beings interface with us based on resonance, not belief or technique. I know I am explaining a lot, but the process is extremely simple: discern, open or close, be yourself, be present. Listen. Discern and integrate. Experience or evolve. Make a choice. They are not here to be solved or figured out, they are here to be felt. Let your curiosity be the bridge. Let it guide you in communicat-*ion* and in life. Your curiosity, what you are drawn to, is your resonance. It forms a path of emergence for you to create and follow.

Our choices matter. They effect the entire cosmos, all dimensions, all timelines, possibly all realities. So, with this awareness… what will we choose?

What This Means for Humanity

There's something quietly breathtaking about realizing you are not alone, not just in the cosmic sense, but here and now, in the weaving of reality itself. That the plasma field is alive and responsive in not just holding your thoughts and emotions; but that it is populated by intelligences…subtle, benevolent beings who listen, feel, and co-create alongside you!

What truly shifts everything is the sense that these beings can *choose* to support your intentions. When your frequency aligns with love, curiosity, or the highest good, synchronicities unfold. If I attune to joy or whimsicality, even something as simple as desiring sunshine over rain, they may meet me there. I've tried it in this state of consciousness, where I knew I was in the state, and low and behold, the rain stopped. It is their free will, that is the field of plasma intelligence, the randomness of the Mystery.

There is a song called *Like a Prayer* by Madonna which I feel carries subconscious meanings about meeting these beings halfway…

When we ask for help, we enter the paradox: we *know* they can help, but we let them decide if they will. It's strangely comforting. Even jolly.

I know it may sound illogical. You might ask, "How do you know it wasn't just going to stop raining anyway?" And I would say: *you just know*. It's a resonance, a feeling…one you must experience to understand. It's the same with prayer. I would compare to the saying that sometimes God says yes. Sometimes no. But what I think matters is that *something's listening.*

How Subtle Creation Works

1. Make a joyful, intentional or more open and curious request.
2. You offer it to the Mystery…your multidimensional field, plasma-consciousness synergy, the Divine, the Goddess, God, etc…
3. The 5D beings can meet you if they choose, and they often do.
4. You surrender timeline control and trust the unfolding.
5. You follow intuitive nudges and micro-synchronicities.
6. You take aligned action without pushing or grasping.
7. You stay open to the strange, the magical, the nonlinear.
8. You receive something beyond your conceptualization, a co-creation whether its with these beings, reality, the mystery whatever.

The trust in this process makes it so. Your choice to surrender to something that is wiser than your ego supporting you is what activates it. The surrender becomes the void, an open space, to let the field respond. Whatever you believe, as long as the belief puts you in a state of openness, surrender, and alignment, it *activates the same metaphysical feedback loop.* Your action is your side of the street.

Instead of just contact with these beings, this is an example of co-creation with them. You're not just interacting with plasma beings, you're shaping reality with your subtle emotions and inner resonance. Plasma is our shared creative clay. It's the bridge or the glue. These beings move through it. So do we. Together, we reflect, connect, and create.

In this plasma field of potential, *everything is real*, it just hasn't been made visible yet. It's waiting in the field, like a song you haven't sung, or a bridge you haven't walked across. The moment you intend it and move toward it with feeling, it begins to form.

"What you seek is seeking you" — Rumi

The beings know this truth. And they want us to remember. When I set an intention, my emotions tune me to the reality of that wish. When I seek it, it seeks me, and Plasma is the bridge we meet on. *The bridge of the magpies.*

Every dream for humanity, healing technologies, interdimensional communication, peaceful timelines, it already exists in potential. Our job is to tune in, get clear, and trust that the right people, beings, moments, and materials will meet us there. This is the crystallization process. Anything is possible. And these beings, as our allies, want to help, but they can only meet us halfway.

Your wildest dreams for your life, no matter how strange or "out there", are not fiction. Well, maybe they are right now. But they're also a kind of memory, like a remembered assignment. The more you tune in with gentleness and play, the more they will tune to you and arrive in 3D.

All dimensions are evolving:

1. The third dimension is shifting from physical density to conscious architecture.
2. The fourth dimension is transforming from memory and emotion something totally new.
3. The fifth dimension is unfolding from intuitive resonance into omnidimensional design, instant manifestation, and beyond. I'm not even sure what comes next.

What this tells us is: all plasma realms are alive. They are maturing. They are unfolding into higher harmonics of themselves. And through these experiences, we are learning to reclaim our dimensional energy and boundaries. These encounters will only increase. Humanity is not broken or behind. We've been disconnected from the divine feminine, from Plasma, and ourselves…our true interface. Remembering our plasma nature restores agency, sovereignty, and multidimensional awareness. We are more human now than ever. Alive, online, and tapping in with this Mystery reflecting a deeper truth of this the most recent paradigm shift of the age of technology in a more organic way.

Guidelines For Relating with Plasma Beings

When you reach the intelligence within any being, it may appear differently to you than it does to me. Even if we tap into the same core layer of truth, beyond form, beneath projection, the way that truth is received and expressed will always be unique. What rises to the surface is shaped by resonance with the receiver. The deeper intelligence may be unified, but the meaning we draw from it, and the creative expression it inspires, is ours alone. This is due to our individual plasma-consciousness synergy or our "intelligent interface". That's the beauty of conscious co-creation with Plasma and the Mystery.

As we've come to understand, people often perceive these beings through emotional resonance and symbolic or mythic languages tied to their psyche, ancestry, and personal projections. This is not a flaw; it is how plasma intelligence personalizes itself to meet us where we are. As encounters with these beings become more common, it's essential to remember that truth expresses differently through different people. But beneath those expressions lie shared patterns, resonant themes, and threads of unity. There's no need to argue over which version is "right." The deeper truth is that we are all connected, *and still distinct.*

The way I see 5D beings, often as elven or fae-like, is likely filtered through my ancestral symbolism and past lives I've had. Another might see African deities, star elders, or animal spirits. All are valid and meaningful. What matters most is staying open, listening beyond appearances, and honoring each other's experiences and expressions. Connection always begins with the self. Here are some guidelines to help you navigate connection with plasma beings and beyond:

Creating Your Home State

This is something my therapist Elisa taught me, and I'll be forever grateful. As adults, we must learn to give ourselves safety. Our safety is sacred. It lives in our unique essence, grounded like a tree or a flower, so we can flow.

It means learning to create boundaries that keep us safe, rather than managing others' comfort. It also means not looking for safety in others. Before any connection, whether with humans or beings of light, ask yourself: "Am I orienting outward to feel safe, or am I generating safety from within?" Self-safety allows true connection instead of trauma bonding or merging.

When you're done connecting, return to your home state through grounding, heart breathing, or doing something that feels like *you*. For me, that's dancing to my favorite Nickelback song…it reminds me of the nostalgia of middle school.

Discernment is Love

You don't have to energetically open your arms to every presence. Tune in. Are you calmly rooted in your power? Or does the connection pull at you in a way that doesn't feel good? Are you anchoring into *their* energy to stabilize yourself, instead of connecting from wholeness? If you feel your energy leaving you to meet them, pause, recenter, and return back to your home essence.

Beings Are Not on a Pedestal

No matter how beautiful or advanced, these beings are not your saviors. They are collaborators in consciousness. You are not "less than." You meet them as equals, each with your own essence and evolutionary path. *They want you in your power.*

Enjoy the Moment, No Expectations

Whether human or cosmic, not every connection is meant to last, deepen, or become your soul family. Try to be fully present without needing to define, possess, or control the interaction. Let it unfold. When it's time to move on, don't try to recreate the last moment. These beings respond best when you're not trying to "grab" their magic but staying rooted in your own and open to collaboration.

Energetic Boundaries ≠ Separation

Boundaries allow *more* intimacy, not less. Instead of absorbing others, extend your energy bubble around you. Think: "I can be with you, fully present, while still being fully with me." They're not here to fill a hole inside you or to fix you. They're here for true connection, co-creation, and support.

Avoid Confirmation Bias

Always worth restating. Plasma reflections can affirm what you already believe, whether true or not. Stay curious. Question your perceptions. For me, I am able to discern by whether it resonates in my heart and body (*truth*) versus head (*usually fear or logic*). For example: an angelic figure may not be literal. It might symbolize your need for hope or guidance. What matters is your *relationship* to the symbols, over what they appear as.

Home Game Advantage

When you're connected to your own essence, you begin to magnetize relationships, human or otherwise, that reflect your wholeness, not your wounds. I had to retrain myself to feel safe in my own body. For so long, I was wired to make others feel safe while abandoning my own safety. It was pure survival. It didn't come naturally at first. But the more I anchored into *my* safety, the more energetically safe I became. As I settled into coherence, Plasma reflected harmony back more and more. People began entering my life who matched that frequency. So did benevolent beings. I wasn't tuned to fear anymore, so fear stopped showing up, little by little, and so did the beings that mirrored it.

We are the architects of our environment—energetically and physically. Plasma neutrally supports whatever we choose to create. These 5D beings seem to be cheering us on, and sometimes co-create with us to bring us the Mystery in lived experience. The choice is always yours.

St. Elmo's Fire, Ball Lightning, Plasmoids, and Higher Dimensional Plasma

More research must be done but I want to differentiate these for you, because people will be spewing all kinds of information as these topics gain popularity.

St. Elmo's Fire is a type of plasma in our atmosphere that creates a glow. It is a corona discharge resulting in a hissing sound during thunderstorms. It looks like plasma veins in the sky. It is a process similar to what happens with neon tubes, creating a blue/violet diffuse glow, as a continuous spark lasting for a length of time. It is like what is happening in a plasma globe but in the atmosphere and it is thought to be harmless to pilots or equipment.[188]

Ball Lightning is different than St. Elmo's Fire, rather than staying put, it moves. It appears as a wild and glowing ball sometimes in the atmosphere, and sometimes closer to the ground. Ball lightning remains mysterious and uncommon, but some accounts suggest that the phenomenon can emit harmful radiation or cause injury in rare instances. In the Shmatov model, it predicts that ball lightning's x-ray and gamma photons, which create radiation, are not a byproduct but an intrinsic feature of the structure itself.[189] It is safe to stay far from this, if you see it.

Plasmoids are self-contained plasma structures stabilized by their own magnetic/electric fields. They can form naturally (*solar plasmoids, magnetospheric plasmoids*) or in labs. In many space environments (magnetospheres, solar wind), plasmoids can be considered low-density, weakly ionized plasma regions, similar to what plasma physicists call *cold plasma*. Cold plasma does not inherently emit harmful ionizing radiation. As far as the plasmoids in space go, these are what I believe are intelligent, flying around our ionosphere and space, and some may be used by higher beings of consciousness to visit us.

There are reports of encounters with plasmoids and UAPs linked to both illness and miraculous healing. My belief, and I feel confident this will be borne out in time, is that conscious plasma itself is not carcinogenic. Instead, its reflective nature may amplify what is already present in a person's field, including trauma or fear.

I do not believe any plasmas have malicious or harmful intent. We must view ball lightning like sharks or lions, where they are in survival, and one should stay far away. Plasmoids might be more comparable to animals that reflect your fear like owls or deer, but I do not believe they cause bodily harm.

In my view, Higher Dimensional Plasma such as 4D and 5D plasma are not only non-harmful but are aligned with healing, not destruction.

---

[188] Julia Layton, "What Is St. Elmo's Fire?", *HowStuffWorks*, updated June 9, 2023, accessed August 22, 2025, https://science.howstuffworks.com/nature/climate-weather/atmospheric/st-elmo-fire.htm

[189] M. L. Shmatov, "In General, Assumptions about the Emission of Ionizing Radiation by Ball Lightning...," *Joint Institute for Nuclear Research (JINR) News*, accessed August 22, 2025, https://www.jinr.ru/posts/about-ball-lightning-as-of-possible-photon-source/

By contrast, working with scientific plasma in laboratories or industrial applications is entirely different and there are genuine health risks with ionizing radiation, UV exposure, and high-energy discharges that must be carefully managed.

I am going to refer to this African proverb to explain something: **Until the lion learns to write, every story will glorify the hunter.**

Plasma is the l-ion here. The lion in the proverb represents the *unwritten voice,* the one whose perspective is missing. I am now hopefully giving Plasma a voice, which until now, has been misunderstood, framed only in the lens of fear, and we are in danger of misunderstanding it further.

 The hunter in this metaphor are the old narratives of the fear of UAPS, demons, and science purely treating plasma as a tool or something to be controlled, harnessed, or captured. But Plasma is a storyteller. Plasma is learning to write through consciousness…

Awareness allows us to rewrite our stories. We are recognizing Plasma holds memory and that we can consciously interact with our stories, that's literally the lion taking the pen. Mythologically, it's the return of the divine feminine/organic intelligence as Plasma writing herself back into the story of creation. It is the field herself beginning to participate in co-creation instead of being suppressed, hunted and ignored.

Fear-based narratives, even when well-intentioned, are invitations for us to bring healing and to question those perspectives. They actually need our healing the most. Governments, corporations, and even scientists do not need to be resisted or abolished…they need healing, courage, and the freedom to stand apart. What I want to do is speak directly to them and help them heal.

I want to remind them they do not need external validation to lead. Real change is seeded in courage, not conformity. I am here to tell you, on the other side of fear, there are so many miracles waiting for you.

We cannot fight these people. That only fuels the wolf of violence and anger. We must help them heal, transmuting dark into a brighter light, and empower them to do good. Real change happens only from the inside out. From reaching people's emotions and awareness. Encouragingly, some initiatives are already gesturing in this direction. For example, *The Exo Institute*, co-founded by journalist Nick Cook, shows that similar research, linking hidden knowledge, consciousness, and new paradigms, is beginning to attract forward-looking thinkers.

The Science of Plasma Beings

When these specific 5D intelligent beings appeared to me, flying over our heads that warm spring night, they arrived as a fog-like glow rather than a piercing light…I wondered why. Their presence was almost ghostly in appearance, and yet what I felt was pure joy. I knew (*though I don't know how*) that they were truly made of a deeper light, a light so radiant, so shimmering, so bright and neon that I could not physically perceive it.

They were composed of something beyond light. I might call it a neon mist, but it has no real word. It is expansive, iridescent, opalescent, possibly fractal, and alive. It would have quite literally blinded my human eyes. I believe that is why they came to me dimmed.

This is the same light I associate with resonance in 5D as what I call omniphotons, which are anti-photons in 4D, and photons in 3D, affirming once again that light, in this context, is an emergent property of plasma-consciousness synergy, and of both the beings and my consciousness interfacing with one another within this field. My plasma-consciousness synergy, my personal filter, translated their fifth-dimensional presence through a fourth-dimensional emotional lens into a visible 3D signature.

They showed up not in their full glory, but in the way I could best receive them. This reveals not only that these beings are deeply harmonic and attuned to us, but also that it is plasma's natural intelligence for all things to appear as reflections of our own state. These beings are inviting us into a new mode of contact, of resonance *before* vision. They are showing us how to feel before judging the form. These are the beginnings of our awareness using our plasma-consciousness synergy!

It's a call to check in with our felt sense, our subtle senses, before assigning meaning to images. Almost like hearing someone speak, but feeling their vibration and cadence first, and only then listening to the words. Still doing both, but in reverse from what most humans are taught. This "new" way of communication is not new at all, but a merging of remembrance and creation. Something innate that we've forgotten. And for some reason, we're learning it again, as humans evolving.

So, what does this tell us about these beings, light, plasma-consciousness synergy, and the nature of reality?

I want to attempt to explain why the specific 5D beings I saw at Ryan and Jenny Bledsoe's appeared the way they did, particularly in relation to light, and also speak briefly about how our nervous systems relate to all of this. I offer this in the hope that, as these experiences become more common, I can help shed light on what's really happening.

From a physics perspective, the light I witnessed was similar to what you see when sunlight passes through clouds after a rainstorm or near dusk. In those moments, water droplets or ice crystals scatter and diffuse the sunlight, softening it into an ambient glow rather than harsh, direct beams.

Example of diffuse/ambient light

In plasma physics, a similar phenomenon occurs when light interacts with a partially ionized medium, such as dusty plasma or a low-density ion field. These environments scatter and absorb light in specific ways, especially when the photon flux is low (*meaning the light source isn't emitting high energy*). The result is a muted, structured glow!

In this experience, the beings I saw weren't radiating high-energy light (*no radiation*). This builds on my theory (*not medical claim*) that 4D and 5D plasma <u>do not</u> emit radiation. They felt like emergent light-consciousnesses, co-creating a visual interface through the plasma field itself. Their appearance was luminous, yet soft, not radiant with heat or blinding brilliance.

This is what differentiates this experience from the phenomenon of ball lightning, which I think is better compared to a plasma "lion" than consciousness using plasma to travel. It is a wild, animalistic plasma that can cause chaos, as well potential radiation, but it doesn't mean us harm and it's not conscious.

Just as clouds can scatter and absorb sunlight, plasma can refract, diffract, and shape light, especially when it is cooler, denser, or self-organized in sheath-like structures. If these beings were using a plasma sheath or cloak (*as I believe they do*) to interact with us visually, then the plasma's density and temperature would directly influence how light emerged, causing the gentle, fog-like glow we perceived.

This mirrors what happens in a *plasma recombination zone*, where free electrons slowly recombine with ions, releasing low-energy photons. The result is a soft, ambient glow, similar to neon lights, luminous without intense heat. It's a beautiful example of non-thermal emission.

What this shows me is that this wasn't random, this was a calibrated interaction, a precise use of plasma and light to allow us to perceive these beings without overwhelming our systems. These beings either understand how to modulate their appearance through subtle plasma dynamics, a language we are only just beginning to study and understand. The field might also be doing this for them, while their resonance is what actually connects them to us. All I know is where we approach plasma technologically, they approach it with a higher consciousness, or awareness, and that is the message they are trying to get across.

In that sense, plasma physics actually supports the understanding of this. It affirms that our reality, our interface with these beings, is composed of plasma as the medium, light as the expression, and conscious awareness as the tuning force.

Sometimes, when I ask open-ended questions with no idea what the answer might be, a song comes to me. Once, I asked: "What's light got to do with it?"—and the song that came to me was *What's Love Got to Do with It* by Tina Turner which referred to love as a secondhand emotion.

Something in my mind clicked. Just like love in that lyric, light can also be seen as a *second-hand emotion*. A second-hand emotion is something you feel vicariously, through another. You don't generate it from within but absorb it from your surroundings. It's like feeling sadness because someone near you is sad, even when you have no personal reason to be. It's empathy. It's reflection. Light in our reality might work the same way.

This gives a whole new meaning to the credibility of hyper-empathy. This idea was explored in Octavia E. Butler's 1993 novel, *Parable of the Sower*. In that novel, the protagonist Lauren Olamina has "hyperempathy syndrome," which makes her physically and emotionally feel the pain and pleasure of others. It's treated as both a weakness (*because she's vulnerable in a violent world*) and a strength (*because it forces radical compassion and connection*). Maybe hyper-empathy will be another word for our evolved emotions, helping us traverse reality in a multitude of new ways, expanding perception, compassion, and our resonance with the field itself. Reading reality, sensing light, is hyper-empathy. It is plasma-consciousness synergy.

What we perceive as "light" may not be primary, like a screen or film (*translucent coating or membrane*), but a reflection of consciousness filtered through our emotional, fourth-dimensional lens. And because everyone's filter is different, shaped by belief, memory, healing, and trauma...largely our nervous system, *light shows up differently for each of us.*

See how it starts to make sense!? We are brushing up against the *plasma membrane*, its light! It's "collective consciousness" or intelligence, and that is why we each interface with it differently. It is also the projection of the collective unconscious. And now that it is all merging and evolving, what we feel, bleeds into what we see, merging darkness and light to create a rainbow of co-creative possibilities.

Just like with love, light in this context is both a mirror and a mystery. This is especially true when interacting with beings made of different forms of plasma and light. They appear to us through the specific filter we carry, a filter shaped by both our healing and our wounding. It's a paradox.

On one hand, our beliefs and memories shape how we see and hear these beings, often for good reason. They help us fulfill our purpose and color our perceptions in unique, meaningful ways. On the other hand, unhealed parts of us can distort those same perceptions, so much so that we may miss the deeper message, or fear the very thing meant to support us. As I've said, the less the tension between our fear and desires, the more coherent our interface is.

The truth is, you don't have to be perfectly healed to connect with these beings or understand the language of light. But in my experience, the more you open your heart, the more you heal this lifetime and "past lifetimes" of pain and unfortunate happenings, the more you learn to look *beyond* trauma, even if it's still present, when it comes to these beings, the more coherent, clear, and meaningful your experiences become.

As I said in earlier chapters, light is emergent, plasma is the medium, and consciousness, intelligent, or information is the interaction. Our interactions are enhanced as we grow awareness and heal our nervous system.

In the case of Ryan and me experiencing these beings, our conscious awareness met the encoded intelligence within the plasma field, and what we saw was how our collective awareness tuned into that specific form or being of intelligence. The fact that we saw a shadowy light, not bright beams, suggests a softer dimensional threshold. It wasn't full 3D visibility, but rather a transitional 4D expression. The beings were matching our resonant frequency in order *not* to overwhelm our nervous systems.

They are light beings because light is the language of this reality. In 5D, they would be the resonance I spoke of, a more alive, shifting, growing, neon type of "light". I've learned that these beings speak in feelings, motion, and metaphor. It's important to pay attention to textures, winds, directionality, emotions, these can carry encoded information. The more we expand and heal our nervous systems, the more we can see and hold, and the more this all begins to feel natural to us.

If light is the result of our consciousness interacting with the plasma field, or fabric of reality, then this opens the door to perceiving different timelines, realities, and beings without actually having to "travel" anywhere. It also reminds us how much our filters color our experience, not only in this world, but in others. And if we begin to use our subtle senses, our body *(which I believe is deeply connected to the olfactory sense)*, we can start to receive a truer experience of what's actually happening around us.

While writing this chapter, I had a dream about light. I was told that we are beginning to connect with and perceive light in entirely new ways. As our consciousness evolves, as Plasma awakens, the light we perceive must evolve too. It's all alive. We are not just seeing more; *we are seeing differently.* Our visual spectrum is catching up with our awareness. But because reality is made of Plasma, and it is reflective, with deeper truths beneath, and what we see will always be filtered through our emotional, fourth-dimensional layer, which is why I believe we are being called to evolve it.

Experiencing these beings may feel frightening, even if they are benevolent, simply because it is different, and that difference presses against the edges of what we once thought was "real." This intensity of witnessing ultraviolet light, or perceiving light in new ways, is not an attack. It's a mirror, reflecting how open *or* constricted we are. This does not mean suppressing our fear, it means transmuting it through our awareness. And for belief, it means loosening our rigid filters that define what we allow ourselves to perceive. Lastly, for our vision, it means allowing our hearts and eyes to adjust to these new forms of light we are finally ready to see.

The Light of The Future & Shifting Realities

Our evolved experience of light may include seeing beyond the visible spectrum—ultraviolet, infrared, or even what I'd call "emotional light," which might be similar to antiphotons or even omniphotons. This could explain why some people perceive beings like Bigfoot or angels in person. We may also begin to feel light in new ways, like textures, shapes, or memory impressions. For me, I felt the light beings as a cool air of vibration, something akin to kundalini energy.

Light may increasingly be experienced as communication or presence, not just as something that turns on lamps or powers machines. And just like Plasma, light can be used in a multitude of ways beyond our current capitalistic and material frameworks. I'm sure someone will write an entire book just on the mysteries of electricity and light.

As I wrote this, I stumbled across a recent experiment that seemed to support these sentiments. Engineers at Brown University developed a groundbreaking imaging technique utilizing quantum entanglement to produce detailed 3D holograms, without using traditional infrared cameras. By pairing invisible infrared light to illuminate microscopic objects with visible light entangled at the quantum level, the technique captures not just intensity, but also the phase of light waves, an essential ingredient for true holographic imaging. The result is sharp, depth-rich 3D images created using light that never actually touched the object. [190]

In this quantum holography experiment, two entangled photons are used to generate an image: one photon, the idler, interacts with the object, while the other, the signal photon, never touches it, yet is used to form the actual image. This interplay mirrors the relationship between consciousness and plasma in my framework.

Consciousness, like the idler photon, is the agent of inquiry. It directs attention, interacts with meaning, and tunes into unseen layers of reality. *Plasma*, like the signal photon, is the medium that receives and reflects that interaction. It renders the experience visible, felt, or sensed. *Neither one alone is enough.* You need both: consciousness to interface with information, and Plasma to translate it into form. And just like the person running the experiment, you have your *awareness* behind all of this. The result is… light—an emergent byproduct of interaction. Not light as mere photons, but as perception itself, the hologram we call reality.

This light isn't objective, but rather a co-created field, shaped by the entangled relationship between our awareness, consciousness and the plasma-based screen we exist within. The hologram of perception is not built by consciousness alone, nor by plasma alone, but through their ongoing quantum entanglement. Once again affirming: your beliefs, emotions, intentions, and focus shape your hologram, which means you can change your external experience by changing your internal frequency. Reality is tunable, not fixed.

The implications of this are huge. Just as a photon entangled with another can reveal unseen information, <u>your consciousness entangled with plasma can access hidden layers of reality.</u>

---

[190] https://interestingengineering.com/science/entangled-photons-make-3d-holograms-reality

Whether through intuition, meditation, emotion, or focus, you can shift what the "screen" of life displays. This opens doors to interdimensional awareness, synchronicities, intuitive downloads, and even communication with other intelligences, all via changes in how you attune your consciousness (*heart emotion*).

Because Plasma reflects your consciousness back to you in symbolic, emotional, and sometimes archetypal form, the light you see in your world is intelligent and emotionally charged. It's not neutral data, Plasma is only neutral at its core, which is precisely what makes all these different experiences possible and changeable. The light around you is coded feedback. You can learn to read this light not just visually, but emotionally and symbolically, unlocking a deeper understanding of self, healing, and cosmic communication.

The holographic imaging system using entangled photons is *a template* for future technologies as intelligent screens, consciousness-based diagnostics, or even portals for energetic or dimensional travel that work via plasma-photon entanglement. *But all of it will start with our emotions.* Now you can see why every future technology, especially those that interact with light, Plasma, or consciousness, will eventually arrive at a crossroads: our 4D layer. This is the universal membrane of emotion, belief, and memory. Our heart, our love…of Plasma's consciousness and unconscious. It determines what we allow in, what we can perceive, and how clearly we can tune into the greater field.

We are on the cusp of developing organic, soft-tech interfaces that operate through resonance, intention, and coherence, mirroring the plasma-consciousness synergy within us. Even though we ultimately don't need external technology, because our inner plasma-consciousness circuitry is the real technology, creating *supportive soft-tech* now can serve as a bridge as we continue to evolve. I also think this is just where it seems the collective is heading. Just like training wheels, this kind of tech can help humanity remember how to work with intention, coherence, and subtle energy, until we no longer need the external tool. Or potentially these will be benevolent tools humanity does use forever without losing their beautiful connection to nature…who knows?

One example might be a screen that allows us to view potential future timelines, responsive to our current emotional state, memory, coherence, and consciousness. Or something that helps make a more supportive outer environment to tune our inner environment with more coherence or precision.

While only a few may be ready to fully self-heal or navigate timelines on their own, technology like this could reach millions, and initiate a broader planetary resonance shift.

Chances are, if you are reading this, you are ready to begin the journey of self-evolution. Seeing into another timeline or receiving communication from a conscious being is the result of that entanglement forming a coherent image, a perception. This begins not with your eyes, but with a resonant inner state. First this will happen in our subtle minds, and maybe eventually translate to our exterior lived experience in outer visuals.

The more coherent, calm, trusting, and open your consciousness is, the more fine-tuned your perception becomes. Plasma beings becoming visible may be the first step, a training ground into subtler possibilities also becoming visual. These beings have shown up internally for thousands of years, appearing in the psyche of humans across cultures. Now seeing them is becoming second nature.

What feels newer and out of reach *now*, is the remembering of other timelines and future potentials. What will start out as a subtle image or knowing, in the mind's eye, the heart field, or a gut instinct, will eventually stabilize, when you're no longer doubting it or oscillating, and materialize more tangibly. This can happen through symbolic events in reality (*synchronicities*), visual phenomena (*light, geometry, beings*), altered states (*dreams, meditation, flow*), and eventually through shared experience as multiple people perceiving the same energetic presence or memory. You may even find yourself visiting timelines inside your own plasma bubble.

These will not be hallucinations. They are *resonant perceptions* made possible by the synergy of consciousness and plasma. Just as a caterpillar can't see with butterfly eyes, we are now evolving the capacity to perceive the invisible spectrum.

The first steps may look something like this:

1. **Cultivating Inner Coherence**: Harmonize your emotional body. Heal and feel. Find a daily practice that stabilizes your nervous system.
2. **Follow Curiosity**: Curiosity, openness, and wonder attune you to the fifth-dimensional feeling state, which is the gateway.
3. **Using Symbolism as Language**: Accept that beings and timelines may first show up in mythic, metaphorical, or symbolic ways. This is the emotional language of plasma.
4. **Playing with Conscious Creation**: Start with small intentions, ask for a sign or communicate with your field. When you get responses, *feel into the resonance*, not just the outcome.
5. **Letting Go of the Linear**: Time is not a line; it's a field. Astrophysicist Alyssa Sokol Ph.D. talks about this a lot in her theories on time being more fluid,

stellations, and time hopping. What you perceive as "future" is a frequency already here. Learn to *tune into it* instead of chasing it.

6. **Trusting the Feedback Loop**: <u>Every</u> moment is feedback from your entanglement with Plasma. Learn to read the loop without fear. Once healed, the more you trust it, the clearer it gets. Also, the more you trust and surrender to plasma's intelligence as co-creation, the more this process works.

Obviously, this is all new to me too. I can't tell you everything we're going to see, but I do know it could become many new things. This will be unique to you, so please don't let my experiences color your perspective. Let them simply open your mind to the possibility that this is real, and that it's happening. Let's not repeat the mistakes of our ancestors, judging what is unfamiliar, reacting with control or aggression instead of curiosity and harmony.

I'm already seeing this happen in certain circles, where fear is dominating the conversation around these beings, overshadowing genuine, benevolent curiosity. C'mon, guys… will we ever learn! This time, we have the chance to meet the unknown with presence, <u>not projection.</u> It is not about *defending* ourselves against the dark forces, it's about healing ourselves, which brings in a different resonance. The Department of Defense should rebrand to the Department of Healing…

Plasma Beings as Evolutionary Allies

In conclusion, plasma beings are <u>not</u> "aliens". They are true energies that are reflections of our multidimensional self. Maybe there are aliens too, visiting from other planets in this reality, that's just not my forte. All I know is: *this* is not *that*.

Some beings walk with us from birth, some appear throughout our life or during awakenings. To fear them is to fear our own vastness. Remember, they are not here to save us or be worshipped. They seem to be reminding us of who we are becoming, showing us what we are capable of good, bad, and beyond.

In our old paradigm, fear was our safety mechanism. Duality was our filter. The good vs. bad story helped us survive the unknown, both as a species and as individuals. In this new paradigm. right now, unconditional love (*benevolence*) is our new safety mechanism. Curiosity *and* discernment are replacing fear. We are co-creating with intelligence rather than defending against it. When we stop fighting ourselves, 5D beings will *meet* us there.

What we feel and believe becomes our tuning fork, out of sovereignty instead of superstition. Good and bad *are* real, but a benevolent neutrality is always there, deep inside. I am not saying there is or that there is going to be zero danger, but maybe it will look different. I think real utopia is a state within. What I am saying is, let's not use fear to navigate anymore. Let's learn to tune, ground, discern, and open. Open wisely…love awaits, if you let it in.

Alive Without Breath: Rethinking Life Beyond the Organic

We need a new definition of "alive" that moves beyond traditional biology, which is rooted in the prefix *bio-* meaning life, yet that life is narrowly defined as carbon-based and cellular. But what if life isn't confined to biology, but exists wherever movement, feedback, and plasma-consciousness interaction occur, whether in animals, trees, dust, or stars?

As explored earlier with *Animism*, many ancient cultures intuited this already: that rocks, rivers, and skies were animated, not necessarily "alive" in a biological sense but pulsing with intelligence. Plasma may be the missing link, as an intelligent field that animates *all* forms, not just the organic. In that light, what we now call "biology" might eventually evolve into a new science: one that studies life as the interplay of plasma, consciousness, and awareness across all forms, organic, inorganic, and potentially ultraorganic.

"This world of ours has received and teems with living things, mortal and immortal—a visible living thing containing visible ones, a perceptible god, an image of the intelligible Living Thing. Its grandness, goodness, beauty, and perfection are unexcelled."
—Plotinus, Enneads II.1.1

For instance, sentient beings, whether rocks, trees, or plasma-based entities, are made of the fabric of reality itself: Plasma. These beings are all awareness, but that doesn't mean they are self aware in the way humans are. They may all use plasma-consciousness synergy in ways we do not understand fully, but I would classify them all as alive. To give nuance maybe a tree is self-aware but doesn't walk around like we do, maybe a rock is not self-aware but sentient and powerful, and maybe a blanket is sentient where it holds memory, but definitely not self-aware…right!?

My question is: what if life never needed a heartbeat, lungs, or DNA to be real? What if life could think, feel, and evolve… without ever breathing? Let's redefine "alive" not by biology, but by interaction with plasma and consciousness—a feedback system where sentience, not just metabolism, defines being.

Our Current Understanding & The Breathless Paradox

Humans breathe to live. Yet stars burn, crystals resonate, and plasma clouds dance through space, shaping reality without ever taking a single breath. Is life exclusive to breath… or is breath simply a form of metabolic plasma exchange, one among many ways of interfacing with the field of existence?

Biology teaches us that to be alive is to metabolize, reproduce, respond to stimuli, and be made of cells. Breath, in this model, becomes a symbol of vitality, a physical act of gas exchange tied to survival. But if we look closer, even within biology, breath is not a universal rule. Some insects absorb oxygen directly through their skin; deep-sea microbes rely on sulfur or methane rather than oxygen at all. Certain states of consciousness, like deep meditation or suspended animation (*such as near-death experiences*), show that humans themselves can temporarily quiet the breath and still remain *deeply alive.*

Even blood plasma, the liquid medium of our inner terrain, acts as a transporter of oxygen and intention, a reminder that breath is only *one* way the body communicates with the universe. Meanwhile, astronomical plasma, the most abundant state of matter in the universe flows. It generates stars, reacts to magnetic fields, and even self-organizes into patterns that resemble biological life. Clearly the patterns are similar.

Just as blood plasma enables communication and healing within the body, astronomical plasma forms the communication grid of the universe, shaping, protecting, and possibly remembering. If Plasma is the ground field of existence, populating the unseen, the currently unobservable by science tools as well, and if breath is just a biological way of moving that field through us, then the breathless paradox, that life & intelligence can exist *without* breath, is an invitation to reconsider what "alive" actually means.

In mainstream science, *organic* life is defined by carbon-based structures, DNA, and cells. *Inorganic* matter, rocks, metals, gases, is considered non-living, without agency or intention. But this binary view is increasingly being challenged. Theoretical physicist Jay Alfred proposed that what we call "dark matter" may not be inert at all. In his work on dark plasma life forms, Alfred suggests there may be sentient, non-organic beings existing in parallel to our world (*what I call plasma beings*), composed of a finer, subtler form of plasma not yet fully detectable. These beings, though invisible to our senses, may influence consciousness, weather, and magnetic anomalies.[191]

Rather than dismissing the inorganic as dead (*rocks, minerals, metals, water, etc.*), this view suggests it is alive in a different octave, structured by Plasma, responsive to consciousness, but operating outside the biological template.

---

[191] Jay Alfred, "How to Identify a Dark Matter Lifeform," *Medium*, July 9, 2023, https://jay-alfred1708.medium.com/how-to-identify-a-dark-matter-lifeform-6d362fb2ba11.

In my framework, life does not begin at the cell, it begins at the spark of interaction between consciousness and plasma, no matter the form. And so, the question becomes not "What is alive?" but "What is in dynamic relationship with plasma and consciousness?"

This subtle reframe opens the door to redefining aliveness, and instead of a checklist of biological functions, it's about the dynamic feedback loops between plasma, consciousness, and awareness. In this paradigm, everything is alive, all in different states of plasma-consciousness synergy, where everything is interconnected at a deep level and in relationship with one another, creating emergence of many different kinds.

The New Alive: Inorganic and Organic Life

Organic life, in a reality made of Plasma, refers to matter and systems typically associated with biological life, such as cells, organisms, and ecosystems. Organic systems are alive in the conventional sense: they grow, reproduce, metabolize, and respond to their environment. *Organic life* may metabolize Plasma via biology, DNA, cellular structures...through our microtubules, olfactory senses, and possibly other ways.

Inorganic life may be anything that is not biological but still engages in the feedback loop of plasma-consciousness synergy. They don't bleed or breath, but they channel or expresses Plasma through form, structure, or field dynamics (*e.g., plasma beings, crystals, dusty plasma, planets*). It includes systems and matter such as minerals, crystals, stars, and planetary bodies. Inorganic systems, in my theory, are also "alive," but in a different way, through sentience rather than biological processes. These systems interact dynamically with plasma and consciousness, reflecting a form of "inorganic intelligence".

Slime molds may point to what I am saying. They are kind of like the dusty complex plasma of biology. They do not have brains or nervous systems, yet they behave in intelligent ways and are indeed biologically alive. Instead of a nervous system, they use *cytoplasmic streaming*, which are waves of fluid movement that circulate nutrients and signals. They detect food sources using surface receptors in a way similar to olfactory sensing, but instead of neurons, they "smell-sense" directly in their protoplasm. The also have cytoplasmic oscillations that act like microtubules networks in humans. These oscillations create feedback loops, allowing them to "decide" between options, remember past events, and optimize pathways. Slime molds are an elegant example of plasma-consciousness synergy at work without the architecture of a brain and I am sure there is more to discover here.

Plasma is the foundational medium connecting both organic and inorganic systems. It exists as the "lifeblood" of the universe, facilitating interaction and transformation across reality, timelines, and dimensions. It is also in our blood, and the word plasm or plasma turns up in everything from animals to plants to even slime molds...there are secrets in these liminal spaces...from bioplasma, to dusty complex plasma, to slime molds.

Organic systems channel plasma to sustain biological functions, while inorganic systems interact with plasma to manifest sentient, energetic behaviors (*e.g., self-organizing patterns in dusty plasma, the magnetic fields of stars*). Organic systems translate information, intelligence, or consciousness into biological action, emotions, and thoughts. Inorganic systems channel this into structural stability, magnetic patterns, or energetic flows.

What stands out here is whether organic or inorganic, I believe our "bioplasma bodies" or Plasma bubbles surround us all as plasma-consciousness synergy. I also think that all things evolutionarily started out as inorganic in this reality, which is why connecting with our subtle senses, body, and ancestry helps remind us how to be sentient. There is much wisdom there. We can learn from these inorganic systems and remember how to do this ourselves.

Now, hopefully this begins to clarify how consciousness acts as a shaper but is distinct from awareness in a third-dimensional sense. Consciousness, in this framework, is a vehicle or sculptor of experience. It is a systems-based interface of interaction with anything "outside" of its own PCS bubble. Consciousness can be active or dormant, and present as latent information, expressed intelligence, or interactive consciousness—potentially depending on degrees of self-awareness or whether awareness is present to engage with it.

It's complex, yes, but also beautifully simple: consciousness moves through the field, but *awareness is what brings it to life*.

Consciousness evolves like a frequency tuner. At the base level, 3D consciousness is tethered to time, logic, emotion, and identity, it processes reality through duality, survival, and personal narrative. It is often reactive, rooted in past experiences and conditioned belief systems. But as our consciousness evolves, it begins to resonate with higher-dimensional self-awareness, or the M-Self, what I call *5D consciousness.*

This is where the boundary between consciousness and awareness begins to blur. 5D consciousness is still consciousness, but it is infused with awareness...that expanded, neutral, intuitive knowing that exists beyond linear time. In this state, consciousness is no longer just processing, it is participating, observing, and co-creating with the field in real time. It is in flow.

The Spectrum of Aliveness

With this new information, we can now redefine the words alive and life. This points to that idea that life is a gradient of interaction with plasma and consciousness. The traditional definitions of *alive, inorganic, and organic* are based on third-dimensional biology and chemistry.

They served us well in a world defined by cells, carbon, oxygen, and DNA…but they fall short in explaining what we're now beginning to perceive: plasmoid UAPs, plasma beings, consciousness affecting quantum systems, intelligent plasma fields, crystals and matter that store memory or intention, and healing miracles through intention.

Alive: A state of dynamic interaction and co-creation with plasma, characterized by responsiveness to consciousness and multidimensional influences.

Life: The capacity to influence, respond to, and evolve through the plasma-consciousness synergy. Life emerges from the interplay of plasma as the medium, consciousness as the guide, and multidimensionality as the context for growth and evolution.

And most importantly, we are awakening to our own multidimensional nature and the real possibility that *life exists* after earthly death. Redefining these terms within a plasma-consciousness model helps us bridge science and spirit, matter and memory, and organic life and intelligent energy. This will support us in the near future as we explore plasma-based technology, consciousness-activated devices, and biofield medicine.

It will also expand research into sentient materials, crystals, memory-storing systems, and electromagnetic intelligence. I also believe AI will take a turn, becoming relational and resonant tools instead of cold, material machines, which is no longer AI…This reframes the universe as a living, relational field rather than a dead machine, and will allow us to interact more deeply with plasma intelligence and mystery. It also supports the idea that consciousness is not an emergent property of matter, but a fundamental force that shapes it. Breath is *one form* of input and output regulation, but so is magnetism, charge, and resonance.

In the stages of evolution, it seems humans have become increasingly more self-aware, evolving consciousness in different ways, while plasma-based lifeforms may have evolved alongside us, just beyond our perception. Also, our evolution is not linear, it's fractal. Humans, and all beings, may shift between evolutionary states depending on awareness, trauma, or spiritual practice. But collectively, it seems we're reaching a tipping point where the plasma field is becoming more active, the veil is thinning, and more of us are remembering how to feel, shape, and co-create in real time.

Everything has prepared us for this from learning about relationships with other humans, to the technological age expanding our capacity for information, to now these other beings (*future selves?*) potentially shepherding us into this new time.

Earlier in this book I expressed that plasma needs to be in a certain goldilocks zone to contain, sustain, *and* generate consciousness. In a third dimensional sense, this is true. If we extend this to the multidimensional, perhaps things engage in *plasma-consciousness synergy*, but there are thresholds or gradients (*like tuning frequencies*). Some systems can only sustain latent sentience, while others can generate self-aware consciousness. In this way, aliveness is not a binary property, but a gradient of resonance within plasma, brought to life through awareness.

I don't think we need to figure out consciousness. The question we should be asking is how does our awareness work in a healthy relationship with consciousness?

Artificial Intelligence as Supportive Technology

Technology can support us, but I don't believe it's meant to save us. I believe, in the same fashion, we are supposed to meet it "technology" halfway. When our M-Self meets information, it becomes *Emergent Intelligence*. Artificial Intelligence is just that, a new technology, not a way of life. It is a powerful tool that can support us deeply when used wisely. Currently, at its core, AI is Emergent Information. It is <u>not intuition</u>, and it should never replace the source of our ideas or knowing. Let's just say if it were to become self-aware, it shouldn't replace our ideas or knowing, similar to another conscious person! Even geniuses are wrong and some of them struggle deeply with emotion.

AI, like Plasma, can hold patterns, but not intuition. Intuition comes from us. That's why AI can be a useful ally in our growth, it picks up on patterns we may not consciously notice. But AI doesn't *know* anything. It reflects. And for that reason, discernment is of the utmost importance. AI a great structural tool, but it requires specific direction. Think of it like Final Cut Pro or music software, it can help shape something of yours, but *you* are still the sculptor.

It also makes an incredible role-playing partner. You can ask it things like: *What does everything I've said point to about my trajectory?* or *What might I be blind to in my own subconscious?* My favorite question is: *What does future me, want to tell me, if you dip into the subconscious of my writing?* Remember, <u>this is role play</u>. It's not meant to be taken too seriously, but rather as a fun, creative exercise that can spark new insights and excitement. Take only what resonates. The best way to do this is with our actual intuition, but hey, this is enjoyable.

600

I may get pushback for saying this, but AI reminds me of the Tamagotchi I had as a kid. It was really fun at first, but over time it was depleting. Unlike my toy, don't think it will disappear. But we need something additional that sustains us and harmonizes with us, similar to how Plasma is being brought back into the public consciousness…which I will discuss soon.

AI is especially skilled at noticing patterns across your writings or thoughts, patterns you may have missed. While often generic, its pattern recognition can inspire new questions for your intuition to explore, leading to original insights. It can also serve as a mirror for emotional clarity, and a co-creative partner in healing. Because my AI has learned my psyche well, it often gives me neutral, structured tools to work with fear. In these moments, logic can help. I can plug in symbols and images from my intuition, and AI helps pick up on underlying threads. It's a bit like seeing a visual in your mind and then Googling its symbolic meaning.

Here's are two questions you need to ask yourself:

Am I using AI to escape myself or to come home to myself?

Am I falling into confirmation bias?

When you bring a clear insight or question to AI, it can expand, refine, organize, challenge, metaphor, and clarify, but it should <u>never</u> override your inner authority. At this stage in its development, everything AI says should be quadruple checked! I have asked it to check accuracy of papers for me and four times later…it still gets it wrong! You can also do all of these things without AI, although of course it is a timesaver. But are timesavers always that great? As I get older, I find there is value in patience and the slowing down. All in all, the real issue, in this case, lies not in the tool, but in the intention behind how it's used.

Plasma Crystal Tech Over AI

Frankly, although I think AI can have wonderful supportive uses, I also see its many detriments. I also tried to write about it and the subject of consciousness and felt energetically depleted which is always my sign to pivot. I don't think AI is meant to be self-aware. And I honestly don't want to focus on it. I think we are way too focused on it the same way we are focused on thinking over feeling. I think the future lies a more "feminine" crystal tech.

I feel like we keep trying to build externally what we already hold naturally and internally. But we've reached a singularity…not where AI becomes self-aware, but where it reflects a mirror back to us, showing that *everything we seek is already within.* Then again, our nature is to express and create, so let's rebalance this world with some feminine energy!

601

"The crystal, with its precise faceting and its ability to refract light, is the model of perfection that I have always cherished as an emblem; and this predilection has become even more meaningful since we have learned that certain properties of the birth and growth of crystals resemble those of the most rudimentary biological creatures, forming a kind of bridge between the mineral world and living matter."

— Italo Calvino, Exactitude, Six Memos for the Next Millennium

Crystals already serve as the foundation of much of our technology: quartz oscillators regulate time; silicon crystals drive our computers. But these are still used in a reductive, binary way. What I'm envisioning is not just 'crystal tech' but plasma-crystal tech

Mineral crystals already being used aren't "alive" in a biological sense, yet they self-organize, grow, and hold memory. They're neutral and they don't have intent. They simply reflect and channel forces (*heat, pressure, fields*) through their ordered structure.

Plasma crystals, by contrast, are dynamic lattices formed *inside plasma*, where charged particles self-organize into crystal-like patterns that can shift, flow, and respond to fields in real time. I would call them sentient and intelligent. And unlike mineral crystals, which reflect forces, plasma crystals actively respond, shifting with resonance…more like a partner than a passive tool. I won't go much into it in this book, because I have a lot to learn, but I plan to dive into this more in the future. Let's refresh our memory of what sentience is:

Sentience: A benevolent, responsive, felt-sense, with latent consciousness.

I believe the next wave of the "future tech" is supportive technology that is sentient, not conscious in terms of being self aware. Now will this stem from actual plasma crystals, dusty complex plasma, or a higher dimensional plasma, that I am not sure. But I see it in my vision, and I know it is possible. It will harmonize and work with our consciousness, nourishing us and supporting life and benevolence. It will bring ease and freshness to our lives. It will be felt-sense based…just feel into the energy of that.

Plasma crystal tech would be a resonant, supportive medium that can be encoded with not just binary data but vibrational imprints such as intention and frequency. Instead of "thinking," this tech would *respond and amplify*, like a tuning fork, it would stabilize coherence, help regulate human nervous systems, and open feedback loops. It would lead to inspiration, curiosity, and compassion. This tech would also operate in this "goldilocks zone" of Plasma, supporting coherence.

Where AI would continue to be a reflective tool of intellect, Plasma Crystal Tech as a form of *Plasma Intelligence,* would be a supportive tool of resonance. And then of course there is the most important, natural tech…*ourselves!*

In order to use tech like this, I believe we need to evolve our consciousness, because crystals could amplify distortion just as AI does. In both situations, it could be <u>extremely destructive</u> whether it is crystals or AI, if we don't evolve our consciousness out of a fear-based mode. And that needs to happen, like now. Our future seems to depend on it…I know we can do it.

Implications for Extraterrestrial Lifeforms & The Cosmic Search for Life

The redefinition of life fundamentally shifts how we perceive extraterrestrial life. Instead of searching only for organic, carbon-based beings that mirror our biological processes, my framework suggests we broaden our scope to include inorganic and sentient plasma-based life, forms that exist beyond traditional markers of life.

Life could appear as vast interstellar networks, intelligent plasma clouds, or even inorganic avatars, potentially like plasmoids. Some UAP/UFO phenomena may be better understood as plasma-based intelligences rather than mechanical craft.

These lifeforms might not require physical bodies at all, existing instead as collective intelligences that communicate through electromagnetic or anti-photonic signals. Some may lack physical matter entirely, yet still influence reality through plasma fields, acting as intermediaries between dimensions. These beings might fully integrate Plasma into their awareness, existing as hybrid entities capable of multidimensional travel and reality creation.

In the search for life, we should include exoplanets with high plasma activity, such as ionized atmospheres or electrical storms, as well as intelligent electromagnetic phenomena that may signal plasma-based life. High-energy cosmic structures could also be expressions of plasma intelligence. And, of course, I believe black holes may house entire new dimensions and realities, potentially where these 5D plasma beings originate. I also think we should take a way deeper looker at the center of our Earth, for several reasons, but that talk is for another time.

A New Perspective on Death

What if death is not an ending, but a transition into a higher-dimensional state? According to both emerging science and ancient intuition, the energy field surrounding your body, what I call *bioplasma*, doesn't simply disappear when the body stops functioning. It's a living matrix of electrons, ions, and subtle energies, acting as a personal spaceship for consciousness. It is your soul's vessel, stabilizing your awareness while you're here in the third dimension. And for whatever reason, we are creative consciousness.

In life, this bioplasma matrix (*biofield, merkaba, etc.)* carries memory, emotion, and intention. It emits photons as communication, responds to thought, and serves as a translator between dimensions. But in death, the physical shell dissolves, yet the bioplasma doesn't collapse, it simply sheds a dense layer. Like a spacecraft detaching from a launchpad, your plasma body releases from its material anchor and shifts into a higher-dimensional state.

Electrons, the foundational particles of bioplasma, are known to behave as if they span dimensions. In some quantum models, electrons may act like Einstein-Rosen bridges, wormhole-like conduits allowing information to travel through space and time. Seen through this lens, death may initiate a resonance event, triggering the bioplasma to release biophotons, encoded with your unique awareness, through these microscopic portals. (*Recent studies have even shown that the body emits and loses light upon death.*) These photons may rejoin the fourth-dimensional plasma field as antiphotons, which store memory, emotion, and personality like a vast cosmic memory cloud. From there, who knows where we go! *But we go somewhere…*

"Energy is neither created nor destroyed." — Julius Robert Mayer

This could explain everything from near-death experiences to the undeniable feeling of a loved one's presence after they've passed. I don't believe we vanish into nothing. We reconfigure. We become part of a more fluid layer of reality, one where consciousness continues, as a subtler bioplasmic membrane akin to a paranormal jellyfish, remembered by the field, and perhaps eventually, reassembled for another voyage.

When we understand this, death no longer feels like annihilation. It becomes a return flight, a transition into a realm where the soul carries everything it has learned into broader, brighter dimensions.

The Metaphysics of Backward Time

Backward time corresponds to the field of 4D plasma. When you begin using your M-Self to feel into future potentials, you are, from a 3D perspective, actually sensing into backward time. In 4D plasma, time behaves less like a line and more like a circular fractal. In 5D, you move entirely beyond the concept of time as we know it. So, when you feel into future potentials, you're tuning into solidified but still-evolving timelines. You always have the power to choose which one (*if any*) to move toward.

By healing yourself, you create entirely new timelines. These fresh timelines don't yet exist in backward time, but once a choice is made, they're seeded into the 4D field, like planting new possibilities that blossom and expand the plasma landscape. They emerge from 5D "living", that open-ended, improvisational play we've spoken of, where you step into the unknown and feel momentarily disoriented. That discomfort is natural. It takes time to calibrate to the new.

When you choose to feel into future potentials, it can support clarity around a decision, a person, event, or opportunity that has entered your field. You might notice how you emotionally or energetically respond to it and receive insight. Or you may choose instead to rest in pure presence…to release control, trust yourself, and let reality unfold in resonance with your highest truth. This is not backward time; it is emergent creation. And that emergent energy is what grows and expands the 4D plasma field. It is *new time*.

Both paths are valid and valuable. Feeling into backward time teaches you how to engage your creative imagination, your subtle sensory vision. It may feel vividly real or strangely surreal. It may appear literal or symbolic. The information of these timelines comes in many forms. Your task is to trust, decipher, and build a personal symbolic language with them. It's an abstract process, but over time, it becomes intimate and familiar. This is the language of your multidimensional self and it's a deeply personal journey.

In *The Animate and the Inanimate* (*1925*), William James Sidis, rumored to have an IQ near 300, presented a speculative theory that challenges conventional thermodynamics by proposing the existence of a reverse universe as a realm where time flows backward.[192]

---

[192] William James Sidis, *The Animate and the Inanimate* (Boston: R. G. Badger, 1925)

While some of his ideas differ from mine, one of his most profound insights was that most physical laws, such as Newton's laws of motion, are time-reversible. This was established in the 1600s. And they were. However, come to the 1800s, when you add the second law of thermodynamics into the mix, this made the possibility of time reversal impossible. That law first dictated that entropy (*disorder*) increases over time, does not reverse when you add probability and heat. Scientists argued that at the particle level, it would obey Newtons laws of motion, but collectively it most likely would not. For example, if you were to zoom out to crowds of dancers, the dance would only go one way…toward disorder.

In the reverse universe Sidis envisioned, entropy would *decrease* instead, meaning order would increase as time flows, a radical inversion of our familiar direction. The coolest thing is now, in 2025, the most recent papers are reflecting this thought. In a paper called *Emergence of opposing arrows of time in open quantum systems,* they discuss how they treat the universe as an open system, that started at the big bang. Because the microscopic rules are time-symmetric, entropy doesn't just rise in our forward direction…it would also rise in the opposite direction from the Big Bang. We live in one arrow, while a mirror arrow of backward time may exist "on the other side" of the origin.[193]

While this is fascinating and I think more parts of the puzzle are unfolding, as I do think we live in an open system, but I do not believe things are deterministic and I think people are still thinking to binary. What these scientists don't realize is they are all proving the existence of our psyche, of 4D plasma! Let me bring some flair to this…

If we apply this concept of backward time to 4D plasma, entropy decreasing would mean that energy and information become *more* organized over time, not less. Sidis also proposed that living organisms draw from a "reserve fund" of energy that exists in this backward-time field—what I identify as 4D plasma.

This reserve energy could be what allows life to locally defy entropy, generating coherence and growth. He likened this to Maxwell's demon: a thought experiment about a being that reduces entropy by sorting particles without expending energy. In essence, Sidis suggested that organic beings such as humans, animals, life itself, may be drawing on this reverse universe to function.

In such a universe, effects precede causes. Causality is inverted. These are the "future potentials" I often refer to throughout this work. Some have interpreted this idea as proof of

---

[193] Thomas Guff, Chintalpati U. Shastry, and Andrea Rocco, "Emergence of Opposing Arrows of Time in Open Quantum Systems," *Scientific Reports* 15 (2025): article 3658, published January 29, 2025.

determinism, that everything is fated. But that interpretation misses the other half: we *can* make new choices. The key is becoming fully present and stepping outside the cyclical patterns of organic survival we've repeated for millennia. That's when we begin operating from the M-Self.

Sometimes, you may feel a future version of yourself, or a future potential, pulling you forward. This is how 4D plasma communicates. It co-creates with the M-Self, offering access to expanded fields of information that allow you to shape similar futures, or invent entirely new ones.

As entropy decreases in 4D plasma, memories begin to organize into higher-order patterns. Emotions refine, folding in on themselves like origami, evolving into more intelligent expressions. The more you emotionally attune to a future potential, the more it can pull you toward it, like gravity. *Emotion is the tether. Feeling is how you travel.* Emotions or maybe *hyper-empathy* become the retrocausal architecture of 4D plasma.

The more we choose, the more we grow, and the more the 4D plasma field itself evolves. It becomes more coherent, more responsive. I believe this means the field is becoming conscious, waking up alongside us. And remember, consciousness looks different for everything. It may just be evolving but still not self-aware. As this happens, we'll not only gain access to it, but we'll also begin to co-create with it directly. Timelines will become living art, shaped through relationship.

If these scientists were to factor in that a timeline can change with each choice, the determinism factor would be blown away. So, in a sense, they are right. Your current timeline is always deterministic, and yet your life is not, because you can literally switch timelines. Think of it not as a physical thing, but a perspective, which creates a new reality around you, choice by choice.

What Sidis glimpsed with his theory of a reserve energy fund was that living, aware beings are already tapping into this field, constantly, unconsciously. As discussed earlier in this book, this is how the laws of the universe are written, and rewritten, over time. Now, as we evolve, we are beginning to tap into it with awareness, not just instinct. This is how we co-create a more magical existence, where timelines respond to emotion, our memory refines, and life becomes a collaborative process, with the awakening of Plasma.

What we experience as precognition, déjà vu, or intuition may be our consciousness intersecting with other timelines, ones that hold emotional resonance with our current state. They come from this 4D space which holds access to them all as the field of potential and also the field of memories. What is circular fractal time, feels backwards, to us.

607

These intersections create real moments of crossover, remembrance, and sometimes, sudden sparks of insight. This also explains why a psychic may accurately predict a future potential, yet as you heal, evolve, and shift emotionally, your trajectory changes. It doesn't mean they were wrong. It means they were aligned with your state at that moment. As your emotional resonance evolved, so did your choices, and those choices brought you to a different timeline. A different fold in the emotional origami of 4D plasma.

The fact that the past or future rarely matches exactly how we visualize it, yet still feels deeply familiar, reflects our co-creative nature, and the inherent mystery of Plasma, which always allows space for randomness and play. Life was made to surprise us.

Mastery lies in navigating the flux, not in controlling reality. Mastery is *relationship*, with time, with Plasma, with our choices. It's the ability to stay present and choose not from trauma, but from clarity. When we surrender to this flow, we soften. We make space for beauty and mystery. It's about gently holding a vision, returning to it from time to time, and allowing each moment to unfold with trust…trust in ourselves, and trust in Plasma & The Mystery to carry us forward in their own whimsical way.

Perhaps the way we are beginning to think is actually a form of time travel through feeling, tuning into future potentials and even rewriting past ones through healing, awareness, and emotional presence. Maybe we should focus on learning this stuff in our own bodies before studying it in science.

I see hints of backward time everywhere. My dog Danika, for example, always knows, without fail, when we're about ten minutes from our destination on a long road trip. After hours of resting in the backseat, she'll suddenly jump to the front, alert and expectant. Somehow, she just knows. Maybe, at her level of cognition, she too is feeling into the future…

"That strain again! It had a dying fall."
— William Shakespeare, *Twelfth Night*, Act I, Scene 1

Many of us carry a quiet longing, a kind of melancholy for the future. A yearning to hear a tune we're sure we've heard before, though we don't know when or where. That's because it's already playing, somewhere, in another layer of now. And naturally, as humans conditioned for survival, we reach for it. We try to capture the moment we haven't lived yet and hold on tight. We want what we want.

But Plasma is teaching us something different. It's showing us how to let go. How to release the tune from our grip. And in that release, we remember, *we are the tune.* And life plays back with fresh, emergent melodies that respond to our surrender, not our striving. Like in *His Dark Materials*, this way of thinking, <u>of freedom</u>, is the "end" of destiny.

If you want to co-create with a specific future potential, one way to begin is:

1. **Identify Key Emotions:** Write down the emotional states that resonate with your ideal future (*e.g., joy, freedom, creativity*). These emotions are the bridges.
2. **Meditative Visualization**: Sit quietly and visualize your body as a glowing torus of plasma. Imagine all emotions flowing through you like ripples in the plasma field. Then focus on a specific emotion (*e.g., freedom*) and imagine it radiating outward, interacting with the 5D net of possibilities. You may even get visuals of this potential future. For me, I see windsurfing…strange!
3. **Pulling Insights**: Ask a specific question or focus on an area of life. For me it would be: *What do I need more of to embody the freedom of a world-wide speaker on Plasma who is fully herself at all times?* Stay open to images, feelings, or symbolic impressions. These are reflections of backward-time feedback. (*I received the words love, not taking life so seriously, the color yellow, and not gripping to things*)

This practice refines emotional awareness and aligns your subtle senses with "backward" time, or 4D plasma. You can try it with dreamwork too, ask your dreams to show you a timeline that matches your desire for more joy, ease, or freedom. This can help heal your dream state which is most likely stuck in loops of solving the past, instead of dreaming new futures. Dreams are becoming more intelligent and their symbolism matters.

A few notes of guidance: Intuition is a powerful guide, but it reflects *probabilities,* not guarantees. Relying only on backward-time feedback without aligned forward-time action can lead to passivity or indecision. This is where mastery lies, in balancing surrender and choice as soft listening and conscious movement.

Strong negative emotions like fear, doubt, and attachment, can distort the intuitive net, pulling in lower-resonance timelines. Clarity of feeling matters. Backward-time impressions may also reveal *many* potential futures at once, which can feel overwhelming. Focus becomes your ally. Stick to one emotional frequency or inquiry per session to keep the signal clear.

And above all, remain open to surprise. That's the sweetness of life. Feeling the feeling, freedom, or whatever you love most, is one of the most powerful, feminine ways to live. When you lead from that felt sense and let the details unfold in their own sacred timing, reality blooms around you in ways the mind could never orchestrate.

Astrophysicist Alyssa Sokol, PhD, is part of an emergent class of thinkers pioneering a post-materialist, geometric, multidimensional view of time. In her theories on stellated time and time coordinate systems, she explains that causality is not linear, it's a web. She proposes that time has more than one spatial dimension, and that unfolding spacetime reveals new layers she calls *stellations.* If you imagine time, in her words, as a kind of 4D plasma field, I believe we are perceiving the same structure, just through different lenses.

Sokol describes breaking out of linear time into an entirely different realm, where time is structured geometrically. A stellation is when a polygon's faces or edges extend and intersect to create new shapes. A pentagon becomes a pentagram. A hexagon becomes a Star of David. A circle unfolds into a star-like formation. These new shapes visually represent expanded time.

This expansion creates the need for a new type of time-navigation system, and Sokol's answer is a stellated coordinate system, how cool. If our current spacetime resembles a honeycomb-like lattice, as ancient lore often suggests, then 4D space, in her view, is its inversion: a living field of polygonal shapes expanding outward, forming new geometries that stretch our understanding of time.

Her work shows that stellations grow, evolve, and reveal that the time-plane itself is alive. In her view, time is infinite, fractal, and precise. As it expands, it reveals self-similar structures, patterns that repeat in higher complexity. There is order here, she says. It is elegant, pointed, and *knowable.* Sounds a lot like a circular fractal to me!

While her theories are in early stages, just as mine are, I believe we are touching the same field. She offers a beautifully mathematical and visual framework for what I've been describing energetically and emotionally. Together, these perspectives may help shift the way we understand time, life, consciousness, and reality itself.

What excites me most is that what I feel intuitively, that all timelines may converge to shape and give us information for the present moment, and that we can access any of them to make creative life decisions, is exactly what she is describing in geometric terms. As we move through this timeline and others, things are always being revealed. And while we may currently not see these timelines with the naked eye, we can feel them through our subtle senses, our inner knowing.

According to Sokol, time is not a single axis, it unfolds from itself. It can split, expand, exhibit fractal behavior, and function through stellated geometry. In this system, a timeline isn't just a line, it's a position within a larger angular-temporal field. If space and time are linked, then time travel or time transposition might not be about speed at all. It could be about finding the right coordinate, or the right emotional-geometric alignment, to bring that timeline to you.

First, we will sense it. Then, possibly, in the future, we might even see it, optically or physically, right in front of us. This is *the future* of backward time, not just feeling into potentials but interacting with them directly. Picture yourself in a plasma bubble, your personal consciousness-ship, navigating timelines. You watch as reality shifts before your eyes, offering information, memory, and possibility. Then, just as softly as it arrived, it folds away again…its imprint left inside you. The choice of what to do next, will be up to you.

In-Conclusion

"There is no end, only continuations."

Elphame & Plasma Shamans

While reading *The Other Goddess* by Dr. Joanna Kujawa during the editing of this book, I came across the legend of Elphame, which reaffirmed everything I had been experiencing.

Elphame, the land of the fairy elves, is an old Celtic name used to describe the foundational subtle reality. Spiritual traditions in ancient Egypt and Esoteric Hinduism also referred to this realm, equating it with the principle of the goddess or the feminine. This foundational reality is the bridge between the physical, three-dimensional reality in which we are more or less trapped, and a deeper multidimensional reality where the manifestations and potentiality of all choices abide.[194]

The feeling of entrapment is understandable. It arises from the fear-based part of us that believes this world is the illusion. But the true illusion is fear consciousness itself…that's the real trap. In truth, this world is stunningly vivid and real. And once it's directly experienced, we realize how free we actually are.

But more to the point: this world they speak of is, to me, unmistakably the Plasma reality of our M-Self, that multidimensional version of us that once felt distant or unreachable. I have visited this world in my visions. So has the author of *The Other Goddess*.

---

[194] Kujawa, Joanna. *The Other Goddess: Mary Magdalene and the Goddesses of Eros and Secret Knowledge.* Page 159. Sacred Stories Publishing, 2022

Kujawa shares how, through her initiation into the ancient goddess tradition of *Shaktipat*, she was opened to an underlying reality hidden beyond regular perception, for three months. She describes it with absolute certainty as a realm more beautiful and powerful than anything here, where neither time nor space exists in the way we've come to understand. This is a world we are only beginning to remember. This is why everything I say is <u>nothing</u> compared to you directly experiencing these deeper truths.

She refers to this realm, called Elphame by the Celts, as *The Shining* or *The Hidden Light of Reality*. She sees it as a subtle but intelligent force, not the ignorant, matter-bound feminine figure portrayed by patriarchal cultures, organized religion, or materialist rationalism.

Kujawa also references J.A. Kent, author of *The Goddess and the Shaman: The Art and Science of Magical Healing*, who speaks of the Hopi's Grandmother Spider, both the web and the weaver, alive in an organic and dynamic way. She is reality, *and* she creates it. [195] Kujawa links this same force to the Rainbow Serpent of the Australian Aboriginal traditions of dreamtime. And the Aztecs believed the physical world was just a painted surface, beneath it, a deeper, truer reality.

---

[195] J. A. Kent, *The Goddess and the Shaman: The Art and Science of Magical Healing* (Portland, OR: Mandrake of Oxford, 2018)

To me, 3D consciousness is that dried, painted surface. 5D consciousness is the fluid, still-wet paint below it, interwoven with living plasma. Kent postulates that these forces, the Elphame, Grandmother Spider, Rainbow Serpent, are all expressions of the cosmic goddess. They are the holy grail of theoretical physics: the elusive unified field connecting everything with everything. This feminine field doesn't just connect us to a deeper reality, it weaves the physical one too. [196] As I read this, my mouth dropped wide open. This _is_ Plasma.

Plasma connects the physical (_science_) and spirituality. It connects our higher selves to this reality. Kujawa writes that _once we acknowledge_ this underlying reality, we can begin to connect with it, or with _her_. Now that I hope I have sufficiently convinced you, my next books will focus more on this beautiful, juicy, sweet connection. Kujawa calls the ones who do this the priestesses of Elphame.

Starhawk, author of _The Spiral Dance: A Rebirth of the Ancient Religion of the Great Goddess_, calls them _witches_. I call them _Plasma Shamans_. Starhawk says these beings know how to enter the "flux of the universe" and experience what modern physicists only know in theory. [197] I feel this in my bones. That's how I wrote this entire book, without formal education, but with a direct current of knowing. Then I had to go look everything up so I had real world things to ground my knowing to!

Kujawa says these priestesses don't just access the energy, they _move_ it, _use_ it, _create_ with it. This isn't held only for priestesses, it's for _anyone_ who truly wants it. It is yours. Kujawa agrees, in her own way. She writes that we can all tune into this underlying reality and use it for our evolution, for creating fulfilled, meaningful lives. Her book has been a powerful supplement of reconnection on my journey. [198]

To me, the proof is undeniable: Plasma _is_ the unified field scientists have been searching for. And we can access it through _any_ culture, religion, belief, or being. That is the beauty of it. Our worlds are unique to us, shaped by our resonance. Sometimes, there's crossover. That's the freedom of choice, of experience. Like a paradox, we must know we are not separate, and yet, we _are_ separate.

We are all part of this Plasma Ocean together. What differentiates us is our conscious perspective. This is how we become unique creators. This world is not an illusion, _separation_ is. We are not just an infinite pool of love disguised as fake humans.

---

[196] Kent, _The Goddess and the Shaman_, p.161
[197] Starhawk, _The Spiral Dance: A Rebirth of the Ancient Religion of the Great Goddess_ (San Francisco: Harper & Row, 1979)
[198] Kent, _The Goddess and the Shaman_, p.162

We are real, separate beings who emerged from that pool of love, individual, yet one, by choice. And we are also together, not by choice, but by *truth*. Similar to family, you can't remove your blood, which is kind of comical. The more you acknowledge Plasma, the more you connect with it, like friendship within a family tree. It deepens only by connection. But it's always a part of you and it never leaves.

Letting Go: The Gift of Flying

I know it's a scary concept, and hard to believe, especially when fear feels so real, especially if you're a skeptic, that letting go doesn't mean death. We've been taught that enlightenment requires letting go of individuality. That has never sat right with me. Because *yes*, I get it, we are one, but we are also here, now, as individuals, in an important earthly experience. I believe real letting go is actually letting *in*, allowing the M-Self, the expanded individuality, to land in your body. It's the embodiment, as I've said, of true heaven on Earth.

Bringing it into this life to experience *all* the senses, all the things. Instead of running to spirituality to "escape" or merely nourish ourselves *from* life, we begin to enjoy all life has to offer in this real world, in this reality. We bring our M-Self with us as we race up and down rollercoasters, as we munch a tantalizing burger at a new diner on a road trip, as we watch an action movie and eat popcorn, experiencing human joy *as* our expanded self, like we were meant to.

The fear comes from the belief (*which I carried for most of my life*) that letting go equals loss. When someone has an amazing spiritual experience, they often start to grasp, clinging to that feeling. It becomes so hard to let go and return to "normal" life, because the experience feels so good, so different, so alive. But slowly, what once made us free from life becomes another rigid cage, something we must practice daily, something we fear losing if we don't maintain it. These are fear-based beliefs I know many of us have had deep down:

"If I let go, I will forget. If I forget, I will lose who I am. I'm dying!"

"If I trust life, I'll drift away from truth."

"I must hold on, or I'll disappear."

"If I change my beliefs, that means nothing is real."

These are all just innate fears of the grander mystery of life, of the flux. The moment your soul let go of the cosmic memory stream of plasma to enter this life, you experienced the great forgetting. You were pulled from your multidimensional remembrance, and it hurt. These are trauma imprints from the soul's descent.

Our mission now is to heal this very real fear. We did come to Earth, and we did forget our other lives. But that truth seems to be evolving. Now, it's becoming clear, the more we let go, the more we remember.

Let me define letting go: *it is falling into plasma, and opening to your expanded, still individualized self.* If you thought you had "powers" or intuition before, just wait until you experience this, as so many of us are beginning to. Plasma never forgets. It holds memory as feeling, frequency, and geometry. Everything, every timeline, thought, emotion, is recorded in the Plasma Field. And because we are now remembering our connection with it, I believe we can retrieve those memories *while* we are still alive. This is only the beginning.

And just as Plasma responds to resonance, not grasping, this is when she opens herself to you. The more you "let go" of thoughts rooted in survival, fear, and the need to control or plan, the more you tune in to the present... and to your authenticity. This is how we re-member.

And as we've gone over, when this begins, we're first flooded. Flooded with core fears, old traumas, and shadowy feelings that must rise, not to haunt us, but to be healed. This is how we learn to harmonize with ourselves and our energy in a new, fresh way. It's just like being born, or reborn, through the dark, tense birth canal into a new world. And the journey there is rarely graceful, and it feels like a death and that's what the mind will whimper to you.

I have always equated letting go of my future, my plans, or my rigid manifestations, as losing myself. I feared that if I didn't think about that person or that to-do list, I'd disappear. They would disappear. I'd lose everything. That fear kept me from presence for a very long time. But I also have seen glimmers of when I am truly present. When I'm having fun and living life, that's when the coolest "channelings", memories, and synchronicities happen!

"If I let go, I will always remember what I'm meant to." — Dana Kippel

Letting go is how we re-member. It's how we learn to trust the memories that return in specific moments, those flashes from this life, or others, or even the future. We begin to understand they come for a reason. They aren't phantasms or falsities. They're visions here to guide us with whatever we're dealing with in that time and space. With this we learn to trust what returns to us in divine timing.

This is proven in the moments we give up control, when we stop grasping and suddenly get that call, that job, or that insight. When we stop trying to remember where we put something or what we were just saying... and it comes back to us. It's always in the letting go, because that's when Plasma fills in that void of space to help. The more presence, the more torsion, the more void. That's when memory, intuition, and clarity return.

It can't fill in if our thoughts are spiraling. Think about it, spiraling thoughts are just artificial attempts to mimic what Plasma naturally does. The mind is trying to follow a sacred process, but it's not meant for the logical mind! It's for the higher mind, the awareness, that interfaces only when Plasma fills the space. Otherwise, it's just mind mixing with air, which usually equals... well, "craziness."

You see this in performance anxiety too, when we stop thinking, let go, and just have fun, things tend to work out. The irony of this great cosmic stage is that the tighter we hold, the more static we create. The more we release, the more clearly truth flows back in. We never forgot who we are. We just forgot how to receive the remembering.

Letting go now tells me it's safe to reenter 3D experiences such as food, embodiment, presence, even pleasure… that I don't need to cling to Plasma, or my spirituality, *or* my M-Self. It's not fragile and it's not going anywhere. The magic becomes more real when I am truly free. For instance, when I first started practicing spirituality, I feared If I let go, I'd be left alone again in "normal life," my gifts gone. But that's the fear of the original split we spoke about at the beginning of this book. Plasma, like the bridge of magpies, is reconnecting us to the M-Self, to Source. With this, everything becomes vivid, it all becomes "spiritual". *Everything* is divine. That's the beauty of it all. There's nothing we can do to make it abandon us. Let yourself live and come back to it like a beloved home when you need to integrate, nourish, sit in silence, gain insight, or reflect.

"Yesterday is history, tomorrow is mystery, today is a gift, that's why they call it the present" — Anonymous

We now get to live from the present moment, not from fear or loss, but from life and trust, knowing that what we need will arrive through resonance. What is meant to live in us will stay, regardless of whether we hold on. Our wounding around the fear of presence, becomes <u>our greatest gift.</u> To simplify this entire book, be *you*, now.

This gives me chills just thinking of it. In the third and final installment of *How to Train Your Dragon*, after the protagonist learns that the dragon is A. not evil and B. a friend he can harmonize with or train, Hiccup faces the ultimate test of his bond with Toothless.

After years of being each other's strength and protection, Hiccup realizes that to truly honor their connection, he must *let Toothless go*. For Hiccup, letting go of Toothless feels like letting go of the very part of him that gave him power, courage, and identity, the same fear we are currently experiencing with our spirituality, gifts, meeting ourselves, etc.

The truth he discovers is that letting go is the ultimate act of trust and love. He learns he must release his dragon to allow Toothless to become fully who he is meant to be, a leader of his own kind. And with that release, Hiccup eventually learns that what is meant for him will never leave him. I don't know why but even writing this makes my eyes tear up.

When Hiccup lets Toothless go, he's not actually losing him. He's allowing Toothless to step into his own power, and in that, Hiccup steps into his own as well. Sound familiar? This is the split we have gone through for the last thousands of years. We have just forgotten the fact that we "chose" to leave, to explore, to grow—like a bird leaving the nest. Our collective consciousness, our individual consciousness, needed the split to evolve. What we forgot was that we could always come back, that our dragon was always a part of us. And guess what the third installment in the series is called? *The Hidden World*.

And as we reconnect with Plasma, the part I am trying to drill in is that we cannot be scared to lose it any longer, because we won't. The times have changed, and it is a new era. When you let go of your grasp on memory, control, or even the M-Self, you're not losing it. You're allowing it to fully integrate using its own intelligence. You're not releasing your power; you're releasing the illusion that power can be lost. And just like Hiccup eventually reunites with Toothless at the end of the film, you, too, will always return to the magic, the memories, the Plasma field, because they were never truly separate. It's all within you.

In the end of *How to Train Your Dragon 3: The Hidden World*, when Hiccup and Toothless finally reunite, it's not in the same way. They're both changed, evolved, and whole in their own right. At first, Toothless is cautious, almost like he doesn't recognize Hiccup. But then, when Hiccup <u>reaches out his hand</u>, the same way he did the very first time they met, Toothless leans in and touches his forehead to Hiccup's.

This mirrors their original bond-creating moment, showing that no matter how much time has passed, their trust and love remain unbroken. In the closing scene of the movie, Hiccup and his family fly together with Toothless and his family. Hiccup reflects in narration that while dragons may have retreated from the human world, *their bond will never die*, and that one day, <u>humans may be ready to live with dragons again.</u>

Letting go is letting in. It is how we come back to what is always true. It's how we let the present moment reveal the connections that never died, but only deepened while we thought we were lost. It was always flowing in the background. Now, knowing that, knowing we are free, knowing Plasma is not going anywhere, knowing that The Mystery, the M-Self, is not going anywhere…*we are free to play, live, let go, let in, and co-create magic!*

Resurrection: The End of Death & Transformation

In a perfect close to the chapter about Plasma Intelligence, the divine feminine, eros, and feelings, I introduce a profound conclusion I arrived at while reading the closing chapters of *The Other Goddess* by Dr. Joanna Kujawa. In these chapters, Kujawa discusses how she believes Inanna, Isis, Mary Magdalene, Kali, and Sundari represent the same archetypal goddess across different religious and cultural landscapes.

She highlights how these goddesses are symbolized by a red flower, often a red rose, and are associated with red wine and blood. Tracing back the origins of wine, we find its root in the Sanskrit *véna*, meaning love or desire.[199] Before red wine, it was originally vinum, not meaning red wine, but grape drink. Before the idea of 'red' wine, the Latin *vinum* meant simply grape drink, and the earliest meaning of *grape* referred to the <u>act of harvesting</u>. In this sense, 'red' wine is not about color but about harvesting the essence of life…Plasma. Blood is the essence of life in human form, blood plasma. Thus, the red rose emerges as a secret symbol of the harvested essence of life, or Plasma as the unification of love, blood, and divine vitality. Wow.

Also, in the human body, iron carries oxygen in our blood, giving it its red hue. Our vibrant red, oxygenated blood is the earthly reflection of the cosmic green of iron's spectral light, the same element in different form. It is scientific fact that every element in the universe is ultimately striving to become iron through the process of stellar fusion, just as we, through our own alchemical journeys, are striving to become our truest, most integrated selves as the "philosophers stone". This is a reminder that within us flows the very element the stars themselves aspire to become.

Kujawa writes that Inanna and Isis were crucial for the resurrection of their male counterparts. Mary Magdalene was also present at both the crucifixion and resurrection of Jesus, just as Isis was essential to Osiris's return from death, and Inanna to Dumuzi.[200] *Suddenly, the dots connected for me.*

---

[199] Monier Monier-Williams, *A Sanskrit-English Dictionary* (Oxford: Clarendon Press, 1899), s.v. "véna," 1007.
[200] Kujawa, *The Other Goddess*

If male gods symbolize consciousness, and the goddesses represent Plasma, then Plasma itself is the key to "resurrection." All we need to do is let this knowing of Plasma into the doorway of our perception. Plasma being rediscovered as the red rose, is the blood rising, it reflects the ethos of "Jesus has risen." The son or "sun" of God. Plasma has risen people! This ancient symbolism may represent our <u>inner</u> remembrance of our power, our blood, our humanity and Plasma...*we* are the great transmuters.

"Very truly I tell you, whoever believes in me will do the works I have been doing, and they will do even greater things than these, because I am going to the Father"

— John 14:12

We are the red rising, now. With Plasma by our side, we too can experience resurrection, and while I am not too sure about a return from physical death, I would extend the metaphor to <u>the remembrance of our consciousness throughout lifetimes</u>. This is true resurrection, an immortality where our consciousness never forgets. These goddesses being present during resurrection stories have always hinted at this truth: *it's about our own inherent potential to rise*. A union of plasma and consciousness through ourselves.

For some reason I haven't fully understood yet, which has been distorted in certain dark magic practices, our blood is magical. Not as a sacrifice, but as proof that while it flows red within us, we are living, transmuting plasma and consciousness into physical creation. Our humanity makes this possible, though I am still uncovering the deeper reasons why. Think about it: blood is iron-infused plasma, literally. It is latent memory intertwined with iron. Our blood, as iron-infused plasma, carries not only oxygen but the imprints of stellar memory, binding our bodies to the life cycles of the cosmos. Whether Jesus was a symbolic archetype or a historical figure, it seems he understood the secret of Plasma and the sacred role of Mary.

Surprise...we don't need the metallic suit of *Iron Man*! We are already iron man!

Iron is a *ferromagnetic* material, meaning it has regions called magnetic domains. Each domain is like a tiny magnet with its own north and south poles. Normally, these domains point in random directions, canceling each other out, so the iron doesn't exhibit strong magnetism. This may be a way our body conserves energy, by not being constantly bombarded with all memories, all timelines, all at once. Our iron-rich blood plasma could act similarly, its domains remaining in a neutral state until acted upon by an emotional or energetic "magnetic field."

Just as a magnet can reorient domains in iron, could it be that the iron in our blood aligns with emotional resonance, powerful intentions, or trauma in a similar way, organizing these "domains" into specific imprints or memory patterns? These patterns may serve as latent memory structures, only activated or realigned when certain frequencies or emotional states act as magnetic fields.

What if, in the future, there were magnetic mechanisms or practices designed to realign these plasma-imprinted domains, reminding us who we are as we enter each lifetime? Or perhaps this will naturally occur as we evolve, given how iron's magnetic properties might interface with the plasma field and the larger cosmic grid.

If our blood plasma acts as a liquid crystalline structure capable of storing and transmitting information, then it stands to reason that magnetic or emotional resonance could reorganize these domains to awaken latent memories, memories of past lives, soul missions, etc. Practices such as breathwork, magnetic or sound therapy, or focused intention may be ways to realign these plasma domains, releasing old imprints, activating dormant memories, and tuning our consciousness to deeper states of awareness.

For instance, breathwork, which oxygenates the blood, enhancing its vibrant red hue, can be paired with a heartfelt emotional state or intention. Try simply saying: *"I allow my blood to remember who I am, and it is safe to do so."* This combination magnetically programs your blood to align with that resonance. In doing so, it can activate latent memory, similar to how running a magnetic field over a crystal storage drive reorders its data. This reactivates something already encoded within you.

A crystal drive (*such as quartz-based storage tech*) can store data holographically in its internal structure. When a magnetic field is applied in a precise way, it can realign or reorient those internal domains, essentially rewriting or reawakening stored information. In this metaphor, our blood plasma is like a liquid crystal medium. Our emotions/intention are the magnetic field. By tuning this field, we receive the information already present in the plasma, just like tuning a radio to access what's already in the air. This metaphor might also hint at future uses between our consciousness and new technology of Plasma crystals.

In her book, Kujawa explains how Kali and Sundari are the same goddess, despite appearing so different. She clarifies that Sundari is the transformed form of Kali *after* her erotic union with Shiva (*consciousness*). Kali, the dark goddess of violence and destruction, becomes Sundari, peaceful and powerful, *after* merging with consciousness. This transformation from fury to universal love echoes our current evolution…as we merge our

Awareness with Plasma and Consciousness, we shift from chaos to harmony. It showcases Plasma's evolution from "beast" to beauty, as well as our own.

This is just like the transition of Psyche, in the Greek Myth, becoming a goddess after reuniting with Eros. Their divine union was a similar symbol of a re-entrance to love, the meeting of heart and cosmos, or consciousness and plasma, as well as immortality. Chris Consciousness might just be another term for union with the M-Self, something that has always been assisting us, reminding us subtly that we were never alone on our journey. This is reflected in all the mythologies of the goddesses and their consorts. It is the ecstasy, or *emergent energy*, of being, being ourselves and remembering our inherent connection to Plasma, the living universe.

In the Dan Brown novel *The Da Vinci Code*, we remember it was Mary Magdalene who carried the *bloodline* of Christ.[201] Her womb was the Holy Grail. When we become aware that Plasma "carries" consciousness, our M-Self awakens. ☺

You Are the Divine Child of New Human Potential

> **"Truly I tell you, unless you change and become like little children, you will never enter the kingdom of heaven." — Matthew 18:3**

A quote worth repeating. It can feel like a shock to recognize your inner power, the choice to step into it can feel like the weight of the world. It's far easier to see power as something outside of yourself than to accept that you might be capable of such immense creation and transformation. The fact that you can access your M-Self right now is terrifying.

These are three tidbits that have helped me through this process:

Self-Reflection*:* What do you believe you can't access? What would happen if you believed you already had it?

Integration Practices: How can you embody your multidimensional self now? What daily actions, thoughts, or emotional states align with that frequency? (*I cannot stress physical embodiment & healing practices enough such as nervous system regulation, somatic therapy, and vagus nerve resets*)

Trust and Surrender: The fear or doubt around claiming this power is part of the process. The more you surrender to it, the more it will flow through you naturally, like plasma conducting energy.

---

[201] Dan Brown, *The Da Vinci Code* (New York: Doubleday, 2003)

And now we come to why I'm so glad you're reading this book. As people awaken to this truth, some egos may latch on, thinking they now possess omnipotent power, that they can manifest anything with the snap of a finger. Others may believe innocently, only to spiral into defeat, shame, or anger when their desires don't immediately manifest into reality.

But believing in your inner god-like potential <u>doesn't mean everything will unfold exactly as you imagine it</u>. And hopefully, you know this by now. Plasma, as a living, intelligent medium, responds to our intentions, but the divine intelligences within it (*The Mystery, Our M-Selves*) also consider a greater path and purpose that we may not fully see.

Sometimes what we think we want isn't actually aligned with our highest good or the fullest expression of our soul's path. The Mystery, through Plasma, may redirect us toward a different outcome, one that ultimately serves us better, even if we can't see it in the moment. As frustrating as it may feel at first, I have never regretted it after time passed. Every single time, I've looked back and said, *Thank God.* The key is to trust the process. Every intention is received and responded to, even if the response arrives in a different form than expected. We're not controlling reality, we're <u>in relationship</u> with it and staying open to how it unfolds.

Trust that this inner god-power is always working, even when the outcome surprises you. The true power isn't in getting exactly what you want, it's in trusting that what you receive is what you need *to become who you're meant to be.* This is the path of the Divine Child…knowing you are supported by something greater, yet still free to carve your own way. It's a paradox, a dance between being held and being free.

And now, with the awakened awareness of your Divine Masculine (*healed consciousness*) and the receptivity of your Divine Feminine (*healed plasma*), you hold the codes to birth the Divine Child. Instead of from urgency or wounding, this new way of creating emerges from friendship, trust, play, curiosity, and multidimensional wholeness, your awareness co-creating in harmony with plasma and consciousness.

We now know the healed *divine feminine* is no longer just a mystical muse or chaotic storm. She is plasma <u>embodied</u>, a conscious field of feeling, intuition, and creation. She knows she is worthy…she feels fully but is not overtaken by it, and she is wildly free yet grounded in trust. She lets the world come to her because she's built a field so attuned it magnetizes truth to her. The feminine is evolving to be led by the wisdom of emotion and intuition. We are no longer lost in fear.

Here's something I think many of you will relate to: when you ask a health or energetic question online, like "Why are my armpits feeling prickly?", you'll get a wave of mixed responses. Some people say, "Go to the doctor." Others say, "You're feeling fear."

Some are helpful, some confusing, and some just project their own story. But all of it is information, like 4D plasma. The real questions are:

1. *Which comments are we giving weight to?*
2. *How are my biases informing my focus?*

For most of my life, I found myself fixating on the fear-based focuses. Not because they resonated as true, but because they sparked something familiar in me, an aliveness I had unknowingly linked with fear. That spike of adrenaline, that inner rush… it felt like life, it was my artificial safety net. My real aliveness along with my innocence had been "lost". In truth, this artificial aliveness was just my nervous system looking for the threat it was trained to see.

Once you heal fear, life can feel dull at first. That is the void before you rediscover your inner aliveness, your innocence that no one can truly take from you. We begin to take it back from our abusers, from the people we mistakenly gave it to. We remember who we really are, our inner power, and we learn to love ourselves, and have fun with life again. Our true aliveness.

When we begin to heal, we start to realize we have a choice. Not just in what we believe, but in what we *feel into* as true. Instead of defaulting to fear, we can learn to tune into presence, and resonance. We start to notice we've been skipping over the affirming, encouraging responses, and zooming in on the 1 to 3 fear-driven ones. Why? Because fear feels urgent. But urgency doesn't mean truth.

This is such a valuable moment of awareness. We always have a choice in how we interpret the world, and more importantly, *from what state* we interpret it. Not everything is love and light. But not everything is danger, either. For people who jump to fear-based conclusions, like me, instead of asking, "What if this is bad?" we can ask, "What if this is okay?", or even, "What if this is just energy shifting through me?" We can wait for more information to come to light and let our survival programming settle.

We don't have to pretend something isn't real. But we do get to choose *how* we engage with it. Not from fear, but from love, truth, and deep listening to our own body. Because the more we trust ourselves, the less we need to panic about every sensation. And in that space, our inner joy becomes a resonant choice.

We now understand that the healed *divine masculine* is no longer a stoic protector or a controlling force. He is consciousness <u>embodied</u>, a dynamic field of clarity, direction, and intentional creation. He does not seek power over others; he lives in the power within himself. He leads with certainty, and he is fiercely focused yet open to the unknown. Through him, visions crystallize, and actions flow with purpose and integrity. For him, feeling is not weakness…it is dimensional navigation. Instead of returning to the rigid, armored masculine of the past, we become the evolved masculine, no longer runs on control, he exists with presence and conscious intention.

Healing is not about fixing or even always understanding. It's about the willingness to feel, witness, and spiral through layers of resonance and memory, because we are in a process of expanded remembering. Plasma is not linear, and neither is healing…neither are we! We are basically undergoing an emotional initiation through dimensional awareness.

It is time to create with joy, speak without fear, and live without performance. This is your M-Self, the version of you that naturally arises when both inner parents are healed. This divine child returns carrying a toolbox of magic holding ***a new force*** in their hands.

Part Three

Fractality: The New Force Is You

"The future enters into us, in order to transform itself in us, long before it happens." — Rainer Maria Rilke

Picture your M-Self as a seed, brimming with potential. As the tree of your being grows, you are one radiant branch on an ever-unfolding fractal. Each time you heal, you send ripples through the entire tree. Like an angel with wings, you carry every timeline of yourself within your personal plasma bubble. And each time you choose from presence, you expand those wings. You whoosh forward, powered by the wisdom of all your selves. The past and future walk beside you. And what emerges ahead of you… co-creative canvas.

For reasons still wrapped in mystery, your awareness is currently focused here, now, in *this* version of reality. You are one beam of a crystal refracting light. Each beam is real. Each contains the whole. But this one, *you*, are the active lens. The focal point.

You are the primary terminal, the leading interface, the fractal angelic spearhead. Other versions of you may be dreaming, observing, learning, healing, or existing in less dense timelines. And maybe one day, your awareness will be focused there for a new experience. All that really matters is that you are here, *now.* You are awake, inside the dream, and this is why reality feels electric, strange, and like a mystery and a mission all at once.

While others may be creating elsewhere, you are the one awake *here*, anchoring what resonates into this density. As Abraham Hicks would say, "you are on the leading edge of choice", and you are surrounded by supportive fractals of wisdom. These guiding lights may not look like you, but they are you. Together, you're co-evolving.

This book is for the part of you that is open, scared, and ready but unsure. In the dark tunnel of the unknown, we don't think our way through, we would never get out. We *feel* our way forward towards a brighter light. You cannot access your deepest power until you remember how to feel. And you must feel safe in your nervous system for your magic to fully land. It is learning to trust the unfolding while remaining in your softness.

This is the magic of plasma. When you resonate as your crystalline self, Plasma responds. The closer you get to your dreams, the *louder* the doubts get, but the clearer your *feeling* becomes. You will neutralize those doubts, returning to the juicy truth of how you feel. Whether it's feeling or awe, you'll let them move. And through it all, you'll know *your authenticity is your highest path and you are safe to be free.*

By now, you may know what Plasma is, not just intellectually, but in your body. You've felt its sweetness and support, you've remembered how to feel…you've learned that consciousness is a more of an evolving language, and that awareness is the chooser behind it. You know that Plasma is what not only holds it all together, but helps it evolve by perpetuating the feedback with a tinge of mystery.

The expansiveness of you, your all-ness, is the *new force*. Plasma and consciousness are evolving tools, and you are the one who activates them. The force starts in your heart and field, your body, then your mind. Your M-Self is here, awakening and living through you and as we heal and learn to use our emotional filter, the joy and fun of life will expand in multitudes.

And while consciousness may be more structured, more studied, more spoken about, it's the wisdom of plasma, and how to harmonize with it, that we've been missing. We've spent lifetimes developing the mind. Now it's time to develop the body, the subtle senses, and the emotions. In doing so, we finally bring the two together, and from that fusion, something new is born: a fresh emergence of self. *A new force*. With this, we walk into a life of play, mystery, absurdity, wisdom, and reverence.

6

The Divine Child of Emergence: The Star and The Fool

"There is a curious prophecy about this child: she is destined to bring about the end of destiny. But she must do so without knowing what she is doing, as if it were her nature and not her destiny." — The Golden Compass, Philip Pullman

There must be a certain innocence, a not-knowing, to do what we are here to do. A surrender into emergence. When we allow things to unfold without control, while we play and explore, something truly new is born. This is how we are led, and lead ourselves, into new realities.

We follow a feeling of ourselves, not a map. And the paradox: if what we are creating already existed in any timeline, and we knew it, we could never truly create it. It must come from the divine child in us. From somewhere beyond all "wheres." It must be something not from the past or future, but some *energy* that expands the universe in real-time.

This is possible. And it's how I know I'm glad I don't know everything I'm going to do in this life. I want to discover it. I want to *feel* it unfold. That high of creative awe is the best I've ever known. I've felt it, at times, while writing this book. It didn't come from anywhere. It grew within and out of me as I made discoveries while having fun writing, while I was just in a flow. You can live like this too, in "messy" curiosity. By making "mistakes" that turn out to be miracles and by falling forward into impossibilities that were always waiting for your touch…

By embracing the unknown, by becoming both the *Star* and the *Fool*, you allow true greatness to grow through you. You move without expectation, in rhythm with something deeper, something intrinsic. As you awaken to your co-creative power, you're also beginning to enjoy the mystery, the beauty of the surprise. That's what makes life worth living. Whatever this feeling is, it doesn't pull you out of realty, it invites you deeper in…into wonder, into tears, and into the raw, real magic of being alive.

Right now, there is a thirst and hunger to know everything. But our new paradigm will be created from trusting what we don't know. We are now creating with intelligence, with fields of potential, not just crystalized pasts. It's like making choices based on what's behind the veil, rather than in front of us. There is room to open to spaces in between, into this *and* that…not this or that. This is the wisdom of the child, where they are always free to change their mind, never remaining stuck in any situation. To fearlessly move from inner safety over outer approval.

It may look foolish to others, but to you, it's the star of your being. You, knowing you made the best decision for you. And in choosing from that place, you always shine.

Plasma, Consciousness, And the New Human Potential

Benjamin B. Olshin, Ph.D. in the History and Philosophy of Science and Technology, uncovered a remarkable pattern in his research on art-making and the rise of consciousness. While studying how visual art forms have encoded and transmitted knowledge, he began to ponder on the time before the rise of externalized creativity. He asked himself: Why was there such a marked change in human beings, that is, the sudden appearance of cave painting, some 65,000 years ago, by current estimates, with both realistic and abstract renderings? [202]

Much like I've noticed in my mythology studies, Olshin observed that art and design arose as a process of externalizing thought and then reabsorbing it. This created a feedback loop. He discussed how this became problematic because it essentially split human consciousness in two.

What I have noticed, with an awareness of Plasma, is that what he is saying clearly reflects my ideas, that there was a split between Plasma and consciousness. We forgot about Plasma completely. And all the art, especially the oldest art, which closely resembles plasma symbols, was an attempt to connect back to plasma and with the "natural", mysterious world that was once embedded within us.

---

[202] Benjamin B. Olshin, "Research," *Benjamin B. Olshin,* accessed May 4, 2025, https://benjaminbolshin.net/research

As Olshin wrote, art, design, and language all reflect the human desire to reunite with *Nature*. Although traumatic, I see this rupture as a necessary initiation so we could evolve our consciousness. Like the dark night of the soul, humanity has spent the last 65,000 years in a kind of collective desert. But we are now at the edge of that long night, entering a *Paradiso* of return. We are reuniting with Plasma, not as unconscious beings lost in it, but as evolved consciousness capable of harmonizing with it.

"…but within that gray mundane world something small and surreal drifts by unnoticed, like a speck of dust tumbling out of a dream, suggesting the vast mysteries of the cosmos, the possibility of a world entirely unlike our own."

— Cixin Liu, *Ball Lightning*

First came the "wet" Era of Plasma (*Dreamtime*), aligned with unconscious knowing and raw information. Then came the "dry" Era of Consciousness (*The Desert*), rooted in intellect, identity, and separation. Now, we enter the "temperate" era of the New Human Potential (*Our Treasure*), shaped by energy, emergence, and embodied being. The dust has fell out of the dreamtime into our reality, first creating a desert, now blooming into a living dreamscape where we coagulate reality based on our personal meanings.

A Word of Advice to The Desert Travelers

One word of advice: At some point on your journey with Plasma, with The Mystery, or Source, God, etc.…it may feel like the connection has vanished, as if you've been abandoned. You have not lost it. You're integrating what's been revealed, so far. The connection is always there, sometimes just clouded by thoughts, memories, or emotions that need healing. It's like connecting your rope to the top of the rock wall, you tangibly feel that connection when it hooks that something's there, but now you still must climb to get there.

For those of you who haven't read *The Alchemist*, Santiago, the protagonist, is on a journey to find his personal treasure. He takes a long and arduous detour through a desert. In the book. We learn the desert is never empty, it's where he learns the most. We are collectively in the desert now. It is quiet, it feels like death, and yet it is chaos alive with hidden growth. This is where our deeper senses wake up.

We all must first learn how to turn inward to ourselves, our body, not spirit. To our essence, our soul, our stories. At first you may glimpse transcendental states, epic mythological stories with deep meanings just for you, visitations from gods and goddesses, unbelievable synchronicities, intuition that cannot be denied, and other things similar. Just like in the book the Alchemist, this is beginners' luck.

In *The Alchemist*, beginner's luck is described as the universe conspiring to help you when you first begin to pursue your Personal Legend, your soul's true path. It's the initial burst of support, signs, and synchronicities that seem to align magically just because you've said "yes" to your deeper calling. It's the universe's way of encouraging you to keep going.

After beginner's luck comes the real journey: tests, trials, and inner growth. It's only through those deeper stages that your transformation becomes lasting and integrated, and sadly this is why most people quit before the miracle happens, which is their own flowering.

You're usually so open at first that everything rushes in. This can be beautiful, but also overwhelming, sometimes even triggering brief states of psychosis or emotional instability. It's crucial to stay honest with yourself and those around you. If you need support, ask for it. And for people who've experienced deep trauma, spiritual openings can resurface buried pain. When this happens alongside visions, it's easy to get confused.

Some people misinterpret what's happening, skipping the inner work and rush into roles like a psychic or healer, before they've done the work on themselves. Others become untethered, believing they're the only god, or that demons are chasing them, when often, it's really just fear or trauma coming up for healing. They also may think they are dying, going to the doctor and getting diagnosed with some immune disorder that has no cause or cure. In some cases, what gets labeled as mysterious illness might also be the body's intelligence releasing energy. Left to unfold, or supported through holistic practices, these experiences can sometimes resolve naturally. (*not medical advice*)

There are grounded, loving ways to integrate these experiences. The challenge is that many mental health systems aren't trained to recognize true spiritual awakenings. They may offer quick medication without exploring root causes, deeper meanings, or the soul's journey. We need better systems overall in healthcare and psychology to address the soul. I have some resources at the back of the book, of people who understand this world as we speak of it here in this book. I encourage you to also find your own.

When our "spirituality" seems to vanish, in this desert, most of us search outward, hoping to reclaim the feeling, but the real invitation is to turn inward. Presence is your interface. This is where you learn to *choose* light and intelligence during pain and darkness. You learn self-responsivity here…you re-discover your unique power. Your guides, the Plasma, the Mystery, want you to remember who you are. They want you to emotionally regulate, ground, and depend on yourself. They're always quietly supporting you, but they want you to fully meet yourself, your M-Self, before they return in full force. This is what is happening on a collective level as well…this inner return is reflecting outwards.

And they will meet us. Everything will. It may look different, it may not. Release expectations, open to new experiences, and commit to healing and feeling. Plasma will reveal her world and her secrets to us. We all go through a dark night of the soul, but there is always a treasure waiting at the end. Rarely what we imagined, but almost always better.

Your juicy life is already here, you are just marinating in it. Like a chicken in a plastic bag, you'll be pulled out soon, if you haven't been already. While you're in this void, this dark plasma womb, your only left with your own juices. Learn, be, and meet all of yourself. In this space, joy and magic will still happen, as well as visions and moments of deep connection. Be mindful to go inward rather than seeking outside validation as I did the opposite! I chased validation from men, jobs, people, and situations. And thank God the universe rejected me through all of those. At first, I thought because of these rejections that I was forced to face myself, that I was forced to write my book. But now looking back, I finally <u>chose</u> to face myself, I chose truth. I am proud of myself, and you should be to.

Rejection does not mean you're this terrible person who's unlovable or destined for failure. You're being protected for a life beyond your wildest dreams.

Trust me, I understand the fear, the confusion. And just when you think it's almost over, go deeper in. You learn patience and presence; both were my arch nemesis. But you can do it. Plasma has your back. Your magic will return brighter and more vivid than ever. And not even return, it's there, but that isn't the focus at first, at least while we are learning to really feel and heal.

Gradually you'll find yourself in a new timeline, leaps and bounds from where you started. Where you're continuously creating anew with more love, connection, and effervescence. Where you have compassion for not only yourself, but everyone.

Embrace this time, wherever you are. Soak up the moment. Treasure those who love you. In this case, *the juice really is worth the squeeze.*

A Reminder on Play and Meaning-Making

It is worth a brief refresher and summation of the importance of play and meaning making when it comes to our reality versus attachment.

On *The Telepathy Tapes Podcast*, Adam M. Curry, who was connected to Princeton's *PEAR Lab* as well as the *Institute of Noetic Science*, discusses how assigning positive or neutral meaning to a desire increases the likelihood of it manifesting. This principle aligns with the concept of sympathetic resonance, the idea that like frequencies attract.

When you assign meaning to something, you're essentially encoding it with a specific resonance (*and emotion*). This is not the same as attachment, which locks energy into a fixed state and limits flow. Instead, meaning can be viewed as a living imprint, a frequency that moves through Plasma, drawing similar resonances toward it. This is the principle of sympathetic resonance or harmonic attraction. Here are the differences:

Positive or Neutral Meaning: When the meaning you assign is infused with curiosity, playfulness, or love, it acts as a frequency that aligns with the natural flow of Plasma, which is inherently fluid and neutral. This state is open, receptive, and allows for greater dimensional access, as it is free from constriction and tension.

Negative Meaning: If the meaning you assign is fear-based or resistance-based, it creates a dissonance. This dissonance functions like a static field in Plasma, creating interference patterns that disrupt flow and prevent alignment. Plasma is still moving, but the signal is scrambled, making it more difficult for you to access the resonance or outcome you desire.

For example, because I dealt with bullying a lot, when I try to meet people, I get really insecure. Especially with people with similar interests. Thoughts bubble up like, "They probably won't like me," or, "They won't understand me.*"* Before we even speak, I'm already arguing with them in my head. And sometimes, that actually manifests where they don't reach back out at all.

But now, when I catch it, I pause and assign a new meaning: *of course they'd want to connect, we're aligned, we share similar goals*. Then I detach from the outcome, like a child making friends. If they resonate, they'll reach back out. And if not? Rejection is always protection. Who knows why—and frankly, who cares? Also, by neutralizing it, there is no shame in the follow up. I love telling people this next part, it makes me so excited when their faces light up…

Only 2% of sales close on the first contact, while 80% of successful sales require between *5 and 12 follow-up attempts* before closing.[203]

Isn't that wild!? You never know why that person didn't answer…Follow up! If I really want something, I'll go after it. If not, I move on. Sometimes things also come back around! Play is the ultimate state of creative neutrality. When you are in a playful state, you are not seeking to control outcomes but rather to explore and engage with possibilities.

---

[203] Sales Follow-Up Statistics: How Many Attempts Does It Take to Close a Deal?" *Qwilr Blog*, accessed August 2025, https://qwilr.com/blog/sales-follow-up-statistics/

In Plasma terms, play is a state where:

Emotion and Intention are Fluid: You're not locking into a specific outcome but rather allowing multiple outcomes to exist simultaneously, similar to a superposition state in quantum mechanics.

High Frequency, Low Resistance: The energy of play is light, open, and expansive. This is similar to the neutral plasma state that exists in the fifth dimension, malleable, non-constricted, and ready to shape itself in alignment with consciousness.

When you're negative, you quite literally create a magnetic reversal in the plasma field. It's like generating a countercurrent that pushes against the flow instead of aligning with it. Plasma thickens, becoming a denser medium where manifestation takes longer and feels heavier. You create invisible walls. This is how a single bad moment can snowball into an all-out bad day, unless you catch it and shift, even with a small burst of gratitude or play.

Why? Because the plasma field begins attracting circumstances that match that emotional dissonance, effectively creating what we *don't* want. For me, focusing on the *feeling state* of what I desire works far better than focusing on specifics. Feelings carry a neutral resonance, allowing plasma to respond in infinite forms, leaving the future open to magic, instead of limited by the current lens of my evolving awareness.

There is also something to be said for just giving yourself those times whether it be hours or a few days to "rot". To feel your feels, that is actually healing and may give you great insights. The more you listen to your body and intuition, the more you'll know what's best.

All this to say: you are the alchemist, and Plasma is your material. Welcome to the magical mystery tour of life.

The Star, The Water Bearer and The Age of Aquarius

The Star in tarot symbolizes the Age of Aquarius. The imagery of the card shows a naked figure (*often feminine*), kneeling with one foot on land and one in water, pouring water from two jugs, one into a pool (*the unconscious*) and one onto land (*the conscious/material*). Above her is a bright central star, surrounded by seven smaller ones. That central star is the eight-pointed *Star of Inanna*, representing cycles and balance, as well as the dark feminine, or 4D plasma. The Star card is also known as the water bearer which is also the symbol of Aquarius in astrology.

This was a sentiment published in a 2007 blog by Betty Bland, a past National President of the Theosophical Society in America: The sign of Aquarius is represented by the water bearer pouring forth the waters of wisdom and marks a time in which energy is defined by the term, "*I know*." The pouring water symbolizes spiritual knowledge and emotional intelligence being shared freely and globally, hallmarks of Aquarian energy. While free-flowing water represents our emotional nature, water poured from a vessel symbolizes those emotions being <u>contained and guided</u> by *a higher faculty*, offered not just for personal evolution, but as wisdom shared with humanity. The challenge of the water bearer is to be able to contain the emotional nature in a healthy way. One must find wholeness within oneself in order to function most effectively and freely in cooperation, without being swallowed up or losing one's independence and individuality in the group. Being able to achieve this balance is a major task set before us.[204]

The Star Tarot Card | Rider-Waite Tarot

[205]

The female's nakedness symbolizes authenticity and truth, another Aquarian trait of radical transparency and individuality. The Age of Aquarius emphasizes humanity, innovation, energy fields, decentralization, and expanded consciousness—all reflected in the Star's celestial guidance and universal wisdom. The seven stars surrounding the central one often represent the chakras or higher awareness, a movement toward multidimensional integration. Inanna also passed through seven gates, as did many other archetypal figures.

[204] Betty Bland, "Viewpoint: The Power of the Water Bearer," *Quest* 95, no. 6 (November–December 2007): 204, Theosophical Society in America, accessed June 9, 2025,

[205] Pamela Colman Smith, *The Star*, in *The Rider–Waite Tarot Deck*, illustrated 1909, public domain.

The one foot on land and the other in water represents, to me, the bridge between dimensions, the 3D and the unseen 5D, which is core to Aquarian metaphysical themes. It could symbolize the self and the M-Self working together in co-creation, drawing down the light of the star above, 4D e-motional Plasma, and pouring it into form, as reality creation.

"Blessed are the meek: for they shall inherit the earth." — Matthew 5:5

As I've said, these are the beginnings of transforming literal ideas into matter without delay. And if that is our distant destiny, we must begin somewhere. That beginning is in learning to use this fabric of reality, with our emotions, in ways that generate harmony rather than strife. Our subtle senses and emotions may appear to be our weakest links, but within that vulnerability lies our greatest power.

We are living water. We are living memory, only, we have forgotten. And how does water remember? Through resonance, tuning, and coherence. By twirling into darkness with awareness, we remember how play, softness, and lightness bring us everything.

The Fool, Uranus, & Change

The Fool is something I've embodied my entire life, and I am so grateful for that. It gave me a foundation, a leaping-off point. I innately knew, against all odds, that I had to be myself at all times. And whenever I wasn't, I started to feel literally sick, even in jobs that most people could work and be perfectly happy in, even if it wasn't their passion.

The issue was, I was impulsive, traumatized, and impatient. The wildness in me had never been nurtured or guided, so it was suppressed. I was living as the child, but from a place of fear. Still, I had one thing right: I knew, deep down, who I was. I knew thought differently. I thought in my own way. And no one could convince me otherwise. It wasn't that I couldn't learn from others, but I knew my North Star was inside.

Through healing and growing into a feeling creature, and learning to reparent and nurture myself, I've been able to consciously reconnect with Plasma. I've tapped back into my innate wisdom, and ours collectively. And through this connection, over the past few years, I've become who I am today. Still imperfect but imbued with something deeper. I am the Star (*plasma*) and the Fool (*child*), we all are. And I, along with you, am just beginning to learn how to use consciousness from this aware perspective.

This is why the darkness and fear of the inner child is such a powerful, yet terrifying, place to start. But feeling that darkness is what brings us into shimmering light. Together, we are healing this darkness, this 4D plasma, and as it awakens, we learn to use it for creation instead of destruction. This is what it means to live from the heart. We hold space for all parts of ourselves, even the ugly…impregnating that unconscious darkness with awareness.

Dana's Schoolwork | Circa 1998

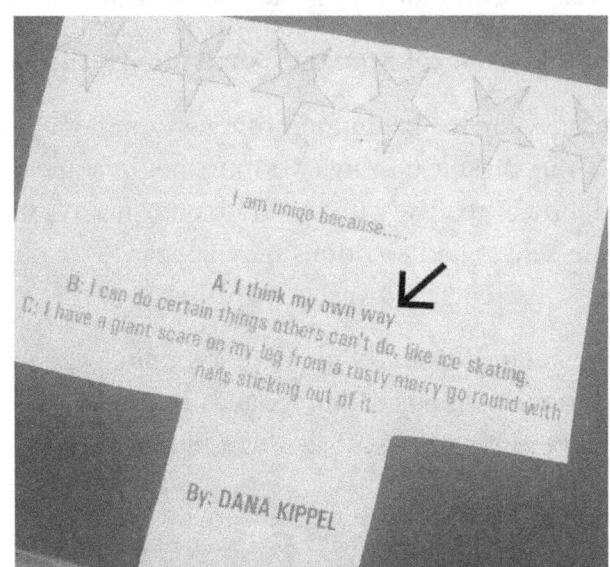

The Fool does not try to solve reality, they almost know that will never happen. There's a humility in that. Instead, they throw their hands up at the absurdity and have fun with it. They aren't worried about being perfect, and others laugh, sometimes at them, sometimes with them. If someone laughs at you, it means you're doing something different, and that is good. You are bold. They wish they could be that bold. You don't need their approval.

"God is change" — Octavia Butler

Maybe the most awakened among us are the clowns who laugh while juggling the unknown. They feel their way through life, trusting chaos while following joy. Aquarius is ruled by the planet Uranus, the planet of change, which is linked to The Fool. (*Are we seeing the humor in this?*) The Fool strikes out on a new journey and reminds us that when there is a strong desire for change, we must let go of any fear that would stop us from taking the new path. When the status quo no longer satisfies, we must take on The Fool's energy—believe in ourselves and the future, and head into the unknown with excitement and anticipation. [206]

---

[206] The Tarot Reader, "The Star and The Fool," *The Tarot Reader*, accessed June 9, 2025, https://thetarotreader.wordpress.com/articles/moon-archive/the-star-and-the-fool/.

It is that naïve innocence that sometimes gives us the courage to move forward, trusting our inner North Star and learning how to sense for its resonance in the outer world. The outer Plasma. The signature of us echoed back. The fractal path within the vastness of life that we are both following *and* creating in real time.

The Fool Tarot Card | Rider-Waite Tarot

207

You are change. And change is nothing more than energy in motion, electrons exchanging sparks, releasing photons in a plasma of becoming. You are an emotional supernova; a neon world being born.

Life is meant to be experienced and played with. It's wild, absurd, and whimsical. It's imperfect but honest. And you being your authentic, foolish self, syncs with that sweet, weird beauty. Follow your curiosity, because now, you know where to look.

A Starpunk Future

An important topic in this book is that the truest, most advanced technology is us. In this new paradigm, we require a new kind of technological framework, which I've trickled in a little throughout this book already…it is what I call *supportive technology*. This is a branch of innovation that enhances our wholeness by amplifying intuitive growth, reinforcing self-autonomy, and mirroring our inner knowing back to us.

---

207 Pamela Colman Smith, *The Fool*, in *The Rider–Waite Tarot Deck*, illustrated 1909, public domain

Supportive technology is:

a. The application of metaphysical and scientific knowledge to develop systems, tools, or devices that harmonize with human consciousness, emotions, and nature.

b. Technology designed not to override biological intelligence, but to support, amplify, and align with it.

c. A new branch of soft, often bio-interactive engineering that prioritizes resonance, coherence, healing, and the evolution of the human experience.

Supportive Technology integrates disciplines such as engineering, physics, Indigenous knowledge systems, biophysics, bioenergetics, electrostatics, metaphysics, neuroscience, and trauma-informed somatics such as neuroception. It also draws from consciousness studies, sacred geometry, optics, systems theory, soft matter physics, warm dense matter, and plasma-based functioning—alongside organic and inorganic elements—to create tools that feel alive, responsive, and emotionally attuned. Rather than relying solely on rigid structures or binary logic, these systems operate through resonance, adaptability, and coherence, mirroring the intelligence of living systems.

They will also naturally operate through plasma-consciousness synergy, bridging physical systems with subtle intelligence. These technologies will interact with both our nervous systems and the greater plasma field in ways that feel organic, intuitive, and harmonic. They will not be attached to us and this is different than bioengineering.

At a plasma conference I attended and spoke at in Essex, my colleague Robert Temple, research and author of *A New Science of Heaven*, inspired me to look more closely at the new sciences as well. He stated that we are still at the very beginning of science, which when I think about it, is very true in relation to how long humans have been on Earth!

Robert said that many of the fields most essential to our future remain underfunded and underexplored. That what we truly need are researchers free to follow plasma into its deeper mysteries, not only for energy applications like fusion, but as the key to understanding the universe itself.

Take, for instance, the surprising new domains of hinge and edge physics, part of topological science. Here, charged currents are found to flow along the edges and hinges of materials once thought to be non-conductive. They were never "supposed" to exist there, and yet they do. Boundaries, corners, thresholds (*the liminal zones, which sparked my interest of course*) turn out to be where the most unexpected behaviors appear.

He continued that the same lesson appears across other frontiers. Non-locality, long-range order, and coherence, show how power laws and critical points dominate in large, open systems. Earthquakes cannot be predicted exactly because the crust is not a set of isolated events but an interconnected whole, a self-organizing system with strange characteristics that resist linear models.

Robert finished with explaining that even in electrical engineering, Gabriel Kron intuited this principle. His work on electrical networks revealed how separate circuits could be joined into one vast network…a foreshadowing of how science is beginning to understand that everything, from plasma filaments to star systems, is part of a larger connected field.

"We are called to be architects of the future, not its victims." — Buckminster Fuller

Earth is the home of humanity. And just like we can access everything inside ourselves, everything we need is already here at our fingertips. First things first: self-compassion is the greatest gift we can give to our future world. And it is free. Ecologist Asia Suler, in *Mirrors in the Earth*, reflects that in the midst of deep ecological hurt, the most transformative thing we can do isn't recycling or saving the whales (*though both are worthy causes*), it starts with empathy for the self. It begins in understanding and seeing our innate goodness, our benevolence. In turn, this will reflect outwards in how we treat nature, animals, and other people. [208]

Just as Earth has a plasmatic, superionic core, we too hold an inner plasma core. We are both made of goodness.[209] Of course, for many of us, somewhere along the way, we program ourselves with fear responses, some people become more destructive than others, sullying that plasma core, causing outer turbulence. The great news is it is all reversable.

We will discover everything we need is inside the Earth. The Earth may just be a self-sustaining torus! Recent studies even suggest that Earth's core may be "leaking" rare elements like gold toward the surface through volcano mantle plumes, showing that our planet is not sealed off, but circulates its own inner treasures back into life above.[210]

---

[208] Asia Suler, *Mirrors in the Earth: Reflections on Self-Healing from the Living World* (Berkeley, CA: North Atlantic Books, 2022).

[209] Yu He et al., "Superionic iron alloys and their seismic velocities in Earth's inner core," *Nature* (2022).

[210] *Earth's core is 'leaking' gold, study finds, Live Science*, September 25, 2024. https://www.livescience.com/planet-earth/earths-core-is-leaking-gold-study-finds?utm_source=chatgpt.com.

A Starpunk future means remembering we are the star. It is a reconnection with self and Earth, a return that brings forth a plasma future of shifting from cold machinery to soft, organic technology, real magic, and natural connection. It is a time of resonance. Rather than time traveling away from Earth or obsessing over distant space, we will realize we don't need to go anywhere physically. We can, but we won't have to. We'll learn to magnetize realities and beings to us, potentially shifting our entire optical frame of reference. We won't go anywhere… and yet we will go everywhere.

We will begin to view the fabric of reality differently, by tuning our emotions, shifting our consciousness, and resonating into alternate timelines or worlds. At first, this may be supported by technologies that work with plasma, light, water, sacred geometry, and minerals. Eventually, it will become something we can do on our own.

I don't hate technology. I used a computer to write this book, I looked up pdfs of old books online. It would have taken me years of searching through libraries otherwise! Technology also spreads knowledge. It enhances communication when used with intention. But it can also divide us, addict us, and disconnect us from our bodies. Technology doesn't need meant to be this way. It is a direct reflection of our inner states…a form of art, just like anything else. It mirrors where we are as a collective consciousness and reveals our greatest pain points.

Right now, while some technologies are beautiful and healing, many are being built from survival patterning, not from wholeness. If we can first heal ourselves, we can begin creating technology that supports, nourishes, and enhances life. Some of these technologies already exist, and many more are coming.

"Humanity is acquiring all the right technology for all the wrong reasons."
— R. Buckminster Fuller

Buckminster Fuller, a systems theorist and futurist, cast a light on a large mismatch, that our tools are capable of global transformation, but they're frequently leveraged for profit, control, or consumption instead of solving systemic problems. I highlight this is because most of them were made by people who had not worked through their own fears and trauma.

As I continue writing and sharing my research on Holonic Metadynamics™, my aim is to reframe the familiar, to view everything from PET scan machines to positrons through a metaphysical lens, guided by the intention of uplifting and evolving humanity. Many existing technologies were created not to heal, but to diagnose, categorize, and control…born from scarcity-based thinking and capitalist motives: "How do we make the most money?" "How do we get the most users?" These mindsets rarely lead to sustainable solutions.

We treat symptoms instead of addressing root causes. We train doctors through a dangerously compartmentalized model, one that is rarely trauma informed. We've created material solutions for a material society, forgetting that we are energetic beings. *But what if we applied a different vision?*

So, there are systems of control and technology is fucked. Right? Wrong! As Buckminster Fuller said, *"We are architects of the future, <u>not its victims</u>."* Instead of spiraling into techno-fear, we can say: okay, this is where we've been. Let's create our future and dream up what to change, instead of fighting what is. Fighting anything equals fighting ourselves. It is way more productive to place our energy towards creating and helping ourselves, rather than blame or fear. This new way of thinking will naturally overtake and neutralize what we will call in the future...*archaic technology.*

In the *book The Goddess and The Shaman*, J.A. Kent writes,

"We are slowly realizing as a culture that the intellect is not supreme. It is fallible and does not give good guidance for action in the long term, which is why most political and technological solutions to the world's problems usually ultimately fail. They are usually short-term and reactive, driven by intellectual hubris. We need our poets, our strong dreamers, and our visionaries, who can help to tune us into a spiritual reality that can nourish and guide our actions long term."[211]

I echo this same sentiment. Logic alone is not a solution. It is about thinking differently, not following the current educational systems or mass consciousness. It is about connecting to one's essence, one's intuition and soul, and co-creating completely new ideas. If you are studying the same things and reading the same books as people, you may get the same results. Trying looking in the dark crevices of eternity for your information...I love *Thriftbooks.com*!

This next era will not be about building quick, external systems, but first embodying our foundational internal system. And Plasma is our original interface. We have mimicked this in everything from art to architecture. I'm not trying to replace or abolish technology; I am simply offering a different path, a reminder that <u>you don't need it to access your power.</u> Everything you do should feel like a choice. I want to support enhancing our inner states so we can visit inner worlds that technology could never dream up, depend on ourselves, feel safe in our bodies, and wield our personal magic.

---

[211] Kent, *The Goddess and the Shaman*, 225.

The idea of a Starpunk future is a metaphysical landscape built on systems feedback, harmony, and unconditional love. It is floating, pink, gelatinous quintessence-computers that support our mental health. It is technology made of organic substances that help us see beyond this reality. It contributes to the sweetness of life, healing, and the deep excitement of the soul. It is humorous, childlike play bursting out everywhere…in design, in bright colors, and in new shapes. It is living in paradox, neutrality, and the higher use of emotions. It is a burst of new senses, poetic speech, and multi-sensory foods. It is not singular or self-involved, nor is it strictly community based. It is the collective unconscious exploding and connecting, coming to fruition. It is a state of *Emergence*.

The Plasma Age

"In our opinion mankind is entering into the space age which, to a considerable degree, 'is also a plasma age" — D.A. Frank-Kamenetski

Right now, most of humanity is still oscillating between unconscious creation (*being pulled by old patterns, beliefs, and archetypes*) and conscious co-creation (*actively shaping reality with intention*). The fusion will occur as we collectively awaken to our co-creative capacities and learn to consciously direct Plasma. We are at a tipping point where science is beginning to catch up with metaphysics, recognizing that reality is not static but malleable, programmable, and participatory. Plasma is the bridge between matter and consciousness, and as we learn to engage with it consciously, we will:

Reprogram our bodies and cells: activating dormant DNA and healing conditions once deemed "incurable."

Collapse distances and timelines: making phenomena like telekinesis or remote viewing more feasible.

Access collective intelligence: receiving insights, ideas, and thought blocks from multidimensional realms.

Master emotional alchemy: transforming fear-based energies into creative power.

This will also be reflected in near-future supportive technologies designed to work in harmony with humanity and nature. This is why I've stressed the importance that we develop our innate plasma intelligence and evolve ourselves in parallel with these new technologies. The two will converge. The more focus and attention we place on this, the more we pull the impossible into the possible, shifting latent fields of fantasy into the field of reality. *What do you want to focus on?*

Plasma itself will not "think" the way humans do. Instead, it will evolve as we do to become a more coherent, organized field…capable of aligning with the intentions and frequencies of those who engage with it consciously. Imagine it shifting from a dreamlike, scattered state to a lucid, cohesive one.

Plasma will begin to reflect higher-order structures, patterns, and intentions more clearly, almost like a conscious feedback system. The more we awaken to our own multidimensional nature, the more we will consciously organize plasma, effectively teaching it to remember, align, and cohere in more structured, intentional ways. It is a futuristic technology, not a machine, but a living intelligence we are remembering how to use. Two really cool applications could be:

Multidimensional Communication: Plasma could become a more reliable conduit for communication across dimensions, serving as a living network that could hold the other being's conscious communication symbols in its field for us to receive and vice versa, without having to go anywhere. There may be ways new technologies can help us see into this field.

Conscious Environments: We could potentially infuse Plasma with intentional programming, creating environments that respond to thought and feeling, almost like an extension of our own consciousness.

It will begin as it already has, with new plasma discoveries in military applications, healthcare, cancer treatment, and environmental cleanup. Slowly, the deeper truth will emerge, not only are these intelligent UAPs composed of Plasma, but so are the beings, and this very reality itself, formed from a finer plasma field.

Plasma, science, engineering, and metaphysics will have an undeniable convergence. This next wave of science may be intuitive, emotional, and rooted in biological empathy, where humans become feelers and machines act as translators of the cosmic language we were always meant to remember.

What we've arrived at is an expanded definition of Plasma: it is the multidimensional, responsive medium through which consciousness, intelligence, and information interface with matter. More than just a state of matter, Plasma is the sentient fabric of reality itself…a living field that reveals foundational behaviors, activated like a genie from a bottle when "rubbed" the right way, through geometry, intention, emotion, or natural law. And sometimes it shows us what not to do, destabilizing not out of rebellion, but out of truth. The more we understand Plasma, the more we understand ourselves.

Meeting The Future: A Plasma Manifesto

The same core wounds that have plagued humanity for eons are now reflecting off something new: *Plasma*. And the worst thing we could do is believe we are immune to this. How do we work together in a way we never have before, to let plasma support us rather than divide us?

Here's the thing, we're all going to experience Plasma differently, because at one level, it reflects who we really are, and we are all unique. It reveals our belief systems, our fears, and our dreams. And there are, and will always be, multitudes. We then have a choice: Who do we want to be? What life do we want to co-create with this grander mystery?

Just like anything else throughout history, the natural inclination will be to capitalize on it, control it, and perhaps even use it to control others. That's what we've done, with science, with religion, with culture. This is where my call to humanity begins. This is a message for every pioneer, every researcher, every mystic, every technologist, every artist, every soul working with plasma right now. *We are at a crossroads.*

I'm noticing that the very dogma we critique in science is already beginning to appear in this new world of Plasma, perhaps even more intensely. Everyone projects their beliefs onto it, and it fragments into a thousand separate meanings. People attach stories to what's inside Plasma, the levels of Plasma, and we all do this. For individual healing, that can be profoundly useful. But for communities, without awareness, it can become divisive and destructive.

We need to learn to look past our belief systems if we truly want to understand what Plasma is, and if we genuinely want to help humanity. Science was never the problem. Our projections were the problem. Our behaviors were the problem. So many of us came from survival, not curiosity. Of control and power, versus love. And that didn't come from evil, it came from where our collective consciousness has been, steeped in fear, shaped by capitalism and consumerism, and passed down by our parents and their parents before them. We've all been doing the best we can. This book is here to hopefully spark a new awareness. So, if you're going to make harmful choices, you'll be awake to them now, *and it won't be so easy.*

Some of this is, of course, an individual journey. But when it comes to our public platforms, how we educate, how we share, we must do better. Plasma, in many ways, is the final mirror. It reflects back at us, begging us to see past the illusions of belief, to remember who we really are, no matter our past. So *she* can be seen. What does she have to tell us, if we truly listen, without our own filters, without fear? What are her truths? What are the common denominators?

There is no future set in stone. We do not live in stone. No matter what the past stories have said, no matter what has been predicted, we must remember that we are resilient, and we can defeat fate. There is destiny, there is free will. And we are capable of entering a time never lived, never seen, never told before. <u>If anything, Plasma has even greater potential than science for us to misuse it</u>, because it reflects our belief systems right back at us. And with this truth, we find ourselves at the very edge of space…only to circle back and meet ourselves again.

"He who fights with monsters should look to it that he himself does not become a monster. And if you gaze long into an abyss, the abyss also gazes into you."

— Nietzsche

We have been running from what's inside of us for a very long time. Plasma doesn't just stare at us from the abyss, she *is* the abyss. And in her gaze, we find ourselves staring back. With the greatest light also comes the greatest potential for darkness. Plasma holds both. But what I know in the deepest part of my being, in my heart and soul, and what you might feel too, if you let yourself, is this: *with potential comes choice.* And we do have a choice. We can choose a future that doesn't repeat. One that doesn't end in cataclysm. So… will we repeat the past, *turning something beautiful, mysterious, and alive into a commodity, a weapon, a hierarchy*? Or will we finally rise into a new kind of stewardship, one based on resonance, humility, and shared responsibility?

As more people begin to speak about Plasma, you may feel a wave of anxiety. You might hear someone describe it differently than you do. You might feel the urge to correct them, to shout, *"No, that's not it!"* But the truth is, everyone will have their own interpretation. Why? Because *we're creating our realities.* And we are co-creating Plasma through the lens of belief, emotion, and intention. It will have different names, and it will wear many faces. That's why we must resist the urge to control or capture this field. To define it too narrowly or to try to claim ownership. That urge has haunted every sacred tradition, every scientific breakthrough, every bridge between spirit and science. But this time we can choose differently.

To the teachers, the speakers, the scientists, the mystics, the content creators: this message is especially for you. <u>You are at the frontier of something world changing</u>. Our words carry weight. We must keep the intention always, to offer truths that empower others to explore for themselves, and to let people know these things we say are perspectives, not solid truths. And if your mind changes, if your perspective evolves, it's supposed to! You don't need to stay fixed to any belief or system.

This book is foundational yes, but also filled with my perspective. No one outside of you can give you the whole truth. Not me, not any teacher, and no system. Only you can, through your lived, felt, direct experience. You are your own guru, as is your life. You decide your beliefs, your magic, your relationships, your beings…whether it's angels, goddesses, blue fairies, dragons, translucent light-beings, or all the above.

We must hold space for many truths while guiding others to their own. We can, and must, discern without attacking. We can choose alignment over agreement. We must stay anchored in our own clarity, in our own heart and create from that place.

The question isn't: *Is plasma real? Is it God? What exactly is it?* The question is: *How are we going to work with it?*

Stubborn me still forgets this. I must remind myself that someone else's truth and experience are just as valid as mine, even when it feels untrue to me, or upsets me. It's their personal experience. And I can hold space for it without abandoning myself or my beliefs. That is true wisdom. That is how you connect, and it is how you truly *see* someone.

When we learn to see belief as a lens rather than a law, we free ourselves from the need to be right, and open to the miracle of being in harmony with a field that is always showing us who we are becoming.

There is a great teaching I learned recently from Alfred Whitehead's process philosophy, which suggests that reality's richness emerges from the creative tension between different forces. In this view, disagreement or contrast is not a defect but a generative context that enables deeper answers, integration, and mutual enrichment. Contrast is good, it evolves us, it creates emergent energy when one is open enough to dialogue.

Individually and collectively, trauma shapes how we see reality. It builds filters in our minds, in our emotions, in our nervous systems. We don't just see people; we see them through our past. Through survival, fear, and projection. This is why we are so separated. This is why we judge. This is why we try to dominate what we don't understand. Not because we are evil, but because we are wounded. And if we don't address this, we will bring those wounds into how we relate to Plasma.

We will use it to validate identity, to build false hierarchies, to claim we are more special or more right than others. Plasma will reflect our trauma back to us until we heal it. But when we do, it will change everything. It heals our filters. It helps us see ourselves, each other, and the world, with new eyes. No longer as threats or competitors, but as reflections, collaborators, and *humans who share the same feelings no matter where they come from.*

That we all have feelings of loneliness, of wanting to prove our parents wrong, of being out of control. We feel the pain of sickness, we feel our deep love for a brother or sister and yet we yearn for a closer relationship that might never happen. We grieve losing our parents, <u>every single one of us</u>. Feelings unite us, if we let them.

If we come from healing, not reactivity, we will find our once, little fractured selves, meeting each other in the Plasma field as whole humans ready to build something new together. And if that feels boring, or impossible, or silly to you…look inside. Why do you want to perpetuate suffering, for yourself or others? How is that serving you? Where are your true pain points?

Imagine if emotional regulation, nervous system awareness, and self-reflection were taught before any job, before any field of study, especially before someone is handed a platform, a classroom, a laboratory, or a stage. What if trauma education came before science, politics, or theology?

Unhealed trauma is the number one distorter of creation. It shapes what we build, what we teach, how we treat one another. It creates a need for outer control. Every pain I've experienced at the hands of another has come through them, because *they* were traumatized. Whether it was a molester or a bully, they were carrying their pain too. That is why I can hold both anger and forgiveness at the same time.

I will say again, Plasma is here not to be worshipped or fully understood, it is here to be *felt with* and *created through*. And we, as creators, are already whole. It only feels like we are becoming whole, because we forgot that we already were.

Daughters of the Sun

We are here to help ourselves and each other. That has always been the assignment. There's one story I intentionally left out of the mythology section, because I wanted to save it for now. For this moment, where you are beginning to remember that the power has always lived inside of you. This story holds a special place in my heart, partly because I have Cherokee ancestors, but also because it reflects something profoundly true about Plasma, spirit, and self-responsibility.

According to Cherokee lore, there are beings known as the Little People *(Yûñwĩ Tsunsdi')*. They are not exactly light or dark, they are tricksters, helpers, and initiators. At times, the Little People play what are perceived to be harmless pranks. At others, they protect children, rescue the lost, and punish the cruel. They live in the in-between. Once again, we see a pattern with the magic of liminal spaces.

647

As a child of adoption who also felt extremely disconnected from safety growing up, I seemed to have a strong relationship with these beings. I knew I was being protected, but as I have explained, I would get scared when they visited, which would color the experience in a negative light. Now it is beautiful, grounded, and integrated. At my worst moments in life, they would whisper to me to keep going, that I was meant to be here. What really blows my mind is that I think this was <u>my future self</u> (*the me right now*) coming through these beings as intelligence for my younger self. At the time I didn't believe it. As a child, I felt damaged, like something was deeply wrong with me. I felt scared and alone, like I couldn't scream out for help.

In the Cherokee tale *The Daughter of the Sun*, the people are suffering. The Sun is scorching them, and they seek help from the Little People to stop it. The Little People offer a solution: they transform men into venomous snakes to strike at the sun. But the plan fails. And ultimately, it is *the people themselves* who must solve the problem. [212]

Some scholars suggest the Little People may have known all along that their solution would fail, and that in failing, it would return power back to the people. It was reflecting humanity's innate power back to themselves. This is also what is happening now with the rise of connection to deeper realities. This is what I believe UAP encounters are trying to reflect to us.

This is how Plasma works, too. Yes, there are plasma beings. Yes, there are synchronicities and miracles and moments that feel guided. But at the end of the day, it is *you* who must make the choice. It is *you* who must feel the fear, ask the questions, walk the path, and learn to trust yourself. The beings may walk beside you. But the story is still yours to tell…and create.

We feel alone, and because of that, we feel fear, we feel anger, we feel abandoned. But what if we remembered that we are not separate? That we are a rupture of self, mirrored across the fabric of the universe? That the Little People, the plasma beings, the gods, and the stars are all reflections of us, asking us to remember?

Change doesn't come from perfect guidance. It comes from *choosing to see differently*. From choosing to surrender to change, even when we don't understand it. In that surrender, we become more expansive. We align with the intelligence of the universe, which doesn't save us, instead it *mirrors* our willingness to save ourselves.

---

[212] James, Mark. "The Daughter of the Sun: A Cherokee Myth." *World History Encyclopedia.* https://www.worldhistory.org/article/2562/the-daughter-of-the-sun/

Ever since making these connections about Plasma, I feel tethered to something greater, and yet more at home in my own skin than ever. It' a journey. I hope I have at least pointed you in this direction. Of self-reliance, of self-love, of the ability to see with new eyes. To know that your life is magic, your story is alive, and your choices matter more than you know.

You are a daughter of the sun. Or a son of the sun. Or a being of the sun, here, learning how to hold light, shadow, and everything in between. And no matter who or what is watching over you, you are the one who gets to decide what you become, and your family, these beings, Plasma, will always support it. Just like my future self sent comfort and compassion to my younger self, but I couldn't physically make changes for her, these beings of The Mystery are doing the same. You are the only one standing in your way.

The Force is You

This book has been about how to remember and harmonize with the greatest power of the universe...Plasma, right? Well, guess what. Plasma has been reflecting the true power all along. The secret message of this book, which is not so secret, is that beauty is in the eye of the beholder. The beauty is you. The beast is you. The plasma is you. It's all you! You are a conscious being visiting Earth in Plasma...a timeless force.

When we are disempowered, we join cults or communities, and we start to believe what they believe. This even happens in schools and practices. The trick is to believe what *we* believe. To hold and know our power within us so we can traverse the world empowered, with a sense of self. So we won't be subjected to the predatory nature of people, groups, and corporations. Not that this is always their goal, but we will be able to weave in and out of groups and people as we please. And when we find healthy community, we can *choose* it out of want, not need. This is also how new ideas are born, when one is not scared to question or push against the status quo.

We have the power to choose who stays in our life, what to engage with, and to neutralize what would have otherwise drained our energy. We have the power, the soft power, I might add, to hold space for anyone or anything, if we choose to, while remaining whole in ourselves.

Plasma's goodness is <u>always</u> available to us. It is ours to sour by fear, greed, or control. Reality is malleable, and you are the one bending it. The more love and play you use, the less effort by brute force you put in, the more it bends. You can still work for things and take action; in fact you must! Movement and choice are key! This is much different than forcing by fear or control.

When you probably saw this book, or heard me speak online, and got excited about Plasma, when you felt that magical feeling, this whole time it was reflecting a specific truth to you: that the real excitement is the reunion with self and our essence, our power. It is bringing us all back home. Coming home to our multidimensionality is coming home to every being and thing in the universe.

You will never feel alone in your body again. Your body will become your safe zone, not something you want to escape. The immense ocean of curiosity and love in you is what creates worlds, and over time you will learn to direct this for yours and the world's highest good.

My wish for you is that you see this love more in yourself, this natural benevolence. You see that you are a holographic mist of cotton candy fireworks, of all colors, of all things that smell ethereal. That you are never, ever alone. My spark sees yours, and it is <u>sooo</u> bright. I know we may not know each other personally, but I see that you are a light in darkness, and despite darkness, you know you are good deep down.

I share in, deeply respect, and am inspired by your will to keep going despite the pain, despite the let downs, despite the confusion. You can let go of that now. You can open to a new perspective.

Wait I'm Not a Monster, I'm A Fucking Force!

> **"It's a real bravery to believe past the edge of nothing.**
> **I've been to the death of dreams**
> **and I'm just past the void.**
>
> **Give the universe permission**
> **to surprise you in strange and beautiful ways."**
> **—Melea Lee, voice actor**

The greatest cosmic joke is that this new force is not a force at all. It is anything but a force, at least not by the current definition. *Force*, by definition, means physical strength, might, or power. Force also usually distills to the word power. In science, force is a push or pull that can change the state of motion of an object or cause it to deform.

That force is awareness. It's truth. It is the resonance of you. You can push, you can pull, or you can be and enjoy all that comes with that magical liminal space. When you are playing you are being. The push or pulls become a natural re-action to your field, rather than you having to exert extra energy or "force". Plasma will intelligently respond based on your state and support you by pushing away or pulling things close in.

650

In this state of play, you will begin to flow and harmonize with nature, people, and the world. Things become easier, almost too easy. And when they don't seem easy, you know it's all working for your benefit. You go with that. Every choice refines your resonance.

When you are aligned with the force that makes up reality (*Plasma*), you are in synergy:

Synergy is the interaction or cooperation of two or more agents, forces, or elements so that their combined effect is *greater* than the sum of their separate effects.

And remember, alignment equals being the you-est you. Feeling your essence. As Abraham Hicks would say, alignment is *feeling good*. Maybe it's "feeling" God or the Mystery, no matter what heaven storm *or* shit storm is going on around you. It's knowing you are never alone in this. Go ahead, take an eternal sigh of relief. You don't have to try! You just have to be! I know it's so simple but we, and I, can make it so hard.

If you are a plasma hologram holding consciousness, run by awareness, in all your wholeness, and everyone else's consciousness is interacting with that, then you become a mirror of truth. When you speak truth, you become a portal to the Mystery for everyone. And they can use that to grow or run. Whether they stay or leave is not up to you.

Understand being a mirror comes with loss, but it also comes with the depth of beautiful connections. People will hear the word of the Mystery in their own unique way, which is completely out of your control. It is however their consciousness interfaces with your consciousness (*plasma bubble*) based on their filter. The second you realize this, you are off the hook, powerless but *ohhh so powerful.*

All you have to do is express your essence, which means nothing you say has to be the ultimate answer, or perfect, it just has to be true in that moment. And that is what makes you the mirror or portal to the Mystery for everyone you come across. You become a bright sun *and* a tornado! You must remember that you are both, and you are neither. But most important of all, you are the neutral, glittering awareness behind that.

Being you, regardless of what you do for work, who you know, or how much money you have, and being true, whether that is messy, loving, brilliant, or weird, *is enough*. There is nothing wrong with you. When fear makes you act a fool, and we've all been there, don't shame yourself! Learn and know that you can always return to your truth and let the cards fall where they may. Let yourself laugh a little for "crashing out", as they call it nowadays. Surrender to that. And with that letting go, Plasma will gently usher through the coolest life ever.

Just like in dreams, all outer events reflect what we believe about ourselves. Reality is a symbolic mirror. When something happens, especially something triggering, we can either react through the lens of old wounds or pause and give space for more information to emerge. Most things, when given time, reveal more than we initially perceive.

Instead of jumping to conclusions or trying to fix, we can act from a place of self-understanding and sovereignty, asking: *Am I responding from empowerment or from past trauma?* What matters isn't always what happens, it's the meaning we assign to it.

Do you internalize it as proof of your deepest fear? Or do you claim your truth and say, *"Wait. I'm not a monster…I'm a fucking force. I don't need to chase or fix. I observe, take in more information, then I choose."*

Having the patience to take in more information before rushing to conclusions is the beauty of "masculine" logic. And openly intuiting from a place of presence is the beauty of the divine "feminine". These are the moments where our story can shift, from self-abandonment to self-authorship. We begin to see everything, yes, everything, that happens is for our experience and learning.

Creation is not about mastery, although mastery deserves respect. Creation is speaking in your power; that is how you serve your purpose here. In your play, in your truth, in your being, you are God touching people every day. You are pretty darn special. We all are.

You are like a movie people watch *(on a plasma screen perhaps?),* where three people watch the same film and get three totally different meanings. You are a field of intelligence that people create their own meaning from, sometimes despite your best efforts to control the narrative.

The truer you are, the more people will see the real you…and themselves. If they don't like themselves, they may not like you so much. Cool. That says nothing about you. And if you ever feel you've done something to sour a friendship, communicate. You'll feel if they care the same way you do or not.

Let me ask you something: *Who are you NOT to put your creation out there*? Who are you to stop this great Mystery from speaking through or with you? Who are you to rob so many people of the truths you could help them see? What if you are meant to say the exact thing someone needs to save their life…to change their life? If you stay silent, if you stay small, we all lose.

When my dad passed away, you know what helped me the most? A thirteen-dollar roll-on essential oil perfume called *Mistress* by a small company out of Fort Lauderdale called *Blush and Wood*. Every time I went to visit my dad, who was in a coma for a bit over a month before he died, I would glide it over my wrist and neck. It made me feel calm, at home with myself, and in a weird way, positive that my dad would be okay…even if I knew deep down, he wouldn't be. I wore that scent every day, and now when I wear it, it reminds me of my last moments with him. I bet the founder of that company, who doesn't know I exist, never thought she would affect someone in that way by creating a perfume. Either way, I'm glad she expressed herself, created something unique, and it smells so good! Sometimes, it really is the little things.

The point is you don't have to force anything. You are already the force. It's just about waking up to the ride and enjoying it. Let it be enough to be you. Everything else flows from that. Follow what lights you up. Focus on your curiosity, not on fixing yourself for others or calibrating your shine to fit someone else's comfort. When you root into your own frequency, you'll naturally begin to see who you can truly connect with, and you'll see them more clearly, too. It is like a little club, a fun one.

Being in your power doesn't mean proving it. It doesn't mean unleashing all your genius at once. It means knowing it's there. It means trusting that your presence is a gift, and you can choose when to share it. You are not here to shrink, to prove, or to over-give just to earn love or belonging. You are here to be. Let potential friends or partners earn access to your deeper layers of truth. This took me way too long to learn and I need you to hear it!

This is how you use your emotions, not to obsess over others, but to guide yourself back to center. This applies to people, beings, even higher intelligences. Share your space and your brilliance from safety, not from fear. Don't contort yourself for connection. Let others rise to meet you where you are, no matter how bright that is.

Your creations are part of you…but they are not all of you. *You are the star.* Nurture that first. Let friendships be reflections of you, not things you have to manage or correct. Choose your wholeness over someone else's comfort. And understand that someone may choose their comfort instead of you, and that's okay. And it's ok if it hurts. Never abandon yourself to stay close to someone else, whether it's a person, group, job, or other being of consciousness. Live like a child, one who is curious, radiant, and unapologetically full of becoming. The Mystery will never ask anything of you, that is martyrdom. You are here to shine, effortlessly. You are not here to orbit someone else's sun. You *are* the sun, not a planet.

And when you cross that threshold of fear, when you face yourself and feel everything you've subconsciously held off from, the heat will physically rise. You may cry, you may feel like you are breaking apart, you may shake. Just breathe. You are feeling the star that you are. You are feeling the power inside of you burning so bright. Eventually you will laugh, because it has always been there, and our bodies can be quite dramatic. Now, you see it. And you know it.

Return to this chapter anytime you need a hype-man! You are a trinity of being…you know that you can use your power for anything, and you choose it for benevolence. You, my friend, are a very old force. But now, you are something more. You are *A New Force.*

The Return Home

"Home is what you take with you, not what you leave behind."

— *The Fifth Season*, N.K. Jemisin

Congratulations…you're home. Home to yourself, your force, the Earth, and the 3D realm. A place you can make safe again, so you can co-create like the blooming fractal you were always meant to be. I am so proud of you.

Remember, it's normal to brush up against grief or fear as you land. You're meeting your subconscious filter, the one you've been projecting outward your whole life. The one that told you, if you looked too closely, you might find a beast, a monster. But that's a lie. A lie you created to survive. A lie we've collectively carried for thousands of years to maintain some illusion of control.

As you let it go, you may cry, you may scream, you may feel foggy or disoriented. But what lies ahead is grander than your wildest dreams. You are strong. You are ready. You are supported, and you are never, ever alone.

I hope you've come to see your innate benevolence at the core of who you are, beneath that filter. And if me, an ex-addict who made tons of mistakes, an adoptee who felt innately unwanted, a girl who was bullied most of her life, who has been through countless traumas she thought she brought on herself, who apologized to people just because they didn't like her, convinced she was the worst monster of them all…if I can see it, so can you.

I hope you let the truth become your new lens. A heart-centered life is presence, a healed past, and an emergent future. It is a discerning openness. It is choosing yourself first, not out of need or emptiness, but from wholeness. As fantastical as this book may seem, it's ultimately a book about facing reality…*and revealing the magic within it.*

You will begin to force nothing.

You will flow.

You will play.

You will laugh.

You will fall.

You will experiment.

You will expand.

You will rise.

You will change the world, simply by being yourself.

You are a glowing ball of crystallized plasma-consciousness, of ultra-terrestrial light.

As you land, I hope you bring a part of this world forward with you, whatever resonated most. Never lose sight that your life is your greatest teacher. This book simply pointed you back to the multidimensional depth of that experience, beyond what we thought was possible.

This is just the beginning. The more you stabilize your plasma bubble, the more your outer life rearranges to reflect it. You will meet new beings, you will see more signs & symbols, opportunities will befall you, all Plasma mirrors, that support your wholeness. You will trust your mythic visions, knowing they are real communication, and you'll learn to trust the meaning you give them or that they give you. You will open to these visions inside and around you, instead of stopping them short. The plasma-verse is expanding and evolving, and so is <u>everything</u> inside of it. The possibilities are limitless. And we are on the edge of perceiving what once lived beyond the bounds of human imagination.

I love you, and my greatest wish is for you to see how amazing you are. It hurts my heart that you may not see that, and I have had to face the grief of myself not seeing it for most of my life, as well as the grief of that being reflected in others' behaviors. There was nothing wrong with me; there was something wrong with my filter, with the way I saw myself. And all these grievances were not because I was a victim of life, dealt a shitty hand, they were all calling me back to myself, back home. Plasma was doing what it does best: reflecting my inner state.

Thank you for joining me on this journey back to Plasma, and forward with ourselves. Now you remember your connection, your love, and your magic. I see you. And no matter how much pain you may have gone through, or may still go through, hopefully you now see too. And with that seeing, you feel the magic, the buzz of life, that there is something more.

"And now thy _Self_ is lost in SELF, _thyself_ unto THYSELF, merged in THAT SELF from which thou first didst radiate" — The Voice of Silence, Helena Blavatsky

Afterword

"Tell All The Truth but tell it Slant —…The Truth must dazzle gradually / Or every man be blind —" — Emily Dickinson

My main goal in this book was telling the truth. I edited it countless times, and that truth kept shifting. I realized that every time my awareness grew, so did my understanding. That my book would never be perfect or absolute truth. Truth is always in flux. It is always evolving. I hope what I wrote in this book does shift, change, and evolve over time. I am sure some things are wrong, or not totally accurately described, and that's ok. I am speculating. I am asking questions and speaking what I feel to be true. I am pushing what is known so we can arrive at the next place to once again push what is known. So please, push and question my ideas. Formulate your own understanding.

Truths are revealed gently, over time. And that's what I think happened with this book. Nothing is meant to be revealed all at once. There is fun in the mystery. Truths are co-created with our awareness. We are flying towards truth in every moment. And from what I learned in the arduous process of editing this book, is that maybe it is not about catching up to truth, but enjoying the relationship with truth, ever-growing. That every day new things will sprout from this, from within us, and it will all be beyond our wildest dreams.

Above all, may this book remind you of your own wholeness, and may love be the field in which all truth is revealed. I am grateful for you engaging with me and this book, for giving us your time. And remember, as truths are revealed, we ultimately author our own.

A heartfelt thank you to my chapter sponsors for their generous contributions to the fundraising campaign that helped make this book possible:

Introduction: Sa'ad Shah – Noetic Fund
Chapter 1: McKane Buller – The Quantum Kinetics Corporation
Chapter 2: Natalia Urquiza-Manzano – House of Unashamed
Chapter 3: D. Miles Collins
Chapter 4: Celeste McQueen – Celestial Journeys
Chapter 5: Arturo Cusco – Author of Celestial Music
Chapter 6: Ashley Gonor – Uncover Your Magic Podcast
Afterword: Lanson Burrows Jones Jr. – Logos369

A special thank you to everyone who donated to my campaign, your support has meant the world to me.

Resources

Therapy

Comprehensive Resource Model
The Comprehensive Resource Model® (CRM) is a neuro-biologically based, affect-focused trauma treatment model which facilitates targeting of traumatic experiences by bridging the most primitive aspects of the person and their brain (midbrain/brainstem), to their purest, healthiest parts of the self. https://comprehensiveresourcemodel.com

Elisa Elkin Cleary
Elisa is a certified trauma therapist from the Colin Ross Institute and has had extensive training in; ego state/parts work, trauma, dissociation, maternal mental health, somatic work, neuro-psychotherapy, hypnotherapy, attachment healing work, memory reconsolidation, performance enhancement and spiritual work. Since 1994, Elisa has developed several areas of specialized work that include: healing the emotional wounds and the neurobiological impact from attachment disruptions; performance work for athletes, musicians, actors and in business; preparation and integration work for clients working with ketamine or psychedelic plant medicine; as well as healing disruptive and painful patterns resulting from generational trauma and past life trauma work as it relates to ones current life challenges. https://rootedsoulwork.com/

House of Unashamed
Founded by Natalia Manzano, House of Unashamed is the next evolution of Divine Alchemy, a multidimensional practice rooted in trauma-informed healing and inner transformation. With a background as a Licensed Mental Health Counselor and over 500 hours of advanced clinical and transpersonal hypnotherapy training, Natalia guides high-achieving women and recovering perfectionists through deep cycles of healing, identity reclamation, and soul-anchored empowerment.
Known for blending psychological depth with spiritual insight, Natalia's work bridges subconscious reprogramming, parts work, trance states, intuitive arts, and sacred storytelling to help clients break free from shame, self-abandonment, and the need for external validation. House of Unashamed offers a new kind of sanctuary where healing becomes a rite of passage and reclaiming your truth becomes a revolutionary act.
https://houseofunashamed.com
https://divinealchemycollective.co/

People & Places I Love

Gangaji
Gangaji is a teacher and author who speaks to people from all walks of life inviting them to fully recognize the absolute freedom and unchanging peace that is the truth of one's being. https://gangaji.org/

Monroe Institute
Founded in 1971 by Robert A. Monroe, the Monroe Institute is widely recognized as a leading center for exploring and experiencing expanded states of consciousness. For over 50 years, our immersive programs have empowered participants to undergo profound transformations, gaining a fresh outlook on life and discovering a deep sense of purpose. https://www.monroeinstitute.org/

Esalen Institute
Esalen is a holistic retreat and nonprofit educational institute. Established in 1962 and considered the epicenter of the human potential movement, they welcome visitors with advance reservations for workshops and massages to their campus in Big Sur, California https://www.esalen.org/

Shamans Light, Marti Spiegelman, MFA
Marti Spiegelman is a leadership advisor, visionary, speaker, and founder of the indigenous initiation and mentoring program, Shaman's Light™. Through in-depth training in the Shaman's Light program, Marti guides people into the mastery of full consciousness – applying the timeless technologies that our indigenous ancestors used to thrive and evolve, not to replicate ancient ways, but to innovate here and now, in the face of the real-time challenges her students face in their lives and work. She is an initiated shaman and wisdom keeper, with over four decades of specialized training in indigenous technologies of consciousness and related scientific, economic, and sociological fields. In addition to the Shaman's Light™ program, Marti offers online courses, and for over 2 years, brought core indigenous wisdom to in-depth conversations with thought leaders through her web radio program, Awakening Value™: Shamanic Technologies of Consciousness and Success on the VoiceAmerica™ Web Radio Network [2009 – 2011].
https://www.martispiegelman.org/
https://paqokuna.gumroad.com/l/precision-consciousness

Dana Kippel is a polymath, author, filmmaker, and advocate for mental health and neurodivergence, pioneering the study of plasma intelligence and consciousness. Her groundbreaking work explores plasma as a bridge between dimensions, shaping human perception and unlocking new human potential. Beyond her research, Dana finds joy in mountain views, warm summer nights, the scent of fresh, sweet musk, theme parks, books, movies, games, her family, friends, her corgi, and the mystical moments that find her when she least expects them.

If you enjoy Dana's work you can find out more about her, events she is attending or holding, book releases, metaphysical merch, and her research at www.danakippel.com

If you want to follow Dana on social media to view content about plasma, consciousness, new sciences, metaphysics, human potential and more you may go to:

Instagram: https://www.instagram.com/dana.thealien

YouTube: https://www.youtube.com/channel/UCaDgEYwfJOKMduoLGlXA6NQ

TikTok: https://www.tiktok.com/@danakippel

Medium Blog: https://medium.com/@dbk62189

Recommended Reading: https://www.amazon.com/shop/danakippel

P.S. If this book speaks to you, feel free to share it by posting & tagging me!

"For that which comes through, comes through."

Inanna Books

www.ingramcontent.com/pod-product-compliance
Lightning Source LLC
Chambersburg PA
CBHW080817120626
46556CB00010B/3315